ELEMENTS OF STRUCTURAL DYNAMICS

ELEMENTS OF STRUCTURAL DYNAMICS

A NEW PERSPECTIVE

Debasish Roy

Indian Institute of Science, Bangalore, India

G Visweswara Rao

Engineering Consultant, Bangalore, India

WILEY

A John Wiley & Sons, Ltd., Publication

This edition first published 2012

Registered office

John Wiley & Sons Ltd, The Atrium, Southern Gate, Chichester, West Sussex, PO19 8SQ, United Kingdom

For details of our global editorial offices, for customer services and for information about how to apply for permission to reuse the copyright material in this book please see our website at www.wiley.com.

Library of Congress Cataloging-in-Publication Data

Roy, Debasish Kumar, 1946-
Elements of structural dynamics : a new perspective / Debasish Roy, G Visweswara Rao.
p. cm.
Includes bibliographical references and index.
ISBN 978-1-118-33962-6 (hardback)
1. Structural dynamics. I. Gorti, Visweswara Rao. II. Title.
TA654.R69 2012
624.1′71–dc23

2012011742

A catalogue record for this book is available from the British Library.

Print ISBN: 9781118339626

Typeset in 10/12pt Times-Roman by Laserwords Private Limited, Chennai, India
Printed and bound in Singapore by Markono Print Media Pte Ltd

Contents

Preface **xi**

Acknowledgements **xv**

Introduction **xvii**

General Notations **xxi**

1 Structural Dynamics and Mathematical Modelling **1**
1.1 Introduction 1
1.2 System of Rigid Bodies and Dynamic Equations of Motion 2
1.2.1 Principle of Virtual Work 2
1.2.2 Hamilton's Principle 3
1.2.3 Lagrangian Equations of Motion 4
1.3 Continuous Dynamical Systems and Equations of Motion from Hamilton's Principle 6
1.3.1 Strain and Stress Tensors and Strain Energy 7
1.4 Dynamic Equilibrium Equations from Newton's Force Balance 11
1.4.1 Displacement–Strain Relationships 11
1.4.2 Stress–Strain Relationships 13
1.5 Equations of Motion by Reynolds Transport Theorem 13
1.5.1 Mass Conservation 15
1.5.2 Linear Momentum Conservation 16
1.6 Conclusions 17
Exercises 17
Notations 18
References 19
Bibliography 19

2 Continuous Systems – PDEs and Solution **21**
2.1 Introduction 21
2.2 Some Continuous Systems and PDEs 22
2.2.1 A Taut String – the One-Dimensional Wave Equation 22
2.2.2 An Euler–Bernoulli Beam – the One-Dimensional Biharmonic Wave Equation 23
2.2.3 Beam Equation with Rotary Inertia and Shear Deformation Effects 27

2.2.4 *Equations of Motion for 2D Plate by Classical Plate Theory (Kirchhoff Theory)* 29
2.3 PDEs and General Solution 36
2.3.1 *PDEs and Canonical Transformations* 36
2.3.2 *General Solution to the Wave Equation* 38
2.3.3 *Particular Solution (D'Alembert's Solution) to the Wave Equation* 38
2.4 Solution to Linear Homogeneous PDEs – Method of Separation of Variables 40
2.4.1 *Homogeneous PDE with Homogeneous Boundary Conditions* 41
2.4.2 *Sturm–Liouville Boundary-Value Problem (BVP) for the Wave Equation* 42
2.4.3 *Adjoint Operator and Self-Adjoint Property* 42
2.4.4 *Eigenvalues and Eigenfunctions of the Wave Equation* 45
2.4.5 *Series Solution to the Wave Equation* 45
2.4.6 *Mixed Boundary Conditions and Wave Equation* 46
2.4.7 *Sturm–Liouville Boundary-Value Problem for the Biharmonic Wave Equation* 48
2.4.8 *Thin Rectangular Plates – Free Vibration Solution* 53
2.5 Orthonormal Basis and Eigenfunction Expansion 56
2.5.1 *Best Approximation to $f(x)$* 57
2.6 Solutions of Inhomogeneous PDEs by Eigenfunction-Expansion Method 59
2.7 Solutions of Inhomogeneous PDEs by Green's Function Method 64
2.8 Solution of PDEs with Inhomogeneous Boundary Conditions 68
2.9 Solution to Nonself-adjoint Continuous Systems 69
2.9.1 *Eigensolution of Nonself-adjoint System* 69
2.9.2 *Biorthogonality Relationship between* $\mathbf{L}$ *and* $\mathbf{L}^*$ 70
2.9.3 *Eigensolutions of* $\mathbf{L}$ *and* $\mathbf{L}^*$ 73
2.10 Conclusions 74
Exercises 75
Notations 75
References 77
Bibliography 77

3 Classical Methods for Solving the Equations of Motion 79
3.1 Introduction 79
3.2 Rayleigh–Ritz Method 80
3.2.1 *Rayleigh's Principle* 84
3.3 Weighted Residuals Method 85
3.3.1 *Galerkin Method* 86
3.3.2 *Collocation Method* 91
3.3.3 *Subdomain Method* 93
3.3.4 *Least Squares Method* 94
3.4 Conclusions 95
Exercises 95
Notations 96
References 97
Bibliography 97

4	**Finite Element Method and Structural Dynamics**	**99**
4.1	Introduction	99
4.2	Weak Formulation of PDEs	101
	4.2.1 Well-Posedness of the Weak Form	103
	4.2.2 Uniqueness and Stability of Solution to Weak Form	104
	4.2.3 Numerical Integration by Gauss Quadrature	107
4.3	Element-Wise Representation of the Weak Form and the FEM	111
4.4	Application of the FEM to 2D Problems	113
	4.4.1 Membrane Vibrations and FEM	113
	4.4.2 Plane (2D) Elasticity Problems – Plane Stress and Plane Strain	115
4.5	Higher Order Polynomial Basis Functions	118
	4.5.1 Beam Vibrations and FEM	118
	4.5.2 Plate Vibrations and FEM	120
4.6	Some Computational Issues in FEM	121
	4.6.1 Element Shape Functions in Natural Coordinates	122
4.7	FEM and Error Estimates	124
	4.7.1 A-Priori Error Estimate	124
4.8	Conclusions	126
	Exercises	126
	Notations	127
	References	129
	Bibliography	129
5	**MDOF Systems and Eigenvalue Problems**	**131**
5.1	Introduction	131
5.2	Discrete Systems through a Lumped Parameter Approach	132
	5.2.1 Positive Definite and Semi-Definite Systems	134
5.3	Coupled Linear ODEs and the Linear Differential Operator	135
5.4	Coupled Linear ODEs and Eigensolution	136
5.5	First Order Equations and Uncoupling	142
5.6	First Order versus Second Order ODE and Eigensolutions	143
5.7	MDOF Systems and Modal Dynamics	145
	5.7.1 SDOF Oscillator and Modal Solution	146
	5.7.2 Rayleigh Quotient	153
	5.7.3 Rayleigh–Ritz Method for MDOF Systems	155
5.8	Damped MDOF Systems	156
	5.8.1 Damped System and Quadratic Eigenvalue Problem	157
	5.8.2 Damped System and Unsymmetric Eigenvalue Problem	158
	5.8.3 Proportional Damping and Uncoupling MDOF Systems	159
	5.8.4 Damped Systems and Impulse Response	160
	5.8.5 Response under General Loading	161
	5.8.6 Response under Harmonic Input	161
	5.8.7 Complex Frequency Response	163
	5.8.8 Force Transmissibility	165
	5.8.9 System Response and Measurement of Damping	167
5.9	Conclusions	173
	Exercises	173
	Notations	175

	References	177
	Bibliography	177
6	**Structures under Support Excitations**	**179**
6.1	Introduction	179
6.2	Continuous Systems and Base Excitations	181
6.3	MDOF Systems under Support Excitation	185
6.4	SDOF Systems under Base Excitation	191
	6.4.1 Frequency Response of SDOF System under Base Motion	192
6.5	Support Excitation and Response Spectra	196
	6.5.1 Peak Response Estimates of an MDOF System Using Response Spectra	197
6.6	Structures under multi-support excitation	198
	6.6.1 Continuous system under multi-support excitation	199
	6.6.2 MDOF systems under multi-support excitation	202
6.7	Conclusions	203
	Exercises	204
	Notations	205
	References	206
	Bibliography	206
7	**Eigensolution Procedures**	**209**
7.1	Introduction	209
7.2	Power and Inverse Iteration Methods and Eigensolutions	210
	7.2.1 Order and Rate of Convergence – Distinct Eigenvalues	212
	7.2.2 Shifting and Convergence	213
	7.2.3 Multiple Eigenvalues	215
	7.2.4 Eigenvalues within an Interval-Shifting Scheme with Gram–Schmidt Orthogonalisation and Sturm Sequence Property	216
7.3	Jacobi, Householder, QR Transformation Methods and Eigensolutions	220
	7.3.1 Jacobi Method	220
	7.3.2 Householder and QR Transformation Methods	224
7.4	Subspace Iteration	231
	7.4.1 Convergence in Subspace Iteration	232
7.5	Lanczos Transformation Method	233
	7.5.1 Lanczos Method and Error Analysis	235
7.6	Systems with Unsymmetric Matrices	237
	7.6.1 Skew-Symmetric Matrices and Eigensolution	245
	7.6.2 Unsymmetric Matrices – A Rotor Bearing System	246
	7.6.3 Unsymmetric Systems and Eigensolutions	253
7.7	Dynamic Condensation and Eigensolution	260
	7.7.1 Symmetric Systems and Dynamic Condensation	262
	7.7.2 Unsymmetric Systems and Dynamic Condensation	264
7.8	Conclusions	268
	Exercises	268
	Notations	269
	References	272
	Bibliography	273

8	**Direct Integration Methods**	**275**
8.1	Introduction	275
8.2	Forward and Backward Euler Methods	281
	8.2.1 Forward Euler Method	281
	8.2.2 Backward (Implicit) Euler Method	284
8.3	Central Difference Method	286
8.4	Newmark-β Method – a Single-Step Implicit Method	289
	8.4.1 Some Degenerate Cases of the Newmark-β Method and Stability	292
	8.4.2 Undamped Case – Amplitude and Periodicity Errors	295
	8.4.3 Amplitude and Periodicity Errors	295
8.5	HHT-α and Generalized-α Methods	297
8.6	Conclusions	303
	Exercises	305
	Notations	305
	References	306
	Bibliography	307
9	**Stochastic Structural Dynamics**	**309**
9.1	Introduction	309
9.2	Probability Theory and Basic Concepts	311
9.3	Random Variables	312
	9.3.1 Joint Random Variables, Distributions and Density Functions	314
	9.3.2 Expected (Average) Values of a Random Variable	315
	9.3.3 Characteristic and Moment-Generating Functions	317
9.4	Conditional Probability, Independence and Conditional Expectation	317
	9.4.1 Conditional Expectation	319
9.5	Some oft-Used Probability Distributions	319
	9.5.1 Binomial Distribution	320
	9.5.2 Poisson Distribution	320
	9.5.3 Normal Distribution	321
	9.5.4 Uniform Distribution	322
	9.5.5 Rayleigh Distribution	322
9.6	Stochastic Processes	323
	9.6.1 Stationarity of a Stochastic Process	323
	9.6.2 Properties of Autocovariance/Autocorrelation Functions of Stationary Processes	325
	9.6.3 Spectral Representation of a Stochastic Process	325
	9.6.4 $\boldsymbol{S_{XX}(\lambda)}$ as the Mean Energy Density of $\boldsymbol{X(t)}$	327
	9.6.5 Some Basic Stochastic Processes	328
9.7	Stochastic Dynamics of Linear Structural Systems	331
	9.7.1 Continuous Systems under Stochastic Input	331
	9.7.2 Discrete Systems under Stochastic Input – Modal Superposition Method	337
9.8	An Introduction to Ito Calculus	338
	9.8.1 Brownian Filtration	340
	9.8.2 Measurability	340
	9.8.3 An Adapted Stochastic Process	340
	9.8.4 Ito Integral	341

9.8.5	*Martingale*	342
9.8.6	*Ito Process*	343
9.8.7	*Computing the Response Moments*	352
9.8.8	*Time Integration of SDEs*	357
9.9	Conclusions	360
	Exercises	361
	Notations	363
	References	365
	Bibliography	366
Appendix A		**367**
Appendix B		**369**
Appendix C		**375**
Appendix D		**379**
Appendix E		**387**
Appendix F		**391**
Appendix G		**393**
Appendix H		**399**
Appendix I		**407**
Index		**413**

Preface

To the best of the authors' knowledge, this book is one of the few attempts at a top-down approach to the subject of structural dynamics. Thus, unlike the oft-treaded route followed in most texts, we depart from introducing the single-degree-of-freedom (SDOF) oscillator and its response features in the first chapter and rather start with the basic principles of linear momentum balance of isotropic and linearly elastic systems, which in turn yield the governing equations of motion in structural dynamics. Whilst an SDOF oscillator is commonly viewed as, and rightly so, as one of the simplest building blocks for the instruments of mathematical modelling needed for an insightful understanding of the subject, it is generally far removed (at least from a functional perspective) from what we usually qualify as structural dynamic systems of engineering interest. Accordingly, despite its advantages in an appropriately sequenced bottom-up exposition, the SDOF system hardly corresponds to what we often perceive and loosely describe as 'real-life' systems. This oft-practiced mode of instruction could leave an inquisitive student somewhat flummoxed until about the last chapter, where he finally gets to connect things up and realise the importance of what he studied in the preceding chapters. Our motivation in writing this treatise has mainly been to upend this scheme of instruction and fashion the discourse that starts directly with an attempt to mathematically model and computationally treat the 'real-life' structural systems, whilst not sacrificing the important feature of a suitably graded exposition with gradually increasing complexity. It is from this perspective that the book can hopefully lay some claim to novelty, mainly aimed at holding the interest of a student right from the start and all through afterwards. In particular, we provide below an overview of the contents in the form of a chapter-wise breakup and include, at appropriate places, some suggestions as to what could be covered within an introductory course covering a semester at an advanced under-graduate or an early graduate level.

The book begins with a prologue providing a few introductory remarks on the significance of 'dynamic' analyses of structural systems. The necessity to have valid mathematical models for both the external loadings and systems is highlighted. Chapter 1 lays down the principles underlying the mathematical models describing the dynamic equations of motion for continuous structural systems. The reader is familiarised with the resulting partial differential equations (PDEs) derived via different routes viz., the familiar Hamilton's principle involving energy expressions, Newton's force balance yielding the state of stress inside a continuum from a solid mechanics point of view and Reynolds transport theorem using the laws of conservation of mass and linear momentum balance. Chapter 2 brings into focus the insight one derives from the knowledge of an assemblage of analytical methods for solving linear PDEs governing the dynamics of a class of continuous systems. Concepts on free vibration solutions based on an eigenanalysis are introduced in the chapter for both self-adjoint and non self-adjoint systems. The elementary notion of Fourier series expansion of a function in a Hilbert space setting is formalised. This is followed by the use of such expansions, based on eigenfunctions, to solve a few dynamic systems governed by inhomogeneous PDEs with both homogeneous and inhomogeneous boundary conditions. The expansion forms a basis for mathematically breaking up a continuous system into a sequence of SDOF oscillators that are more easily inverted. Given the general intractability of analytical approaches for solving PDEs of motion of complex (large) systems, Chapter 3 describes approximate projection methods that attempt to solve the equations via either the extremization of a functional or orthogonalization of certain residuals. Rayleigh–Ritz method and weighted residual methods like Galerkin, falling under this class of approximate methods, are covered in this chapter. An exposition of these classical methods brings into relief the notion of semi-discretization of a continuum mathematical model leading to its discrete, finite-dimensional counterpart (with finite degrees of freedom, usually known as multi-degree of freedom (MDOF) systems). A piecewise (i.e. spatially localised) application of these methods sets the stage for a brief, yet very relevant, introduction to the finite element method (FEM), which forms the subject of Chapter 4.

Analysis of MDOF systems governed by coupled ordinary differential equations (ODEs) that in turn result from the semi-discretization of a continuous system (by either a classical method or the FEM) is dealt with in Chapter 5, with particular attention to solutions of linear time invariant (LTI) systems. Eigenvalue analysis, being the prime mover to obtaining free and forced vibration solutions of linear undamped or damped MDOF systems, is described in some detail in this chapter. In view of their practical importance in the seismic qualifications of structures/equipments, Chapter 6 is devoted to the problem of structural dynamic analysis under support excitations in both time and frequency domains. Here again, the top-down approach of treating a continuous structure first, only to move to the discrete MDOF model later and to the simplest building block of an SDOF oscillator at the last stage, is followed in order to be in sync with the overall expositional format of this book. The response spectrum method vis-à-vis time history methods and a practitioner's preference of the former for design purposes are also briefly touched upon.

Like Chapter 5, Chapter 7 has again a tilt to the broad area of computational structural dynamics in that it is exclusively devoted to a family of eigenvalue extraction techniques for the generalized eigenvalue problem, starting from simple power and inverse iteration methods to the powerful Lanczos method. Use of shifting for better convergence and the role of multiple eigenvalues in obtaining an eigensolution are discussed. Methods suitable for both symmetric and unsymmetric systems are covered along with illustrative examples. Whilst the fundamental aspects of these techniques are always highlighted, the issue of their numerical implementation in the context of large systems is also not lost sight of. Consistent with this approach, algorithms are provided for each of the eigensolution techniques so as to facilitate their implementations on a computer. Direct integration schemes to solve coupled ODEs represented by MDOF systems are presented in Chapter 8 with specific applications to linear structural dynamic systems. The necessary attributes of an efficient integration scheme – consistency, stability and convergence and numerical dissipation are discussed at length in this chapter along with supportive examples.

Chapter 9 briefly introduces what is somewhat loosely referred to as stochastic structural dynamics (or random vibration), which provides a useful tool in propagating the input (forcing) uncertainty, as in earthquake or wind induced structural vibrations, through the equations of motion to characterise the response uncertainty. In laying out the methodology, we have here emphasised a modern and insightful treatment of the subject based on the elements of stochastic calculus (Ito calculus in particular).

How To Read This Book

If used as a textbook for a one-semester course of Structural Dynamics at the undergraduate level, it is possible to start with and cover Chapter 1 in full, followed by parts of Chapter 2 (for instance, a few 1D examples from Sections 2.2–2.4 restricted to only simply supported beam equations as examples; Sections 2.5 and 2.6). Whilst Chapter 3 may be entirely covered, Chapter 4 on the FEM may only be used to introduce the basic notion of semi-discretization (again emphasising 1D examples and without covering the error estimates). This may be followed by a nearly full coverage of Chapters 5 and 6, whilst only touching upon the basic eigensolution approaches in Chapter 7 (e.g. Section 7.2 in full and the Jacobi method as well as its convergence in). Finally, Chapter 8 may be barely touched to introduce only the Euler and Newmark methods. In case of a one-semester long post-graduate course, the above scheme may be generally followed (possibly with the addition of examples on Timoshenko beam and 2D plate bending in Chapter 2, finite element error estimates in Chapter 4) and Chapter 9 may also be used towards the end to introduce the notion of stochastic processes and consequently the fundamental aspects of Ito calculus in evaluating the response moments.

However, if used as a reference, a reader (depending on the background) may choose to go through the chapters in this book in a selective, non-sequential manner to gain additional insights into the subject.

Acknowledgements

Writing this book has taken some effort spanned over a time interval of about one and half years, a significant part of which should have deservedly gone to our family members to whom we express our gratefulness not entirely unmixed with a sense of apology.

Introduction

Dynamics is an inherent nature of a system reacting to its environmental forces. The dynamic response may be described by some quantitative measures of the system behaviour (e.g. displacements, stresses, etc.) evolving with time. As time progresses, the transient phenomena may disappear and it is possible that the system attains a steady state. The transients disappear typically due to the internal damping present in structural dynamic systems. The steady state may be independent of time in which case the system is said to have reached the static equilibrium position. For example, a bridge/ship deck under an impact loading may undergo oscillations and eventually reach a steady state. This may also be the case with such time-invariant (static) loads acting on the system forever, once the initial perturbations die down. Certain time-varying loads of sustained nature may be periodic in time so that the system dynamically responds with (some of) the response measures also exhibiting periodicity, especially following some initial cycles of transients. For example, a rotating system such as a generator or a pump motor in a power plant experiences such periodic loads due to the inherent rotor mass imbalances. This may in turn cause excessive vibrations in the adjacent structural components/equipments. One needs to take care of the resulting response magnitudes of both the primary/secondary systems that may exceed the safe allowable limits. On the other hand, loads such as those caused by seismic events or bomb blasts may be of characteristically shorter duration. If the systems need to be designed to withstand such shock loads without a major failure, the interest for an analyst lies in mitigating or damping out the vibration levels, if not suppressing them altogether.

All the environmental loads for a structural system may not be as easily amenable for mathematical representation, which, however, constitutes a pre-requisite for a subsequent analysis to be carried out on the system. One needs to use the right blend of skill and experience in (approximately) modelling the environmental loads. The exercise is all the more fraught with difficult in that environmental loads are inherently of uncertain characteristics, that is they are often not modelled deterministically. In particular, some of them may admit modelling through what are referred to as stochastic processes. For example, wind, wave or earthquake excitations could be interpreted as instances of a stochastic process (Crandall and Mark, 1963; Lin, 1967). Stochastic processes must be treated within the framework of probability theory (Papoulis, 1991), with random variables as the basic building block. In any case, the subject of structural dynamics is relevant and may lay its claim to some level of completeness only if the excitations are properly modelled.

From the viewpoint of analysis of a structural system – a building, or a mechanical component such as a crank shaft or a space structure – it is imperative that we reduce the system into a mathematically (and computationally) viable form. This in turn requires

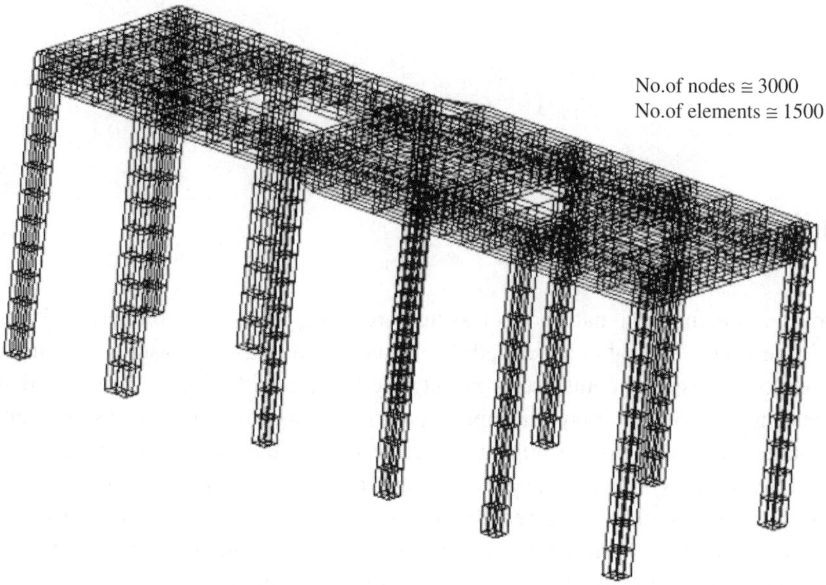

Figure I.1 Turbo-generator foundation (Indian Standard 2974 (Part 3): 1992); finite element model with 8-noded solid elements (Courtesy of Cranes Software International Ltd., Bangalore, India)

efficient tools, say to decompose the generally complex system on hand into components/elements whose mathematical models are individually far easier to handle and analyse. The mathematical models of a truss, an Euler–Bernoulli beam, a shear beam, a plate a shell or a solid body as a continuum (which typically form system components) are fortunately derivable from Newton's principle of momentum balance combined with the constitutive equations relating stresses and strains. Despite the apparent complexity with a structural system in its original form, it often admits decomposition into such simpler elemental sub-systems. A skillful approach is nevertheless warranted in carving out an acceptable model comprising of different types of these known elemental systems and finally realise reasonable results. In many cases, it is a trade-off between the effort put in the modelling and the availability of computational facilities to treat the model. With the advent of high capacity personal computers/workstations, it is no wonder that the modelling ability has of late grown manifold with the result that one is now enabled to model and solve for the response of complex structural dynamic systems with lesser effort. However, a need is always present for an adequately informed, yet intuitive, approach whilst arriving at the mathematical models.

Shown below, by way of an example, is a possible mathematical model of a turbo-generator foundation (Figure P1) that carries the turbine generator set and other auxiliary structural components. Forming this model, by all means, seems to be a formidable task. Figure P2 shows yet another case of what could possibly be labelled a 'complex structural system' – a high voltage transmission line tower with numerous bolted steel angle members. A major difficulty in the modelling exercise is to resolve the question on what constitutes a fine enough discretization to arrive at converged numerical solution for a structural system with continuous mass and stiffness distributions. It also possible that the solution corresponding to too fine a discretization may have high numerical pollution. An

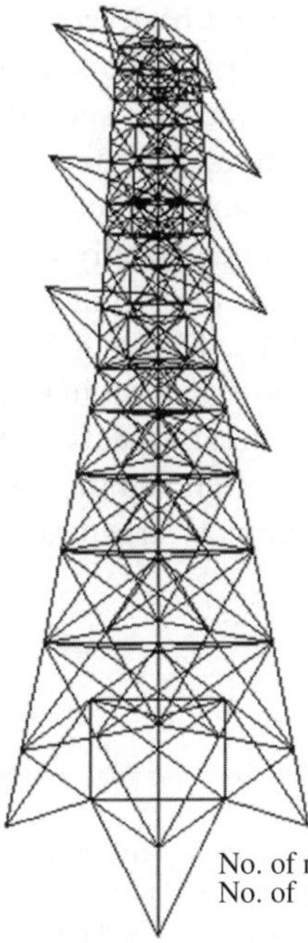

Figure I.2 Transmission line tower with two-noded truss elements (Indian Standard 802 (part 1/ Sec 1): 1995; finite element model (Courtesy of Cranes Software International Ltd., Bangalore, India)

appropriate reduction of a continuous, infinite-dimensional system to a finite dimensional one (model reduction/semi-discretization) is again an issue that is mostly resolved through professional experience with such problems and the scientific information gleaned from this book alone might prove quite inadequate to meet this challenge. Nevertheless, only professional experience, not assisted by a suitable scientific basis, is likely to spring nasty surprises especially in somewhat uncharted territories. For instance, it is readily possible to conceive of a structural system which, whilst yielding strictly positive definite stiffness matrices, may not admit a unique solution under a certain class of boundary conditions. No amount of jugglery with discretization, based on just professional skills, is going to make the computed solution unique in these scenarios.

Whilst it may not be the only available method to do so, the finite element method (FEM) (Zienkiewicz and Cheung, 1967; Hughes, 1987; Cook, Malkus and Piesha, 1989) is nevertheless the most widely used route to accomplish such model reduction. It is basically a discretization tool with a wide range of simple elemental systems from which one can make an informed choice in building up the finite dimensional mathematical model of a system on hand.

In any case, the subject of structural dynamics which, in recent times, has encompassed diverse fields of interest – from structural system identification, health assessment or monitoring, control, nanostructures and opto-electro-structural systems – is too wide in scope to be covered in a single volume, especially one such as this dealing only with the elementary building blocks of the subject. Specifically, this book is devoted to linear structural dynamics with emphasis only on the basic concepts of vibration theory under deterministic and stochastic excitations, whilst not losing sight of the recent links the subject has with computational mechanics aided by such discretization strategies as the FEM. For an analyst or a designer, a structural system or component is often a 3D continuum, which could possibly admit a dimensional reduction to 2D or even 1D continuum based on some 'engineering' (as against mathematically rigorous) approximations. In our setting, the mathematical model representing the vibratory state of a continuous system is often a partial differential equation (PDE). Deriving these equations of motion from the basic principles of continuum mechanics and solving them even for simple systems provide a great insight into some of the generic aspects of the system behaviour. It also helps in understanding the need for discretization and the philosophy behind the construction of discrete models, especially for complex structural systems which are otherwise not amenable to an analytical solutions.

References

Cook, R.D., Malkus, D.S. and Piesha, M.E. (1989) *Concepts and Applications of Finite Element Analysis*, 3rd edn, John Wiley & Sons, Inc., New York.

Crandall, S.H. and Mark, W.D. (1963) *Random Vibration in Mechanical Systems*, Academic, New York.

Hughes, T.R.J. (1987) *The Finite Element method*, Prentice Hall International, Inc.

Lin, Y.K. (1967) *Probabilistic Theory of Structural Dynamics*, McGraw-Hill, New York.

Papoulis, A. (1991) *Probability, Random Variables and Stochastic Processes*, 3rd edn, McGraw-Hill, Inc.

Zienkiewicz, O.C. and Cheung, Y.K. (1967) *The Finite Element Method in Structural and Continuum Mechanics*, McGraw-Hill, London.

General Notations

A	Area of cross-section
$\boldsymbol{A}$	Matrix operator (linear transformation)
$\boldsymbol{B}$	Partial differential operator in strain-displacement relationship
c	Damping coefficient in an SDOF oscillator
$\boldsymbol{c}$	Matrix of the elasticity constants in stress-strain constitutive relationship
$\boldsymbol{C}$	Damping matrix
$\mathbb{C}$	The field of complex numbers
$C^m(a, b)$	Functions that, together with their first m derivatives, are continuous on (a, b)
E	Young's modulus
E_λ	Eigenspace corresponding to the eigenvalue λ
$\boldsymbol{f(t)}, \boldsymbol{f(x, t)}$	External forcing functions
F	Deformation gradient
F_b	Body force
F_s	Surface force
$h(t)$	Impulse response function of an oscillator
$h_j(t)$	Impulse response function of jth mode
$H(\lambda)$	Complex response function (Fourier transform of $h(t)$)
I	Moment of inertia
$\boldsymbol{I}$	Energy functional, identity matrix
J	Determinant of F
$\boldsymbol{k}$	Stiffness of an SDOF oscillator
$\boldsymbol{K}$	Stiffness matrix
l	Length parameter
L	Lagrangian, $T - V$
$\boldsymbol{L}$	Differential operator
m	Lumped mass in an SDOF oscillator, mass density per unit length for 1D systems, mass density per unit area for 2D systems
$\boldsymbol{M}$	Mass matrix
N	Size of the matrix
$\boldsymbol{N}$	Null space operator
$\mathbb{N}$	The natural numbers
$\boldsymbol{q}(t)$	Vector of generalised coordinates

$\mathbb{R}$	Set of real numbers (real line)
$\mathbb{R}^+$	Non-negative real numbers
$\mathbb{R}^n$	n-Dimensional Euclidean space
$\mathbb{R}^{n\times m}$	$n \times m$ Matrices with real elements
t	Time variable
T	Kinetic energy
$\hat{u}(\boldsymbol{x}, t)$	Approximate solution
$\boldsymbol{u}(\boldsymbol{x}, t)$	Displacement field variable $(= (u_1, u_2, u_3)^T)$
$U(.)$	Unit-step function
V	Potential energy
$\boldsymbol{x}$	Coordinate axes
X	Coordinate axis
Y	Coordinate axis
Z	Coordinate axis
$\mathbf{Z}$	The integers
$\delta(.)$	Dirac delta function (see Equation 2.180)
δ_{ij}	Kronecker delta (see Equation 2.167)
$\delta\Omega$	Boundary domain
ε	Strain tensor
ε_{ij}	Strain components
λ, λ_j	Eigenvalues
(λ_i, u_i)	Eigenpair of eigenvalue and associated eigenvector
μ	Frequency (eigenvalue) parameter
υ	Poisson's ratio
ξ	Damping ratio of an SDOF oscillator
ρ	Mass density
$\rho(.)$	Rayleigh quotient
$\boldsymbol{\sigma}$	Stress tensor
σ_{ij}	Stress components
ω	Radian frequency
ω_n	Natural frequency of an SDOF oscillator
ω_d	Damped natural frequency of an SDOF oscillator
Δ	Laplacian operator: $\Delta f = \sum_j \frac{\partial^z f}{\partial x_j^2}$
$\Phi_j(\boldsymbol{x}), j = 1, 2, \ldots$	Orthonormal eigenfunctions of $\boldsymbol{x}$ (for a continuous system)
$\boldsymbol{\Phi}$	Matrix of eigenvectors $(= \{\boldsymbol{\Phi}_n\})$
$\boldsymbol{\Lambda}$	Diagonal matrix containing eigenvalues on the diagonal
Ω	Material domain
£	Laplace transform
∇	Gradient of a function: $\nabla f = \left(\frac{\partial f}{\partial x_1}, \ldots, \frac{\partial f}{\partial x_n}\right)$
$\Vert.\vert$	Norm

1

Structural Dynamics and Mathematical Modelling

1.1 Introduction

Structural dynamics finds wide application in all areas of engineering – civil, mechanical, aerospace, marine and many others. Excellent text books and treatises initiating a newcomer to structural dynamics are available in Meirovitch (1967), Craig (1981), Clough and Penzien (1982). For an academic learner and a practising engineer, it is particularly important to assimilate the basic concepts, judiciously apply them whilst assessing or predicting the performance of the structure and interpret the results in a coherent and meaningful fashion. The material presented in this chapter is mainly aimed at serving this purpose. The mathematical rigour in the presentation is thus kept at a level consistent with the above aim. The effort has been to emphasise the basic concepts whilst keeping in mind their usefulness from a practical perspective.

Dynamic equations of motion are derivable based on variants of Newton's principle of force balance and they typically appear in the form of partial differential equations (PDEs). Any structural system with varying stiffness and mass distribution over its volume may be (mathematically) modelled as a continuum having, in principle, infinite degrees of freedom (*dof* s). The degrees of freedom generally refer to the unknown displacements at any spatial (material) point in the system. For a solid body modelled as a continuum, there exist 3 *dof* s – for example three orthogonal translations – at a material point. However, for certain cases wherein the measure (length or span) of the material domain along one or two spatial dimensions is much less compared to the other(s), the mathematical model can be considerably simplified through a dimensional descent, that is by appropriately averaging over such dimensions leaving only the dominant dimension(s) in the reduced model. This is, for instance, the case with a plate (or a beam), which is derivable from the 3D continuum model through such a descent by averaging across the so-called thickness dimension(s). For such dimensionally reduced continuum models, an additional set of three rotation degrees of freedom at every material point needs to be introduced in order to capture the post-deformed structural orientations across the eliminated dimensions. The governing PDEs of motion (in the retained space variables and time) are typically derived

Elements of Structural Dynamics: A New Perspective, First Edition. Debasish Roy and G Visweswara Rao.

based on the principles of virtual work and variational calculus (Hughes, 1987; Reddy, 1984, Humar, 2002).

In practice, further simplification is achieved by reducing the governing PDEs of motion into a system of ordinary differential equations (ODEs) through a process called discretization, wherein the spatial variable(s) in the governing equations is (are) removed, leaving time as the only independent variable. The process of discretization typically involves expressing the dependent response function in a (piecewise smooth) series involving known functions of the space variable(s) (often referred to as the shape functions) and unknown functions of time (often called generalized coordinates). This finally yields, upon some form of averaging or integration over the space variable(s), a system of ODEs of motion. The finite element method (FEM) is probably the most widely employed tool for such discretization based on spatially localised and piecewise smooth shape functions. Pending further elaboration of these ideas on discretization, some of which will be taken up subsequently in this book, we presently focus in this chapter on the basic principles of deriving the equations of motion based on virtual work.

1.2 System of Rigid Bodies and Dynamic Equations of Motion

To start with, it is instructive to consider a few simple cases of rigid-body dynamics, wherein the governing equations are in the form of ODEs.

1.2.1 Principle of Virtual Work

It is known that for a single particle, the dynamic equilibrium is characterised by the equation of motion as:

$$F_i = m\ddot{x}_i, \; i = 1, 2, 3 \tag{1.1}$$

where the vector $\{x_i\}$ denotes translations in the three global directions. The above equation is due to Newton's second law of motion with $\ddot{x}_i$ denoting the acceleration of the particle and F_i the external force in the ith global direction. If the inertial term is also treated as a form of (fictitious) force acting on the particle, we arrive at D'Alembert's principle by restating Equation (1.1) as:

$$F_i - m\ddot{x}_i = 0, \; i = 1, 2, 3 \tag{1.2}$$

Generalising Equation (1.2) for a system of particles (Figure 1.1) with mass m_j, $j = 1, 2 \ldots . N$, one obtains the corresponding equations of motion for dynamic equilibrium as:

$$\sum_{j=1}^{N} (F_{ij} - m_j\ddot{x}_{ij}) = 0, \; i = 1, 2, 3 \tag{1.3}$$

At this stage, we may consider the configuration (solution) at dynamic equilibrium to extremise (minimise) some (energy) functional. Then, the so-called principle of virtual work may be stated as:

$$\sum_{j=1}^{N} (F_j - m_j\ddot{x}_j) \cdot \delta x_j = 0, \; j = 1, 2 \ldots N \tag{1.4}$$

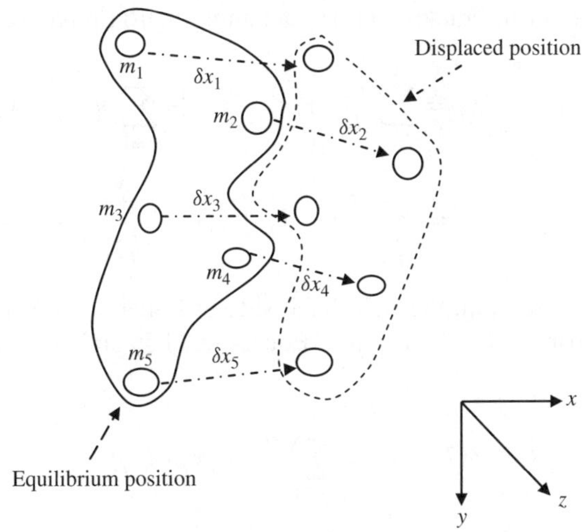

Figure 1.1 Particles in equilibrium and imposed virtual displacements

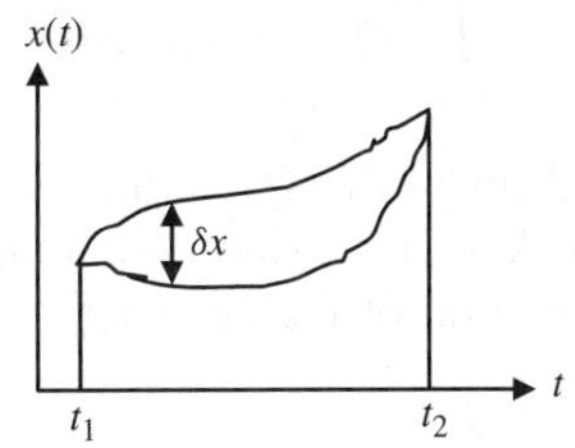

Figure 1.2 Vanishing of virtual displacement, δx at end time instants t_1 and t_2

Here, $\{F_j\}$ is the force vector with scalar components resolved in the three global directions and $\{\delta x_j\}$ is a vector of virtual displacements, which are $C^1(t_1, t_2)$ with $\delta x_j(t_1) = \delta x_j(t_2) = 0$ (Figure 1.2), imposed on the system in equilibrium (Figure 1.1). In particular, the virtual displacements may be interpreted as admissible variations (i.e. possible solutions of the governing second-order ODEs) of the dynamic equilibrium configuration. Moreover, they must vanish at the boundaries of the time interval of interest. It is worthwhile to note that the virtual variation, denoted by $\delta x(t)$, must be distinguished from the functional increment $dx(t) = x(t)dt$ in time. Now, the scalar product on the left-hand side of Equation (1.4) may be interpreted as the virtual work done by the forces at equilibrium, which, according to the principle of virtual work, is zero.

1.2.2 Hamilton's Principle

The applied force vector F_j in Equation (1.4) is conservative, it is derivable from a potential V as:

$$\sum_{j=1}^{N} F_j \cdot \delta x_j = -\delta V \tag{1.5}$$

Further, the second term in Equation (1.4) containing inertia forces, can be rewritten as:

$$\sum_{j=1}^{N} m_j\ddot{x}_j \cdot \delta x_j = \sum_{j=1}^{N} \frac{d}{dt}(m_j\dot{x}_j \cdot \delta x_j) - \sum_{j=1}^{N} m_j\dot{x}_j \cdot \delta\dot{x}_j$$

$$= \sum_{j=1}^{N} \frac{d}{dt}(m_j\dot{x}_j \cdot \delta x_j) - \delta \sum_{j=1}^{N} \frac{1}{2} m_j\dot{x}_j \cdot \dot{x}_j \tag{1.6}$$

Noting that the second term on the right-hand side of Equation (1.6) is the kinetic energy T of the system of particles, we substitute Equations (1.5) and (1.6) in Equation (1.4) to obtain:

$$\delta(T - V) = \sum_{j=1}^{N} \frac{d}{dt}(m_j\dot{x}_j \cdot \delta x_j) \tag{1.7}$$

Integration of Equation (1.7) over the interval (t_1, t_2) together with the fact that the variational and integration operators commute yields:

$$\delta \int_{t_1}^{t_2} (T - V)dt = 0 \tag{1.8}$$

In deriving the above result, vanishing of $\delta x(t)$, at the two ends of the interval of integration is utilised. Equation (1.8) is known as Hamilton's principle, which states that, for a conservative system, the first variation of the Lagrangian $T - V$ remains stationary with respect to the virtual displacement.

1.2.3 Lagrangian Equations of Motion

At this stage it is convenient to introduce the concept of generalized coordinates, which constitute a set of linearly independent response variables (e.g. displacements or their linear combinations) that, along with their first time derivatives (called generalized velocities) enable recasting the governing ODEs of motion in the state space form, that is as a system of first order ODEs. For example, the dynamics of a pendulum (Figure 1.3) of mass m can be described in term of the position of the centre of mass of the bob in Cartesian coordinates x, y and their instantaneous velocities. However, since $\sqrt{x^2 + y^2} = l$, the number of generalized coordinates is one in this case. Accordingly, it is more convenient

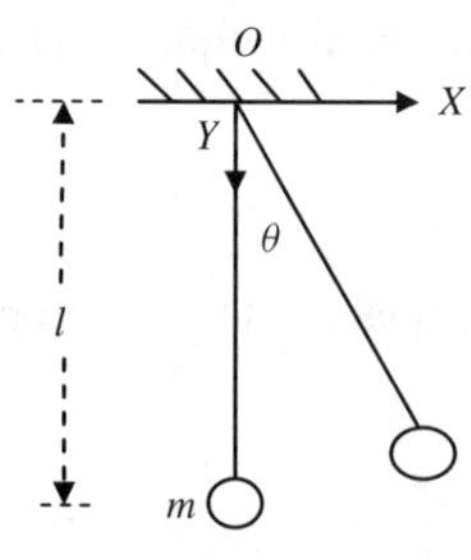

Figure 1.3 Generalized coordinates; dynamic motion of a pendulum, $\theta = q_1$

to describe the motion in terms of the angle θ, around the point of suspension 'O', and its time derivative $\dot{\theta}$. Note that the position coordinates x and y can be expressed in terms of θ as $x = l\sin\theta$ and $y = l\cos\theta$.

Thus, in general, all displacements x_j in a (discrete) dynamical system can be expressed in terms of these generalized coordinates, $q_i, i = 1, 2, \ldots n$ as:

$$x_j = x_j(q_1, q_2, \ldots\ldots\ldots, q_n) \tag{1.9}$$

From Hamilton's principle, considering $T = T(q_k, \dot{q}_k, t)$ and $V = V(q_l, t)$ with $k = 1, 2 \ldots, n$, we have the following expressions:

$$\delta T = \sum_{k=1}^{n}\left\{\frac{\partial T}{\partial q_k}\delta q_k + \frac{\partial T}{\partial \dot{q}_k}\delta \dot{q}_k\right\},\ \delta V = \sum_{k=1}^{n}\left\{\frac{\partial V}{\partial q_k}\delta q_k\right\} \tag{1.10}$$

Here, δq_k denotes virtual displacement of the generalized coordinate q_k. The above two expressions, when substituted in Equation (1.8), yield:

$$\int_{t_1}^{t_2}\sum_{k=1}^{n}\left\{\frac{\partial T}{\partial q_k}\delta q_k + \frac{\partial T}{\partial \dot{q}_k}\delta \dot{q}_k - \frac{\partial V}{\partial q_k}\delta q_k\right\}dt = 0 \tag{1.11}$$

Integrating the second term in Equation (1.11) by parts results in the following:

$$\begin{aligned}\int_{t_1}^{t_2}\frac{\partial T}{\partial \dot{q}_k}\delta \dot{q}_k dt &= \int_{t_1}^{t_2}\frac{\partial T}{\partial \dot{q}_k}d(\delta q_k) = \left\{\frac{\partial T}{\partial \dot{q}_k}\delta q_k\right\}_{t_1}^{t_2} - \int_{t_1}^{t_2}\frac{d}{dt}\left(\frac{\partial T}{\partial \dot{q}_k}\right)\delta q_k dt\\ &= -\int_{t_1}^{t_2}\frac{d}{dt}\left(\frac{\partial T}{\partial \dot{q}_k}\right)\delta q_k dt\end{aligned} \tag{1.12}$$

Whilst deriving Equation (1.12), the vanishing of δq_k at the ends of the interval of integration and the commutation of the differentiation operator with that for the virtual displacement have been made use of. In view of Equation (1.12), Equation (1.11) may be recast as:

$$\int_{t_1}^{t_2}\sum_{k=1}^{n}\left\{-\frac{d}{dt}\left(\frac{\partial T}{\partial \dot{q}_k}\right) + \frac{\partial T}{\partial q_k} - \frac{\partial V}{\partial q_k}\right\}\delta q_k dt = 0 \tag{1.13}$$

Since $\delta \boldsymbol{q} \neq \boldsymbol{0}$ we have:

$$\frac{d}{dt}\left(\frac{\partial T}{\partial \dot{q}_k}\right) - \frac{\partial T}{\partial q_k} + \frac{\partial V}{\partial q_k} = 0,\quad k = 1, 2 \ldots\ldots n \tag{1.14}$$

Equation (1.14) is the Lagrange equation of motion for the system in dynamic equilibrium. It represents a set of ODEs in the n generalized coordinates.

Example 1.1: Equation of motion for a pendulum

We derive the Lagrange equation of motion for the pendulum as in Figure 1.3. Towards this, we first express T and V in terms of the generalized coordinate θ and its time derivative $\dot{\theta}$ as:

$$T = \frac{1}{2}ml^2\dot{\theta}^2,\ V = mgl(1 - \cos\theta) \tag{1.15}$$

$$\frac{\partial T}{\partial \dot{\theta}} = ml^2\dot{\theta},\ \frac{d}{dt}\left(\frac{\partial T}{\partial \dot{\theta}}\right) = ml^2\ddot{\theta}\ \text{ and }\ \frac{dV}{d\theta} = mgl\sin\theta \tag{1.16}$$

Thus, the governing ODE of motion is given by:

$$\ddot{\theta} + \frac{g}{l}\sin\theta = 0 \tag{1.17}$$

Note that Equation (1.17) is nonlinear in θ.

1.3 Continuous Dynamical Systems and Equations of Motion from Hamilton's Principle

Hamilton's principle in Equation (1.8) described for a discrete dynamical system, consisting of a system of particles, can also be generalized for a continuous system. To this end, let us consider a deformable continuous body with body force F_b and imposed boundary constraints as well as surface traction F_s, shown in Figure 1.4. Let the material domain of the body be denoted as Ω so that $\partial\Omega$ denotes the boundary (considered as Lipschitz). If $\boldsymbol{u}(x, y, z) = (u_1, u_2, u_3)^T$ represents the displacement vector function, the kinetic energy T is given by (Einstein's summation convention is consistently used in the rest of the chapter):

$$T = \left(\frac{1}{2}\right)\int_{\Omega} m\dot{u}_i\dot{u}_i dV,\ i = 1, 2, 3 \tag{1.18}$$

Expression for the potential energy V consists of two parts, namely, the strain energy (stored due to material deformation) and the work done by the external forces. The contribution V_1, from the external forces, is given by:

$$V_1 = -\int_{\Omega} F_{bi}u_i dV - \int_{\partial\Omega} F_{si}u_i dS \tag{1.19}$$

where F_{bi} and F_{si} are the body and surface traction forces in the ith global direction with $i = 1, 2$ and 3 that are equivalent to X, Y and Z axial directions.

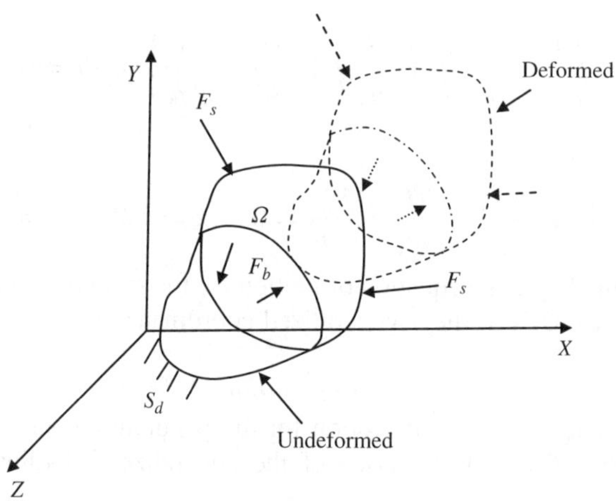

Figure 1.4 Undeformed (thick line) and deformed (dashed line) configurations of a continuous system; F_b – body forces, F_s – surface forces and S_d – specified displacements

1.3.1 Strain and Stress Tensors and Strain Energy

The strain energy V_2 is a function of the strains within the body. In Figure 1.4, both the undeformed and deformed configurations are shown. The deformation, resulting due to the application of the external forces, causes strains and hence stresses in the body. By the word 'deformation' (as distinguished from the rigid-body movement), we imply the class of translations of material points such that the Euclidean distance between a pair of such points changes. Strains are typically pointwise measures of rates of changes in such deformations as functions of the space variables. Accordingly, a natural starting point in the mathematical description of strains would be to consider a pair of material points and study their relative orientations in the undeformed and deformed configurations. Thus, if $\mathbf{x}$ and $\boldsymbol{x}'$ denote a material point in the undeformed and deformed configurations, respectively, (see Figure 1.7) and $\boldsymbol{u}(\boldsymbol{x}) := (u_1, u_2, u_3)^T$ denotes the material displacement vector, we have:

$$\boldsymbol{x}' = \boldsymbol{x} + \boldsymbol{u} \tag{1.20}$$

Use of the chain rule of differentiations immediately leads to:

$$dx_i' = \frac{\partial x_i'}{\partial x_j} dx_j \tag{1.21}$$

We refer to $\boldsymbol{F} = \left[\frac{\partial x_i'}{\partial x_j}\right]$ as the deformation gradient tensor such that $d\boldsymbol{x}' = \boldsymbol{F}d\boldsymbol{x}$ (see Bibliography for tensors and their notations). Using Equation (1.20), one may also write $\boldsymbol{F}$ as:

$$[\boldsymbol{F}]_{ij} = \boldsymbol{\delta}_{ij} + \frac{\partial u_i}{\partial x_j} \tag{1.22}$$

where $\boldsymbol{\delta}_{ij}$ denotes the Kronecker delta defined as $\delta_{ij} = 1$ for $i = j$ and zero otherwise. Since $\boldsymbol{F}$ is assumed to be an isomorphism between the deformed and undeformed configurations, its inverse exists and we may also write $d\boldsymbol{x} = \boldsymbol{F}^{-1}d\boldsymbol{x}'$. Now, if ds and ds' denote the distances between a pair of infinitesimally separated points in the undeformed and deformed configurations, respectively, we have $(ds)^2 = dx_i dx_i = (d\mathbf{x})^T d\boldsymbol{x}$ and $(ds')^2 = dx_i' dx_i' = (d\boldsymbol{x}')^T d\boldsymbol{x}' = (d\boldsymbol{x})^T \boldsymbol{F}^T \boldsymbol{F}(d\boldsymbol{x}) = dx_i [F]_{ji} [F]_{ij} dx_j$. The Green–Lagrange strain tensor ε is now defined through the following equation:

$$(ds')^2 - (ds)^2 = 2dx_i [\varepsilon]_{ij} dx_j \tag{1.23}$$

Accordingly, ε may be readily expressed in terms of $\boldsymbol{F}$ as:

$$\boldsymbol{\varepsilon} = \frac{1}{2}[\boldsymbol{F}^T \boldsymbol{F} - \boldsymbol{I}] \tag{1.24}$$

Alternatively, in terms of the scalar components, we have:

$$[\boldsymbol{\varepsilon}]_{ij} = \left(\frac{1}{2}\right)\left\{\frac{\partial u_i}{\partial x_j} + \frac{\partial u_j}{\partial x_i} + \frac{\partial u_k}{\partial x_i}\frac{\partial u_k}{\partial x_j}\right\} \tag{1.25}$$

Note that the last term on the right-hand side is nonlinear in $\boldsymbol{u}$. For dynamical systems undergoing deformations associated with small strains, a linearised form of $\boldsymbol{\varepsilon}$ may be readily obtained by dropping the nonlinear term:

$$[\boldsymbol{\varepsilon}_L]_{ij} = \left(\frac{1}{2}\right)\left\{\frac{\partial u_i}{\partial x_j} + \frac{\partial u_j}{\partial x_i}\right\} \tag{1.26}$$

Consistent with the definition of the strain tensor above, stresses in a body are also pointwise defined functions of the space variables. Typically, stress at a material point within the body is a measure of the traction per unit area of a cross section passing through that point. The concept of stress at a material point is useful in evaluating the traction acting along any cross section passing through the material point. In order to formalise this concept, it is convenient to consider in succession (i.e. not simultaneously) three mutually orthogonal sections passing through the material point and conduct the following thought experiment. Consider one such square section and resolve the associated traction (not necessarily normal to that section) into three orthogonal components, which are then divided by the area of the square. In the limit of this square approaching a point, the three stress components at the material point corresponding to that section are recovered as the limits of the associated traction components divided by the area (provided the limits exist). Repeating this thought exercise for the other two orthogonal sections, one finally recovers nine stress components at a material point (Figure 1.5). If these thought experiments are conducted with cross sections in the deformed configuration, one obtains the so-called Cauchy stress tensor σ, given by:

$$\sigma = [\sigma_{ij}] = \begin{bmatrix} \sigma_{11} & \sigma_{12} & \sigma_{13} \\ \sigma_{21} & \sigma_{22} & \sigma_{23} \\ \sigma_{31} & \sigma_{32} & \sigma_{33} \end{bmatrix} \tag{1.27}$$

where the integer subscripts 1,2,3 correspond to the triad $(x_1, x_2, x_3) := (X, Y, Z)$. Considering a specific stress component σ_{ij}, we observe that the first subscript is the integer index corresponding to the normal direction $\boldsymbol{x}_i$ to the section over which the component is acting and the second subscript denotes the direction $\boldsymbol{x}_j$ along which the component is acting. Thus, the stress tensor is a second-order tensor. Alternatively,

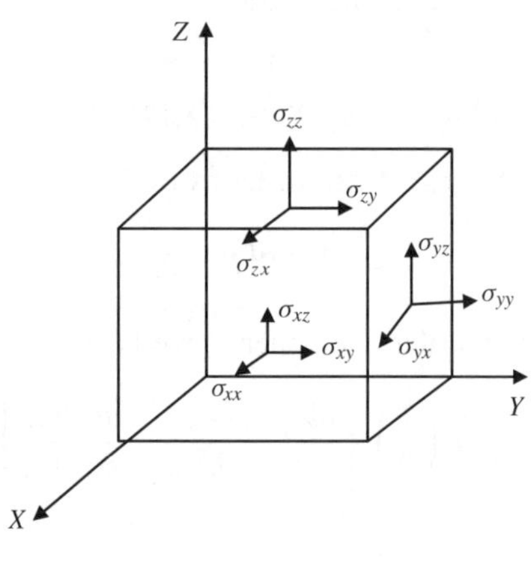

Figure 1.5 Stress at a material point in a body

σ_{ij} may also be interpreted as a bilinear transformation involving the first derivatives with respect to x_i and x_j. Whilst the diagonal elements σ_{11}, σ_{22} and σ_{33} are normal stresses, and the off-diagonal ones σ_{12}, σ_{13}, σ_{21}, σ_{23}, σ_{31} and σ_{32} are shear stresses. Satisfaction of the equation of angular momentum at the material point requires that σ is symmetric, thereby reducing the number of unknown stresses to six. Following the Voigt convention, σ may sometimes be represented in the form of a vector as $\sigma = (\sigma_{11}\ \sigma_{22}\ \sigma_{33}\ \sigma_{23}\ \sigma_{31}\ \sigma_{12})^T$. However, since the deformed area *da* is generally not known a priori, other forms of stress measures such as the first and second Poila–Kirchoff stress tensors are defined with respect to the undeformed configuration. Thus, it is appropriate to understand the definitions of these two stress measures as well.

The first Poila–Kirchoff stress tensor $[T_{ij}]$ relates the forces in the deformed configuration to the areas in the undeformed configuration. Here we invoke Nanson's relation between deformed and undeformed areas, namely, $da\vec{n} = F^{-T} J\, dA\, \vec{N}$, where $J := \det(F) \cdot \vec{n}$ and $\vec{N}$ are the unit normal vectors to the deformed and undeformed areas *da* and *dA*, respectively. The first Poila–Kirchoff stress tensor is defined through the following equation:

$$\sigma_{ij}\, da\vec{n} = T_{ij} dA\, \vec{N} \Rightarrow \sigma_{ij} F^{-T} J\, dA\, \vec{N} = T_{ij} dA\, \vec{N} \Rightarrow T_{ij} = \sigma_{ij} F^{-T} J = J \sigma F^{-T} \quad (1.28)$$

Note that the first Poila–Kirchoff stress tensor is not symmetric. This is an obvious disadvantage in numerical computations. The second Poila–Kirchhoff stress tensor $[S_{ij}]$ relates forces in the reference configuration to areas in the reference configuration. To derive the stress sensor, it is required to relate the forces in the deformed and undeformed configurations. If dP and dP' are the forces in the undeformed and deformed configurations, respectively, it can be easily observed that:

$$\begin{aligned} dP' &= \boldsymbol{F} dP = S \cdot \vec{N} dA \Rightarrow F^{-1} \cdot \sigma \cdot \vec{n}\, da = S \cdot \vec{N} dA \\ &\Rightarrow F^{-1} \cdot \sigma \cdot F^{-T} J\, dA\, \vec{N} = S \cdot \vec{N} dA \\ &\Rightarrow S = J\, F^{-1} \cdot \sigma \cdot F^{-T} \end{aligned} \quad (1.29)$$

It can be observed that the second Piola-Kirchhoff stress tensor is symmetric. It follows that the strain energy is given by:

$$V_2 = \int_\Omega W_{ij} dV, \quad W_{ij} = \sigma_{ij} \varepsilon_{ij} \quad (1.30)$$

Substitution of expressions in Equations (1.18), (1.19) and (1.30) in Equation 1.8 gives:

$$\delta \int_{t_1}^{t_2} (T - V) dt = \int_{t_1}^{t_2} \left\{ \int_\Omega \left(m\dot{u}_i \delta \dot{u}_i - \sigma_{ij} \delta \varepsilon_{ij} - F_{bi} \delta \dot{u}_i \right) dV + \int_{\partial\Omega} F_{si} \delta u_i dS \right\} dt = 0 \quad (1.31)$$

The first integral on the right-hand side of Equation (1.31) takes the form:

$$\begin{aligned} \int_{t_1}^{t_2} \int_\Omega m\dot{u}_i \delta \dot{u}_i dV dt &= \int_\Omega \int_{t_1}^{t_2} m\dot{u}_i d(\delta u_i) dV = \int_\Omega \left\{ m\dot{u}_i \delta u_i \Big|_{t_1}^{t_2} - \int_{t_1}^{t_2} m\ddot{u} \delta u_i dt \right\} dV \\ &= -\int_\Omega \left(\int_{t_1}^{t_2} m\ddot{u} \delta u_i dt \right) dV \end{aligned} \quad (1.32)$$

In the above simplification, vanishing of the virtual displacements at the two ends of the interval is utilised. The second term in the volume integral on the right-hand side of Equation (1.31) can be simplified as follows:

Considering the symmetry of the two-dimensional stress tensor ($\sigma_{ij} = \sigma_{ij}$) and integrating second term in Equation (1.31) by parts, one gets:

$$\int_{\Omega} \sigma_{ij}\delta\varepsilon_{ij}\, dV = \int_{\Omega} \sigma_{ij}\delta\left(\frac{\partial u_j}{\partial x_i}\right) dV = \int_{\Omega} \sigma_{ij}\frac{\partial(\delta u_j)}{\partial x_i} dV$$
$$= \int_{\partial\Omega} n_i \sigma_{ij} \delta u_j dS - \int_{\Omega} \frac{\partial \sigma_{ij}}{\partial x_i}\delta u_j dV \tag{1.33}$$

Here, $n_i,\ i = 1, 2, 3$ are unit normals in respective global directions. Now, in view of Equations (1.32) and (1.33), Equation (1.31) reduces to:

$$\delta\int_{t_1}^{t_2} (T - V)dt = \int_{t_1}^{t_2} \left[\int_{\Omega}\left(\frac{\partial\sigma_{ij}}{\partial x_j} - m\ddot{u}_j + F_{bj}\right)\delta u_j dV\right.$$
$$\left. + \int_{\partial\Omega} (F_{sj} - n_i\sigma_{ij})\delta u_j dS\right] dt = 0 \tag{1.34}$$

Since the virtual displacement δu_j is arbitrary, the dynamic equations of motion along with the natural boundary conditions are obtained from Equation (1.34) as:

$$\frac{\partial\sigma_{ij}}{\partial x_j} - m\ddot{u}_j + F_{bj} = 0 \tag{1.35}$$

$$F_{sj} - n_i\sigma_{ij} = 0 \tag{1.36}$$

Equation (1.35) represents the dynamic equations of motion of a continuous system in the form of a PDE. Equation (1.36) represents the natural boundary condition that has emerged from the application of the principle of virtual work. Dynamic equilibrium equations for the case of large deformation can be similarly derived (Geradin and Rixen, 1994) using the nonlinear strain displacement relationship in Equation (1.25).

Example 1.2: Equations of motion for an axial loaded bar

If the simple case of an axially loaded bar (Figure 1.6) is considered, the expressions for kinetic and potential energies are given by

$$T = \left(\frac{1}{2}\right)\int_0^l m(x)\left(\frac{\partial u}{\partial t}\right)^2 dx, \quad V = \int_0^l EA(x)\left(\frac{\partial u}{\partial x}\right)^2 dx \tag{1.37}$$

E and A in Equation (1.37) stand for Young's modulus, the material constant and the area of cross section of the bar.

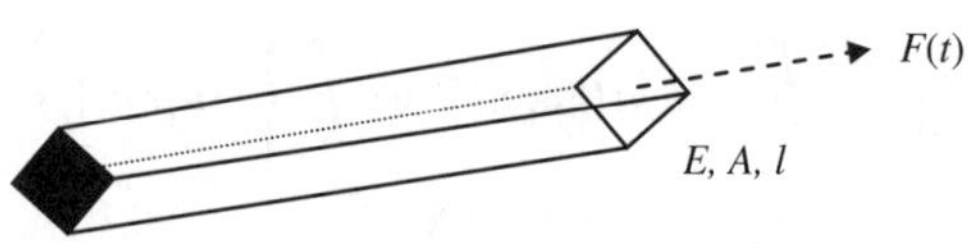

Figure 1.6 Axially loaded bar as a continuous system

In writing the expression for V, we utilise the fact that the stress and strain tensors are one-dimensional and are given by:

$$\varepsilon_x = \frac{\partial u}{\partial x} \quad \text{and} \quad \sigma_x = E\varepsilon_x = E\frac{\partial u}{\partial x} \tag{1.38}$$

The dynamic equation of motion for the vibrating bar follows from the equilibrium Equation (1.35) as:

$$\frac{\partial}{\partial x}\left(EA\frac{\partial u}{\partial x}\right) + F = m\ddot{u} \tag{1.39}$$

The PDE (Equation (1.39)) represents a one-dimensional wave equation. One can derive in a similar fashion dynamic equations of motion for other continuous systems. Such derivations further appear in the later chapters as and when they are needed. Table 1.1 describes the dynamic equations of motion for some of the familiar continuous systems.

1.4 Dynamic Equilibrium Equations from Newton's Force Balance

The dynamic equation of motion in Equation (1.35) can be derived from the solid mechanics point of view also. The state of a stress at any point inside a continuum is shown in Figure 1.5. It is familiar that the stress tensor σ consists of six components and six corresponding components of strains constitute the strain tensor ε. Thus:

$$\boldsymbol{\sigma}^T = (\sigma_{xx}, \sigma_{yy}, \sigma_{zz}, \sigma_{xy}, \sigma_{xz}, \sigma_{yz}) \tag{1.40}$$

$$\boldsymbol{\varepsilon}^T = (\varepsilon_{xx}, \varepsilon_{yy}, \varepsilon_{zz}, \varepsilon_{xy}, \varepsilon_{xz}, \varepsilon_{yz}) \tag{1.41}$$

1.4.1 Displacement–Strain Relationships

Since strain at a point is the change of displacement per unit length, it follows:

$$\begin{aligned} &\varepsilon_{xx} = \frac{\partial u}{\partial x}, \quad \varepsilon_{yy} = \frac{\partial u}{\partial y}, \quad \varepsilon_{zz} = \frac{\partial u}{\partial z}, \\ &\varepsilon_{xy} = \frac{\partial u}{\partial y} + \frac{\partial v}{\partial x}, \quad \varepsilon_{xz} = \frac{\partial u}{\partial z} + \frac{\partial w}{\partial x}, \quad \varepsilon_{yz} = \frac{\partial v}{\partial z} + \frac{\partial w}{\partial y} \end{aligned} \tag{1.42}$$

In matrix form Equation (1.42) can be written as:

$$\boldsymbol{\varepsilon} = \boldsymbol{B}\boldsymbol{U}, \quad \text{with} \quad \boldsymbol{U}^T = (u, v, w) \text{ and}$$

$$\boldsymbol{B} = \begin{bmatrix} \frac{\partial}{\partial x} & 0 & 0 \\ 0 & \frac{\partial}{\partial y} & 0 \\ 0 & 0 & \frac{\partial}{\partial z} \\ \frac{\partial}{\partial y} & \frac{\partial}{\partial x} & 0 \\ \frac{\partial}{\partial z} & 0 & \frac{\partial}{\partial x} \\ 0 & \frac{\partial}{\partial z} & \frac{\partial}{\partial y} \end{bmatrix} \tag{1.43}$$

Table 1.1 Equations of motion for typical cases of continuous systems

Transverse vibrations of a uniform taut string	Torsional vibrations of a beam	Flexural vibrations of a uniform beam	Flexural vibrations of a uniform plate
		(With rotary inertia effects neglected)	(With rotary inertia effects neglected)
Energy expressions:	Energy expressions:	Energy expressions:	Energy expressions:
$T = \frac{1}{2}\int_0^l m\left(\frac{\partial y}{\partial t}\right)^2 dx$	$T = \frac{1}{2}\int_0^l I\left(\frac{\partial u}{\partial t}\right)^2 dx$	$T = \frac{1}{2}\int_0^l m\left(\frac{\partial u}{\partial t}\right)^2 dx$	$T = \frac{1}{2}\int\int m\left(\frac{\partial w}{\partial t}\right)^2 dxdy$
$V = \frac{1}{2}\int_0^l T_0\left(\frac{\partial y}{\partial x}\right)^2 dx$	$V = \frac{1}{2}\int_0^l GJ\left(\frac{\partial^2 u}{\partial x^2}\right)^2 dx$	$V = \frac{1}{2}\int_0^l EI\left(\frac{\partial^2 u}{\partial x^2}\right)^2 dx$	$V = \frac{1}{2}\int\int D\left(\frac{\partial^2 w}{\partial x^2} + \frac{\partial^2 w}{\partial y^2}\right)^2 dxdy$
The governing PDE for vertical deflection $y(x,t)$:	The governing PDE for rotational dof $u(x,t)$:	The governing PDE for transverse deflection $u(x,t)$:	The governing PDE for vertical deflection $w(x,t)$:
$T_0\frac{\partial^2 y}{\partial x^2} + m\frac{\partial^2 y}{\partial t^2} = F(x,t)$	$GJ\frac{\partial^2 u}{\partial x^2} + I\frac{\partial^2 u}{\partial t^2} = T(x,t)$	$\frac{\partial^2}{\partial x^2}\left(EI\frac{\partial^2 u}{\partial x^2}\right) + m\frac{\partial^2 u}{\partial t^2} = F(x,t)$	$D\left(\frac{\partial^4 w}{\partial x^4} + \frac{\partial^4 w}{\partial y^4} + 2\frac{\partial^4 w}{\partial x^2 \partial y^2}\right) + m\frac{\partial^2 w}{\partial t^2} = F(x,y,t)$
T_0: Tension in the String	where	m: mass per unit length	m: mass per unit area
m: mass per unit length	I: Polar mass moment of inertia	E: Young's modulus of elasticity	D: Flexural rigidity $= \frac{Et^2}{12(1-\nu^2)}$
	G: Shear modulus	I: Flexural moment of inertia	E = Young's modulus
	J: Torsional constant		ν = Poisson's ratio
			t = plate thickness

1.4.2 Stress–Strain Relationships

The generalized Hooke's law for an anisotropic material is given by:

$$\boldsymbol{\sigma} = \boldsymbol{c}\boldsymbol{\varepsilon} \tag{1.44}$$

$\boldsymbol{c}$ is the matrix of material constants (see, for instance, Chapter 2 for an explicit form of $\boldsymbol{c}$ for beams and plates) and in the linear case it is a symmetric matrix with $c_{ij} = c_{ji}$. Referring to the infinitesimally small block in Figure 1.5, equilibrium of forces, say in the X direction, is given by:

$$d\sigma_{xx} dy\, dz + d\sigma_{yx}\, dx\, dz + d\sigma_{zx}\, dx\, dy + F_x = m\ddot{u}\, dz\, dy\, dz \tag{1.45}$$

However, $d\sigma_{xx} = \frac{\partial \sigma_{xx}}{\partial x} dx, d\sigma_{yx} = \frac{\partial \sigma_{yx}}{\partial y} dy$ and $d\sigma_{zx} = \frac{\partial \sigma_{zx}}{\partial z} dz$. Thus, the equilibrium Equation (1.45) takes the form:

$$\frac{\partial \sigma_{xx}}{\partial x} + \frac{\partial \sigma_{yx}}{\partial y} + \frac{\partial \sigma_{zx}}{\partial z} + F_x = m\ddot{u} \tag{1.46a}$$

The equilibrium equations for the Y and Z directions follow in a similar fashion:

$$\frac{\partial \sigma_{xy}}{\partial x} + \frac{\partial \sigma_{yy}}{\partial y} + \frac{\partial \sigma_{zy}}{\partial z} + F_y = m\ddot{v} \tag{1.46b}$$

$$\frac{\partial \sigma_{xz}}{\partial x} + \frac{\partial \sigma_{yz}}{\partial y} + \frac{\partial \sigma_{zz}}{\partial z} + F_z = m\ddot{w} \tag{1.46c}$$

Equation (1.46) is identical to Equation (1.35) with the only difference that in the later case, the equations are written in abbreviated form with respect to global direction variables $i, j = 1, 2, 3$ which are equivalent to the X, Y and Z directions.

1.5 Equations of Motion by Reynolds Transport Theorem

Reynolds transport theorem, as applied to a solid body with temporally varying volume $V(t)$ and surface $S(t)$, helps in obtaining (Jog, 2002; Malvern, 1969) the balance laws (−law of conservation of mass, momentum and energy balance – as it moves ('transports') with a velocity v(t)). If there are no material sources (i.e. if material is neither destroyed nor produced), $V(t)$ clearly corresponds to the 'material volume' and let $V_0 := V(t_0)$ denote the initial volume. Thus, consider the initial and deformed configurations of a deformable body as shown in Figure 1.7. The displacement field variable is given by $\boldsymbol{u} = \boldsymbol{x} - \boldsymbol{X}$.

$\boldsymbol{X}$ and $\boldsymbol{x}$ are the position vectors of generic points respectively located in the initial and deformed configurations and this mapping is assumed to be one-to-one. The coordinates in V_0 and V are, respectively, referred to as the material and spatial coordinates. The deformation gradient, $\boldsymbol{F}$ is the gradient of the vector field $\boldsymbol{x}(\boldsymbol{X}, t)$ and is given by:

$$\boldsymbol{F} = \begin{bmatrix} \frac{\partial x_1}{\partial X_1} & \frac{\partial x_2}{\partial X_1} & \frac{\partial x_3}{\partial X_1} \\ \frac{\partial x_1}{\partial X_2} & \frac{\partial x_2}{\partial X_2} & \frac{\partial x_3}{\partial X_2} \\ \frac{\partial x_1}{\partial X_3} & \frac{\partial x_2}{\partial X_3} & \frac{\partial x_3}{\partial X_3} \end{bmatrix} = \nabla \boldsymbol{x} \tag{1.47}$$

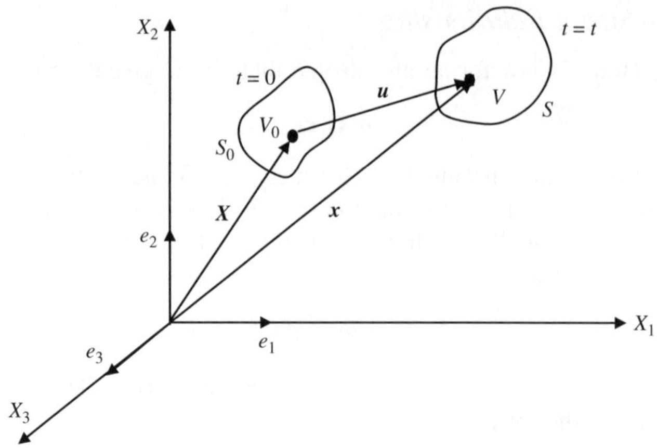

Figure 1.7 Initial and deformed configurations of a deformable body

Since F is an identity matrix at $t = t_0$, $J(=\det F)$ is unity in the initial configuration and remains a function of time as the body deforms. The infinitesimal volume elements in the initial and deformed configurations are related by $dV(t) = J\, dV_0$. Reynolds transport theorem facilitates an understanding of how a field variable $f(\boldsymbol{x}, t)$ changes with time as the body deforms. Here, the field variable (scalar, vector or tensor) is assumed to be continuous and differentiable in both the arguments. Defining the volume integral $\int_{V(t)} f\, dV$, its rate of change is given by $\frac{d}{dt}\int_{V(t)} f\, dV$. Note that since the limits of integration are functions of time, it is not possible to take the time derivative inside the integral directly. Reynolds transport theorem utilises the 3D analogy of the Liebnitz rule of integrating $1D$ integrals and gives:

$$\frac{d}{dt}\int_{V(t)} f\, dV = \int_{V(t)} \left\{ \frac{df}{dt} + f(\nabla \cdot v) \right\} dV \tag{1.48}$$

The above result follows from the following steps:

$$\begin{aligned} \frac{d}{dt}\int_{V(t)} f\, dV &= \frac{d}{dt}\int_{V_0} \bar{f} J dV_0 \\ &= \int_{V_0} \frac{d(\bar{f}J)}{dt} dV_0 \\ &= \int_{V_0} \left(J\frac{df}{dt} + \bar{f}\frac{dJ}{dt} \right) dV_0 \\ &= \int_{V_0} \left(\frac{d\bar{f}}{dt} + \frac{\bar{f}}{J}\frac{dJ}{dt} \right) J dV_0 \end{aligned} \tag{1.49}$$

Here, $\bar{f} = \bar{f}(\boldsymbol{X}, t)$, indicating the representation of f in the undeformed coordinates. Since $\frac{1}{J}\frac{dJ}{dt} = \nabla \cdot v$ (Appendix A) in Equation (1.49), we finally have from Reynolds transport theorem:

$$\frac{d}{dt}\int_{V(t)} f\, dV = \int_{V(t)} \left\{ \frac{df}{dt} + f(\nabla \cdot v) \right\} dV \tag{1.50}$$

The above equality can also be written as:

$$\frac{d}{dt}\int_{V(t)} f\,dV = \int_{V(t)} \left\{\left(\frac{\partial f}{\partial t} + v \cdot \nabla f\right) + f(\nabla \cdot v)\right\} dV$$

$$= \int_{V(t)} \left\{\frac{\partial f}{\partial t} + \nabla \cdot (f v)\right\} dV \tag{1.51}$$

By the divergence theorem, the right-hand side of the above equation takes the form:

$$\int_{V(t)} \left\{\frac{\partial f}{\partial t} + \nabla \cdot (f v)\right\} dV = \int_{V(t)} \frac{\partial f}{\partial t} dV + \int_{S(t)} f(v \cdot \vec{n}) dS \tag{1.52}$$

$\vec{n}$ is the outward directed unit normal to $S(t)$. Thus, we have an alternative form of the Reynolds transport equation as:

$$\frac{d}{dt}\int_{V(t)} f\,dV = \int_{V(t)} \frac{\partial f}{\partial t} dV + \int_{S(t)} f(v \cdot \vec{n}) dS \tag{1.53}$$

1.5.1 Mass Conservation

The mass conservation (in the absence of mass sources or sinks in the material volume) requires that the masses in V_0 and $V(t)$ are the same. Thus, we have:

$$\int_{V_0} \rho_0(\boldsymbol{X}) dV_0 = \int_{V(t)} \rho(\boldsymbol{x}, t) dV$$

$$= \int_{V_0} \bar{\rho}(\boldsymbol{X}, t) J dV_0 \tag{1.54}$$

$\bar{\rho}(\boldsymbol{X}, t)$ is the mass density in the undeformed configuration and $\rho_0(\boldsymbol{X}) = \bar{\rho}(\boldsymbol{X}, 0)$. By the localisation theorem (Appendix A), we get the Lagrangian version of mass conservation in terms of the material coordinate, $\boldsymbol{X}$ as:

$$\rho_0(\boldsymbol{X}) = J\bar{\rho}(\boldsymbol{X}, t) \tag{1.55}$$

The Euler version of the conservation of mass (i.e. in spatial coordinates) is given by:

$$\frac{d}{dt}\int_{V(t)} \rho\, dV = 0 \tag{1.56}$$

If we let $f(x, t) = \rho(x, t)$ in Equation (1.51) and utilise the transport theorem, we have:

$$\int_{V(t)} \left\{\frac{\partial \rho}{\partial t} + \nabla \cdot (\rho v)\right\} dV = 0 \tag{1.57}$$

Since $V(t)$ is arbitrary, by localisation theorem, we obtain the differential form of mass conservation law, valid at any point inside the continuum:

$$\frac{\partial \rho}{\partial t} + \nabla \cdot (\rho v) = 0 \tag{1.58}$$

1.5.2 Linear Momentum Conservation

The governing equations of motion for many structural systems often result from linear momentum balance. This requires that the rate of change of linear momentum is equal to the net forces acting on the material volume. The net force consists of the body force f_b acting on the volume (prescribed per unit volume) and the traction forces acting on the surface (prescribed per unit area). The surface tractions correspond to the surface stresses denoted by τ, the stress tensor. The linear momentum balance equation is given by:

$$\frac{d}{dt}\int_{V(t)} \rho v\, dV = \int_{V(t)} \rho f_b dV + \int_{S(t)} \tau \vec{n} dS \tag{1.59}$$

Following the similar steps as in Equation (1.49), we simplify the left-hand side of the above equation as:

$$\begin{aligned}\frac{d}{dt}\int_{V(t)} \rho v\, dV &= \frac{d}{dt}\int_{V_0} \bar{\rho}\bar{v} J\, dV_0 \\ &= \frac{d}{dt}\int_{V_0} \rho_0 \bar{v} dV_0 \quad (\text{since } J\bar{\rho}\,(\boldsymbol{x}, t) = \rho_0(\boldsymbol{X}) \\ &= \int_{V_0} \frac{d}{dt}\left(\rho_0 \bar{v}\right) dV_0 \end{aligned}\tag{1.60}$$

Since $\rho_0\,(X)$ is independent of time, the above equation further simplifies to:

$$\frac{d}{dt}\int_{V(t)} \rho v\, dV = \int_{V_0} \rho_0 \frac{d}{dt}\,(\bar{v})\, dV_0 \tag{1.61}$$

Changing the integration on the right-hand side of the above equation back to $V(t)$, we obtain

$$\frac{d}{dt}\int_{V(t)} \rho v\, dV = \int_{V} \rho \frac{dv}{dt} dV \tag{1.62}$$

Now, we write the momentum Equation (1.59) as:

$$\int_{V(t)} \rho \frac{dv}{dt} dV = \int_{V(t)} \rho f_b dV + \int_{S(t)} \tau \vec{n} dS \tag{1.63}$$

By divergence theorem, $\int_{S(t)} \tau \vec{n} dS = \int_{V(t)} (\nabla \cdot \tau) dV$ and the above equation can be written as:

$$\int_{V(t)} \left(\rho \frac{dv}{dt} - \nabla \cdot \tau - \rho f_b\right) dV = 0 \tag{1.64}$$

Using the localisation theorem, along with the fact that $V(t)$ can be chosen arbitrarily (e.g. as any subset of the volume of the body), we finally obtain Cauchy's equations of motion as:

$$\rho \frac{dv}{dt} - \nabla \cdot \tau - \rho f_b = 0 \tag{1.65}$$

The above equations of motion are identical to those in Equations (1.35) and (1.46) derived by application of Hamilton's principle and Newton's force balance, respectively. The reader can find an application of linear momentum balance in Chapter 7 in deriving the equations of motion.

1.6 Conclusions

In this chapter, a brief overview of the mathematical modelling of structural dynamic systems is presented with a view to familiarising the reader with the moorings of the subject. Thus, whilst systems of rigid bodies are governed by ODEs, flexible structural systems with continuous mass and stiffness distributions are typically represented by PDEs – linear or nonlinear. Irrespective of the method followed in obtaining these equations of motion (say, via Hamilton's principle, Newton's force balance or the linear momentum balance), one perceives a close inter-relationship amongst them. This is further highlighted in some of the later chapters, wherein additional applications of these methods in deriving the governing PDEs appear as needed. Since this book deals only with the dynamics of linear systems, the next chapter starts with a categorisation of linear PDEs and a focused treatment of the analytical solution methods including separation of variables, eigenexpansion, Green's function, and so on. Specific issues like a self-adjoint differential operator and its implications (in the context of such solutions) are explained in the following.

Exercises

1. Consider a vehicle-mounted equipment shown in Figure 1.8. The vehicle is modelled as a mass–spring–damper system with mass m, spring stiffness k and damper with damping coefficient c. The rigid vertical shaft of length l and mass M representing the equipment is mounted over the vehicle and has stiffness α against rotation about O. Use as (generalized) coordinates the horizontal movement of the vehicle and the rotation θ of the shaft. Assuming θ to be small, write the equation of motion under base excitation $g(t)$.
2. Derive the equation of motion for the lateral (transverse) vibrations of a taut string supported on spring mounts (N_s in number) as shown in Figure 1.9. Assume that the initial tension in the string is H and the supporting springs are nonuniformly spaced.

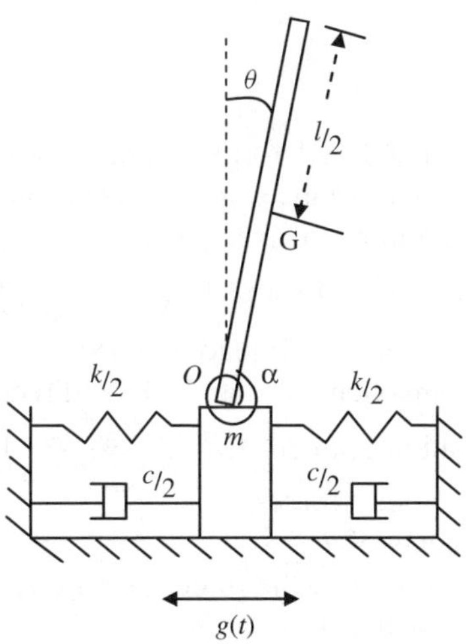

Figure 1.8 Vehicle-mounted shaft under base excitation; G – mass centre of the shaft

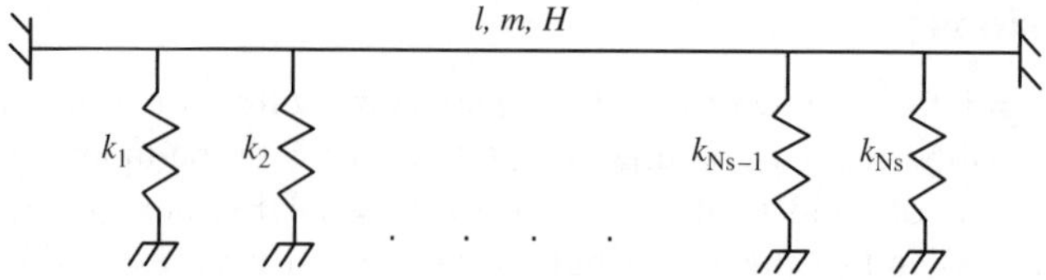

Figure 1.9 A taut string supported on springs

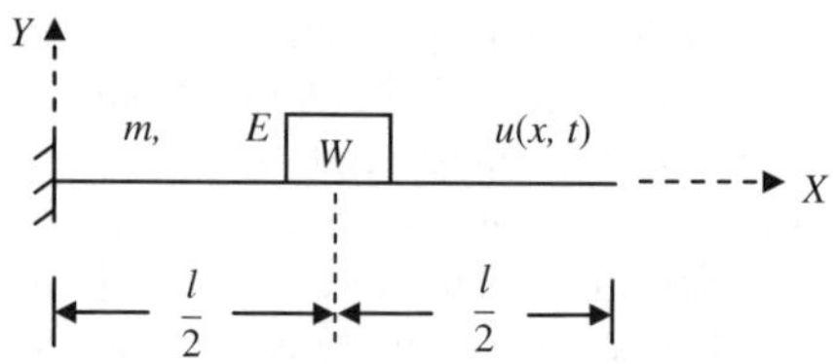

Figure 1.10 A cantilever with a weight at centre

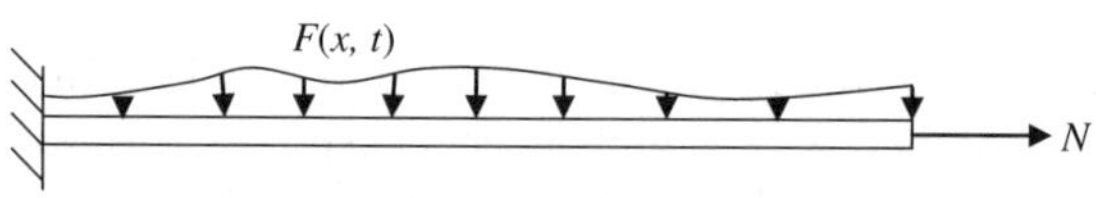

Figure 1.11 A cantilever beam under a constant axial force (prestressed beam)

3. Derive the equation of motion for the lateral vibrations of a cantilever beam (Figure 1.10) with a weight $W = Mg$ at its centre. Assume uniform properties (material/geometric) for the beam.
4. Apply Hamilton's principle to obtain the equation of motion of a uniform cantilever subjected to combined transverse loading $F(x, t)$ and an axial force, N applied at the free end (Figure 1.11). Neglect the shear and rotary inertia effects.

Notations

da, dA	Deformed and undeformed areas, respectively
ds, ds'	Infinitesimal distances in the undeformed and deformed configurations, respectively
D	Flexural rigidity of a plate $\left(=\frac{Et^2}{12(1-\mu^2)}\right)$
$f(\boldsymbol{x}, t)$	Field variable (see Equation 1.48)
F_x, F_y, F_z	Force components in the X, Y, Z directions
$\boldsymbol{F}$	Deformation gradient tensor $\left(=\left[\frac{\partial x_i'}{\partial x_j}\right]\right)$
F_b	Vector of body forces
F_{bi}, $i = 1, 2, 3$	Components of F_b
F_i, $i = 1, 2, 3$	External forces in the three global directions – X, Y and Z
F_s, F_{si}	Surface traction force
g	Acceleration due to gravity
i, j	Integers

$\vec{n}$, $\vec{N}$	Unit normal vectors
dP, dP'	Forces in the undeformed and deformed configurations respectively
$q_i, i = 1, 2, \ldots, n$	Generalized coordinates
$S(t)$	Body surface area
S_d	Specified displacements on the boundary
$S(t_0)$	Initial surface area time, $t = 0$
$[S_{ij}]$	Second Poila–Kirchhoff stress tensor
t	Time variable, plate thickness
$[T_{ij}]$	First Poila–Kirchoff stress tensor
u, v, w	Components of the vector $\boldsymbol{U}$
$\boldsymbol{U}$	Displacement vector
$V(t)$	Material volume
$V(t_0)$	Initial volume at time, $t = 0$
W_{ij}	Components of work done per unit volume $\left(= \sigma_{ij}\varepsilon_{ij}\right)$,
$x_i, \dot{x}_i, \ddot{x}_i, i = 1, 2, 3$	Translations, velocities and accelerations of a rigid body in the three global directions – X, Y and Z
$\boldsymbol{x}$	Position vector of a generic point in the initial (deforned) configuration
$\boldsymbol{X}$	Position vector of a generic point in the initial (undeforned) configuration
θ	Angle of rotation
$\delta x_j, j = 1, 2, ..N$	Virtual displacements
$\rho(\boldsymbol{x}, t)$	Mass density in the deformed configuration
$\bar{\rho}(\boldsymbol{X}, t)$	Mass density in the undeformed configuration
$\rho_0(\boldsymbol{X})$	$= \bar{\rho}(\boldsymbol{X}, 0)$

References

Clough, R.W. and Penzien, J. (1982) *Dynamics of Structures*, McGraw Hill, Singapore.

Craig, R.R. (1981) *Structural Dynamics – An Introduction to Computer Methods*, John Wiley & sons, Inc., New York.

Geradin, M. and Rixen, D. (1994) *Mechanical Vibrations – Theory and Application to Structural Dynamics*, John Wiley & Sons, Inc. Chichester, UK.

Hughes, T.J.R. (1987) *The Finite Element Method*, Prentice Hall, Englewood Cliffs, New Jersey.

Humar, J.L. (2002) *Dynamics of Structures*, 2nd edn, A. A. Balkema Publishers in Lisse, Exton, PA.

Jog, C.S. (2002) *Foundations and Applications of Mechanics*, Narosa Publishing House, New Delhi.

Malvern, L.E. (1969) *Introduction to the Mechanics of a Continuous Medium*, Prentice Hall Inc., Englewood Cliffs, NJ.

Meirovitch, L. (1967) *Analytical Methods in Vibration*, The Macmillan Company, New York.

Reddy, J.N. (1984) *Energy and Variational Methods in Applied Mechanics*, John Wiley & Sons, Inc., New York.

Bibliography

Danielson, D. A. (2003) *Vectors and Tensors in Engineering and Physics*, Westview (Perseus), New York.

Lovedock, D. and Hanno, R. (1989) *Tensors, Differential forms and Variational principles*, Dover, New York.

Beikmann, R.S., Perkins, N.C. and Ulsoy, A.G. (1996) Free vibration of serpentine belt systems. *Transactions of ASME, Journal of Vibration and Acoustics*, **118**, 406–413.

Forray, M.J. (1968) *Variational Calculus in Science and Engineering*, McGraw-Hill, New York.

2

Continuous Systems – PDEs and Solution

2.1 Introduction

Continuous systems with distributed mass and stiffness parameters are governed by partial differential equations (PDEs). The principal methods to derive these PDEs (in terms of space variables and time) for structural systems are described in Chapter 1. Whilst solutions of many nonlinear PDEs are possible via numerical techniques only, in the case of linear PDEs, there exist a few elegant analytical methods (Meirovitch, 1967, 1970; Kreyszig, 1999), for example methods of separation of variables and the integral transform. These methods typically find application for certain special cases of continuous systems and the resulting analytical solutions are often more insightful of the behaviour of such systems governed by homogeneous/nonhomogeneous PDEs (Appendix B) that commonly arise in structural engineering. For instance, whilst the one-dimensional wave equation $\nabla^2 u = \frac{\partial^2 u}{\partial t^2}$ (with $\nabla^2 = \frac{\partial^2}{\partial x^2}$) (say, under homogeneous boundary conditions) might be looked upon as a (possibly crude) idealisation to the small-amplitude vibratory motion of a uniform string, a solution of this idealised system via the method of separation of variables leads to the so-called Sturm–Liouville eigenvalue problem (Simmons and Krantz, 1925), which in turn provides considerable insights into the possible vibration modes via the associated eigenfunctions. These vibration modes may even be exploited to devise approximate analytical solutions to a class of nonhomogeneous PDEs via the so-called eigen- (or Fourier) expansion methods. For certain linear PDEs over infinite domains, integral transform methods using Fourier cosine/sine transform, Hankel transform, Laplace transform, and so on, may also be effectively used to arrive at analytical solutions. In addition to being insightful, analytical solutions, if available, are often preferred to numerical methods as the former typically yields faster and more accurate solutions (probably without even needing a computer). Accordingly, in this chapter, certain linear PDEs (of 1D and 2D structural systems), amenable to such solutions, are considered and the procedures to obtain the analytical solutions discussed.

Elements of Structural Dynamics: A New Perspective, First Edition. Debasish Roy and G Visweswara Rao.

2.2 Some Continuous Systems and PDEs

In Chapter 1, the equilibrium equations for the longitudinal vibration of an axially loaded 1D bar are derived and shown to be $\frac{\partial}{\partial x}\left(EA\frac{\partial u}{\partial x}\right) + F = m\frac{\partial^2 u}{\partial t^2}$. For a uniform bar, this equation takes the form $EA\frac{\partial^2 u}{\partial x^2} + F = m\frac{\partial^2 u}{\partial t^2}$. Denoting $\frac{EA}{m} = c^2$, the unforced vibration is governed by the PDE:

$$\nabla^2 u = \frac{1}{c^2}\frac{\partial^2 u}{\partial t^2} \tag{2.1}$$

where $\nabla^2(=\frac{\partial^2}{\partial x^2})$ is the Laplacian operator. The PDE in Equation (2.1), known as the one-dimensional wave equation, represents many other phenomena like small-amplitude oscillations of taut strings and torsional oscillations of shafts. The equation also governs phenomena in other disciplines such as acoustics and electrostatics, wherein it, respectively, describes the propagation of sound waves and the transmission of electric signals along a cable. A two-dimensional analogue of the taut string is a taut membrane, which is governed by the same Equation (2.1) with the Laplacian $\nabla^2 = \frac{\partial^2 u}{\partial x^2} + \frac{\partial^2 u}{\partial y^2}$ now being an operator over 2D.

2.2.1 A Taut String – the One-Dimensional Wave Equation

Consider a taut and flexible string of finite length l with initial tension T (Figure 2.1a). Let m be its mass per unit length. The inertia force over an infinitesimal segment of length Δx is given by $-m\Delta x\frac{\partial^2 u}{\partial t^2}$ (Figure 2.1b). In deriving the equation, we presently neglect the self-weight of the string.

The transverse displacement $u(x,t)$ is the dependent variable and is a function of the two independent variables x and t. $\theta = \frac{\partial u}{\partial x}$ is the slope and $\Delta\theta = \frac{\partial\theta}{\partial x}\Delta x$ the small increment in the slope over Δx with the assumption that $\tan\theta \approx \theta$ for small angles. Applying Newton's second law over the length Δx of the string, the equilibrium of forces gives the equation of motion for the string as:

$$-m\Delta x\frac{\partial^2 u}{\partial t^2} + T\left(\theta + \frac{\partial\theta}{\partial x}\Delta x\right) - T\theta = 0 \Rightarrow T\frac{\partial\theta}{\partial x} = m\frac{\partial^2 u}{\partial t^2} \Rightarrow T\frac{\partial^2 u}{\partial x^2} = m\frac{\partial^2 u}{\partial t^2} \tag{2.2}$$

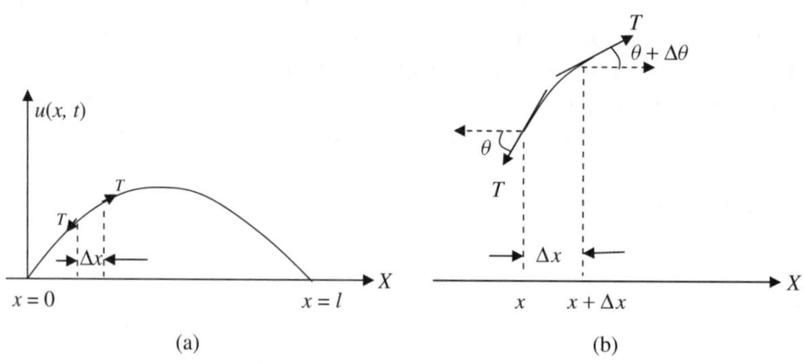

Figure 2.1 (a) Flexible taut string (b) Small element of length dx

In the above equation, if $\frac{T}{m} = c^2$ (with c being a constant and dimensionally representing velocity), the equation is identical to the one in Equation (2.1). The PDE (Equation (2.2)) is accompanied by the associated boundary/initial conditions. In the present case, the conditions are:

$$u(0, t) = u(l, t) = 0 \tag{2.3a}$$

$$u(x, 0) = f(x) \text{ and } u_t(x, 0) = g(x) \tag{2.3b}$$

$f(x)$ and $g(x)$ are the imposed initial displacement and velocity functions. A solution to the wave equation (as its name indicates) physically represents a wave-propagation problem.

2.2.2 *An Euler–Bernoulli Beam – the One-Dimensional Biharmonic Wave Equation*

Structural systems like strings/wires and membranes are perfectly flexible and are incapable of offering resistance to bending (or any external traction in the form of compression). Thus, these structures are predominantly meant to support tensile loads only. In many engineering structures, the external loading (e.g. transversely applied load on a beam) is often resisted through the so-called 'bending' that involves deformation that, for instance, makes a structure 'curved'. In order to support such deformations, internal reaction forces – tensile as well as compressive – are developed. Unlike strings, beams are typical such one-dimensional structures that can support deformations due to bending. Let us thus consider a uniform beam under flexure, of length l, cross-sectional area A and mass density per unit volume ρ, oriented in the X-Z plane as shown in Figure 2.2. Since

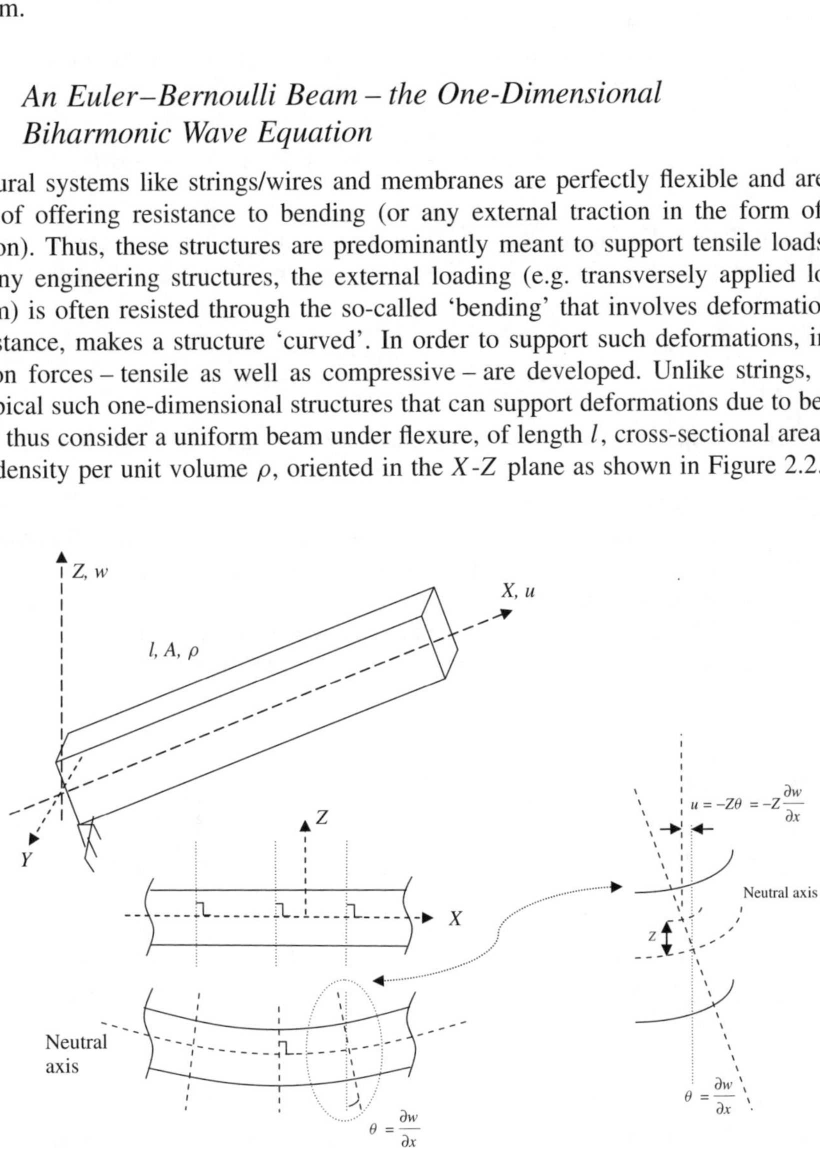

Figure 2.2 Beam under flexure under Euler–Bernoulli thin-beam theory

the cross-sectional dimensions are much smaller than that in the axial direction, a beam is considered to be a one-dimensional elastic body and as such there is a dimensional dissent by order of 2 when compared to a general solid body. It follows that the only stress components on a cross section of the beam are the normal stress σ_{xx} and the shear stress σ_{xz}. If we focus our attention on thin beams and follow Euler–Bernoulli beam theory, the plane cross sections normal to undeformed centroidal axis remain plane and normal to the deformed centroidal axis (Figure 2.2). This implies that the shear strain component $\varepsilon_{xz} = 0$ (or rather negligibly small).

However, in the presence of transverse external loading, equilibrium of forces requires that shear stress component σ_{xz} has to be finite and nonzero. This implies that shear modulus G must be infinitely large.

From Figure 2.2, the normal strain varies across the beam cross section and is related to the transverse displacement $w(x, t)$ as

$$\varepsilon_{xx} = \frac{\partial u}{\partial x} = \frac{\partial}{\partial x}(-z\theta) = \frac{\partial}{\partial x}\left(-z\frac{\partial w}{\partial x}\right) = -z\frac{\partial^2 w}{\partial x^2} = -\boldsymbol{B}w \tag{2.4}$$

The partial differential operator, $\boldsymbol{B} = z\frac{\partial^2}{\partial x^2}$. z is the distance from neutral axis of a generic point in the cross section (Figure 2.3). The stress–strain relationship (Equation 1.44) is given by

$$\sigma_{xx} = \boldsymbol{c}\ \varepsilon_{xx} = -Ez\frac{\partial^2 w}{\partial x^2} = -E\boldsymbol{B}w \tag{2.5}$$

The constitutive matrix $\boldsymbol{c}$ (containing the elasticity constants) in the present case is a scalar and is equal to E, the Young's modulus of elasticity.

To obtain the internal strain energy ($=\int_\Omega \frac{1}{2}\sigma_{xx}\varepsilon_{xx}\, dV$ for a solid body) for the beam under flexure, we require an averaging operation over the two ignored dimensions in the y-z plane. To comprehend this requirement, consider the bending action on a beam segment of infinitesimal length dx at a distance x from origin as in Figure 2.3a, we find that the distributed normal stress σ_{xx} varies linearly across the beam cross section in the z direction, whilst remaining constant at any distance z from the neutral axis and over the

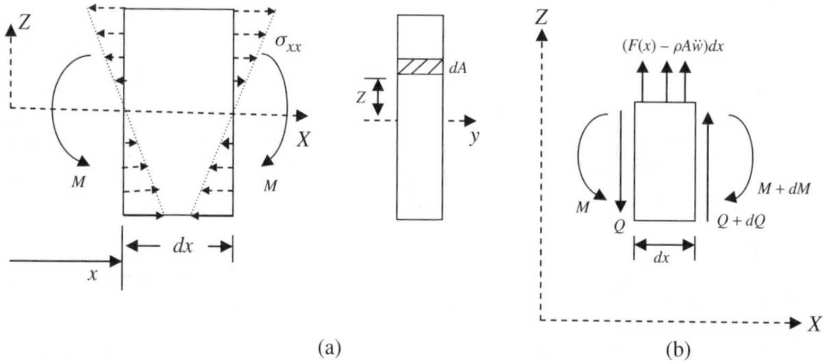

Figure 2.3 Beam under flexure; (a) normal stress σ_{xx} at a beam cross section and (b) free-body diagram on a beam segment of length dx

strip of area dA in the y direction. Thus, the internal strain energy of the beam segment is obtained from Equations (2.4) and (2.5) as

$$\int_{\Omega} \sigma_{xx}\varepsilon_{xx}\, dA\, dx = -E\left(\int_A z^2 dA\right)\left(\frac{\partial^2 w}{\partial x^2}\right)^2 dx = -EI\left(\frac{\partial^2 w}{\partial x^2}\right)^2 dx \tag{2.6}$$

with $I = \int_A z^2\, dA$, the area moment of inertia of the beam cross section.

Incidentally, we can derive the expressions for the stress resultants – moment M and shear force Q for the beam under flexure. From Figure 2.3a, the moment arising out of the normal stress varying across the cross section is obtained as

$$M = \int_A (\sigma_{xx}\, dA)\, z = -E\left(\int_A z^2\, dA\right)\frac{\partial^2 w}{\partial x^2} = -EI\frac{\partial^2 w}{\partial x^2} \tag{2.7}$$

The moment equilibrium of the beam segment (Figure 2.3b) leads to the required expression for the shear force Q. Whilst ignoring the second-order terms, it is given by:

$$-Qdx + \frac{1}{2}\left(F(x,t) - \rho A\ddot{w}\right)(dx)^2 + dM = 0 \Rightarrow Q = \frac{dM}{dx} = -\frac{\partial}{\partial x}\left(EI\frac{\partial^2 w}{\partial x^2}\right) \tag{2.8}$$

Referring back to Equation (2.6) and integrating the incremental internal strain energy over the length l, we get the total internal strain energy. Thus, the total potential energy is obtained as $V = \int_0^l \left[-\frac{1}{2}E\,I\left(\frac{\partial^2 w}{\partial x^2}\right)^2 + F\,w\right]dx$. With the kinetic energy $T = \int_0^l \frac{1}{2}m\,\dot{w}\,\dot{w}\,dx$, where $m = \rho A$, we now can express the energy integral $\mathrm{I}(\cdot) = \int_{t_1}^{t_2}(T - V)\,dt$ for the beam as

$$I(\cdot) = \int_{t_1}^{t_2}\left\{\int_0^l \frac{1}{2}m\,\dot{w}\,\dot{w}\,dx + \int_0^l\left[\frac{1}{2}EI\left(\frac{\partial^2 w}{\partial x^2}\right)^2 - F w\right]dx\right\}dt \tag{2.9}$$

To apply the variational principle on $\mathrm{I}(\cdot)$ in Equation (2.9), let us refer to Equation (1.8) and write the first variation of $\mathrm{I}(\cdot)$ as

$$\delta\int_{t_1}^{t_2}(T - V)\,dt = \int_{t_1}^{t_2}\left[\int_0^l\left\{m\,\dot{w}\delta\dot{w} - EI\frac{\partial^2 w}{\partial x^2}\frac{\partial^2(\delta w)}{\partial x^2} + F\delta w\right\}dx\right]dt \tag{2.10}$$

In Equation (2.10), both space and time derivatives of virtual displacement δw are present and by appropriate reordering of integration operations on the right-hand side of the equation and utilising integration by parts, we can express all variations in terms of δw only. These operations result in:

$$\begin{aligned}\delta\int_{t_1}^{t_2}(T - V)\,dt &= \int_0^l\left[(m\,\dot{w}\,\delta w)_{t_1}^{t_2} - \int_{t_1}^{t_2} m\,\ddot{w}\,\delta w\;dt\right]dx \\ &\quad - \int_{t_1}^{t_2}\left\{\left(EI\frac{\partial^2 w}{\partial x^2}\delta\frac{\partial w}{\partial x}\right)_0^l - \left(\frac{\partial}{\partial x}\left(EI\frac{\partial^2 w}{\partial x^2}\right)\delta w\right)_0^l\right. \\ &\quad \left. + \int_0^l\left(\frac{\partial^2}{\partial x^2}\left(EI\frac{\partial^2 w}{\partial x^2}\right) - F\,\delta w\right)dx\right\}dt\end{aligned}$$

$$= -\int_{t_1}^{t_2}\int_0^l \left(m\ddot{w}+\frac{\partial^2}{\partial x^2}\left(EI\frac{\partial^2 w}{\partial x^2}\right) - F\right)\delta w\,dx\,dt$$

$$-\int_{t_1}^{t_2}\left\{\left(EI\frac{\partial^2 w}{\partial x^2}\delta\frac{\partial w}{\partial x}\right)_0^l + \left(\frac{\partial}{\partial x}\left(EI\frac{\partial^2 w}{\partial x^2}\right)\delta w\right)_0^l\right\}dt \quad (2.11)$$

In obtaining Equation (2.11) we have made use of the fact $\delta w(t_1) = \delta w(t_2) = 0$ (Figure 1.4). Noting that $\delta \boldsymbol{w}$ is arbitrary over the time interval, we obtain the strong form of the equations of motion for the beam under flexure as:

$$m\frac{\partial^2 w}{\partial t^2}+\frac{\partial^2}{\partial x^2}\left(EI\frac{\partial^2 w}{\partial x^2}\right) = F(x,t) \quad (2.12)$$

and the boundary conditions as

$$EI\frac{\partial^2 w}{\partial x^2} = 0 \text{ or } \frac{\partial w}{\partial x} = 0 \text{ at } x = 0 \text{ and } l \quad (2.13)$$

and

$$\frac{\partial}{\partial x}\left(EI\frac{\partial^2 w}{\partial x^2}\right) = 0 \text{ or } w = 0 \text{ at } x = 0 \text{ and } l \quad (2.14)$$

Note that $w = 0$ and $\frac{\partial w}{\partial x} = 0$ are the essential (geometric) boundary conditions, and $EI\frac{\partial^2 w}{\partial x^2} = 0$ and $\frac{\partial}{\partial x}\left(EI\frac{\partial^2 w}{\partial x^2}\right) = 0$ are the natural (force) boundary conditions.

We can also obtain the above equations of motion for the beam by referring to the free-body diagram of the beam segment in Figure 2.3b. The force equilibrium in the z direction gives $dQ + (F(x,t) - \rho A\ddot{w})\,dx = 0$. Utilising the expression for Q in Equation (2.8), we obtain the same Equation (2.12) of the beam vibratory motion as:

$$\frac{dQ}{dx} = -F(x,t) + \rho A\ddot{w} \Rightarrow \frac{\partial^2}{\partial x^2}\left(EI\frac{\partial^2 w}{\partial x^2}\right) + \rho A\ddot{w} = F(x,t) \quad (2.15)$$

For a beam with uniform EI, Equation (2.15) is recast as

$$\frac{\partial^4 w}{\partial x^4} = -\frac{m}{EI}\frac{\partial^2 w}{\partial t^2} + F(x,t) \quad (2.16)$$

As presented above, Equation (2.16) represents an inhomogeneous biharmonic wave equation and Equations (2.13) and (2.14) define homogeneous boundary conditions. A more general form of the biharmonic wave equation, applicable over 2D domains, is written as

$$\Delta(\Delta u) = -\frac{1}{c^2}\frac{\partial^2 u}{\partial t^2} + f(t) \quad (2.17)$$

where $\Delta = \frac{\partial^2 u}{\partial x^2} + \frac{\partial^2 u}{\partial y^2}$ is the Laplacian operator. The above equation governs the vibration of thin plates under bending (Kirchhoff's plate theory, which disregards shear deformation). More specifically, with the assumption of uniform material properties, the PDE representing the vertical vibrations of a rectangular thin plates with uniform material properties is given by Geradin and Rixen (1994).

$$\frac{\partial^4 u}{\partial x^4} + 2\frac{\partial^4 u}{\partial x^2 y^2} + \frac{\partial^4 u}{\partial y^4} + \frac{m}{D}\frac{\partial^2 u}{\partial t^2} = f(x, y, t)/D \tag{2.18}$$

where $D\left(=\frac{Eh^3}{12(1-\nu^2)}\right)$ is a scalar with E being the Young's modulus, ν the Poisson's ratio and h the thickness of the plate. $f(x, y, t)$ is the transverse (vertical) loading acting over the plate. As in the case of the beam equation above, the assumption that the transverse shear strain is negligible is implicit in the PDE (Equation (2.18)). Details of the derivation of the PDE for thin plates are enumerated in a later section.

2.2.3 Beam Equation with Rotary Inertia and Shear Deformation Effects

Timoshenko beam theory considers the effects of rotary inertia and shear deformation that are ignored in the Euler–Bernoulli theory. These two effects may not be negligible, particularly for beams wherein the thickness dimension is not small enough compared to the one along the beam axis. During the bending of such a beam (unlike a thin beam), the effects of rotations of cross sections (see Figure 2.4) as well as the inertia due to this rotation (presently denoted as ψ) must be included. Accordingly, the expression for the kinetic energy, T, takes the form:

$$T = \frac{1}{2}\int_0^l \left(m\dot{w}^2 + mr^2\dot{\psi}^2\right) dx \tag{2.19}$$

r is the radius of gyration given by $r^2 = I/A$ with A being the area of cross-section. The normal strain is related to the rotation angle, ψ as $\varepsilon_{xx} = \frac{\partial u}{\partial x} = \frac{\partial}{\partial x}(-z\psi) == -z\frac{\partial \psi}{\partial x}$ and $\sigma_{xx} = E\varepsilon_{xx}$. Note that, in the present case, $\psi \neq \frac{\partial w}{\partial x}$. The internal strain energy due to this normal strain is given by:

$$\int_\Omega \sigma_{xx}\varepsilon_{xx} dA\, dx = -E\left(\int_A z^2 dA\right)\left(\frac{\partial \psi}{\partial x}\right)^2 dx = -EI\left(\frac{\partial \psi}{\partial x}\right)^2 dx \tag{2.20}$$

As shown in Figure 2.4, the infinitesimal segment dx is deformed by the shear force and the shear strain τ_{xx} is given by:

$$\tau_{xx} = \frac{\partial u}{\partial z} + \frac{\partial w}{\partial x} = -\psi + \frac{\partial w}{\partial x} \tag{2.21}$$

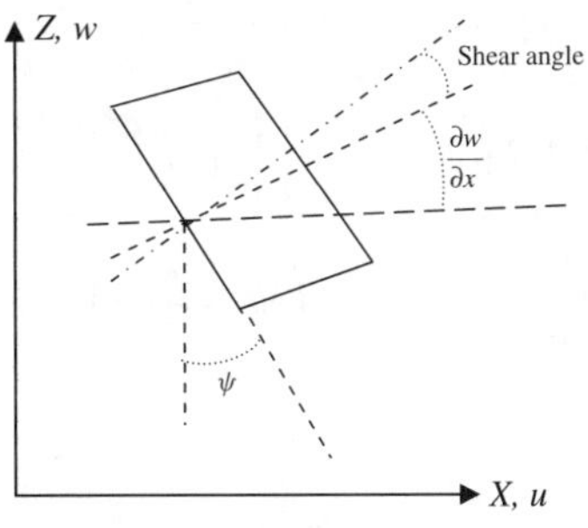

Figure 2.4 Shear deformation at a beam cross section

The internal strain energy due to the shear deformation is $\int_\Omega \sigma_{xx}\tau_{xx}d\Omega = \int_0^l \int_A \sigma_{xx}dA\,\tau_{xx}dx$. The shear force $(= \int_A \sigma_{xx}dA)$ on the cross-sectional area, A, is commonly expressed as:

$$\int_A \sigma_{xx}\, dA = k' AG\tau_{xx} \tag{2.22}$$

k' is known as the cross section reduction factor and $k'A$, the effective area of cross section over which the shear force acts. G is the shear modulus of the material. With the inclusion of shear deformation effects, the total potential energy is given by $V = \int_0^l \left[-\frac{1}{2}EI\left(\frac{\partial\psi}{\partial x}\right)^2 + k'AG\left(-\psi + \frac{\partial w}{\partial x}\right)^2 + Fw \right] dx$ and the energy integral in Equation (2.9) takes the form:

$$I(\cdot) = \int_{t_1}^{t_2} (T - V)dt = \int_{t_1}^{t_2} \left\{ \int_0^l \left(\frac{1}{2}m\,\dot{w}\,\dot{w} + mr^2\dot{\psi}^2\right)dx + \int_0^l \left[-\frac{1}{2}\left(EI\left(\frac{\partial\psi}{\partial x}\right)^2\right) + k'AG\left(-\psi + \frac{\partial w}{\partial x}\right)^2 + Fw \right] dx \right\} dt \tag{2.23}$$

Following the variational principle that renders the first variation of the functional I to zero with respect to the dependent variables w and ψ, we obtain:

$$\delta \int_{t_1}^{t_2} (T-V)\,dt = \int_0^l \left[(m\,\dot{w}\,\delta w)_{t_1}^{t_2} - \int_{t_1}^{t_2} m\,\ddot{w}\delta w\,dt \right] dx + \int_0^l \left[\left(mr^2\dot{\psi}\delta\psi\right)_{t_1}^{t_2} - \int_{t_1}^{t_2} mr^2\ddot{\psi}\delta\psi\,dt \right] dx$$
$$- \int_{t_1}^{t_2} \left\{ \begin{array}{c} \left(EI\frac{\partial\psi}{\partial x}\delta\psi\right)_0^l + \int_0^l \left(\frac{\partial}{\partial x}\left(EI\frac{\partial w}{\partial x}\right) - F\delta w\right)dx \\ -k'AG\int_0^l \psi\partial\psi - \left(k'AG\frac{\partial w}{\partial x}\partial w\right)_0^l - \int_0^l k'AG\frac{\partial w}{\partial x}dw \\ \int_0^l k'AG\frac{\partial w}{\partial x}\partial\psi - (k'AG\psi\partial w)_0^l - \int_0^l k'AG\frac{\partial\psi}{\partial x}\partial w \end{array} \right\} dt = 0 \tag{2.24}$$

Noting that the admissible variations (i.e. the virtual deformations) δw and $\delta\psi$ are otherwise arbitrary over the time interval, we obtain the strong form of the equations of motion, coupled in the two dependent field variables, $w(x,t)$ and $\psi(x,t)$, as:

$$\frac{\partial}{\partial x}\left\{k'AG\left(\psi - \frac{\partial w}{\partial x}\right)\right\} + m\frac{\partial^2 w}{\partial t^2} = f(x,t)$$
$$\frac{\partial}{\partial x}\left(EI\frac{\partial\psi}{\partial x}\right) = k'AG\left(\psi - \frac{\partial w}{\partial x}\right) + mr^2\frac{\partial^2\psi}{\partial t^2} \tag{2.25a,b}$$

with boundary conditions:

$$\left(EI\frac{\partial\psi}{\partial x}\delta\psi\right)_0^l = 0 \Rightarrow \psi = 0 \;or\; EI\frac{\partial\psi}{\partial x} = Moment = 0,\;\; at\; x = 0 \;and\; x = l$$

$$\left(k'AG\left(\frac{\partial w}{\partial x}-\psi\right)\delta w\right)_0^l=0 \tag{2.26a}$$

$$\Rightarrow w=0\ or\ k'AG\left(\frac{\partial w}{\partial x}-\psi\right)=\text{shear force}=0,\ \text{ at } x=0 \text{ and } x=l \tag{2.26b}$$

In the case of a uniform beam, the shear angle ψ can be eliminated from coupled Equation (2.25) and a single PDE in only one dependent variable $w(x,t)$ can be constructed. With no external forcing term, the second equation in Equation (2.25) gives $\frac{\partial\psi}{\partial x}=\frac{\partial^2 w}{\partial x^2}-\frac{m}{k'AG}\frac{\partial^2 w}{\partial t^2}$. Using this derivative and differentiating the first equation once with respect to x, we finally obtain the PDE for a uniform Timshenko beam in terms of the vertical displacement as:

$$EI\frac{\partial^4 w}{\partial x^4}+m\frac{\partial^2 w}{\partial t^2}-mr^2\frac{\partial^4 w}{\partial x^2\partial t^2}-\frac{mEI}{k'AG}\frac{\partial^4 w}{\partial x^2\partial t^2}+\frac{m^2r^2}{k'AG}\frac{\partial^4 w}{\partial t^4}=f(x,t) \tag{2.27}$$

If the rotary inertia and shear deformation terms are neglected, the above equation reduces to the PDE (Equation (2.16)) corresponding to the Euler–Bernoulli beam theory.

2.2.4 *Equations of Motion for 2D Plate by Classical Plate Theory (Kirchhoff Theory)*

A plate may be viewed as a 2D analogy of a beam-type continuous system. In the classical plate theory (Kirchhoff plate theory) involving thin plates, shear deformation effects are ignored, These effects, however, assume significance for thicker plates and are accordingly accounted for, for example in the first-order shear deformation theory (Reissner–Mindlin plate theory). Presently, consider the deformation (assumed to be 'small') of a thin plate subject to the following kinematic assumptions: (i) stress, σ_z in the transverse direction is zero and (ii) any cross section initially normal to the neutral plane (predeformation) remains normal to the plane postdeformation (i.e. shear deformation is negligible). Let the material be homogeneous and isotropic (of the St. Venant–Kirchhoff type; see Appendix B) and the thickness h of the plate uniform. The governing equations of motion for such a thin plate can be looked upon as a statement of balance of externally applied and reactive forces, the latter resulting from the so-called membrane and bending actions. Specifically, in this context, the membrane action of interest is due to the inplane displacements $u(x,y,t)$ and $v(x,y,t)$ resulting from the transverse loading of the plate.

2.2.4.1 Strain–Displacement Relationships

Taking $w(x,y,t)$ as the transverse displacement field at a generic point (x,y), the inplane displacements can be expressed (Figure 2.5) as:

$$u=-z\frac{\partial w}{\partial x},\ v=-z\frac{\partial w}{\partial y} \tag{2.28}$$

In view of Equation (2.28) and the kinematic assumptions considered above, the components of the symmetric and linearised (i.e. with the nonlinear terms removed) Green

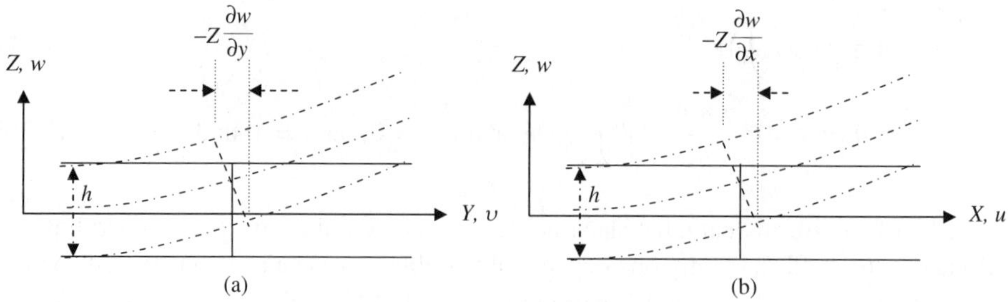

Figure 2.5 Deformation of a thin plate in the (a) X and (b) Y directions and the inplane displacements due to bending

strain tensor, $\varepsilon_{ij} = \frac{1}{2}\left(\frac{\partial u_i}{\partial x_j} + \frac{\partial u_i}{\partial x_j}\right)$ (Equation (1.27)), are presently given by:

$$\varepsilon_x = \frac{\partial u}{\partial x} = -z\frac{\partial^2 w}{\partial x^2}, \varepsilon_y = \frac{\partial v}{\partial y} = -z\frac{\partial^2 w}{\partial y^2}, \varepsilon_z = 0, \gamma_{xy} = \frac{\partial u}{\partial y} + \frac{\partial v}{\partial x} = -2z\frac{\partial^2 w}{\partial x \partial y},$$
$$\gamma_{xz} = \frac{\partial u}{\partial z} + \frac{\partial w}{\partial x} = 0, \gamma_{yz} = \frac{\partial v}{\partial z} + \frac{\partial w}{\partial y} = 0 \tag{2.29}$$

Thus, for the current structure, writing the nontrivial components of the strain tensor in a vector form, the strain–displacement equations are given by:

$$\varepsilon = \begin{Bmatrix} \varepsilon_x \\ \varepsilon_y \\ \gamma_{xy} \end{Bmatrix} = \begin{Bmatrix} -z\dfrac{\partial^2 w}{\partial x^2} \\ -z\dfrac{\partial^2 w}{\partial y^2} \\ -2z\dfrac{\partial^2 w}{\partial x \partial y} \end{Bmatrix} = -\boldsymbol{B}w \tag{2.30}$$

$\boldsymbol{B}$ is the differential operator:

$$\boldsymbol{B} = \begin{Bmatrix} -z\dfrac{\partial^2}{\partial x^2} \\ -z\dfrac{\partial^2}{\partial y^2} \\ -2z\dfrac{\partial^2}{\partial x \partial y} \end{Bmatrix} \tag{2.31}$$

$\frac{\partial^2 w}{\partial x^2}$ and $\frac{\partial^2 w}{\partial y^2}$ are the (linearised) curvatures of the plate in the X-Z and Y-Z planes and the $\frac{\partial^2 w}{\partial x \partial y}$ is the crosscurvature of the plate due to torsional motion about the Z-axis.

2.2.4.2 Stress–Strain Relationships

If $\left\{\sigma_{x,}\sigma_y, \tau_{xy}\right\}^T$ is the stress 'vector' (corresponding to the strain vector defined earlier), the St. Venant–Kirchhoff constitutive equations are given by:

$$\varepsilon_x = \frac{1}{E}\sigma_x \nu(\sigma_y + \sigma_z)$$

$$\varepsilon_y = \frac{1}{E}\sigma_y - \nu(\sigma_x + \sigma_z)$$

$$\gamma_{xy} = \frac{1}{G}\tau_{xy} \tag{2.32}$$

G is the shear modulus given in terms of E as $G = \frac{E}{2(1+\nu)}$. From Equation (2.32), the stress–strain relationship takes the form:

$$\begin{Bmatrix} \sigma_x \\ \sigma_y \\ \tau_{xy} \end{Bmatrix} = \frac{E}{(1-\nu^2)} \begin{bmatrix} 1 & \nu & 0 \\ \nu & 1 & 0 \\ 0 & 0 & \frac{1-\nu}{2} \end{bmatrix} \begin{Bmatrix} \varepsilon_x \\ \varepsilon_y \\ \gamma_{xy} \end{Bmatrix} = \boldsymbol{c} \begin{Bmatrix} \varepsilon_x \\ \varepsilon_y \\ \gamma_{xy} \end{Bmatrix} \text{ with}$$

$$\boldsymbol{c} = \frac{E}{(1-\nu^2)} \begin{bmatrix} 1 & \nu & 0 \\ \nu & 1 & 0 \\ 0 & 0 & \frac{1-\nu}{2} \end{bmatrix} \tag{2.33}$$

2.2.4.3 Energy Expressions

The kinetic energy, T in terms of the vector of velocities, $\dot{U} = (\dot{u}, \dot{v}, \dot{w})^T$ of an infinitesimal volume segment dV is given by:

$$T = \frac{1}{2}\int \rho \dot{U}^T \dot{U} dV$$

$$= \frac{1}{2}\int_V \rho \left[\dot{w}^2 + z^2 \left\{ \left(\frac{\partial \dot{w}}{\partial x}\right)^2 + \left(\frac{\partial \dot{w}}{\partial y}\right)^2 \right\} \right] dV \tag{2.34}$$

ρ is the reference mass density per unit volume. The second term $z^2\left\{\left(\frac{\partial \dot{w}}{\partial x}\right)^2 + \left(\frac{\partial \dot{w}}{\partial y}\right)^2\right\}$ in the above equation represents the rotary inertia effect. As in the case of a beam, this inertia term is significant only in the presence of shear deformation. For thin plates, this term is usually ignored. Taking $m\left(= \int_{-\frac{h}{2}}^{\frac{h}{2}} \rho dz\right)$ as the mass 'density' per unit area, the kinetic energy expression takes the familiar form in terms of only the translation in the Z direction as:

$$T = \frac{1}{2}\int_A m\dot{w}^2\, dA \tag{2.35}$$

With $\sigma = \{\sigma_x, \sigma_y, \tau_{xy}\}^T$, the strain energy, V is given by:

$$V = \frac{1}{2}\int \sigma^T \boldsymbol{\varepsilon} dV = \frac{1}{2}\int \boldsymbol{\varepsilon}^T \boldsymbol{c}^T \boldsymbol{\varepsilon}\, dV \tag{2.36}$$

Substituting expressions for the strain vector $\boldsymbol{\varepsilon}$ and the matrix of elastic coefficients $\boldsymbol{c}$, Equation (2.36) takes the form:

$$V = \frac{1}{2}\frac{E}{(1-\nu^2)}\int_V z^2 \left[\left(\frac{\partial^2 w}{\partial x^2} + \frac{\partial^2 w}{\partial y^2} \right)^2 - 2(1-\nu)\left\{ \frac{\partial^2 w}{\partial x^2}\frac{\partial^2 w}{\partial y^2} - \left(\frac{\partial^2 w}{\partial x \partial y} \right)^2 \right\} \right] dV \tag{2.37}$$

$$= \frac{1}{2}\frac{E}{(1-v^2)}\int_A \left[\left(\frac{\partial^2 w}{\partial x^2}+\frac{\partial^2 w}{\partial y^2}\right)^2\right.$$

$$\left. -2(1-v)\left\{\frac{\partial^2 w}{\partial x^2}\frac{\partial^2 w}{\partial y^2}-\left(\frac{\partial^2 w}{\partial x \partial y}\right)^2\right\}\right] dA \int_{-\frac{h}{2}}^{\frac{h}{2}} z^2 dz$$

$$\Rightarrow V = \frac{1}{2}D\int_A \left[\left(\frac{\partial^2 w}{\partial x^2}+\frac{\partial^2 w}{\partial y^2}\right)^2 - 2(1-v)\left\{\frac{\partial^2 w}{\partial x^2}\frac{\partial^2 w}{\partial y^2}-\left(\frac{\partial^2 w}{\partial x \partial y}\right)^2\right\}\right] dA \quad (2.38)$$

with $D = \frac{Eh^3}{12(1-\nu^2)}$, often referred to as the plate bending rigidity.

With the Lagrangian, $L = T - V$ and applying Hamilton's principle, we have $\delta \int_{t_1}^{t_2} L\left(\frac{\partial^2 w}{\partial x^2}, \frac{\partial^2 w}{\partial y^2}, \frac{\partial^2 w}{\partial x \partial y}, \frac{\partial w}{\partial t}, w\right) dt = 0$. Taking the first variation of the kinetic energy gives:

$$\delta \int_{t_1}^{t_2} T\, dt = \delta \int_{t_1}^{t_2} \frac{1}{2}\int_A m\dot{w}^2 dA\, d = \int_{t_1}^{t_2}\int_A m\dot{w}\ \delta\dot{w}\ dA\ dt \quad (2.39)$$

Integration by parts with respect to time and application of the admissible virtual displacements yields:

$$\delta \int_{t_1}^{t_2} T\, dt = \int_{t_1}^{t_2}\int_A m\ddot{w}\delta w\, dA\, dt \quad (2.40)$$

Assuming the presence of an external (transverse) loading field, $f(x, y, t)$, over the plate, the associated potential energy is given by $V_e = -\int_A f\, w dA$ and its first variation is

$$\delta \int_{t_1}^{t_2} V_o\, dt = \int_{t_1}^{t_2}\int_A f\delta w\, dA\ dt \quad (2.41)$$

Similarly the first variation of the potential energy expression in Equation (2.38) gives:

$$\partial \int_{t_1}^{t^2} V\ dt = \int_{t_1}^{t^2}\int_A D\left[\left\{\left(\frac{\partial^2 w}{\partial x^2}+\frac{\partial^2 w}{\partial y^2}\right) - (1-v)\frac{\partial^2 w}{\partial y^2}\right\}\delta\left(\frac{\partial^2 w}{\partial x^2}\right)\right.$$

$$+\left\{\left(\frac{\partial^2 w}{\partial x^2}+\frac{\partial^2 w}{\partial y^2}\right) - (1-v)\frac{\partial^2 w}{\partial x^2}\right\}\delta\left(\frac{\partial^2 w}{\partial y^2}\right)$$

$$\left. +2(1-v)\frac{\partial^2 w}{\partial x \partial y}\delta\left(\frac{\partial^2 w}{\partial x \partial y}\right)\right] dA\ dt$$

$$= \int_{t_1}^{t_2}\int_A \left[\left(\frac{\partial^2 w}{\partial x^2}+\nu\frac{\partial^2 w}{\partial y^2}\right)\delta\left(\frac{\partial^2 w}{\partial x^2}\right) + \left(\nu\frac{\partial^2 w}{\partial x^2}+\frac{\partial^2 w}{\partial y^2}\right)\delta\left(\frac{\partial^2 w}{\partial y^2}\right)\right.$$

$$\left. +2(1-\nu)\frac{\partial^2 w}{\partial x \partial y}\delta\left(\frac{\partial^2 w}{\partial x \partial y}\right)\right] dAdt \quad (2.42)$$

At this stage, it is significant to obtain the relationships between the stress resultants – moments and shear forces – and the plate curvatures. As in the case of the beam,

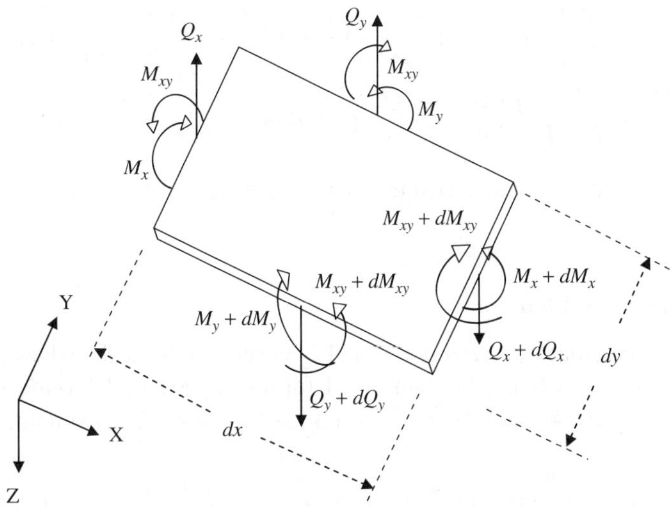

Figure 2.6 Free-body diagram on an infinitesimal plate segment

the moments arising out of the stress components σ_x, σ_y and τ_{xy} varying across the cross sections are obtained as

$$\begin{Bmatrix} M_x \\ M_y \\ M_{xy} \end{Bmatrix} = \int_{-\frac{h}{2}}^{\frac{h}{2}} (\sigma dz)\, z = \int_{-\frac{h}{2}}^{\frac{h}{2}} z\, \boldsymbol{c}\varepsilon\, dz \tag{2.43}$$

Substituting the respective expressions (2.30) and (2.33) for ε and $\boldsymbol{c}$ in the above equation gives:

$$\begin{Bmatrix} M_x \\ M_y \\ M_{xy} \end{Bmatrix} = -D \begin{Bmatrix} \dfrac{\partial^2 w}{\partial x^2} + \nu \dfrac{\partial^2 w}{\partial y^2} \\ \nu \dfrac{\partial^2 w}{\partial x^2} + \dfrac{\partial^2 w}{\partial y^2} \\ (1-\nu)\dfrac{\partial^2 w}{\partial x \partial y} \end{Bmatrix} \tag{2.44}$$

The shear forces Q_x and Q_y are obtained from moment equilibrium conditions (Figure 2.6) as

$$Q_x = \frac{\partial M_x}{\partial x} + \frac{\partial M_{xy}}{\partial y} \text{ and } Q_y = \frac{\partial M_y}{\partial y} + \frac{\partial M_{xy}}{\partial x} \tag{2.45}$$

In view of the moment–curvature relationships, Equation (2.41) is rewritten as

$$\delta \int_{t_1}^{t_2} V\, dt = \int_{t_1}^{t_2} \int_A \left[M_x \delta\left(\frac{\partial^2 w}{\partial x^2}\right) + M_y \delta\left(\frac{\partial^2 w}{\partial y^2}\right) + 2M_{xy}\delta\left(\frac{\partial^2 w}{\partial x \partial y}\right)\right] dA dt \tag{2.46}$$

Applying Green's theorem twice to the above area integral within braces, we get

$$\delta \int_{t_1}^{t_2} V\, dt = \int_{t_1}^{t_2} \left[\int_\Gamma \left\{ (l_1 M_x + l_2 M_{xy})\,\delta\left(\frac{\partial w}{\partial x}\right) + (l_2 M_y + l_1 M_{xy})\,\delta\left(\frac{\partial w}{\partial y}\right) \right\} d\Gamma \right.$$

$$+\int_{\Gamma}\left\{l_1\left(\frac{\partial M_x}{\partial x}+\frac{\partial M_{xy}}{\partial y}\right)+l_2\left(\frac{\partial M_y}{\partial y}+\frac{\partial M_{xy}}{\partial x}\right)\right\}\delta w d\Gamma\Bigg] dt$$
$$-\int_{t_1}^{t_2}\int_A\left(\frac{\partial Q_x}{\partial x}+\frac{\partial Q_y}{\partial y}\right)\delta w\, dAdt \tag{2.47}$$

Here, l_1 and l_2 are the direction cosines of the outward normal at a point on the plate boundary.

2.2.4.4 Rectangular Plate

If the plate is rectangular, $l_1=\cos\theta=1$ and $l_2=\sin\theta=0$ for the edges parallel to the X-axis and $l_1=\cos\theta=0$ and $l_2=\sin\theta=1$ for the edges parallel to the Y-axis. In this case, $M_{xy}=0$, $Q_x=\frac{\partial M_x}{\partial x}$ and $Q_y=\frac{\partial M_y}{\partial y}$ and the Equation (2.47) simplifies to

$$\delta\int_{t_1}^{t_2}V\,dt=\int_{t_1}^{t_2}\left[\int_0^b\left|M_x\delta\left(\frac{\partial w}{\partial x}\right)\right|_0^a dy+\int_0^a\left|M_y\delta\left(\frac{\partial w}{\partial y}\right)\right|_0^b dx+\int_0^b\left|Q_x\delta w\right|_0^a dy\right.$$
$$\left.+\int_0^a\left|Q_y\delta w\right|_0^b dx\right]dt-\int_{t_1}^{t_2}\int_A\left(\frac{\partial Q_x}{\partial x}+\frac{\partial Q_y}{\partial y}\right)\delta w\, dAdt \tag{2.48}$$

Combining Equations (2.40), (2.41) and (2.48), the PDE governing the vibration of a rectangular thin plate may be readily extracted as

$$\frac{\partial Q_x}{\partial x}+\frac{\partial Q_y}{\partial y}+m\frac{\partial^2 w}{\partial t^2}-f=0 \tag{2.49}$$

From Equation (2.48), we also obtain the following boundary conditions – either the bending moment or the slope is zero and either the shear force or the displacement is zero on each edge of the rectangular plate.

$$\begin{aligned}
&M_x(a)=0 \text{ or } \tfrac{\partial w(a)}{\partial x}=0,\quad M_x(0)=0 \text{ or } \frac{\partial w(0)}{\partial x}=0,\\
&M_y(b)=0 \text{ or } \tfrac{\partial w(b)}{\partial y}=0,\quad M_y(0)=0 \text{ or } \frac{\partial w(0)}{\partial y}=0,\\
&Q_x(b)=0 \text{ or } w(b)=0,\quad Q_x(0)=0 \text{ or } w(0)=0,\\
&Q_y(a)=0 \text{ or } w(a)=0,\quad Q_y(0)=0 \text{ or } w(0)=0
\end{aligned} \tag{2.50}$$

2.2.4.5 Plate with Nonsmooth Boundaries (Sharp Edges)

Referring to the plate with an arbitrary boundary, it is required to have the coordinate transformation from the Cartesian to the curvilinear coordinates (along the tangential 's' and normal 'n' directions) at any point on the boundary contour, Γ. From Figure 2.7, we have

$$x=l_1n-l_2s,\; y=l_1s+l_2n \text{ and also } \frac{\partial}{\partial x}=l_1\frac{\partial}{\partial n}-l_2\frac{\partial}{\partial s} \text{ and } \frac{\partial}{\partial y}=l_1\frac{\partial}{\partial s}+l_2\frac{\partial}{\partial n} \tag{2.51}$$

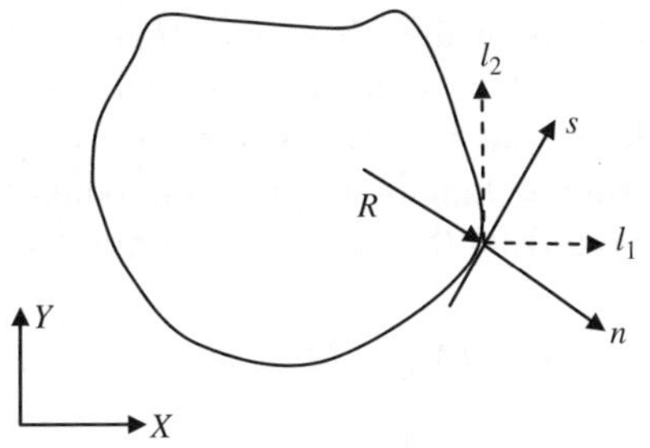

Figure 2.7 Plate with arbitrary boundary

With the above coordinate transformation, the bending moments M_x and M_y and shear force Q_x and Q_y are related to the moments M_n, M_s and M_{ns} and shear force Q_n on the plate boundary as

$$M_n = l_1^2 M_x + 2l_1 l_2 M_{xy} + l_2^2 M_y,\ M_s = l_1^2 M_y - 2l_1 l_2 M_{xy} + l_2^2 M_x,$$
$$M_{ns} = l_1 l_2 \left(M_y - M_x\right) + M_{xy}\left(l_1^2 - l_2^2\right),\ Q_n = \frac{\partial M_n}{\partial n} + \frac{\partial M_{ns}}{\partial s} \tag{2.52}$$

In view of the coordinate transformation and the associated differentiation formulae in Equation (2.51), the two boundary integrals in the first variation of potential energy (Equation (2.47)) take the form:

$$\int_\Gamma \left\{ (l_1 M_x + l_2 M_{xy})\, \delta\left(\frac{\partial w}{\partial x}\right) + (l_2 M_y + l_1 M_{xy})\, \delta\left(\frac{\partial w}{\partial y}\right) \right\} d\Gamma$$
$$= \int_\Gamma \left(-M_n \frac{\partial w}{\partial n} + \frac{\partial M_{ns}}{\partial s} \delta w \right) \partial\Gamma - M_{ns}\delta w|_\Gamma \tag{2.53a}$$

and

$$\int_\Gamma \left\{ l_1 \left(\frac{\partial M_x}{\partial x} + \frac{\partial M_{xy}}{\partial_y} \right) + l_2 \left(\frac{\partial M_y}{\partial y} + \frac{\partial M_{xy}}{\partial x} \right) \right\} \delta w d\Gamma$$
$$= \int_\Gamma \left\{ Q_n + \frac{1}{R}\left(M_n - M_s\right) \right\} \delta w d\Gamma \tag{2.53b}$$

Here, R is the radius of curvature of the boundary at the point under consideration as given by $R = -\frac{l_2}{\partial l/\partial s} = \frac{l_1}{\partial m/\partial s}$.

Using the above boundary integrals in Equation (2.47) and applying the Hamilton's principle to the first variation of the Lagrangian, $L = T - V$, we get

$$\delta \int_{t_1}^{t_2} L\ dt = \int_{t_1}^{t_2} \left[M_{ns}\delta w\ |_\Gamma + \int_\Gamma \left(-M_n \frac{\partial w}{\partial n} + \frac{\partial M_{ns}}{\partial s} \delta w \right) \right.$$
$$\left. + \left\{ Q_n + \frac{1}{R}\left(M_n - M_t\right) \right\} \delta w + \int_A \left(\frac{\partial Q_x}{\partial x} + \frac{\partial Q_y}{\partial y} + m\ddot{w} - f \right) \delta w\ dA \right] dt = 0 \tag{2.54}$$

The term $M_{ns}\delta w\big|_\Gamma$ on the RHS of the above equation stands for variation of twisting moment along the boundary. If the boundary is sufficiently smooth with a continuous first-order derivative, this term vanishes. Whilst the governing PDE for a thin plate with arbitrary boundary is the same as Equation (2.49), the boundary conditions at any point on the boundary to be inferred from the variational formulation are

$$Q_n + \frac{1}{R}\left(M_n - M_t\right) = 0 \text{ or } w = 0$$

$$M_n = 0 \text{ or } \frac{\partial w}{\partial n} = 0$$

$$M_{ns} = 0 \text{ or } w = 0 \qquad (2.55\text{a–c})$$

The first two sets of boundary conditions in the above equation are relevant for any regular point on the boundary, whereas the last one is applicable for any angular point on the boundary with a discontinuous first-order derivative of the contour. The expression $Q_n + \frac{1}{R}\left(M_n - M_t\right)$ stands for a shear force known as the Kirchhoff shear force.

2.3 PDEs and General Solution

Equation (2.1) represents a linear second-order PDE with two independent variables and is a special case of the general linear homogeneous equation:

$$a\frac{\partial^2 u}{\partial x^2} + 2h\frac{\partial^2 u}{\partial x \partial y} + b\frac{\partial^2 u}{\partial y^2} + 2f\frac{\partial u}{\partial x} + 2g\frac{\partial u}{\partial y} + du = 0 \qquad (2.56)$$

Equation (2.56) is analogous to that of a general conic section: $ax^2 + 2hxy + by^2 + 2fx + 2gy + c = 0$. As is known, a conic is an ellipse or a parabola or a hyperbola depending on $h^2 - ab < 0, = 0, > 0$, respectively. If a similar classification is used for the PDE (Equation (2.1)), we compare Equation (2.56) with Equation (2.1) that gives $a = c^2, h = 0, b = -1$ and $f = g = d = 0 \; and \; h^2 - ab > 0$. Thus, the wave equation (Equation (2.1)) is hyperbolic. Similarly, Laplace's equation $\nabla^2 u = 0$ and Poisson equation $\nabla^2 u = f$ are elliptic PDEs encountered in linear elastostatic problems (e.g. plane stress and/or plane strain problems). The heat conduction or the diffusion equation is represented by the parabolic equation $\nabla^2 u = \frac{1}{k}\frac{\partial u}{\partial t}$.

2.3.1 *PDEs and Canonical Transformations*

It is advantageous (in cases where the independent variables are ≥ 2) to reduce the original PDE into a canonical form. For example, consider a special case of the general Equation (2.56) in the form:

$$a\frac{\partial^2 u}{\partial x^2} + 2h\frac{\partial^2 u}{\partial x \partial y} + b\frac{\partial^2 u}{\partial y^2} = 0 \qquad (2.57)$$

Use the canonical transformation $\xi = y + m_1 x$ and $\eta = y + m_2 x$, m_1 and m_2 being arbitrary constants such that the Jacobian $J = \begin{vmatrix} \xi_x & \xi_y \\ \eta_x & \eta_y \end{vmatrix} \neq 0$, where the subscript stands for partial derivative with respect to x and y. Now, we have

$$\frac{\partial u}{\partial x} = m_1 \frac{\partial u}{\partial \xi} + m_2 \frac{\partial u}{\partial \eta}, \; \frac{\partial u}{\partial y} = \frac{\partial u}{\partial \xi} + \frac{\partial u}{\partial \eta},$$

$$\frac{\partial^2 u}{\partial x^2} = m_1^2 \frac{\partial^2 u}{\partial \xi^2} + 2 m_1 m_2 \frac{\partial^2 u}{\partial \xi \partial \eta} + m_2^2 \frac{\partial^2 u}{\partial \eta^2},$$

$$\frac{\partial^2 u}{\partial y^2} = \frac{\partial^2 u}{\partial \xi^2} + 2 \frac{\partial^2 u}{\partial \xi \partial \eta} + \frac{\partial^2 u}{\partial \eta^2} \text{ and}$$

$$\frac{\partial^2 u}{\partial x \partial y} = m_1 \frac{\partial^2 u}{\partial \xi^2} + (m_1 + m_2) \frac{\partial^2 u}{\partial \xi \partial \eta} + m_2 \frac{\partial^2 u}{\partial \eta^2} \tag{2.58a-d}$$

Substituting Equation (2.58) in Equation (2.57) gives the transformed PDE as:

$$\left(a + 2hm_1 + bm_1{}^2\right) \frac{\partial^2 u}{\partial \xi^2} + 2\left\{a + bm_1 m_2 + h\left(m_1 + m_2\right)\right\} \frac{\partial^2 u}{\partial \xi \partial n}$$
$$+ \left(a + 2hm_2 + bm_2^2\right) \frac{\partial^2 u}{\partial \eta^2} = 0 \Rightarrow A \frac{\partial^2 u}{\partial \xi^2} + H \frac{\partial^2 u}{\partial \xi \partial \eta} + B \frac{\partial^2 u}{\partial \eta^2} = 0 \tag{2.59}$$

Now choose m_1 and m_2 so that these constants are the roots λ_1 and λ_2 of the so-called auxiliary equation:

$$a + 2h\lambda + b\lambda^2 = 0 \tag{2.60}$$

Substituting the above condition in Equation (2.59), we obtain

$$2\left\{a + bm_1 m_2 + h\left(m_1 + m_2\right)\right\} \frac{\partial^2 u}{\partial \xi \partial \eta} = 0 \tag{2.61}$$

We have $m_1 + m_2 = -\frac{2h}{b}$ and $m_1 m_2 = a/b$. With these values, Equation (2.61) may be written as

$$\frac{2}{b}\left(ab - h^2\right) \frac{\partial^2 u}{\partial \xi \partial \eta} = 0 \tag{2.62}$$

For the nonparabolic case $(ab - h^2 \neq 0)$, integration of Equation (2.62) gives

$$u(x, y) = F(\xi) + G(\eta) \Rightarrow F\left(y + m_1 x\right) + G\left(y + m_2 x\right) \tag{2.63}$$

In the above equation, F and G are arbitrary functions that are at least twice differentiable. The above solution $u(x,y)$ is analogous to the complementary function (in terms of arbitrary constants) whilst deriving solutions for time-inhomogeneous ordinary differential equations (ODEs). For the parabolic case, we start from Equation (2.59) with the requirement that the transformation retains the parabolicity, $AB - H^2 = 0$. A possible scenario is provided by $H = 0$ and either $A = 0$ or $B = 0$. However, it can be verified that $A = 0 \Rightarrow H = 0$. Hence, setting A $= 0$ gives $m_1 = -\frac{h}{2a}$ and hence $\xi = y - \frac{h}{2a}x$. The other coordinate $\eta(x, y)$ can now be chosen in any way as desired, subject only to the condition that the Jacobian J remains nonzero.

2.3.2 General Solution to the Wave Equation

For the wave equation (Equation (2.1)), Equation (2.63) gives the solution $u(x, t)$ in terms of the arbitrary functions as:

$$u(x,t) = F\left(x + m_1 t\right) + G\left(x + m_2 t\right) \tag{2.64}$$

For this equation, $h^2 - ab > 0$ leads to real roots for m_1 and m_2. In the particular case of a taut string (with $a = c^2, h = 0$ and $b = -1$), we have $m_1 = +c$ and $m_2 = -c$ so that the solution $u(x,t)$ is given by

$$u(x,t) = F\,(x + ct) + G\,(x - ct) \tag{2.65}$$

The arbitrary functions F and G represent two travelling waves, the first propagating to the left and the other to the right, with the same speed c and the wave shapes remaining unchanged as they travel.

2.3.3 Particular Solution (D'Alembert's Solution) to the Wave Equation

The general solution in Equation (2.65) may be of little use unless it is followed up by determination of the unknown functions F and G. This is possible only when the physical phenomenon is represented by adequate boundary and/or initial conditions specific to the problem at hand. If time t is one of the independent variables in a PDE, the functions $u\,(x, 0)$ and $\frac{\partial u(x,0)}{\partial t}$ specified at $t = 0$ are referred to as boundary conditions of the Cauchy type (Appendix B) with respect to t. For the wave equation, let us specify the initial conditions at $t = 0$ in the form:

$$u\,(x,0) = f\,(x) \;\text{ and }\; \frac{\partial u(x,0)}{\partial t} = g(x) \tag{2.66}$$

For the example of the taut string, $f(x)$ is the specified initial displacement and $g(x)$, its initial velocity. Equating the general solution in Equation (2.65) at $t = 0$ and the initial conditions,

$$\begin{aligned} f(x) &= F(x) + G(x) \\ g(x) &= c\left(\frac{dF}{dx} - \frac{dG}{dx}\right) \end{aligned} \tag{2.67a,b}$$

Integration of Equation (2.67b) with respect to x gives

$$\int_\alpha^x g\,(s)\,ds = c\,(F(x) - G(x)) \tag{2.68}$$

Here, α is an arbitrary constant introduced as the lower limit of integration. Combining the above equation and Equation (2.67a), we obtain

$$F(x) = \frac{1}{2}\left(\left\{f(x) + \frac{1}{c}\int_\alpha^x g\,(s)\,ds\right\}, G(x) = \frac{1}{2}\left(\left\{f(x) - \frac{1}{c}\int_\alpha^x g\,(s)\,ds\right\} \tag{2.69}$$

The particular solution with the specified initial data is now obtained from Equation (2.65) as

$$u(x,t) = \frac{1}{2}\{f\,(x + ct) + f\,(x - ct)\} + \frac{1}{2c}\int_{x-ct}^{x+ct} g\,(s)\,ds \tag{2.70}$$

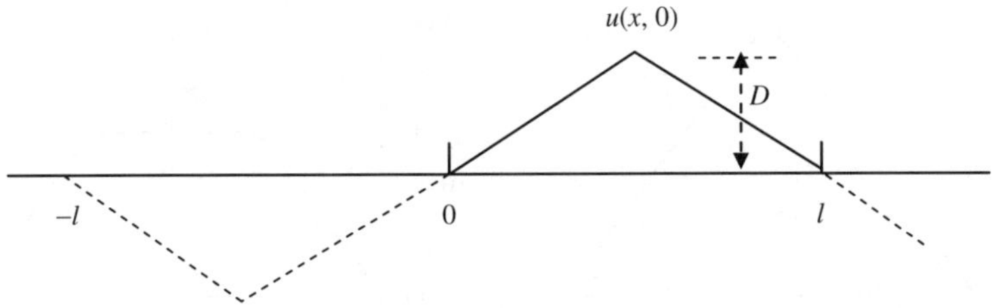

Figure 2.8 Taut string; initial displacement profile. Odd periodic extension

The above equation gives the so-called D'Alembert's solution of the wave equation. The property of the solution of the wave equation is understood by visualising propagation of left and right waves along the characteristic lines $x + ct =$ a constant, towards left and $x - ct =$ a constant, towards right. The propagation continues in t and x, if the spatial domain is infinite (i.e. $-\infty \le x \le \infty$). For a finite spatial domain, on the other hand, the solution represents a stationary standing wave.

Example 2.1: Stationary wave solution of a taut string

The general solution for a taut string is given by Equation (2.65). Suppose that to obtain a particular solution, the initial displacement function $f(x)$ in Equation (2.66) is specified as (Figure 2.8):

$$f(x) = u(x, 0) = 2D\frac{x}{l}, 0 \le x \le l/2 \text{ and}$$

$$= 2D\left(1 - \frac{x}{l}\right), l/2 \le x \le l \tag{2.71}$$

Let the initial velocity $g(x) = \frac{\partial u(x,0)}{\partial t} = 0$. Along with these initial conditions, we impose the boundary conditions $u(0, t) = 0$ and $u(l, t) = 0$. The particular solution given in Equation (2.70) takes the form:

$$u(x, t) = \frac{1}{2}\{f(x + ct) + f(x - ct)\} \tag{2.72}$$

Because of the boundary condition $u(0, t) = 0$, we note from Equation (2.72) that the displacement function $f(x)$ must be odd and from the boundary condition $u(l, t) = 0$ that $f(x)$ must be periodic with period $2l$. Odd periodic expansion of $f(x)$, as a half-range sine series, gives

$$f(x) = \frac{8D}{\pi^2} \sum_{n=1,3,5,\ldots}^{\infty} \sin\frac{n\pi}{2} \sin\frac{n\pi x}{l} \tag{2.73}$$

Upon substituting the above expansion of $f(x)$ in Equation (2.72), the particular solution for the taut string becomes

$$u(x, t) = \frac{4D}{\pi^2}\left\{\sum_{n=1,3,5,\ldots}^{\infty} \sin\frac{n\pi}{2}\left(\sin\frac{n\pi(x + ct)}{l} + \sin\frac{n\pi(x - ct)}{l}\right)\right\} \tag{2.74}$$

The fundamental period ($n = 1$) of oscillation is $T_1 = \frac{2\pi}{\pi c/l} = \frac{2l}{c}$. Figure 2.9 shows a few snapshots (at equal time intervals) of an oscillating steel wire over one half-period of the

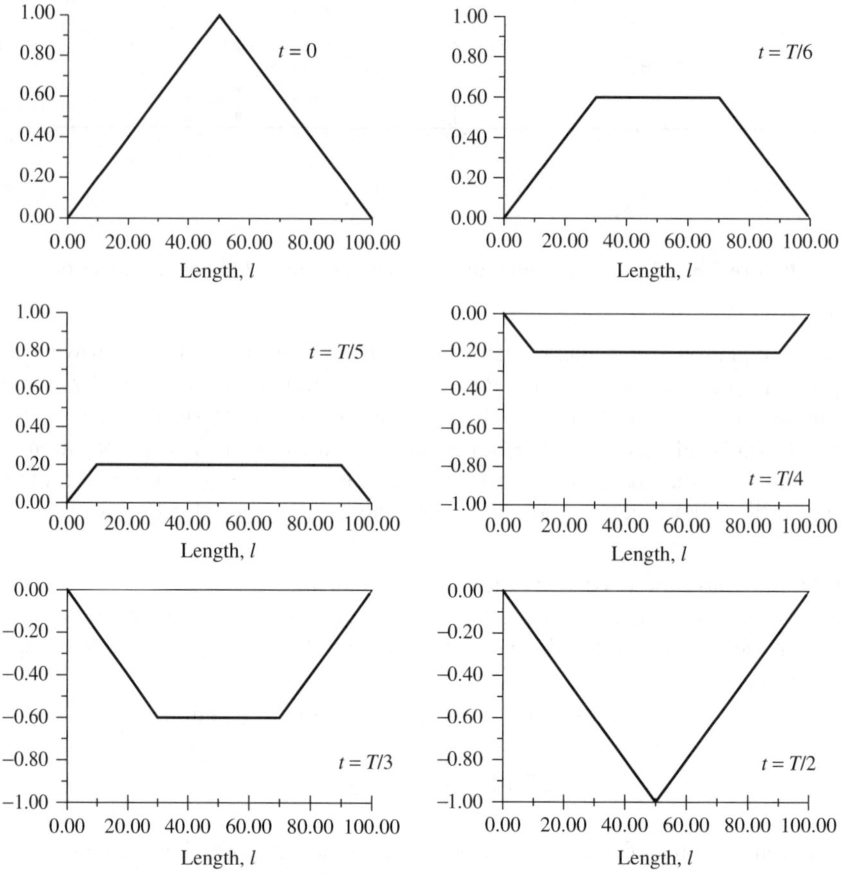

Figure 2.9 Stationary-wave solution of a taut string. Few snap shots in one half-period of fundamental frequency

fundamental. The solution is obtained for the data: length = 100 mm, diameter = 2 mm, mass density = 7850 kg/m^3, tension in the wire = 1 N.

2.4 Solution to Linear Homogeneous PDEs – Method of Separation of Variables

The periodic solution in Equation (2.74) for the wave equation can be rewritten as

$$u(x,t) = \frac{8D}{\pi^2}\left\{\sum_{n=1,3,5,\ldots}^{\infty} \sin\frac{n\pi}{2}\sin\frac{n\pi x}{l}\cos\frac{n\pi ct}{l}\right\} \tag{2.75}$$

The solution in the above form is in variables-separable form in that each term in the Fourier series is a product of two terms – one a function of spatial variable x alone and the other of time t alone. Such a possible form of the solution enables arriving at the method of separation of variables to solve linear homogeneous PDE within a finite domain. The method can be applied to obtain an asymptotically exact solution to a linear homogeneous PDE with homogeneous boundary conditions.

2.4.1 Homogeneous PDE with Homogeneous Boundary Conditions

The basic idea in the method of separation of variables is to write the solution to the PDE as a product of functions, each of which is only dependent upon a single (scalar) independent variable. Thus, consider, once more, the wave equation (Equation (2.1)) with the following boundary and initial conditions:

$$u\,(0,t) = u\,(l,t) = 0,\, t \geq 0$$

$$u\,(x,0) = f(x) \text{ and } \frac{\partial u(x,0)}{\partial t} = g(x) \tag{2.76a,b}$$

A solution is sought in the variable separable form:

$$u(x,t) = \Phi(x)Y(t) \tag{2.77}$$

Substituting Equation (2.77) in Equation (2.1) gives

$$c^2 Y \Phi'' = \Phi \ddot{Y}$$

$$\frac{\Phi''}{\Phi} = \frac{1}{c^2}\frac{\ddot{Y}}{Y} \tag{2.78a,b}$$

Superscript $''$ indicates differentiation with respect to x and an overdot with respect to t. In Equation (2.78b), the left term is a function of x alone and the right term a function of t alone and this can be true if and only if each is equal to a constant. Thus, let

$$\frac{\Phi''}{\Phi} = -k = \frac{1}{c^2}\frac{\ddot{Y}}{Y} \tag{2.79}$$

k is an arbitrary constant. In Equation (2.79) we have two ODEs:

$$\ddot{Y} + kc^2 Y = 0$$

$$\Phi'' + k\Phi = 0 \tag{2.80a,b}$$

The functions $\Phi(x)$ and $Y(t)$ are to be determined with the appropriate boundary/initial conditions. In this respect, $u(x,t) = \Phi(x)Y(t)$ shall satisfy the specified boundary and initial conditions in Equation (2.76). Thus,

$$u\,(0,t) = \Phi\,(0)\,Y(t) = 0 \text{ and } u\,(l,t) = \Phi\,(l)\,Y(t) = 0$$

$$u\,(x,0) = \Phi(x)Y\,(0) = f(x) \text{ and } \frac{\partial u(x,0)}{\partial t} = \Phi(x)\dot{Y}\,(0) = g(x) \tag{2.81a,b}$$

To have a nontrivial solution for $u(x,t)$, $Y(t) \neq 0\ \forall t$ and hence the boundary conditions for the ODE in $\Phi(x)$ are given by

$$\Phi\,(0) = 0 \text{ and } \Phi\,(l) = 0 \tag{2.82}$$

2.4.2 *Sturm–Liouville Boundary-Value Problem (BVP) for the Wave Equation*

Equation (2.80b) along with the boundary conditions in Equation (2.82) constitutes a Sturm–Liouville-type boundary-value problem (Simmons and Krantz, 1925). More generally, a Sturm–Liouville type boundary-value problem is of the generic form

$$\boldsymbol{L}y + \lambda r(x)y = 0, a \leq x \leq b \tag{2.83}$$

with boundary conditions

$$B_1(y) = \alpha_1 y(a) + \alpha_2 y'(a) = 0 \text{ and } B_2(y) = \beta_1 y(b) + \beta_2 y'(b) = 0 \tag{2.84}$$

where $\boldsymbol{L} = \left(p(x)\frac{d}{dx}\right)' + q(x)$ is a second-order linear differential operator (a Sturm–Liouville operator) with $p(x) \in C^1[a,b], q(x)$ and $r(x) \in C^0[a,b]$ and, in addition, $p(x)$ and $r(x) > 0$. α_1 and α_2 in Equation (2.84) are constants (not both zero) and so are β_1 and β_2. Note that any second-order linear operator can be recast into the form of the Sturm–Liouville operator by multiplying with a suitable weight function. A Sturm–Liouville operator is self-adjoint. The self-adjointedness of a differential operator is associated with the concept of an adjoint operator.

2.4.3 *Adjoint Operator and Self-Adjoint Property*

Defining the inner product of $u(x)$ and $v(x)$ (with $u(x)$ and $v(x)$ belonging to a Hilbert space $\mathcal{H}$; (see Appendix B) as

$$< u, v > = \int_a^b u(x)v(x)dx \tag{2.85}$$

The adjoint, $\boldsymbol{L}^* : \mathcal{H}^* \to \mathcal{H}$ ($\mathcal{H}^*$ also being a Hilbert space) of an operator $\boldsymbol{L} : \mathcal{H} \to \mathcal{H}^*$ satisfies $<u, \boldsymbol{L}v> = <\boldsymbol{L}^*u, v>$ for all v in the domain of $\boldsymbol{L}$ and u in the domain of $\boldsymbol{L}^*$. Considering $\boldsymbol{L}$ to be a second-order linear differential operator given by $L(v) = a_2(x)v'' + a_1(x)v' + a_0(x)v$, with boundary conditions prescribed as in Equation (2.56) and integrating by parts twice yields:

$$\begin{aligned}< u, \boldsymbol{L}v > &= \int_a^b v\left(a_2u'' + a_1u' + a_0u\right)dx = \left(a_1 - a_2'\right)uv + a_2uv' \\ &- a_2u'v\,|_a^b + \int_a^b \left[\left(a_2u\right)'' - \left(a_1u\right)' + a_0u\right] v\,dx \Rightarrow < u, \boldsymbol{L}v > - \int_a^b \left[\left(a_2u\right)''\right. \\ &\left. - \left(a_1u\right)' + a_0u\right] v\,dx = \left(a_1 - a_2'\right)uv + a_2uv' - a_2u'v\,|_a^b \end{aligned} \tag{2.86}$$

Substituting the known boundary conditions for $v(x)$, the conditions on $u(x)$ are to be found such that $\left(a_1 - a_2'\right)uv + a_2uv' - a_2u'v\big|_a^b = 0$. It follows from Equation (2.86):

$$<u, \boldsymbol{L}v> = \int_a^b \left[\left(a_2u\right)'' - \left(a_1u\right)' + a_0u\right] v\,dx = < \boldsymbol{L}^*u, v > \tag{2.87}$$

Thus, we have the adjoint operator $L^* = \left(a_2 u\right)'' - \left(a_1 u\right)' + a_0 u$. From Equations (2.82) and (2.87), we obtain the following Green's identity:

$$<u, Lv> - <L^* u, v> = \left(a_1 - a_2'\right) uv + a_2 uv' - a_2 u'v\big|_a^b = P(u, v)_a^b \tag{2.88}$$

Here, $P(u, v) = \left(a_1 - a_2'\right) uv + a_2 uv' - a_2 u'v$. The above procedure can be extended to find an adjoint operator for an nth-order linear differential operator L_n. If L_n is given by

$$L_n = \sum_{j=0}^{n} a_j(x) \frac{d^{n-j}}{dx^{n-j}} \tag{2.89}$$

then the adjoint operator is obtainable as

$$L_n^* = \sum_{j=0}^{n} (-1)^{n-j} \frac{d^{n-j}(a_j(x))}{dx^{n-j}} \tag{2.90}$$

If $L^* = L$, L is self-adjoint and it follows that the domain of L is the same as that of L^*. In the present case, note that if $a_2' = a_1$,

$$L^* u = \left(a_2 u\right)'' - \left(a_1 u\right)' + a_0 u = a_2 u'' + a_1 u' + a_o u \tag{2.91}$$

Thus, with $a_2' = a_1$, above equation shows the self-adjointedness of the operator, L. The operator can be put into the compact form: $L = \left(a_2 \frac{d}{dx}\right)' + a_o \mathbf{I}$, where I denotes the identity operator.

With L being self-adjoint, Equation (2.86) reduces to

$$<u, Lv> - <Luv> = a_2\left(uv' - u'v\right)\big|_a^b \Rightarrow \int_a^b (uLv - vLu)dx = a_2\left(uv' - u'v\right)\big|_a^b \tag{2.92}$$

It follows that the Sturm–Liouville operator, $L = \left(p(x)\frac{d}{dx}\right)' + q(x)I$ in Equation (2.83) is in the self-adjoint form with $a_2 = p$ and $a_o = q$. In addition, for the Sturm–Liouville boundary-value problem with functions $u(x), v(x) \in$ H, L satisfies the boundary conditions in Equation (2.84), that is

$$\alpha_1 u(a) + \alpha_2\, u'(a) = 0,\ \alpha_1 v(a) + \alpha_2\, v'(a) = 0$$
$$\beta_1 u(b) + \beta_2\, u'(b) = 0,\ \beta_1 v(b) + \beta_2\, v'(b) = 0 \tag{2.93a,b}$$

In the matrix vector form, we have:

$$\begin{bmatrix} u(a) & u'(a) \\ v(a) & v'(a) \end{bmatrix} \begin{Bmatrix} \alpha_1 \\ \alpha_2 \end{Bmatrix} = \begin{Bmatrix} 0 \\ 0 \end{Bmatrix}$$
$$\begin{bmatrix} u(b) & u'(b) \\ v(b) & v'(b) \end{bmatrix} \begin{Bmatrix} \beta_1 \\ \beta_2 \end{Bmatrix} = \begin{Bmatrix} 0 \\ 0 \end{Bmatrix} \tag{2.94a,b}$$

Since the pair of constants α_1 and α_2 and β_1 and β_2 the coefficient matrices in the above equations must be singular, that is $u(a)\, v'(a) - v(a)\, u'(a) = 0$ and $u(b)\, v'(b) -$

$v(b)\,u'(b) = 0$. This implies that the right-hand side of Equation (2.92) vanishes and thus, for the Sturm–Liouville boundary-value problem, the self-adjoint operator satisfies:

$$\int_a^b (u\boldsymbol{L}v - v\boldsymbol{L}u)\,dx = 0 \tag{2.95}$$

The Sturm–Liouville (self-adjoint) boundary-value problem, $\boldsymbol{L}y + \lambda r(x)y = 0$ involving a parameter λ assumes significance due to its property that nontrivial solutions $y(x)$, known as the eigenfunctions, exist for specific values of λ, known as the eigenvalues. We now show that these eigenvalues are non-negative, real and the eigenfunctions orthogonal. With $(\lambda, y(x))$ as an eigenpair, we have

$$\boldsymbol{L}y + \lambda r(x)y = 0 \tag{2.96}$$

Taking complex conjugates on both sides of the above equation gives:

$$\overline{\boldsymbol{L}y} + \overline{\lambda r(x)y} = 0 \Rightarrow \boldsymbol{L}\bar{y} + \bar{\lambda}r(x)\bar{y} = 0 \tag{2.97}$$

Overbars in the above equation indicate complex conjugates. It is clear that if (λ, y) is an eigenpair, $(\bar{\lambda}, \bar{y})$ is also an eigenpair. By the self-adjoint property:

$$\int_a^b (\bar{y}\boldsymbol{L}y - y\boldsymbol{L}\bar{y})\,dx = 0 \Rightarrow -\int_a^b \left(\bar{y}\lambda r(x)y - y\bar{\lambda}r(x)\bar{y}\right)dx = 0$$

$$\Rightarrow \left(\lambda - \bar{\lambda}\right)\int_a^b y\bar{y}r(x)dx = 0 \Rightarrow \left(\lambda - \bar{\lambda}\right)\int_a^b |y|^2\, r(x)dx = 0 \tag{2.98}$$

Since $r(x) > 0$ and $|y|^2 > 0$ (y(x) is an eigenfunction), $\int_a^b |y|^2 r(x)dx \neq 0$. Therefore, $\lambda - \bar{\lambda} = 0 \Rightarrow \lambda = \bar{\lambda} \Rightarrow \lambda$ is real. A similar argument shows that the eigenvalues of the Sturm–Liouville (self-adjoint) boundary-value problem are simple in that the dimension of the null space (Appendix B) of $\boldsymbol{L}_\lambda = \boldsymbol{L} - \lambda\boldsymbol{I}$ is one, that is the dimension of $\left\{\Phi : \boldsymbol{L}_\lambda\Phi = 0\right\}$ is one. The orthogonal property of the eigenfunctions is manifest again by the self-adjoint property of the Sturm–Liouville boundary-value problem. To show this, let (λ_i, u_i) and (λ_j, u_j) be two eigenpairs of the boundary-value problem. Then, we have:

$$\int_a^b \left(u_i\boldsymbol{L}u_j - u_j\boldsymbol{L}u_i\right)dx = 0 \Rightarrow -\int_a^b \left(u_i\lambda_i r(x)u_i - u_j\lambda_j r(x)u_i\right)dx = 0$$

$$\Rightarrow \left(\lambda_i - \lambda_j\right)\int_a^b u_iu_jr(x)dx = 0 \Rightarrow \int_a^b u_iu_jr(x)dx = 0 \text{ since } \lambda_i \neq \lambda_j \tag{2.99}$$

Note the similarity between the transpose of a matrix and the adjoint of a differential operator. In the same vein, an analogy may be drawn between a self-adjoint differential operator and a symmetric (Hermitian) matrix $\boldsymbol{A}$ that defines a linear transformation on a vector space. A matrix operator $\boldsymbol{A}$ is symmetric (Hermitian), if $\boldsymbol{A} = \boldsymbol{A}^*$ or, equivalently, if $\boldsymbol{x}^T\boldsymbol{A}\boldsymbol{x} = \boldsymbol{x}^T\boldsymbol{A}^*\boldsymbol{x}$ for any $\boldsymbol{x}$ in the vector space. It is also useful to compare the analogy of the properties of a self-adjoint operator and a symmetric matrix, the latter being known to possess only real eigenvalues (Appendix E). This symmetry is intrinsically carried forward even when we transform the strong form of the system PDEs into an integral form (for instance, a weak form), which is more frequently used with different discretization methods, especially the finite element method (FEM). These methods form the subject of Chapters 3 and 4.

2.4.4 Eigenvalues and Eigenfunctions of the Wave Equation

Referring back to the boundary-value problem (Equation (2.80b)) corresponding to the one-dimensional wave equation, it is evident that $\boldsymbol{L}\Phi = \Phi''$ is self-adjoint with $p(x) = 1, q(x) = 0$. Also, we have $r(x) = 1$ and the eigenvalue parameter $k = \lambda$. Comparison of the respective boundary conditions in Equations (2.82) and (2.84) yield $\alpha_1 = 1$, $\alpha_2 = 0$, $\beta_1 = 1$ and $\beta_2 = 0$. For nontrivial solution $\Phi(x)$ of Equation (2.80b), the possible values of the parameter k gives the eigenvalues of the boundary-value problem. However, since the equation $\Phi'' + k\Phi$ is self-adjoint, k must be positive real. It is evident that for $k = 0$, the solution for Equation (2.80b) is $\Phi(x) = A + Bx$. Substituting the boundary conditions in Equation (2.82), we find that the constants A and B are simultaneously zero and with $\Phi(x) = 0$, only the trivial solution is possible for $u(x,t)$ and $k = 0$ is not an eigenvalue. Similarly, with $k < 0$ and particularly with $k = -\mu^2$, the solution is $\Phi(x) = Ae^{\mu x} + Be^{-\mu x}$. The boundary conditions in Equation (2.82) leads to $\Phi(x) = 0$ that in turn yields again the trivial solution $u(x,t) = 0$. Thus, the possible eigenvalues are with $k > 0$ only. With $k > 0$ and $k = \mu^2$, the solution for $\Phi(x)$ is

$$\Phi(x) = Ae^{i\mu x} + Be^{-i\mu x} \Rightarrow \Phi(x) = C \sin \mu x + D \cos \mu x \tag{2.100}$$

A, B, C and D are arbitrary constants. Boundary condition $\Phi\,(0) = 0$ results in $D = 0$. The other boundary condition $\Phi\,(l) = 0$ gives

$$C \sin \mu l = 0 \tag{2.101}$$

If $C = 0$, only the trivial solution is possible. On the other hand, the equation leads to a nontrivial solution for $\Phi(x)$ provided

$$\sin \mu l = 0 \tag{2.102}$$

Equation (2.102) is the equation in the frequency parameter μ. The possible solutions are $\mu l = n\pi$, $n = \pm 1, \pm 2, \ldots\ldots$ ($n = 0$ is excluded to avoid the trivial solution again). These possible values of μ in turn yield infinite positive eigenvalues as

$$k_n = \mu_n^2 = \frac{n^2\pi^2}{l^2}, \quad n = 1, 4, 9, \ldots. \tag{2.103}$$

We obtain from Equation (2.100) the corresponding eigenfunctions (i.e. vibrating modes) as

$$\Phi_n(x) = C_n \sin \frac{n\pi}{l} x, n = 1, 2, 3, \ldots. \tag{2.104}$$

The orthogonal eigenfunctions $\Phi_n(x)$ are harmonic. μ_n and $\Phi_n(x)$ form a eigenpair of the Sturm–Liouville boundary-value problem corresponding to the wave equation (Equation (2.1)).

2.4.5 Series Solution to the Wave Equation

With the result that k $(= \mu^2)$ is positive and has possible infinite values, the solution to the first ODE (Equation (2.80a)) is given by

$$Y_n(t) = A_n \sin c\mu_n t + B_n \cos c\mu_n t \tag{2.105}$$

In the variables-separable form, the solution to the wave equation (Equation (2.1)) with the boundary conditions in Equation (2.76a) is now given by

$$\begin{aligned} u_n(x,t) &= \Phi_n(x)Y_n(t) = C_n \sin \mu_n x \left(A_n \sin c\mu_n t + B_n \cos c\mu_n t\right) \\ &= \sin \mu_n x \left(A_n \sin c\mu_n t + B_n \cos c\mu_n t\right) \end{aligned} \tag{2.106}$$

In the above equation, the constant C_n is absorbed into the constants A_n and B_n without loss of generality. Since the Equation (2.1) is linear and homogeneous, it follows from linearity (superposition) principle that the solution to the equation is the sum of the denumerable set of solutions, $u_n(x,t)$ given in Equation (2.106). Thus, the series solution to the wave equation can be taken as

$$u(x,t) = \sum_{n=1}^{\infty} \sin \mu_n x \left(A_n \sin c\mu_n t + B_n \cos c\mu_n t\right) \tag{2.107}$$

Substituting the initial conditions in Equation (2.81b) gives

$$\begin{aligned} u(x,0) &= \sum_{n=1}^{\infty} B_n \sin \mu_n x = f(x) \text{ and} \\ \frac{\partial u(x,0)}{\partial t} &= \sum_{n=1}^{\infty} A_n c\mu_n \sin \mu_n x = g(x) \end{aligned} \tag{2.108a,b}$$

The constants A_n and B_n are obtained by expanding the known functions $f(x)$ and $g(x)$ into Fourier sine series over the interval $[0,l]$. Thus

$$A_n = \frac{2}{c\mu_n l}\int_0^l g(x)\sin \mu_n x\,dx \text{ and } B_n = \frac{2}{l}\int_0^l f(x)\sin \mu_n x\,dx \tag{2.109}$$

Equation (2.107) along with Equation (2.109) finally gives the solution to the wave equation by the method of separation of variables.

2.4.6 Mixed Boundary Conditions and Wave Equation

The boundary conditions in Equation (2.76a) are of the Dirchlet type, where the dependent variable is directly specified on the boundary of the domain. The wave equation (Equation (2.1)) representing the axial (longitudinal) vibrations of a rod or bar (Chapter 1) may have Neumann-type conditions where normal derivatives are prescribed on the boundary. For example, for an axially vibrating rod with fixed–free boundary conditions, we have a mixture of both Dirichlet and Neumann boundary conditions given by:

$$\begin{aligned} &u(0,t) = 0,\ t \geq 0\text{: Dirchlet-type boundary condition} \\ &\frac{\partial u(l,t)}{\partial x} = 0,\ t \geq 0\text{: Neumann-type boundary condition} \end{aligned} \tag{2.110a,b}$$

Condition (2.110b) indicates the existence of no stress at the free end of the rod for all $t \geq 0$. The corresponding boundary conditions for the ODE (Equation (2.80b)) in $\Phi(x)$ are obtained as

$$u(0,t) = \Phi(0)\,Y(t) = 0 \text{ and } \frac{\partial u(l,t)}{\partial x} = \frac{\partial \Phi(l)}{\partial x} Y(t) = 0 \tag{2.111a,b}$$

To have a nontrivial solution for $u(x,t)$, $Y(t) \neq 0\ \forall t$ and hence the boundary conditions for the ODE in $\Phi(x)$ are given by

$$\Phi(0) = 0 \text{ and } \frac{\partial \Phi(l)}{\partial x} = \Phi'(l) = 0 \tag{2.112a,b}$$

Equation (2.100) yields $\Phi'(x)$ as

$$\Phi'(x) = C\mu \cos \mu x - D\mu \sin \mu x \tag{2.113}$$

The boundary condition $\Phi\,(0) = 0$ results in $D = 0$ and the condition $\Phi'\,(l) = 0$ gives

$$C\mu \cos \mu l = 0 \tag{2.114}$$

A nontrivial solution for $\Phi(x)$ is possible only if

$$\cos \mu l = 0 \tag{2.115}$$

The equation above gives the possible values for the frequency parameter μ as $\mu_n l = \frac{2n-1}{2}\pi, n = 1, 2 \ldots.$ With $k = \mu^2$, these possible values of μ in turn yield an infinite sequence of positive eigenvalues $k_n = \left(\frac{2n-1}{2}\frac{\pi}{l}\right)^2, n = 1, 2 \ldots\ldots$ The corresponding eigenfunctions are of the form:

$$\Phi_n(x) = C_n \sin \frac{(2n-1)\,\pi x}{2l}, \quad n = 1, 2 \ldots.. \tag{2.116}$$

The mode shapes for the first few eigenvalues (natural frequencies) representing transverse vibrations of the taut string (Equation (2.100)) and the longitudinal (axial) vibrations of the rod (Equation (2.116)) are shown in Figure 2.10.

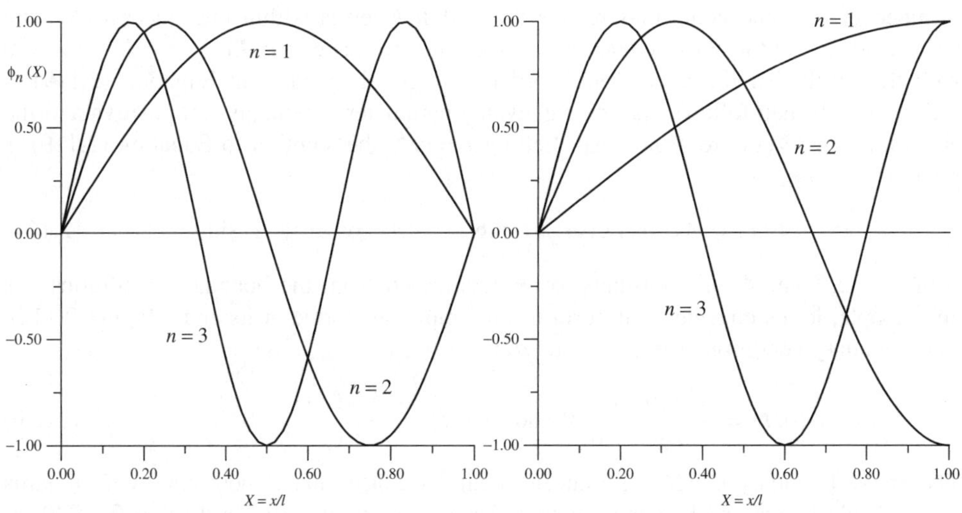

Figure 2.10 First three eigenfunctions (mode shapes) of (a) vertically vibrating taut string and (b) axially vibrating rod

2.4.7 Sturm–Liouville Boundary-Value Problem for the Biharmonic Wave Equation

The Euler–Bernoulli beam PDE (Equation (2.16)) without any external forcing term is rewritten as

$$\frac{\partial^4 w}{\partial x^4} = -\frac{1}{c^2}\frac{\partial^2 w}{\partial t^2}, x \in [0, l], t \in [0, \infty] \tag{2.117}$$

Here, $c^2 = \frac{EI}{m}$. If a solution is sought in the variables-separable form, we have

$$w(x, t) = \Phi(x)Y(t) \tag{2.118}$$

As the variables-separable method leads eventually to a harmonic solution in time, we can directly assume that $Y(t) = Ae^{i\omega t}$ with A being an arbitrary constant. Substitution in Equation (2.117) results in the following fourth-order ODE in the spatial variable:

$$\Phi^{iv} - \mu\Phi = 0 \tag{2.119}$$

Here, $\mu = \frac{\omega^2}{c^2} = \frac{m\omega^2}{EI}$. Equation (2.119) along with appropriate boundary conditions is a Sturm–Liouville boundary-value problem given by $\boldsymbol{L}f = \mu f,\ f \in \mathcal{H}$. The Sturm–Liouville differential operator in this case is given by $\boldsymbol{L} = \frac{d^4}{dx^4}$. If we consider the inner product $< u, \boldsymbol{L}v >$, with $u(x)$ and $v(x) \in \mathcal{H}$, integration by parts gives:

$$\begin{aligned} < u, \boldsymbol{L}v > &= \int_0^l uv''''dx = \left\{uv'''\big|_0^l - u'v''\big|_0^l + u''v'\big|_0^l - u'''v\big|_0^l\right\} \\ &+ \int_0^l u''''v\,dx = \left\{uv'''\big|_0^l - u'v''\big|_0^l + u''v'\big|_0^l - u'''v\big|_0^l\right\} + < \boldsymbol{L}u, v > \end{aligned} \tag{2.120}$$

For the beam under consideration, the boundary conditions corresponding to hinged or clamped or free end conditions reduce the first few terms within the braces {.} on the right-hand side of the above equation to zero. Thus, we have $< u, \boldsymbol{L}v > - < \boldsymbol{L}u, v > = 0$ with the result that the linear fourth order differential operator in Equation (2.119) is self-adjoint. It then follows that the eigenvalue parameter μ remains real positive and the eigenfunctions $\Phi(x)$ are orthogonal. Letting $\mu = \lambda^4$, the solution to Equation (2.119) is given by

$$\Phi(x) = A_1 \cosh\lambda x + A_2 \sinh\lambda x + A_3 \cos\lambda x + A_4 \sin\lambda x \tag{2.121}$$

$A_i, i = 1, 2, 3$ and 4, are constants to be determined from the boundary conditions. As an example, let us consider a uniform beam simply supported at its ends (Figure 2.11a). The boundary conditions for this beam are:

$$w\,(0, t) = \frac{\partial^2 w(0, t)}{\partial x^2} = 0 \text{ and } w\,(l, t) = \frac{\partial^2 w(l, t)}{\partial x^2} = 0 \tag{2.122a,b}$$

The above boundary conditions indicate identical constraints at both ends with regards to the displacement and bending moment. These conditions result in $\Phi(0) = 0, \Phi''(0) = 0, \Phi(l)$ and $\Phi''(l) = 0$. If the boundary conditions at the left end ($x = 0$) are substituted in Equation (2.121), we have $A_1 + A_3 = 0$ and $A_1 - A_3 = 0$ with the result $A_1 = A_3 = 0$.

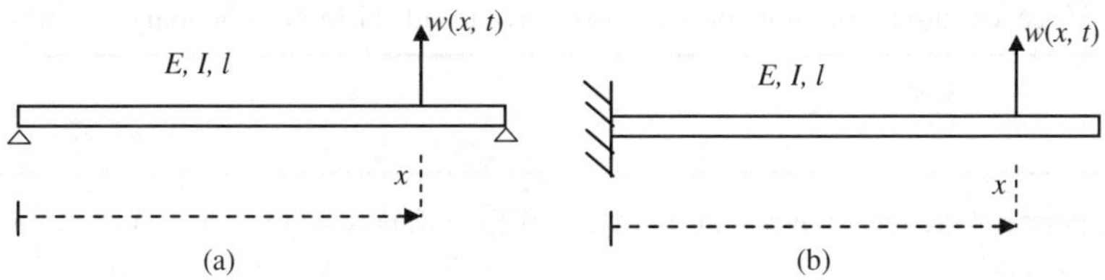

Figure 2.11 A beam. (a) Simply supported case and (b) cantilever (fixed–free)

Thus, $\Phi(x) = A_2 \sinh \lambda x + A_4 \sin \lambda x$. The two boundary conditions at the right end $(x = l)$ give the following matrix equation in A_2 and A_4:

$$\begin{bmatrix} \sinh \lambda l & \sin \lambda l \\ \sinh \lambda l & \sin \lambda l \end{bmatrix} \begin{Bmatrix} A_2 \\ A_4 \end{Bmatrix} = \begin{Bmatrix} 0 \\ 0 \end{Bmatrix} \tag{2.123}$$

For A_2 and A_4 not to be simultaneously zero, it is required that the coefficient matrix be singular, that is $\sinh \lambda l \sin \lambda l = 0$. If $\sinh \lambda l = 0$ and $\sin \lambda l \neq 0$, then $A_4 = 0 = A_2$ which gives a trivial solution for $\Phi(x)$.On the other hand, $\sinh \lambda l \neq 0 \cap \sin \lambda l = 0$ leads to $A_2 = 0$. Here, the possible values for λl are:

$$\lambda_n l = n\pi, n = 1, 2 \ldots.. \tag{2.124}$$

These values of λ_n in turn yield the eigenvalue parameter μ as $\left(\frac{n\pi}{l}\right)^4$. Thus, the mode shape function for the simply supported beam is given by

$$\Phi(x) = A_4 \sin \frac{n\pi}{l} x, n = 1, 2 \ldots.. \tag{2.125}$$

Now consider the fixed–free support conditions as shown in Figure 2.11b. These conditions pertain to a cantilever beam and are described by:

$$w\,(0, t) = \frac{\partial w(0, t)}{\partial x} = 0 \text{ and } \frac{\partial^2 w(l, t)}{\partial x^2} = \frac{\partial^3 w(l, t)}{\partial x^3} = 0 \tag{2.126a,b}$$

The boundary conditions at the left end in Equation (2.126a) refer to the constraints with regards to the displacement and its slope (first derivative) and those at the right end (i.e. at $x = l$) imply vanishing bending moment $(EI\frac{\partial^2 w}{\partial x^2})$ and shear force $EI\frac{\partial^3 w}{\partial x^3}$. These boundary conditions result in $\Phi(0) = 0$, $\Phi'(0) = 0$, $\Phi''(l)$ and $\Phi'''(l) = 0$. Substitution of $\Phi(0) = 0$ and $\Phi'(0) = 0$ in Equation (2.121) gives: $A_1 + A_3 = 0$ and $A_2 + A_4 = 0$. Thus, $\Phi(x)$ can be written as:

$$\Phi(x) = A_1\,(\cosh \lambda x - \cos \lambda x) + A_2\,(\sinh \lambda x - \sin \lambda x) \tag{2.127}$$

Substitution of zero bending moment and shear force at the right end results in the following equations in A_1 and A_2:

$$\begin{bmatrix} \cosh \lambda l - \cos \lambda l & \sinh \lambda l + \sin \lambda l \\ \sinh \lambda l + \sin \lambda l & \cosh \lambda l + \cos \lambda l \end{bmatrix} \begin{Bmatrix} A_1 \\ A_2 \end{Bmatrix} = 0 \tag{2.128}$$

Table 2.1 Free vibration solution for a uniform beam with different boundary conditions

Boundary condition	Results
Simply supported (hinged–hinged) 	Frequency equation : $\sin\lambda l = 0$ Eigenfunction: $\Phi_n(x) = A_1 \sin \frac{n\pi}{l} x$
Cantilever (Fixed–Free) 	Frequency equation: $1 + \cos\lambda l \cosh\lambda l = 0$ Eigenfunction: $\Phi_n(x) = A_1\left[\left(\cosh\lambda_n x - \cos\lambda_n x\right)\frac{\sinh\lambda_n l + \sin\lambda_n l}{\cosh\lambda_n l + \cos\lambda_n l}(\sinh\lambda_n x - \sin\lambda_n x)\right]$
Fixed–Fixed 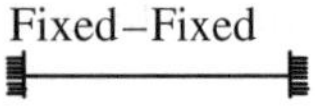	Frequency equation: $1 - \cos\lambda l \cosh\lambda l = 0$ Eigenfunction: $\Phi_n(x) = A_1\left[\left(\cosh\lambda_n x - \cos\lambda_n x\right)\frac{\sin\lambda_n l - \sinh\lambda_n l}{\cos\lambda_n l - \cosh\lambda_n l} + (\sin\lambda_n x - \sinh\lambda_n x)\right]$
Free–Free	Frequency equation: $\cos\lambda l \cosh\lambda l = 1$ (admitting zero frequencies) Eigenfunction: $\Phi_n(x) = A_1\left[\left(\cosh\lambda_n x + \cos\lambda_n x\right)\frac{\sinh\lambda_n l - \sin\lambda_n l}{\cosh\lambda_n l - \cos\lambda_n l} + (\sinh\lambda_n x + \sin\lambda_n x)\right]$

For nontrivial solutions, it is required to impose the condition of singularity of the coefficient matrix, which leads to:

$$1 + \cos\lambda l \cosh\lambda l = 0 \tag{2.129}$$

The above equation is transcendental in the unknown frequency parameter λ and its zeros correspond to an infinite set of values $\{\lambda_n\}$ for λ. For each λ_n, the natural frequency ω_n is obtained as $\omega_n = \left(\lambda_n^2\right)\sqrt{\frac{EI}{m}}$. From Equation (2.128), the coefficient A_2 can be expressed in terms of A_1 using any one of the two equations. In particular, if the second equation is used, we have

$$A_2 = -\frac{\sinh\lambda l + \sin\lambda l}{\cosh\lambda l + \cos\lambda l} A_1 \tag{2.130}$$

With A_2 substituted in Equation (2.127), the eigenfunction for each mode of vibration is obtainable in the form

$$\Phi_n(x) = A_1\left[\left(\cosh\lambda_n x - \cos\lambda_n x\right) - \frac{\sinh\lambda_n l + \sin\lambda_n l}{\cosh\lambda_n l + \cos\lambda_n l}(\sinh\lambda_n x - \sin\lambda_n x)\right] \tag{2.131}$$

The results for the biharmonic wave equation representing the free vibration solution of a uniform beam are given in Table 2.1.

2.4.7.1 Timoshenko Beam PDE – Free Vibration Solution

Consider the homogeneous part of the PDE (Equation (2.27)):

$$EI\frac{\partial^4 w}{\partial x^4} + m\frac{\partial^2 w}{\partial t^2} - mr^2\frac{\partial^4 w}{\partial x^2 \partial t^2} - \frac{mEI}{k'AG}\frac{\partial^4 w}{\partial x^2 \partial t^2} + \frac{m^2 r^2}{k'AG}\frac{\partial^4 w}{\partial t^4} = 0 \tag{2.132}$$

Seeking a harmonic solution in time, we substitute $W(x)e^{i\omega t}$ for $w(x,t)$ in the above equation and obtain:

$$\boldsymbol{L}W = \mu W \tag{2.133}$$

$\boldsymbol{L} = \frac{d^4W}{dx^4} + b\frac{d^2W}{dx^2} + cW$ is a fourth-order differential operator over [0,l] with $b = \left(\frac{m\omega^2}{k'AG} + \frac{mr^2\omega^2}{EI}\right)$, $c = \left(\frac{m\omega^2}{k'AG}\right)\left(\frac{mr^2\omega^2}{EI}\right)$ and the eigenvalue parameter $\mu = \frac{m\omega^2}{EI}$. Since we have $b^2 - c > 0$, the PDE (Equation (2.132)) is hyperbolic. Let us consider the inner product:

$$\begin{aligned} < u, \boldsymbol{L}v > &= \int_0^l u(x)\left(\frac{d^4v}{dx^4} + b\frac{d^2v}{dx^2} + cv\right)dx \\ &= \left\{(uv''')_0^l - (u'v'')_0^l + (u''v')_0^l - (u'''v)_0^l\right\} + b\left\{(uv^i)_0^l - (u'v)_0^l\right\} + < \boldsymbol{L}u, v > \end{aligned} \tag{2.134}$$

The above equation is the Green's identity:

$$\begin{aligned} &< u\boldsymbol{L}v > - < v\boldsymbol{L}u > = P(u, v)_0^l, \text{ with} \\ &P\,(u, v) = \left\{uv''' - u'v'' + u''v' - u'''v\right\} + b\left\{uv^i - u'v\right\} \end{aligned} \tag{2.135}$$

From the boundary conditions in Equation (2.26), we find that $P(u, v)_0^l = 0$ for the commonly encountered end constraints (e.g. those listed in Table 2.1) and the operator $\boldsymbol{L}$ is self-adjoint. Thus, μ remains positive and real. From Equation (2.133), the frequency equation is given by

$$D^4 + bD^2 + (c - \mu) = 0 \tag{2.136}$$

The roots of the above equation are:

$$D^2 = \frac{1}{2}\left(-b \pm \sqrt{b^2 + 4(\mu - c)}\right) \tag{2.137}$$

If $\mu > c$, the roots are real, one positive and the other negative. This condition is equivalent to $\frac{mr^2\omega^2}{K'AG} < 1$ so that the effects of rotary inertia and shear deformation are small. Then, the possible roots for the fourth-order Equation (2.136) can be taken as $D = \pm\lambda_1$ and $\pm i\lambda_2$ with $= \sqrt{\frac{1}{2}\left|-b \pm \sqrt{b^2 + 4(\mu - c)}\right|}$. This yields the free vibration solution as

$$w(x, t) = \left(A_1 \cosh\lambda_1 x + A_2 \sinh\lambda_1 x + A_3 \cos\lambda_2 x + A_4 \sin\lambda_2 x\right)e^{i\omega t} \tag{2.138}$$

The constants A_i, $i = 1, 2, 3$ and 4 are to be determined from the boundary conditions.

The boundary conditions (Equation (2.26)) for the case of a simply supported beam are:

$$w\,(0, t) = w\,(l, t) = 0 \text{ and } EI\frac{d\psi(0, t)}{dx} = EI\frac{d\psi(l, t)}{dx} = 0 \tag{2.139a,b}$$

Given $\frac{\partial\psi}{\partial x} = \frac{\partial^2 w}{\partial x^2} - \frac{m}{k'AG}\frac{\partial^2 w}{\partial t^2}$ for a uniform beam (Equation (2.25a)), the condition on the moments at the two ends take the form: $EI\frac{d^2w}{dx^2} + \frac{m\omega^2 w}{k'AG} = 0$ at $x = 0$ and l. In view of the displacement constraint in Equation (2.139a) at the two ends, the condition on bending moments reduces to $\frac{d^2w(0,t)}{dx^2} = \frac{d^2w(0,t)}{dx^2} = 0$. The last two boundary conditions give $A_1 = A_3 = 0$. The other two conditions at $x = l$ result in the following two homogeneous simultaneous equations:

$$\begin{bmatrix} \sinh \lambda_1 l & \sin \lambda_2 l \\ \sinh \lambda_1 l & -\sin \lambda_2 l \end{bmatrix} \begin{Bmatrix} A_2 \\ A_4 \end{Bmatrix} = \begin{Bmatrix} 0 \\ 0 \end{Bmatrix} \tag{2.140}$$

For nontrivial A_2, A_4, the required singularity of the coefficient matrix gives

$$\sinh \lambda_1 l \sin \lambda_2 l = 0 \tag{2.141}$$

If $\sinh \lambda_1 l = 0$ and $\sin \lambda_2 l \neq 0$, Equation (2.138) imply $A_4 = 0$ and in turn, a trivial solution for $w(x,t)$. It then follows that $\sin \lambda_2 l = 0$ and $A_2 = 0$. This gives the possible values for $\lambda_2 l$

$$\lambda_2 l = n\pi, \ n = 1, 2, \ldots \tag{2.142}$$

The mode shape for each natural frequency, $\omega_{\mathrm{n}} = (\lambda_n^2)\sqrt{\frac{EI}{m}}$ as:

$$W(x) = A_4 \sin \frac{n\pi}{l} x, \ n = 1, 2 \ldots \tag{2.143}$$

Equation (2.143) shows that the mode shape of a simply supported uniform beam is the same as that obtained even without considering the effects of rotary inertia and shear deformation (Equation (2.125)).

Substitution of $D = \pm i\lambda_2 = \pm i\frac{n\pi}{l}$ in Equation (2.136) gives:

$$\left(\frac{n\pi}{l}\right)^4 - b\left(\frac{n\pi}{l}\right)^2 + (c - \mu) = 0$$

$$\Rightarrow \left(\frac{n\pi}{l}\right)^4 - \left(\frac{m\omega^2}{k'AG} + \frac{mr^2\omega^2}{EI}\right)\left(\frac{n\pi}{l}\right)^2 + \left(\frac{m\omega^2}{k'AG} + \frac{mr^2\omega^2}{EI}\right) - \frac{m\omega^2}{EI} = 0$$

$$\Rightarrow \lambda^8\left(\frac{r^2 EI}{k'AG}\right) - \lambda^4\left(\frac{n\pi}{l}\right)^2\left(\frac{EI}{k'AG} + r^2\right) - \left\{\lambda^4 - \left(\frac{n\pi}{l}\right)^4\right\} = 0, \text{ with } \lambda^4 = \frac{m\omega^2}{EI} \tag{2.144}$$

Without accounting for the rotary inertia and shear deformation effects, the above equation (the last two terms within the braces) gives the natural frequencies corresponding to the Euler–Bernoulli beam as before. Ignoring the first term containing λ^8 in Equation (2.144), we obtain an expression for λ^4 as

$$\lambda_n^4 = \frac{\left(\frac{n\pi}{l}\right)^4}{\left\{1 + \left(\frac{n\pi}{l}\right)^2\left(\frac{EI}{k'AG} + r^2\right)\right\}} \tag{2.145}$$

Recognising $\lambda_n^4 = \left(\frac{n\pi}{l}\right)^4$ as the result from the Euler–Bernoulli beam theory for the simply supported case, Equation (2.145) provides a measure of the effect of rotary inertia and

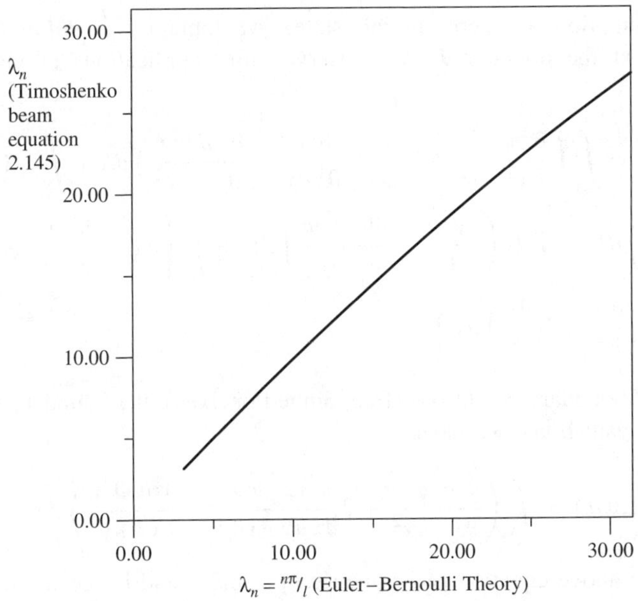

Figure 2.12 Simply supported beam: comparison of natural frequencies form Euler–Bernoulli and Timoshenko beam theories $l = 1$, $E/k'G = 3$, $r^2 = I/A = 0.00021$

shear deformation in reducing the natural frequencies of the beam. That the reduction is more predominant for higher frequencies is illustrated in Figure 2.12.

2.4.8 Thin Rectangular Plates – Free Vibration Solution

Consider a homogeneous isotropic rectangular thin plate (Equation 2.18) of thickness, h The homogeneous part of the PDE is rewritten as:

$$D\left(\frac{\partial^4 w}{\partial x^4} + 2\frac{\partial^2 w}{\partial x^2 \partial y^2} + \frac{\partial^4 w}{\partial y^4}\right) + m\frac{\partial^2 w}{\partial t^2} = 0 \tag{2.146}$$

In an operator form the above PDE is written as

$$\boldsymbol{L}w = \boldsymbol{M}w \tag{2.147}$$

Here, $\boldsymbol{L} = \frac{\partial^4}{\partial x^4} + 2\frac{\partial^2}{\partial x^2 \partial y^2} + \frac{\partial^4}{\partial y^4}$ and $\boldsymbol{M} = -\frac{m}{D}\frac{\partial^2}{\partial t^2}$. Both the operators, $\boldsymbol{L}$ and $\boldsymbol{M}$ are self-adjoint. This is obvious from the following statements:

$$\int_0^T v\mathbf{M}w\,dt = -\frac{m}{D}\int_0^T v\frac{\partial^2 w}{\partial t^2}\,dt = -\frac{m}{D}\left(v\frac{\partial w}{\partial t}\bigg|_0^T - \frac{\partial v}{\partial t}w\bigg|_0^T + \int_0^T w\frac{\partial^2 v}{\partial t^2}\,dt\right) \tag{2.148}$$

With the assumption of zero initial state, we obtain $\int_0^T vMwdt = \int_0^T wMvdt$. Considering next the operator $\boldsymbol{L}$, we observe, after applications of Green's theorem twice that

$$\int_A v\boldsymbol{L}wdA = \int_A \left(\frac{\partial^2 w}{\partial x^2}\frac{\partial^2 v}{\partial x^2} + 2\frac{\partial^2 w}{\partial x\partial y}\frac{\partial^2 v}{\partial x\partial y} + \frac{\partial^2 w}{\partial y^2}\frac{\partial^2 v}{\partial y^2}\right) dA + \int_\Gamma l_1 \left(v\frac{\partial^2 w}{\partial x^2} - \frac{\partial v}{\partial x}\frac{\partial^2 w}{\partial x^2}\right) d\Gamma + \int_\Gamma l_2 \left(v\frac{\partial^2 w}{\partial y^2} - \frac{\partial v}{\partial y}\frac{\partial^2 w}{\partial y^2}\right) d\Gamma + \int_\Gamma \left\{v\left(l_1\frac{\partial^2 w}{\partial x^2\partial y} + l_2\frac{\partial^2 w}{\partial y^2\partial x}\right) - \frac{\partial^2 w}{\partial x\partial y}\left(l_1\frac{\partial w}{\partial x} + l_2\frac{\partial v}{\partial y}\right) d\Gamma\right\} \tag{2.149}$$

For any type of boundary conditions (free, hinged or fixed), the boundary integrals in the above equation vanish and we have

$$\int_A v\boldsymbol{L}wdA = \int_A \left(\frac{\partial^2 w}{\partial x^2}\frac{\partial^2 v}{\partial x^2} + 2\frac{\partial^2 w}{\partial x\partial y}\frac{\partial^2 v}{\partial x\partial y} + \frac{\partial^2 w}{\partial y^2}\frac{\partial^2 v}{\partial y^2}\right) dA \tag{2.150}$$

The RHS of the above equation is symmetric in u and v and hence we also have,

$$\int_A^w w\boldsymbol{L}vdA = \int_A \left(\frac{\partial^2 v}{\partial x^2}\frac{\partial^2 w}{\partial x^2} + 2\frac{\partial^2 v}{\partial x\partial y}\frac{\partial^2 w}{\partial x\partial y} + \frac{\partial^2 v}{\partial y^2}\frac{\partial^2 w}{\partial y^2}\right) dA \tag{2.151}$$

Thus, $\int_A v\boldsymbol{L}wdA = \int_A w\boldsymbol{L}vdA$ and the self-adjoint property ensures that the eigenvalues of the plate are real and the corresponding eigenfunctions are orthogonal. To obtain the eigensolution, assume a harmonic solution of the form:

$$w(x, y, t) = W(x, y)\, e^{i\lambda^4 t} \tag{2.152}$$

Substitution of the above in Equation (2.146) gives

$$\left(\frac{d^4 W}{dx^4} + 2\frac{d^2 W}{dx^2 dy^2} + \frac{dW}{dy^4}\right) - \lambda^4 W = 0 \tag{2.153}$$

Here, $\lambda^4 = \frac{m\omega^2}{D}$. With $\Delta = \frac{d^2 W}{dx^2} + \frac{d^2 W}{dy^2}$, the above equation is recast in the form:

$$\left(\Delta + \lambda^2\right)\left(\Delta - \lambda^2\right) W = 0 \tag{2.154}$$

$$\left(\Delta + \lambda^2\right) W = 0 \text{ and } \left(\Delta - \lambda^2\right) W = 0 \tag{2.155a,b}$$

Each of the above two equations is known as the two-dimensional Helmholtz equation. If we consider $\left(\Delta + \lambda^2\right) W = 0$, and assuming $W(x, y) = W_1(x)W_2(y)$ in a separable form, we obtain

$$\frac{d^2 W_1}{dx^2} W_2 + \frac{d^2 W_2}{dy^2} W_1 = -\lambda^2 W_1 W_2$$

$$\Rightarrow \frac{1}{W_1}\frac{d^2 W_1}{dx^2} = \frac{1}{W_2}\frac{d^2 W_2}{dy^2} + \lambda^2 \tag{2.156}$$

Equating both sides to a constant that must be, say, α^2 results in the two ODEs:

$$\frac{d^2W_1}{dx^2}-\alpha^2 W_1 = 0 \text{ and } \frac{d^2W_2}{dy^2}+\left(\lambda^2-\alpha^2\right)W_2 = 0 \tag{2.157a,b}$$

Handling the Equation (2.155b) also in a similar manner, we finally obtain the linear ODEs:

$$\frac{d^2W_1}{dx^2}\pm\alpha^2 W_1 = 0 \text{ and } \frac{d^2W_2}{dy^2}\pm\left(\lambda^2-\alpha^2\right)W_2 = 0 \tag{2.158a,b}$$

With the possible combinations of the characteristic roots of the above equations, the solution to Equation (2.153) is given by

$$\begin{aligned} W(x,y) = {} & A_1 \sin\alpha x \sin\beta y + A_2 \sin\alpha x \cos\beta y + A_3 \cos\alpha x \sin\beta y \\ & + A_4 \cos\alpha x \cos\beta y + A_5 \sinh\alpha x \sinh\beta y \\ & + A_6 \sinh\alpha x \cosh\beta y + A_7 \cosh\alpha x \sinh\beta y \\ & + A_8 \cosh\alpha x \cosh\beta y \end{aligned} \tag{2.159}$$

Here, $\beta = \left(\lambda^2-\alpha^2\right)^{\frac{1}{2}}$. The constants $A_i, i = 1, 2, .., 8$ are to be computed from the prescribed boundary conditions. In particular, if the boundary conditions pertain to a plate simply supported on all sides, the applicable boundary conditions are: $w = 0$ and $\frac{\partial^2 w}{\partial x^2} = 0$ at $x = 0, a$ and $\frac{\partial^2 w}{\partial y^2} = 0$ at $y = 0, b$. These in turn result in

$$\begin{aligned} & W(x,y) = 0 \text{ for } x = 0, a \text{ and } y = 0, b \\ & \frac{\partial^2 W}{\partial x^2} = 0 \text{ for } x = 0, a \text{ and } \frac{\partial^2 W}{\partial y^2} = 0 \text{ for } y = 0, b \end{aligned} \tag{2.160a,b}$$

With the above boundary conditions, the general solution in Equation (2.159) reduces to

$$W(x,y) = A_1 \sin\alpha x \sin\beta y \tag{2.161}$$

subjected to the following requirement:

$$\sin\alpha a = 0 \text{ and } \sin\beta b = 0 \tag{2.162}$$

The above equation gives

$$\alpha = \frac{n_1\pi}{a} \text{ and } \beta = \frac{n_2\pi}{b}, n_1, n_2 = 1, 2, \ldots, \infty \tag{2.163}$$

The above possible values of α and β (wave numbers in the X and Y directions, respectively) yield the eigenvalues of the simply supported plate in the form:

$$\lambda^4 = \left(\alpha^2+\beta^2\right)^2 \Rightarrow \omega_{n_1 n_2} = \pi^2\sqrt{\frac{D}{m}}\left\{\left(\frac{n_1}{a}\right)^2+\left(\frac{n_2}{b}\right)^2\right\} \tag{2.164}$$

The corresponding orthogonalized eigenfunctions (mode shapes) are given by

$$W(x,y) = A_1\frac{2}{\sqrt{ab}}\sin\frac{n_1\pi x}{a}\sin\frac{n_2\pi y}{b}, n_1, n_2 = 1, 2, \ldots, \infty \tag{2.165}$$

2.5 Orthonormal Basis and Eigenfunction Expansion

Using the orthogonality property (Equation (2.99)), the eigenfunctions, $\Phi_n(x)$ of the Strum–Liouville boundary-value problems can be made orthonormal, that is to satisfy

$$\int_a^b u_i u_j r(x)dx = \delta_{ij} \tag{2.166}$$

δ_{ij} is the Kronecker delta given by

$$\begin{aligned}\delta_{ij} &= 1, \quad i = j \\ &= 0, \quad i \neq j\end{aligned} \tag{2.167}$$

The orthonormality is thus realised by the operation: $\Phi_n(x) = \frac{\Phi_n(x)}{\|\Phi_n\|}$. For example, considering the eigenfunctions $\Phi_n(x)$ of a taut string or of a simply supported Euler–Bernoulli beam, we have for $i = j$

$$< \Phi_i, \Phi_j > = \int_0^l \sin^2\frac{i\pi x}{l}dx = \frac{1}{2}\int_0^l \left(1 - \cos\frac{2i\pi x}{l}\right)dx = \frac{l}{2} \tag{2.168}$$

The orthonormal functions for these examples are given by $\sqrt{\frac{2}{l}}\sin\frac{n\pi}{l}x$, $n = 1,2\ldots$ Similarly, for wave equation corresponding to the axial vibration of a rod, the orthonormal eigenfunctions can be obtained from Equation (2.116) as $\sqrt{\frac{2}{l}}\sin\frac{(2n-1)\pi}{2l}x$, $n = 1,2\ldots$ In general, these eigenfunctions form an orthonormal basis whose linear span is a subspace that in turn is dense (Appendix B) in $\mathcal{H}$. For a finite-dimensional space, we speak of the basis as a subset of the linearly independent (Appendix E) functions that span (Appendix E) the space. Spanning a space implies that every member of the space can be represented by a linear combination of the elements of the basis set (this is even true for infinite-dimensional spaces that are separable; see Appendix B). This is in fact equivalent to saying that a finite-dimensional inner product space is complete (see Appendix B for definition of completeness). Extension of this idea to an infinite functional space $\mathcal{H}$ is, however, not straightforward. More specifically, we can think of a possible functional expansion in terms of the element functions of the basis set for an infinite-dimensional function space that is separable. Nevertheless, complexities arise whilst implementing this idea numerically, as the basis set is infinite (even though, countable) and a finite subset of functions in the basis may not span $\mathcal{H}$. Given that only finitely many terms can be included in the functional approximation, a notion of convergence of such approximations must be made use of. Thus, it may be possible that any member of $\mathcal{H}$ is approximated (within arbitrarily small ε-closeness) by a finite linear combination of the basis functions. Thus, if $\boldsymbol{\Phi} = \{\Phi_n, n \in \mathbb{N}\}$ is the orthonormal set of eigenfunctions, then $\boldsymbol{\Phi}$ is said to be a maximal orthonormal set in $\mathcal{H}$ (Daya Reddy, 1998). By definition, $\boldsymbol{\Phi}$ is maximal, if $< \varphi, \Phi_i > = 0, \forall i$, has $\varphi = 0$ as the only solution in $\mathcal{H}$, implying that no other new member function is orthogonal to the available basis set. Any square integrable function (see Appendix B), $f(x) \in \mathcal{H}$ can be approximated by the sequence of these orthonormal functions, that is if $\mathbf{B} = \{\Phi_I(x), I \in \mathbb{N}\}$ is an orthonormal basis of $\mathcal{H}$, then

$$f = \sum_{\Phi \in B} < f, \Phi > \Phi = \sum_{k=1}^{\infty} c_k \Phi_k \tag{2.169}$$

with $c_k =< f, \Phi >$. The sum in Equation (2.169) is called the generalized Fourier expansion of $f(x)$ in terms of the orthonormal basis functions and c_k, the Fourier coefficients. This is elaborated below.

2.5.1 Best Approximation to f(x)

Suppose we define the partial (finite) sums $S_n = \sum_{k=1}^{n} c_k \Phi_k$ with c_k being the so-called Fourier coefficients and $T_n = \sum_{k=1}^{n} d_k \Phi_k$ with d_k, a set of arbitrary coefficients. Note that span $\{\Phi_k | k = 1, \ldots, n\}$ defines a subspace, which is, by definition, closed and convex (Appendix B).

$$\|f - T_n\|^2 = \int_\Omega (f - T_n)^2 \, dx =< f, f > - < f, T_n > - < T_n, f > + < T_n, T_n > \tag{2.170}$$

Here,

$$< f, T_n > =< T_n, f > =< f, \sum_{k=1}^{n} d_k \Phi_k > = \sum_{k=1}^{n} d_k < f, \Phi_k > = \sum_{k=1}^{n} d_k c_k \tag{2.171}$$

$$\begin{aligned} < T_n, T_n > &=< \sum_{k=1}^{n} d_k \Phi_k, \sum_{j=1}^{n} d_j \Phi_j > \sum_{j=1}^{n} \sum_{k=1}^{n} < d_k \Phi_k, d_j \Phi_j > \\ &= \sum_{j=1}^{n} \sum_{k=1}^{n} d_k d_j < \Phi_k, \Phi_j > = \sum_{k=1}^{n} |d_k|^2 \end{aligned} \tag{2.172}$$

In view of Equations (2.171) and (2.172), Equation (2.170) becomes

$$\begin{aligned} \|f - T_n\|^2 &= \|f\|^2 - \sum_{k=1}^{n} d_k c_k - \sum_{k=1}^{n} d_k c_k + \sum_{k=1}^{n} |d_k|^2 \\ &= \|f\|^2 + \sum_{k=1}^{n} (c_k - d_k)^2 - \sum_{k=1}^{n} |c_k|^2 \end{aligned} \tag{2.173}$$

Two observations are significant from Equation (2.173). One is that with d_k substituted by c_k, we get $\|f - S_n\|^2 = \|f\|^2 - \sum_{k=1}^{n} |c_k|^2 < \|f - T_n\|^2$, which indicates that $\|f - S_n\|^2$ is minimum and thus $f(x)$ is 'best approximated' by S_n using the generalized Fourier coefficients. Moreover, one expects that, as $n \to \infty$, $\|f - S_n\|^2$ can be made small and S_n remains bounded in the associated L^2 norm. This leads to Bessel's inequality for infinite-dimensional $\mathcal{H}$

$$\sum_{k=1}^{\infty} |c_k|^2 \le \|f\|^2 \tag{2.174}$$

By Bessel's inequality, S_n is convergent and thus for every $m, n \in \mathbb{N}$ and any $\varepsilon > 0$, we have

$$\|S_n - S_m\|^2 = \sum_{k=1}^{n} |c_k|^2 - \sum_{k=1}^{m} |c_k|^2 = \sum_{k=m+1}^{n} |c_k|^2 < \varepsilon \tag{2.175}$$

Thus, the sequence, S_n is a Cauchy sequence (Appendix B) in $\mathcal{H}$. Suppose that $\lim_{n\to\infty} S_n = f^*$, then

$$\begin{aligned} < f - f^*, \Phi_m > &= < f - \lim_{n\to\infty} \sum_{k=1}^{n} c_k \Phi_k, \Phi_m > \\ &= \lim_{n\to\infty} \left(< f, \Phi_m > - \sum_{k=1}^{n} < f, \Phi_k > < \Phi_k, \Phi_m > \right) \\ &= \lim_{n\to\infty} \left(< f, \Phi_m > - < f, \Phi_m > \right) \\ &= 0 \end{aligned} \tag{2.176}$$

Equation (2.176) is true for $\forall m$ with Φ_m as the basis, and hence it follows that $f = f^*$ and

$$\lim_{n\to\infty} S_n = f \tag{2.177}$$

Moreover, from Bessel's inequality, we also have Parseval's formula:

$$\sum_{k=1}^{\infty} |c_k|^2 = \|f\|^2, \text{ if only if } \|f - S_n\| \to 0 \text{ as } n \to \infty \tag{2.178}$$

Equation (2.178) is an infinite-dimensional analogue of the Pythagorean theorem. S_n is called the projection of f onto the orthonormal basis B (Figure 2.13). Equation (2.169) thus formalises the elementary notion of a Fourier series expansion of any function $\in \mathcal{H}$ in terms of an orthonormal basis. Since the eigenfunctions of a Strum–Liouville problem form an orthonormal basis, the series expansion (eigenfunction expansion) in Equation (2.169) gains relevance in solving PDEs (particularly if they are inhomogeneous) with boundary conditions that are homogeneous or inhomogeneous.

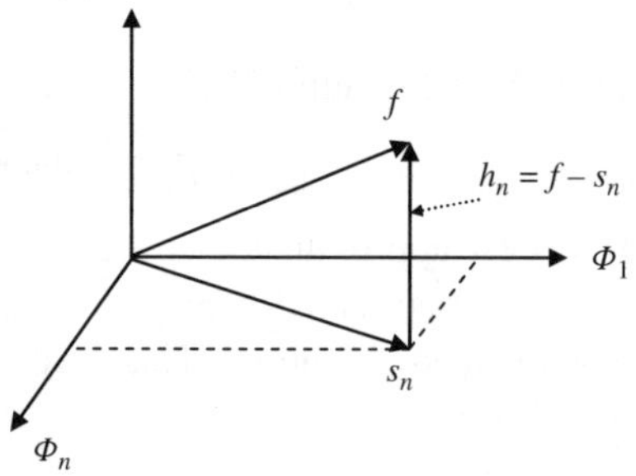

Figure 2.13 Geometric interpretation of S_n as a projection of f onto the subspace spanned by the orthonormal basis set, B

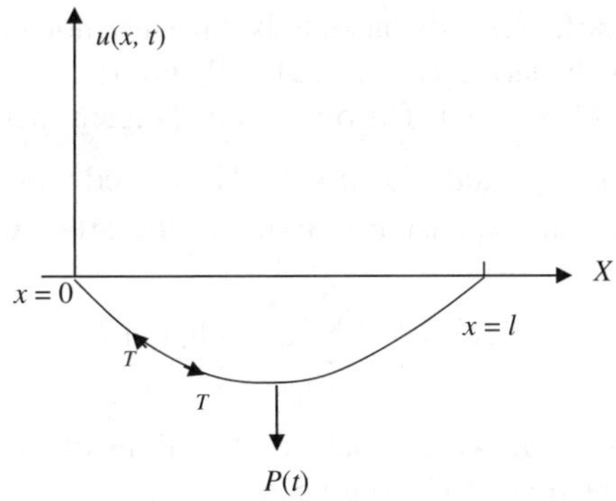

Figure 2.14 Taut string under forced vibration. $F(x,t) = P(t)\delta(x - l/2)$

2.6 Solutions of Inhomogeneous PDEs by Eigenfunction-Expansion Method

Forced structural systems are often governed by nonhomogeneous PDEs. For example, in the presence of external loading, the wave Equation (2.2) (taut string) or the biharmonic Equation (2.16) (Euler–Bernoulli beam) takes the form:

$$m\frac{\partial^2 u}{\partial t^2} = T\frac{\partial^2 u}{\partial x^2} + F(x,t)$$

$$m\frac{\partial^2 w}{\partial t^2} + \frac{\partial^2}{\partial x^2}\left(EI\frac{\partial w^2}{\partial x^2}\right) = F(x,t) \tag{2.179a,b}$$

In the case of a taut string represented by PDE (Equation (2.179a)), the external load function $F(x,t)$, for any given x, may be a harmonically (in time) varying point (concentrated) load acting at $x = l/2$ as shown in Figure 2.14.

For a structural system modelled the 1D beam Equation (2.179b), $F(x,t)$ could be a moving load that traverses over the beam, say, with a uniform velocity, V_s (Figure 2.15).

Example 2.2: Forced vibration solution of the taut string subjected to a point load
Consider the taut string in Figure 2.14 with an external point load, $F(x,t) = P(t)\delta(x - \frac{l}{2})$ where $\delta\,(.)$ is the Dirac delta function:

$$\delta\left(x - x'\right) = 1, x = x' = 0, \text{ otherwise} \tag{2.180}$$

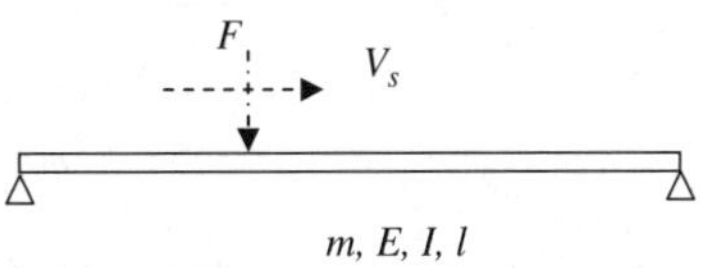

Figure 2.15 A simply supported beam under a moving load

$P(t)$ is the time-varying part, presently taken to be time-harmonic, $\bar{P}\sin\lambda t$. The boundary conditions considered for the string are: $u(0,t)=0$ and $u(l,t)=0$. Let the initial conditions be $u(x,0)=0$ and $\frac{\partial u(x,0)}{\partial t}=0$. The orthonormal eigenfunctions for the taut string as obtained earlier are $\Phi_n(x)=\sqrt{\frac{2}{l}}\sin\frac{n\pi}{l}x,\ n\in\mathbb{N}$. The forced vibration solution is now assumed to be expressible via an expansion in terms of the orthonormal basis functions as

$$u(x,t)=\sum_{n=1}^{\infty}q_n(t)\Phi_n(x) \tag{2.181}$$

$q_n(t),\ n=1,2..\infty$ are the unknown coefficients (functions of time only). Substituting the assumed expansion in Equation (2.179a) gives:

$$\sum_{n=1}^{\infty}\left[\ddot{q}_n(t)+\omega_n^2 q_n(t)\right]\Phi_n(x)=\frac{P\sin\lambda t}{m}\delta(x-\frac{l}{2}) \tag{2.182}$$

where $\omega_n^2=\frac{T}{m}\frac{n^2\pi^2}{l^2}=c^2\frac{n^2\pi^2}{l^2}$. Taking inner products on both sides with respect to any element $\Phi_j(x)$ in the orthonormal basis set, we obtain

$$\sum_{n=1}^{\infty}\left[\ddot{q}_n(t)+\omega_n^2 q_n(t)\right]<\Phi_n(x),\Phi_j(x)>=\frac{P\sin\lambda t}{m}<\delta\left(x-\frac{l}{2}\right),\Phi_j(x)>$$

$$\Rightarrow \ddot{q}_j(t)+\omega_j^2 q_j(t)=\Phi_j(l/2)\frac{P\sin\lambda t}{m},\ j\in\mathbb{N} \tag{2.183}$$

where we have made use of the property $\int_\Omega \delta(x-x^*)\,g(x)dx=g(x^*)$ with $g(,)$ assumed to be bounded continuous. The equation represents an infinite set of uncoupled second-order ODEs in $q_j(t)$, the modal response components of which are also referred to as generalized coordinates of the structural system. Each of these second-order ODEs corresponds to the vibratory motion of an oscillator, often referred to as a single degree of freedom (SDOF) system (Figure 2.16) in the structural engineering parlance. Here, the jth degree of freedom refers to the coordinate $q_j(t)$ that describes the motion of the fictitious uncoupled oscillator for all time $t\in(0,\infty)$. In general, the equation of motion of an SDOF oscillator is given by:

$$M\ddot{q}(t)+Kq(t)=g(t)\Rightarrow \ddot{q}(t)+\omega^2 q(t)=g(t)/M \tag{2.184}$$

M and K are the mass and stiffness parameters of the SDOF system. $\omega\left(=\sqrt{\frac{K}{M}}\right)$ is the natural frequency of the oscillator. Procedure to obtain a solution to a linear time-invariant

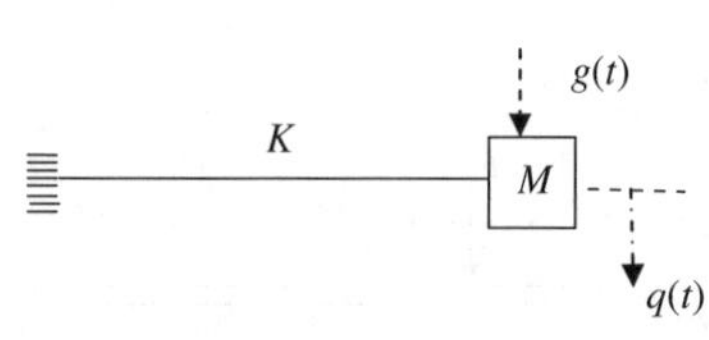

Figure 2.16 Single degree of freedom (SDOF) oscillator

second-order ODE (Equation (2.183) or (2.184)) is straightforward and for most cases of external excitations, solutions can be obtained either by the operator method or the Laplace transform method. For example, adopting the operator method, we may obtain the solution in terms of the complementary function and the particular integral. The former is the solution to the homogeneous part of the ODE and is given by $q(t) = A \sin\omega t + B \cos\omega t$ (A, B being the arbitrary constants of integration). The particular integral is the part of the solution due to the forcing function $g(t)$. It is given by $\boldsymbol{L}^{-1}(g(t))$, where $\boldsymbol{L}(=\boldsymbol{D}^2+\omega^2\boldsymbol{I})$ stands for the differential operator corresponding to the ODE with $\equiv \frac{d}{dt}$. Thus, the total solution to the SDOF system is given by

$$q_j(t) = A \sin\omega t + B \cos\omega t + \boldsymbol{L}^{-1}\left(\frac{g(t)}{m}\right) \tag{2.185}$$

The arbitrary constants A and B are determined from the prescribed initial conditions. Now referring to the modal Equation (2.183) with $g(t) = \frac{\Phi_j(l/2)}{m} P \sin\lambda t$, the solution for the j-th coordinate is given by

$$q_j\,(t) = A_j \sin\omega_j t + B_j \cos\omega_j t + \Phi_j\,(l/2)\,\frac{1}{m}\frac{P \sin\lambda t}{\omega_j^2 - \lambda^2} \tag{2.186}$$

The constants A_j and B_j are to be determined from the initial conditions on each modal coordinate $q_j(t)$. By using the orthogonal property of the eigenfunctions, we have $q_n(0) = u\,(x,0)\int_0^1 \Phi_n(x)dx = 0$ and $\dot{q}_n(0) = \dot{u}\,(x,0)\int_0^1 \Phi_n(x)dx = 0$. Substituting these initial conditions in Equation (2.186) gives $A_j = -\frac{1}{\omega_j}\frac{\Phi_j(l/2)}{\left(\omega_j^2-\lambda^2\right)}\frac{P}{m}$ and $B_j = 0$. Having thus derived the set of solutions (Equation (2.186)) $\{q_j(t) | j = 1, 2 \ldots\}$, the solution to the inhomogeneous PDE of the taut string is obtained from the eigenexpansion (Equation (2.181)). Figure 2.17 shows the string response (i.e. the amplitude spectrum obtained by varying the excitation frequency over the range 0–100 rad/s) at $x = l/2$ under the harmonic excitation. The number of modes considered in the eigenexpansion is 25 and zero initial conditions are assumed in obtaining the solution.

Peaks in the above graph (not shown beyond an amplitude level of 30 units) signify the unbounded response at excitation frequency coinciding with each of the string natural frequencies, ω_n. In physical systems, damping is invariably present and helps in reducing the excessive vibration levels under the so-called resonance conditions $\left(\frac{\lambda}{\omega_n} \to 1.0, n \in \mathbb{N}\right)$.

Example 2.3: Forced vibration solution an Euler–Bernoulli beam subjected to a moving load

The simplest model (Fryba, 1972) for a moving load is one of constant magnitude F, moving with a uniform speed V_s. With such a moving force model, the inhomogeneous PDE (Equation (2.179b)) governing the dynamics of a simply supported beam (Figure 2.15) is given by

$$m\frac{\partial^2 w}{\partial t^2} + \frac{\partial^2}{\partial x^2}\left(EI\frac{\partial w^2}{\partial x^2}\right) = F\delta(x - V_s t) \tag{2.187}$$

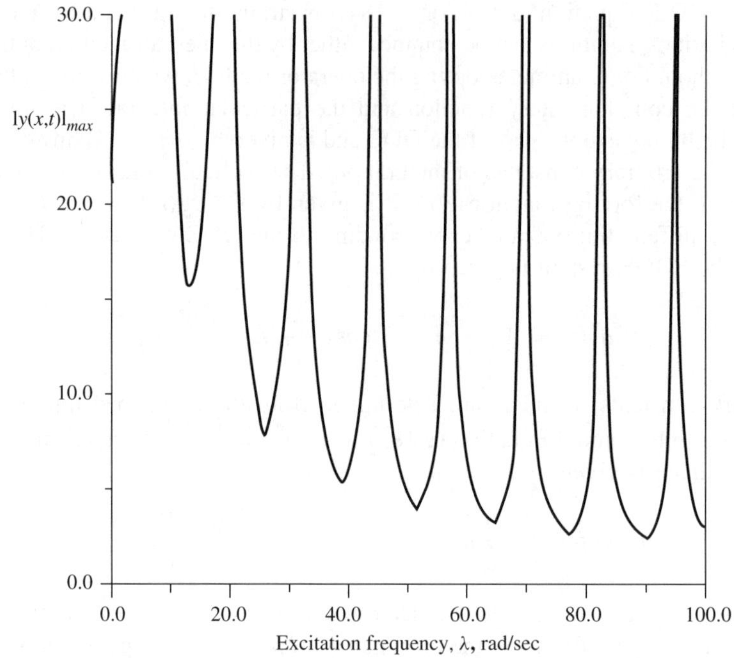

Figure 2.17 Forced vibration response of a taut string under harmonically varying point load, $l = 1000$ mm, radius = 1 mm, mass density = 0.785 E−15 Kg/mm^3, tension = 1 N.

where $F = Mg$ (M is the mass of the moving load and 'g', the acceleration due to gravity). δ (.) is the Dirac delta function (Equation (2.180)). Assuming the beam to be uniform and introducing the nondimensionalised quantities $X = \frac{x}{l}$, $\tau = \frac{V_s t}{l}$, and $W = w/l$, Equation (2.16) takes the form:

$$\frac{\partial^2 W}{\partial \tau^2} + \mu^4 \frac{\partial W^4}{\partial X^4} = \varepsilon P \delta(X - \tau) \tag{2.188}$$

Here, $\mu^4 = \frac{EI}{ml^2 V_S^2}$, $\varepsilon = \frac{M}{ml}$ and $P = \frac{gl}{V_S^2}$. In terms of the nondimensionalised quantities as above, the orthonormalised eigenfunctions for the simply supported beam are given by $\Phi_n(x) = \sqrt{2} \sin n\pi X$ (Section 2.5) and the natural frequencies are given by $\bar{\omega}_n^2 = \frac{EI}{m} \frac{n^4 \pi^4}{l^4}$. Using these orthonormal basis functions in Equation (2.188), the forced vibration solution is again assumed to be an expansion in terms of the orthonormal basis set as

$$u(x, t) = \sum_{n=1}^{\infty} q_n(t) \Phi_n(x) \tag{2.189}$$

$q_n(t), n \in \mathbb{N}$ are the modal responses to be determined. If the initial conditions are specified as $u(x, 0) = 0$ and $\dot{u}(x, 0) = 0$, the corresponding initial conditions for $q_n(t)$ are derivable from the orthogonal property of the eigenfunctions as

$q_n(0) = u(x, 0) \int_0^1 \Phi_n(X)dx = 0$ and $\dot{q}_n(0) = \dot{u}(x, 0) \int_0^1 \Phi_n(X)dX = 0$. Substituting Equation (2.189) in Equation (2.188) gives:

$$\sum_{n=1}^{\infty} \left[\ddot{q}_n(\tau) + \omega_n^2 q_n(\tau)\right] \Phi_n(X) = \varepsilon P \delta(X - \tau) \tag{2.190}$$

In the above equation, $\omega_n \left(= \frac{\bar{\omega}_n}{\frac{V_s}{l}} = \sqrt{\frac{EI}{m}} \frac{n^2\pi^2}{l V_s} \right)$ is the nondimensional natural frequency. Taking inner products on both sides with respect to any $\Phi_j(X)$ in the orthonormal basis, we obtain the uncoupled (modal) equations as:

$$\ddot{q}_j(\tau) + \omega_j^2 q_j(\tau) = \sqrt{2}\, \varepsilon P \sin n\pi\tau,\ j \in \mathbb{N} \tag{2.191}$$

We now obtain the solution to $q_j(t)$ by the Laplace transform method. Using $\mathcal{L} \sin n\pi\tau = n\pi / \left(s^2 + n^2\pi^2\right)$, the solution in the transform domain is given by:

$$Q_j(s) = \mathcal{L} q_j(t) = \sqrt{2}\, \varepsilon P \frac{1}{\left(s^2 + \omega_j^2\right)\left(s^2 + n^2\pi^2\right)} + \frac{s q_j(0) + \dot{q}_j(0)}{s^2 + \omega_j^2} \tag{2.192}$$

Since the initial conditions are zero and $\frac{1}{\left(s^2+\omega_j^2\right)\left(s^2+n^2\pi^2\right)} = \frac{1}{\left(\omega_j^2 - n^2\pi^2\right)} \left(\frac{1}{s^2+n^2\pi^2} - \frac{s}{s^2+\omega_j^2} \right)$, the Laplace inverse transform of $Q_j(s)$ is given by

$$\begin{aligned} q_j(\tau) = \mathcal{L}^{-1} Q(s) &= \frac{\sqrt{2}\, \varepsilon P}{\omega_j^2 - n^2\pi^2} \mathcal{L}^{-1} \left(\frac{1}{s^2 + n^2\pi^2} - \frac{1}{s^2 + \omega_j^2} \right) \\ &= \frac{\sqrt{2}\, \varepsilon P}{\omega_j^2 - n^2\pi^2} \left(\frac{\sin n\pi\tau}{n\pi} - \frac{\sin \omega_j \tau}{\omega_j} \right) \end{aligned} \tag{2.193}$$

The response to a Euler–Bernoulli beam under the moving load is obtained by substituting the above solution $q_j(t)$ in the eigenexpansion in Equation (2.189). Figure 2.18 shows the variation of beam displacement at $X = 0.5$ with respect to the speed, V_s of the moving load and mass ratio, ε. The beam parameters assumed in obtaining the response are $l = 10\,m$, $EI = 1.67 \times 10^6\,\text{N m}^2$ and $m = 78.5\,\text{kg/m}$. The moving load speed is varied in the range 1–100 m/s and ε over the range 0.1–0.5. The peaks in the individual graphs in the figure relate the response to the critical speed of the moving load. The critical speed remains nearly the same for different mass ratios ε.

Figure 2.19 shows the displacement over the length of the beam for two speeds of $V_s = 10$ m/s and 100 m/s. In this figure, the response variation with respect to ε is also shown. For higher speeds, the response peaks at a location away from the centre of the beam. In the above computations, 10 modes are retained in the eigenexpansion (Equation (2.189)), as it is found sufficient for attaining converged response.

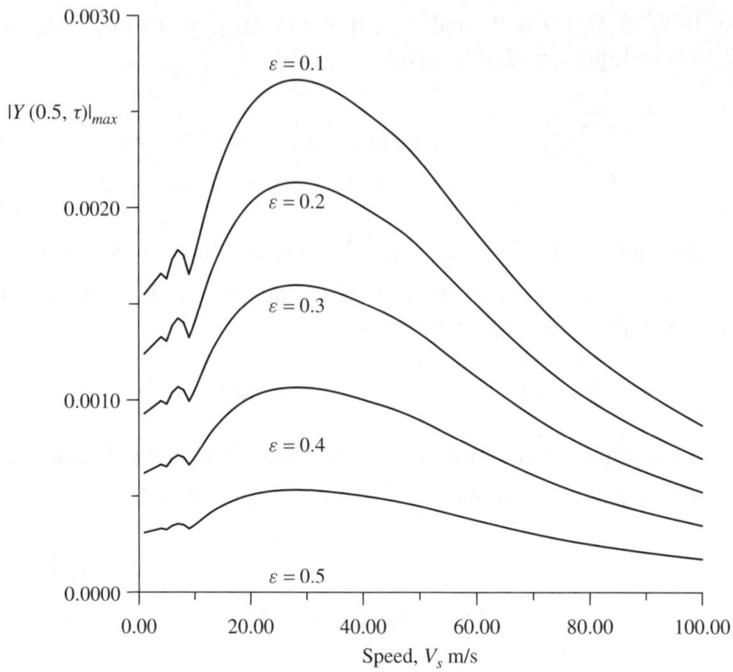

Figure 2.18 Beam response (at $X = 0.5$) vs speed for different mass ratios, ε

2.7 Solutions of Inhomogeneous PDEs by Green's Function Method

Green's function method is useful in solving linear inhomogeneous differential equations including PDEs. Consider the PDE:

$$\boldsymbol{L}u(X) = f(X) \tag{2.194}$$

Here, $\boldsymbol{X}$ is the vector of independent variable. The Green's function method involves finding the inverse of the operator $\boldsymbol{L}$ in the form of an integral operator with a Kernel $G(X;\xi)$ such that

$$u(X) = \boldsymbol{L}^{-1} f(X) = \int G(\boldsymbol{X};\xi)\boldsymbol{f}(\boldsymbol{\xi})d\boldsymbol{\xi} \tag{2.195}$$

In the above equation $\boldsymbol{\xi}$ is a parameter and the Kernel $G\,(\boldsymbol{X};\xi)$ is known as the Green's function for the operator $\boldsymbol{L}$. From this equation, we obtain

$$\boldsymbol{L}\left\{\boldsymbol{L}^{-1} f(\boldsymbol{X})\right\} = f(\boldsymbol{X}) = \int \boldsymbol{L}\left\{G(\boldsymbol{X};\xi)\right\} \boldsymbol{f}(\boldsymbol{\xi})d\boldsymbol{\xi} \tag{2.196}$$

From the property: $\int \delta(\boldsymbol{X}-\boldsymbol{a})f(\boldsymbol{X})d\boldsymbol{X} = f(\boldsymbol{a})$ of the Dirac delta function, $\delta(\boldsymbol{X})$, the above equation implies:

$$\boldsymbol{L}\left\{G\,(\boldsymbol{X};\xi)\right\} = \delta(\boldsymbol{X}-\boldsymbol{\xi}) \tag{2.197}$$

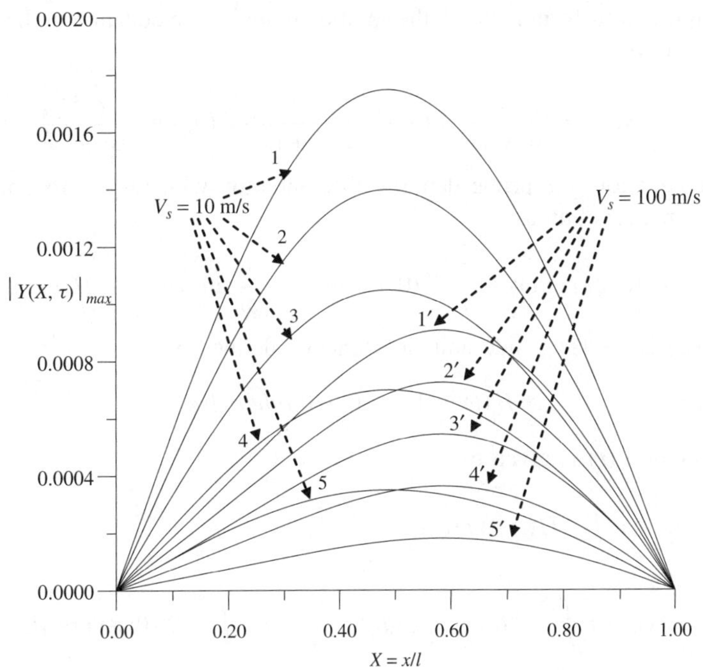

Figure 2.19 Beam response for $V = 10\,m/sec.$ and $100\,m/s$ with $\varepsilon \in [0.1, 0.5]$. Graphs $1, 1' - \varepsilon = 0.1$, $2, 2' - \varepsilon = 0.2$, $3, 3' - \varepsilon = 0.3$, $4, 4' - \varepsilon = 0.4$ and $5, 5' - \varepsilon = 0.5$

As a solution to the above equation, the Green's function $G\,(X;\xi)$ is a singularity solution. Once the function is available, the Equation (2.195) yields the required solution to the original differential Equation (2.194). In practice, difficulty primarily arises in generating the Green's function from Equation (2.197).

Example 2.4: Green's function of a SDOF oscillator
Consider the linear ODE: $lu = f(t)$, with

$$L = \frac{d^2}{dt^2} + \omega^2 \tag{2.198}$$

Let the initial conditions be $u = u_0$ and $\frac{du}{dt} = \dot{u}_0$. The above differential operator, L above belongs to the class of undamped SDOF oscillator equations (Equation (2.184)) and ω is known as the natural frequency of the oscillator. In terms of the Green's function the solution to the above equation is given by

$$u(t) = \int_0^t G(t;\xi) f(\xi) d\xi \tag{2.199}$$

$G\,(t;\xi)$ is the singular solution of the equation:

$$\left(\frac{d^2}{dt^2} + \omega^2\right) G = \delta(t - \xi) \text{ for } 0 \le \xi \le \infty \tag{2.200}$$

Using the Laplace transform method, the solution to the above equation in the transformed domain is given by

$$\bar{G}(s;\xi) = \frac{s}{s^2+\omega^2} G(0;\xi) + \frac{1}{s^2+\omega^2} G'(0;\xi) + \frac{e^{-\xi st}}{s^2+\omega^2} \tag{2.201}$$

In the above equation, the prime denotes differentiation with respect to time. Inversion gives the Green's function as

$$G(t;\xi) = G(0;\xi)\cos\omega t + \frac{1}{\omega} G'(0;\xi)\sin\omega t + \frac{1}{\omega}\sin\omega(t-\xi)\,H(t-\xi) \tag{2.202}$$

Here, $H(t-\xi)$ is the Heaviside unit step function defined by

$$H(t) = 1, t > 0 = 0 \textit{ otherwise} \tag{2.203}$$

Now, the solution $u(t)$ is given by

$$u(t) = \int_0^\infty G(t;\xi) f(\xi) d\xi$$

$$= \cos\omega t \int_0^\infty G(0;\xi) f(\xi) d\xi + \frac{1}{\omega}\sin\omega t \int_0^\infty G'(0;\xi) f(\xi) d\xi$$

$$+ \frac{1}{\omega}\int_0^\infty \sin\omega(t-\xi) H(t-\xi) f(\xi) d\xi$$

$$= u_0 \cos\omega t + u_0 \sin\omega t + \frac{1}{\omega}\int_0^t \sin\omega(t-\xi) H(t-\xi) f(\xi) d\xi \tag{2.204}$$

Since $H(t-\xi) = 0$ for $\xi > t$, the last integral on the *RHS* of the above equation has the limits from 0 to t The first two terms together constitute the so-called complementary solution to the homogeneous part of the ODE (Equation (2.198)) and the last term, the particular integral which is also known as the Duhamel integral or convolution integral.

Example 2.5: Green's function of a simply supported beam

Consider a simply supported beam (Figure 2.15) subjected to a concentrated harmonic force $f(x,t) = f_0 \sin\omega t$. Assuming the solution to the fourth-order PDE (Equation (2.16)) of the beam in the form $w(x,t) = W(x)e^{i\omega t}$ yields

$$\frac{d^4 W}{dx^4} - \lambda^4 W = \frac{f}{EI} \tag{2.205}$$

Here, $\lambda^4 = \frac{m\omega^2}{EI}$. Green's function for the above equation is obtained from:

$$\frac{d^4 G(x;\xi)}{dx^4} - \lambda^4 W = \frac{\delta(x-\xi)}{EI} \tag{2.206}$$

The Laplace transform of the above equation is

$$\bar{G}(s;\xi) = \frac{1}{s^4-\lambda^4}\left\{\frac{e^{-s\xi}}{EI} + s^3 G\,(0;\xi) + s^2 G'^{(0;\xi)} + sG''\,(0;\xi) + G'''(0;\xi)\right\} \quad (2.207)$$

Primes in the above equation indicate differentiation with respect to the spatial variable x. The inverse transform of this equation (Abu-Hilal, 2003) yields:

$$\begin{aligned} G(x;\xi) = {} & \frac{1}{EI\lambda}^{3}\,\Phi_4\,(x-\xi)\,H\,(x-\xi) + \Phi_1(x) G\,(0;\xi) + \Phi_2\,G'(0;\xi) \\ & + \Phi_3\,G''(0;\xi) + \Phi_4\,G'''(0;\xi) \end{aligned} \quad (2.208)$$

where

$$\begin{aligned} \Phi_1(x) &= \frac{1}{2}\,(\cosh\lambda x + \cos\lambda x)\,,\ \Phi_2(x) = \frac{1}{2}\,(\sinh\lambda x + \sin\lambda x)\,, \\ \Phi_3(x) &= \frac{1}{2}\,(\cosh\lambda x - \cos\lambda x)\,,\ \Phi_4(x) = \frac{1}{2}\,(\sinh\lambda x - \sin\lambda x) \end{aligned} \quad (2.209)$$

Noting that the Green's function satisfies the homogeneous boundary conditions associated with the original PDE, we thus have $G(0;\xi)=0$ and $G''\,(0;\xi)=0$. This simplifies the Equation (2.208) as

$$G\,(x;\xi) = \frac{1}{EI\lambda}^{3}\,\Phi_4\,(x-\xi)\,H\,(x-\xi) + \Phi_2 G'(0;\xi) + \Phi_4 G'''(0;\xi) \quad (2.210)$$

To apply the remaining two boundary conditions $G\,(l;\xi)=0$ and $G''\,(l;\xi)=0$, it is needed to have expressions for $G''(x;\xi)$ from Equation (2.21):

$$G''\,(x;\xi) = \frac{1}{EI\lambda}\Phi_2\,(x-\xi) + \lambda\Phi_4 G'(0;\xi) + \frac{\Phi_2}{\lambda} G'''(0;\xi) \quad (2.211)$$

Substituting $x=l$ in the Equations (2.210) and (2.211) and applying the boundary conditions $G\,(l;\xi)=0$ and $G''\,(l;\xi)=0$, we have:

$$\begin{aligned} G'\,(0;\xi) &= \frac{1}{\lambda^2 EI}\,\frac{\Phi_4(l)\Phi_2\,(l-\xi) - \Phi_2(l)\Phi_4\,(l-\xi)}{\Phi_2^2(l) - \Phi_4^2(l)} \\ G'''\,(0;\xi) &= \frac{1}{\lambda^2 EI}\,\frac{\Phi_4(l)\Phi_4\,(l-\xi) - \Phi_2(l)\Phi_2\,(l-\xi)}{\Phi_2^2(l) - \Phi_4^2(l)} \end{aligned} \quad (2.212a,b)$$

With the above expressions of $G'\,(0;\xi)$ and $G'''\,(0;\xi)$, we obtain from the Equation (2.210), the Green's function for a simply supported beam in the form:

$$\begin{aligned} G\,(x;\xi) &= \frac{\sin\lambda x\,\sinh\lambda l\,\sin\lambda\,(l-\xi) - \sinh\lambda x\,\sin\lambda l\,\sinh\lambda\,(l-\xi)}{\lambda^3 EI\sin\lambda l\,\sinh\lambda l}, x \le \xi \\ &= \frac{\sin\lambda\xi\,\sinh\lambda l\,\sin\lambda\,(l-x) - \sinh\lambda\xi\,\sin\lambda l\,\sinh\lambda\,(l-x)}{\lambda^3 EI\sin\lambda l\,\sinh\lambda l}, x \ge \xi \end{aligned} \quad (2.213)$$

2.8 Solution of PDEs with Inhomogeneous Boundary Conditions

The boundary conditions in a structural system may not always be homogeneous (i.e. taking zero values). For example, the fixity locations on the boundary of a continuous system may be subjected to time-dependent motions (Meirovitch, 1967, Warburton, 1976). Some familiar cases in point are seismically excited structures, equipments subjected to induced vibrations from surrounding environment and structures undergoing vibrations during transportation. As a specific example, consider a continuous system modelled as a 1D beam structure (Figure 2.20) subjected to the support excitation. If the support excitation is only in the transverse direction, the governing PDE of motion and the boundary conditions are given by

$$\frac{\partial^2}{\partial x^2}\left(EI\frac{\partial^2 w}{\partial x^2}\right) + m\frac{\partial^2 w}{\partial t^2} = 0,\ w(0,t) = g(t) \text{ and } w(l,t) = h(t) \tag{2.214}$$

We now write the transverse displacement of the beam as a sum of two displacement components in the form:

$$w(x,t) = w_s(x,t) + w_d(x,t) \tag{2.215}$$

The underlying scheme in splitting the response into two components is to choose the function $w_s(x,t)$ of such a form that makes the boundary conditions for the dynamic component $w_d(x,t)$ homogeneous. Substituting the above in Equation (2.214) gives

$$\frac{\partial^2}{\partial x^2}\left(EI\frac{\partial^2 w_d(x,t)}{\partial x^2}\right) + m\frac{\partial^2 w_d(x,t)}{\partial t^2} = -\left\{\frac{\partial^2}{\partial x^2}\left(EI\frac{\partial^2 w_s(x,t)}{\partial x^2}\right) + m\frac{\partial^2 w_s(x,t)}{\partial t^2}\right\} \tag{2.216}$$

The boundary conditions in Equation (2.214) take the form:

$$w\,(0,t) = w_s\,(0,t) + w_d\,(0,t) = g(t) \text{ and} \tag{2.217a}$$

$$w\,(l,t) = w_s\,(l,t) + w_d\,(l,t) = h(t) \tag{2.217b}$$

Now choose $w_s(x,t)$ as:

$$w_s(x,t) = A_1(x)g(t) + A_2(x)h(t) \tag{2.218}$$

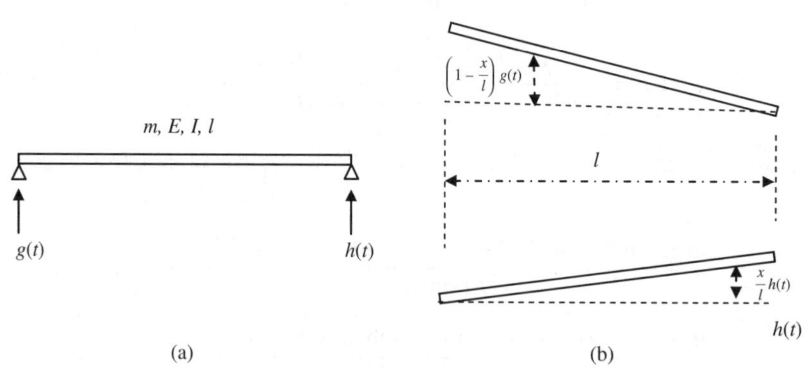

Figure 2.20 (a) Uniform beam under support excitations and (b) pseudostatic support displacements

The above choice of $w_s(x,t)$ is prompted by the requirement: $w_s(0,t) = g(t)$ and $w_s(l,t) = h(t)$ so that the boundary conditions for the dynamic component become homogeneous:

$$w_d(0,t) = 0 \text{ and } w_d(l,t) = 0 \tag{2.219}$$

$A_1(x)$ and $A_2(x)$ can be considered as displacement shapes induced by unit movement at the two supports applied statically and independently. Choice of $A_1(x)$ and $A_2(x)$ is not unique (see Section 6.5). From Figure 2.20, one choice is that $A_1(x) = \left(1 - \frac{x}{l}\right)$ and $A_2(x) = \frac{x}{l}$. $w_s(x,t)$, thus amounts to a displacement component equivalent to that induced by the differential (nonsynchronous) support motions statically and is commonly termed as a pseudo-static component. Substitution of Equation (2.218) along with the displacement shape functions $A_1(x)$ and $A_2(x)$, in Equation (2.216) leads to

$$\frac{\partial^2}{\partial x^2}\left(EI\frac{\partial^2 w_d(x,t)}{\partial x^2}\right) + m\frac{\partial^2 w_d(x,t)}{\partial t^2} = -m\left[\ddot{g}(t) + \frac{x}{l}\left\{\ddot{h}(t) - \ddot{g}(t)\right\}\right] \tag{2.220}$$

The terms on the right-hand side of the above equation represents the effective loading, $P_{eff}(t)$ due to the support excitations. If $h(t) = g(t)$, $P_{eff}(t) = -m\ddot{g}(t)$ and the beam is said to be under uniform support excitation. The nonhomogeneous PDE (Equation (2.220)) along with the homogeneous boundary conditions in Equation (2.219) is now solvable by the eigenexpansion method. A more elaborate description of structural systems under support excitation and their dynamic behaviour is presented in Chapter 6.

2.9 Solution to Nonself-adjoint Continuous Systems

In the case of structural systems such as spinning shafts, axially moving strings/belts, elastic beams with moving inertial loads and pipes conveying fluids, the systems are not self-adjoint. The parameter representing the spinning speed in a rotating system, the travel velocity of a moving load/moving fluid introduces additional terms due to gyroscopic (Coriolis) effects (Meirovitch, 1970) in the governing PDEs. If $\boldsymbol{L}$ is a self-adjoint operator and $\boldsymbol{L}^*$ is its adjoint, we have $< v, \boldsymbol{L}u > = < u, \boldsymbol{L}^*v >$ (Equation (2.87)) with $u \in$ domain of $\boldsymbol{L}$ and $v \in$ domain of $\boldsymbol{L}^*$. For nonself-adjoint systems, $\boldsymbol{L} \neq \boldsymbol{L}^*$.

2.9.1 *Eigensolution of Nonself-adjoint System*

For nonself-adjoint systems, the eigensolution is obtained by considering the solution of both the original and adjoint operators. The following two results pertain to the properties of these two operators:

- eigenvalues of the adjoint operator $\boldsymbol{L}^*$ are the same as those of the original operator, $\boldsymbol{L}$;
- eigenfunctions of the operator, $\boldsymbol{L}$ are orthogonal to those of its adjoint operator.

The following statements validate the above properties of the two operators. Let us consider, for a nonself-adjoint continuous system, u_i and v_j, $i, j \in \mathbb{N}$, the eigenfunctions of $\boldsymbol{L}$ and $\boldsymbol{L}^*$, respectively, that is

$$\boldsymbol{L}u_i = \lambda_i u_i \text{ and}$$
$$\boldsymbol{L}^* v_j = \lambda_j^* v_j \tag{2.221a,b}$$

$\lambda_i, i \in \mathbb{N}$ are the eigenvalues of the operator $\boldsymbol{L}$ and $\lambda_i^*, i \in \mathbb{N}$, the eigenvalues of $\boldsymbol{L}^*$. λ_i and λ_i^* are in general complex. The Equation (2.221b) represents an eigenproblem adjoint to the original one in Equation (2.221a). From Equation (2.221a), we have

$$< v_i, \boldsymbol{L}u_i > =< v_i, \lambda_i u_i > \tag{2.222}$$

By Equation (2.87), we have the property of an adjoint operator as

$$< v_i, \boldsymbol{L}u_i > =< u_i, \boldsymbol{L}^* v_i > \tag{2.223}$$

Subtracting $< v_i, \lambda_i u_i >$ from both sides of the above equation, we get

$$< v_i \boldsymbol{L}u_i > - < v_i, \lambda_i u_i > =< u_i \boldsymbol{L}^* v_i > - < v_i, \lambda_i u_i > \tag{2.224}$$

In view of Equation (2.222), the left-hand side of the above equation vanishes. Thus,

$$\begin{aligned}
&0 < u_i, \boldsymbol{L}^* v_i > - < v_i, \lambda_i u_i > \\
&\Rightarrow 0 =< u_i, \boldsymbol{L}^* v_i > -\lambda_i < v_i, u_i > \\
&\Rightarrow 0 =< u_i, \boldsymbol{L}^* v_i > - < \lambda_i v_i, u_i > \\
&\Rightarrow 0 =< u_i, \left(\boldsymbol{L}^* - \lambda_i\right) v_i >
\end{aligned} \tag{2.225}$$

Equation (2.225) shows that u_i is orthogonal to every function $\left(\boldsymbol{L}^* - \lambda_i\right) v_i$. Zero is also a solution contained in the above equation, which leads to:

$$\left(\boldsymbol{L}^* - \lambda_i\right) v_i = 0 \Rightarrow \boldsymbol{L}^* v_i = \lambda_i v_i \tag{2.226}$$

The above equation reveals that the eigenvalues of the adjoint operator are the same as those of the original operator, $\boldsymbol{L}$. The eigenfunctions are, however, not same.

2.9.2 *Biorthogonality Relationship between* **L** *and* **L***

From the eigenvalue problem in Equation (2.221), we have

$$< v_j \boldsymbol{L}u_i > = \int_\Omega v_j \left(\lambda_i u_i\right) d\Omega = \lambda_i \int_\Omega u_i v_j \, d\Omega \tag{2.227}$$

and

$$< u_i \boldsymbol{L}^* v_j > = \int_\Omega u_i \lambda_j^* v_j \, d\Omega = \lambda_j^* \int_\Omega u_i v_j \, d\Omega \tag{2.228}$$

Equation (2.87) along with Equations (2.227) and (2.228) gives

$$< v_j, \boldsymbol{L}u_i > - < u_i, \boldsymbol{L}^* v_j = 0 = \left(\lambda_i - \lambda_j^*\right) \int_\Omega u_i v_j \, d\Omega \tag{2.229}$$

If $\lambda_i \neq \lambda_j^*$, we obtain the relation between the eigenfunctions of the original operator, $\boldsymbol{L}$ and those of the adjoint operator $\boldsymbol{L}^*$ as:

$$\int_\Omega u_i v_j \, d\Omega = 0 \tag{2.230}$$

If u_i and v_j, $i, j \in \mathbb{N}$ are normalised, we obtain from the above equation the biorthonormality relationship:

$$\int_{\Omega} u_i v_j \, d\Omega = \delta_{ij}, i, j \in \mathbb{N} \tag{2.231}$$

In view of the above equation, the eigenfunctions $\{(u_i, v_j), i, j \in \mathbb{N}\}$ form an orthonormal basis for the nonself-adjoint system and for any function $f \in \mathcal{H}$, the generalized Fourier expansion in Equation (2.169) takes the form:

$$f = \sum_{k=1}^{\infty} c_k u_k, \quad c_k = < v_k f > \tag{2.232}$$

The above expansion can be used for forced vibration solution of a nonself-adjoint system.

Example 2.6: An axially moving string – a nonself-adjoint continuous system and eigensolution

Consider a uniform string of mass, m per unit length, having axial tension H and translating with a constant speed V_s in the axial direction between two fixed supports (Figure 2.21). With the assumption of small-amplitude vibrations, the energy expressions are given by

$$\textit{Kinetic Energy}: T = \frac{1}{2}\int_0^l m\left(\frac{\partial y}{\partial t} + V_s\frac{\partial y}{\partial x}\right)^2 dx \tag{2.233}$$

$$\textit{Potential energy}: V = \frac{1}{2}\int_0^l H\left(\frac{\partial y}{\partial x}\right)^2 dx \tag{2.234}$$

$y(x, t)$ is the transverse displacement of the string and l is the length of the string. Note that the travelling speed induces an additional velocity component $V_s\frac{\partial y}{\partial x}$ and the string velocity at any position x and time t is given by $\frac{\partial y}{\partial t} + V_s\frac{\partial y}{\partial x}$. Hence, the kinetic energy, T contains terms linear in $\frac{\partial y}{\partial t}$, thus rendering the system gyroscopic in nature. From Equations (2.233) and (2.234), the system Lagrangian is given by

$$L = T - V = \frac{1}{2}\int_0^l \left\{m\left(\frac{\partial y}{\partial t} + V_s\frac{\partial y}{\partial x}\right)^2 - H\left(\frac{\partial y}{\partial x}\right)^2\right\} dx \tag{2.235}$$

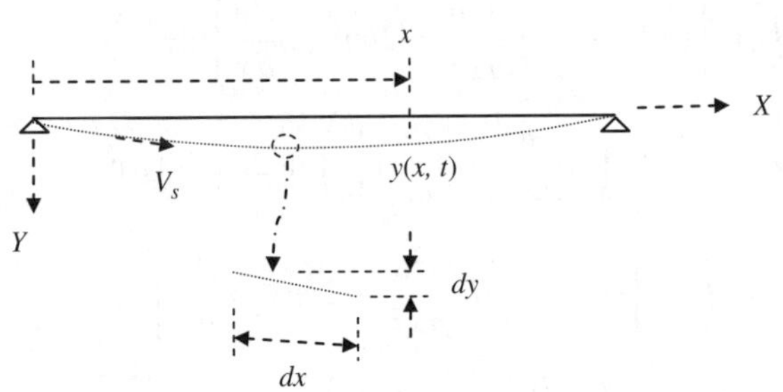

Figure 2.21 Axially moving string and an infinitesimal moving segment

The equations governing the motion of the system are obtained by the Hamilton's principle:

$$\delta \int_{t_1}^{t_2} L\left(y, y', \dot{y}, t\right) dt = 0$$

$$\Rightarrow \int_{t_1}^{t_2} \int_0^l \left\{ m\frac{\partial y}{\partial t}\delta\dot{y} + \left(mV_s^2 - H\right)\frac{\partial y}{\partial x}\delta y' + mV_s\left(\frac{\partial y}{\partial x}\delta\dot{y} + \frac{\partial y}{\partial t}\delta y'\right)\right\} dx\, dt = 0 \tag{2.236a}$$

Integrating each term on the left-hand side of the above equation by parts, we obtain

$$\Rightarrow \int_{t_1}^{t_2} \int_0^l \left\{ m\frac{\partial^2 y}{\partial t^2} + \left(mV_s^2 - H\right)\frac{\partial^2 y}{\partial x^2} + 2mV_s\frac{\partial^2 y}{\partial x \partial t}\right\} \delta y\, dx\, dt$$
$$+ \left(mV_S^2 - H\right)\frac{\partial y}{\partial x}\delta y\bigg|_0^l + mV_s\frac{\partial y}{\partial t}\delta y\bigg|_0^l = 0 \tag{2.236b}$$

Since the virtual displacement δy is arbitrary over the time interval, we obtain the strong form of the equations of motion for the moving string as

$$m\frac{\partial^2 y}{\partial t^2} + 2mV_S\frac{\partial^2 y}{\partial x \partial t} + \left(mV_S^2 - H\right)\frac{\partial^2 y}{\partial x^2} = 0 \tag{2.237}$$

and the boundary conditions as

$$y = 0 \text{ at } x = 0 \text{ and } l \tag{2.238}$$

Assuming a harmonic solution to the PDE (Equation (2.237)) in the form $y(x,t) = Y(x)e^{i\omega t}$ we obtain:

$$\left(mV_S^2 - H\right)\frac{d^2Y}{dx^2} + 2imV_S\omega\frac{dY}{dx} - m\omega^2 Y = 0 \tag{2.239}$$

From the above equation, the linear differential operator $\boldsymbol{L}$ is identified as

$$\boldsymbol{L} = \left(mV_S^2 - H\right)\frac{d^2}{dx^2} + 2imV_S\omega\frac{d}{dx} \tag{2.240}$$

The adjoint operator is obtained by considering the inner product $< u, \boldsymbol{L}v>$:

$$< u, \boldsymbol{L}v> = \int_0^l u\left\{\left(mV_S^2 - H\right)\frac{d^2v}{dx^2} + 2imV_S\omega\frac{dv}{dx}\right\} dx$$
$$= \left(mV_S^2 - H\right)\left(u\frac{dv}{dx}\bigg|_0^l - \frac{du}{dx}v\bigg|_0^l\right) + 2imV_S\left(uv|_0^l\right)$$
$$+ \int_0^l v\left\{\left(mV_S^2 - H\right)\frac{d^2u}{dx^2} - 2imV_S\omega\frac{du}{dx}\right\}dx$$
$$= \left(mV_S^2 - H\right)\left(u\frac{dv}{dx}\bigg|_0^l - \frac{du}{dx}v\bigg|_0^l\right) + 2imV_S\left(uv|_0^l\right) < v, \boldsymbol{L}^*u> \tag{2.241}$$

The above equation identifies the adjoint operator as:

$$\boldsymbol{L}^{*}=\left(mV_S^2-H\right)\frac{d^2}{dx^2}-2imV_S\omega\frac{d}{dx} \tag{2.242}$$

Note that $\boldsymbol{L}^* \neq \boldsymbol{L}^*$ and the system is nonself-adjoint.

2.9.3 Eigensolutions of L and L*

The eigenvalue problem corresponding to the operator $\boldsymbol{L}$ is given by $\boldsymbol{L}Y = \lambda Y$. Using the notation $D = \frac{dY}{dx}$, the auxiliary equation corresponding to $\boldsymbol{L}$ is

$$\left(mV_S^2-H\right)D^2+2imV_S\omega D-m\omega^2Y=0 \tag{2.243}$$

The characteristic roots/eigenvalues of the above equation are obtained as

$$\lambda_1,\lambda_2=-i\frac{\omega}{c\left(1+\frac{V_s}{c}\right)},\quad i\frac{\omega}{c\left(1-\frac{V_s}{c}\right)} \tag{2.244}$$

Here, $c=\sqrt{\frac{H}{m}}$. Note that for a gyroscopic system, the eigenvalues are purely imaginary (Meirovitch, 1967) The eigenfunction of the operator $\boldsymbol{L}$ is given by

$$Y(x)=A_1e^{\lambda_1 x}+A_2e^{\lambda_2 x} \tag{2.245}$$

A_1 and A_2 are arbitrary complex constants. Substitution of the taut string boundary conditions in Equation (2.245) gives

$$A_2=-A_1 \quad \text{and} \tag{2.246}$$

$$e^{\lambda_1 l}=e^{-\lambda_2 l}\Rightarrow e^{i\frac{\omega l}{c\left(1+\frac{V_S}{c}\right)}}=e^{-\frac{\omega l}{c\left(1-\frac{V_S}{c}\right)}}\Rightarrow \omega_n=\frac{n\pi c}{l}\left(1-\frac{V_s^2}{c^2}\right),\ n\in\mathbb{N} \tag{2.247}$$

Equations (2.246) and (2.247) gives the natural frequencies of the axially moving string. With $V_s=0$, ω_n reduces to $\frac{n\pi}{l}\sqrt{\frac{H}{m}}$, the frequencies of a stationary taut string. Figure 2.22 shows the effect of the translating velocity on the first few natural frequencies of the string. As the velocity approaches the phase velocity c $(=\sqrt{\frac{H}{m}})$, the natural frequencies, $\omega_n \to 0$ indicating an instability (a sort of buckling instability) that the translating string undergoes at this critical velocity V_c $(=c)$.

With $A_2=-A_1$ and substituting for λ_1 and λ_2 in terms of ω_n, Equation (2.245) gives the eigenfunctions of $\boldsymbol{L}$ as

$$Y_n(x)=iAe^{\frac{in\pi V_S x}{l}}\sin\frac{n\pi x}{l},\ n\in\mathbb{N} \tag{2.248}$$

In the above equation, 'A' is an arbitrary complex constant. The adjoint eigenvalue problem is given by $\boldsymbol{L}^*Y=\boldsymbol{\lambda}^*Y$. $\boldsymbol{L}^*$ being a complex conjugate of the operator, $\boldsymbol{L}$, the eigenvalues and the eigenfunctions can be obtained as the complex conjugates of those of $\boldsymbol{L}$. Since λ_1 and λ_2 of $\boldsymbol{L}$ are purely imaginary (Equation (2.247)), the eigenvalues of

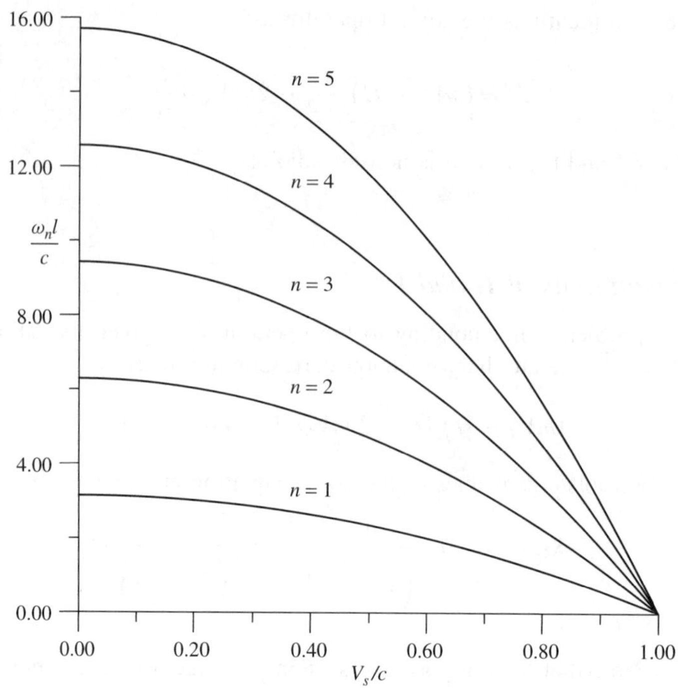

Figure 2.22 Axially translating string. Effect of translation on the natural frequencies

the adjoint operator $\boldsymbol{L}^*$ remain same except that $\lambda_1^* = -\lambda_1$ and $\lambda_2^* = -\lambda_2$. Similarly, the eigenfunction of $\boldsymbol{L}^*$ is given by

$$Y_n^*(x) = -i\bar{A}e^{\frac{-in\pi V_S x}{l}} \sin\frac{n\pi x}{l}, n \in \mathbb{N} \tag{2.249}$$

$\bar{A}$ is the complex conjugate of A.

2.10 Conclusions

Whilst the classical treatment of linear continuous systems, based on eigenexpansion (or a generalized Fourier series), appears to be insightful (owing to their analytical character), elegant and sometimes deceptively simple, the approach lacks generality. It is not only difficult to extend these methods to complex structural systems (for example, a 3D frame consisting of many beams or a plate or slab with cross girders), but also to those with nonuniformity in geometric and/or material properties. Hence, more general solution strategies, typically involving finite-dimensional discretizations of continuous systems (yielding what will be referred to as multidegree-of-freedom systems) through well-known basis functions of the transcendental or polynomial type, are called for. The material in the next two chapters is aimed at introducing some of these concepts, preparing the groundwork for modern computational tools.

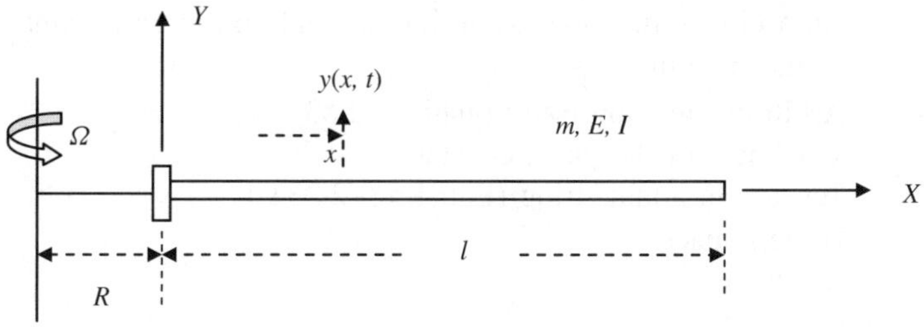

Figure 2.23 Transverse vibration of a helicopter blade modelled as a cantilever beam

Exercises

1. Derive the frequency equation and the eigenfunctions for the transverse vibrations of a uniform beam with fixed–fixed and free–free boundary conditions (see Table 2.1).
2. A simplified model of a helicopter blade as a cantilever beam is shown in Figure 2.23. Assuming a constant rotational speed Ω about the vertical axis at a distance R from the left end (clamped end) of the beam, derive the equation of motion for the transverse (flapping mode) vibrations of the beam.
3. Derive the equations of motion for a circular plate and obtain the free vibration solution for the fully clamped condition (with its perimeter as the boundary). (Hint: Transform the rectangular coordinates into polar coordinates: $x = r\cos\theta$ and $y = r\sin\theta$).

Notations

a	Side of a plate, constant
a_0, a_1, a_2	Constants
$\boldsymbol{a}_j(\boldsymbol{x})$	Coefficient functions in differential operator
A_i, A_n	Constants
b	Side of a plate, constant
B	Orthonormal basis functions
$B_1(\cdot), B_2(\cdot)$	Boundary conditions in BVP (Equation (2.84))
$\boldsymbol{B}$	Partial differential operator, Equations (2.4) and (2.30)
B_n	Constants
c	Parameter in PDE; $c^2=\frac{T}{m}$ for taut string, $c^2=\frac{EA}{m}$ for a uniform bar and $c^2 = \frac{EI}{m}$ for a uniform beam
c_k	Fourier coefficients
C_n	Constants
d_k	Constants
D	$-\frac{dY}{dx}$ (Equation (2.243))
$f(x)$	An arbitrary function (as an initial condition in wave equation)
$f(x, y, t)$	External (transverse) loading field
$F(x)$	Arbitrary function (see Equation (2.63))
$F(x, t)$	Transverse loading

$g(x)$	An arbitrary function (as an initial condition in wave equation)
G	Shear modulus, $\frac{E}{2(1+v)}$
$G(x)$	Arbitrary function (see Equation (2.63))
h	Thickness of the plate, constant
H	Tension in a taut string (Equation (2.234))
$\mathcal{H}$	Hilbert space
J	Jacobian
k	Constant
l_1, l_2	Direction cosines of the outward normal at a point on the plate boundary
$\boldsymbol{L}$	Differential operator (Sturm–Liouville operator)
$\boldsymbol{L}^*$	Adjoint Sturm–Liouville operator
$\mathbf{l}_\lambda$	$=\mathbf{L}-\lambda I$
m_1, m_2	Constants
M	Mass of SDOF system in Figure 2.16
	Mass of the moving load in Example 2.3
	Stress resultant – bending moment in a beam
M	Differential operator
M_x, M_y, M_z	Moments
$n, \bar{n}, n_1, n_2$	Integers
$p(x)$	Coefficient function in Sturm–Liouville operator
$q(x)$	Coefficient function in Sturm–Liouville operator
Q	Shear force in a beam
Q_x, Q_y	Shear forces
r	Radius of gyration
$r(x)$	Coefficient function in Sturm–Liouville operator
R	Radius of curvature of the plate (Figure 2.7)
S_m, S_n, T_n	Partial sums using generalized Fourier expansion
$\boldsymbol{T}$	Tension in a taut string
u	Transverse (lateral) displacement (for taut string)
u_i	Eigenfunction
$u(x)$	Functions
U	Vector of displacements, $\left(= (u, v, w)^T\right)$
v_k	Eigenfunction
$v(x)$	Functions
V_s	Speed of the moving load on a beam (Example 2.3)
	Speed of a translating string (Example 2.6)
w	Transverse (lateral) displacement (for beam)
$Y(t)$	Function of time
$\alpha, \alpha_1, \alpha_2$	Constants
β, β_1, β_2	Constants
$\gamma_{xy}, \gamma_{xz}, \gamma_{yz}$	Shear strains
δw	Virtual displacement
$\varepsilon_x, \varepsilon_y, \varepsilon_z$	Normal strains
$\varepsilon_{xz}, \varepsilon_{xz}$	Normal stress and shear strain'
θ	Slope at any location on a taut string $\left(= \frac{\partial u}{\partial x}\right)$

λ, $\bar{\lambda}$, λ_i	Eigenvalues (overbar indicates complex conjugate)
μ, μ_n	Frequency (eigenvalue) parameter
σ_{xx}, σ_{xz}	Normal stress and shear stress
τ_{xz}	Shear strain
Ψ	Rotation angle
$\Phi(x)$, $\Phi_n(x)$	Eigenfunctions
$\mathbf{\Phi}$	Set of eigenfunctions

References

Abu-Hilal, M. (2003) Forced vibration of Euler-Bernoulli beam by means of dynamic Greens functions. *Journal of Sound and Vibration*, **267**, 191–207.

Daya Reddy, B. (1998) *Introductory Functional Analysis-With Applications to Boundary Value Problems and Finite Elements*, Springer-Verlag, New York.

Fryba, L. (1972) *Vibration of Solids and Structures Under Moving Loads*, Noordhoff International, Groningen, Netherlands.

Geradin, M. and Rixen, D. (1994) *Mechanical Vibrations -- Theory and Application to Structural Dynamics*, John Wiley & Sons, Inc., Chichester, UK

Kreyszig, E. (1999) *Advanced Engineering Mathematics*, John Wiley & Sons, Inc., Singapore

Meirovitch, L. (1967) *Analytical Methods in Vibration*, Macmillan, New York.

Meirovitch, L. (1970) *Methods of Analytical Dynamics*, Macmillan, New York.

Simmons, G.F. and Krantz, S.G. (1925) *Differential Equations-Theory, Technique and Practice*, McGraw-Hill, Inc., New York

Warburton, G.B. (1976) *The Dynamical Behaviour of Structures*, Pergamon Press, Oxford.

Bibliography

Coddington, E.A. and Levinson, N. (1955) *Theory of Ordinary Differential Equations*, McGraw-Hill, New York.

Pryce, J.D. (1993) *Numerical Solutions of Sturm-Liouville Problems*, Oxford University Press, Oxford.

Stakgold, I. (1979) *Green's Functions and Boundary Value Problems*, John Wiley & Sons, Inc., New York.

Wright, A.D., Smith, C.E., Thresher, R.W. and Wang, J.L.C. (1982) Vibration modes of centrifugally stiffened beams. *Transactions of ASME, Journal of Applied Mechanics*, **49** (2), 197–202.

3

Classical Methods for Solving the Equations of Motion

3.1 Introduction

The classical methods that we are going to briefly dwell upon are mostly based on the extremization of certain positive definite functionals, such as the energy integral. These schemes, whilst being important for obtaining useful information on the response of a class of dynamical systems of practical interest, also serve as a precursor to the weak formulation that is employed with the finite element method (FEM). For further motivation, consider a vibrating solid body and the associated Cauchy's equations of motion (Equations (1.35), (1.46) or (1.65)) involving the first (spatial) derivatives of stresses. They are often referred to as the strong form of the governing partial differential equations (PDEs) of motion. An attempt at numerical solutions of these equations (say, through collocation or finite difference) must account for the C^1 continuity of the stress components (and hence the C^2 continuity of the displacement vector) over the domain of interest. This precludes solutions of a large class of problems involving stress discontinuities. Cases in point are the so-called shock tube or shear band problems. Apart from the unsolvability of (or infeasibility of solving) the strong form, there is also the impracticability of point-wise definition of the system parameters necessitated by the strong form. Solutions of the strong form, therefore, cannot also be obtained when the system parameters (material or geometric) are defined piecewise, as in the case of stepped beams or bimaterial interfaces involving jumps in the material properties. Finally, from the perspective of structural design, it often suffices to compute only the stresses that involve first-order derivatives of the response variables (e.g. displacements and/or velocities) vis-à-vis the second-order derivatives demanded for satisfying the strong form.

The classical methods such as the Rayleigh–Ritz method and weighted residual methods obtain solutions within a framework that utilises weakened continuity of the response variables and thus overcome most of the above limitations of the strong form. For instance, the Rayleigh–Ritz method, described below in more detail, is based upon a functional discretization and subsequent minimization (with respect to the unknown coefficients in the discretization) of an energy integral (a potential functional) that is expressed purely

Elements of Structural Dynamics: A New Perspective, First Edition. Debasish Roy and G Visweswara Rao.

in terms of the stresses (or the stress resultants). Thus, this method does not, in principle, require stresses to be C^1 continuous. Another class of methods, which may broadly be categorised as variants of the weighted residuals like the Galerkin method, starts with the system equation of motion itself and aims at approximating the targeted response function. Substitution of this approximation in the strong form yields a residual, which is orthogonalized with respect to weight functions so selected as to drive the integral of the weighted residual to zero. The weighted integral involves integration by parts and as can be intuitively understood, relaxes the strong form and weakens the derivative requirement. We can call these classical methods of solution discrete methods of approximation and through the discretization, the continuum mathematical model is reduced to a discrete one with finite degrees of freedom (dof) that can be conveniently solved for system response either in time or frequency domain. Some of these classical methods are elaborated in this chapter. The FEM, which is the subject matter of the next chapter is indeed known to find a basis from the Rayleigh–Ritz method and the Galerkin method of weighted residuals.

3.2 Rayleigh–Ritz Method

In this method, an approximate solution to the equations of motion of an elastic body (Figure 1.4) is obtained by discretization followed by minimization of a functional. Recall here that the equations of motion are in fact derived in Chapter 1 by the variational principle applied directly to the energy functional $I\left(\boldsymbol{t}, \boldsymbol{x}, \boldsymbol{u}, \boldsymbol{u}_x, \boldsymbol{u}_t\right) = \int_{t_1}^{t_2}(T - V)\boldsymbol{dt}$ where $T = \frac{1}{2}\int \rho \dot{u}_i \dot{u}_i dV$ and $V = \frac{1}{2}\int \sigma_{ij}\varepsilon_{ij} dV - \int f_{bi}u_i dV - \int f_{si}u_i dS$ are, respectively, the kinetic and total potential energy of the elastic body. In the Rayleigh–Ritz method, the energy functional $I(\cdot)$ is first discretized using a linear combination of basis or trial functions for the displacement field variable $\boldsymbol{u}(\boldsymbol{x}, t) := \{u_1, u_2, u_3\}^T$. The discretization is of the form:

$$u_1(\boldsymbol{x}, t) = \sum_{j=1}^{N} U_{1j}^T(\boldsymbol{x}) q_j(t), \quad u_2(\boldsymbol{x}, t) = \sum_{j=1}^{N} U_{2j}^T(\boldsymbol{x}) q_{j+N}(t) \quad \text{and}$$

$$u_3(\boldsymbol{x}, t) = \sum_{j=1}^{N} U_{3j}^T(\boldsymbol{x}) q_{j+2N}(t) \tag{3.1}$$

$\boldsymbol{U}(\boldsymbol{x}) = \left(\boldsymbol{U}_1, \boldsymbol{U}_2, \boldsymbol{U}_3\right)^T$ is the $3 \times N$ matrix of assumed trial functions. These trial functions are required to be admissible in that these functions shall be continuous, linearly independent and complete (Appendix C, also see Appendix B). They shall satisfy the essential boundary conditions over the domain Ω. The trial functions need not satisfy the natural boundary conditions, since this requirement is already contained in the variational formulation adopted by the Rayleigh–Ritz method. $\boldsymbol{q}(t)$ are $3Nx1$ vectors of unknown functions of time that can be treated as the generalized coordinates. The above discretization is the Rayleigh–Ritz approximation that results in new expressions for T and V in terms of the coordinate functions/generalized coordinates $\boldsymbol{q}(t)$. In deriving these expressions for $T$and V, we utilise the strain–displacement relationship $\boldsymbol{\varepsilon} = \boldsymbol{BU}$ and the constitutive relationship between stress and strain $\boldsymbol{\sigma} = \boldsymbol{c\varepsilon}$ with $\boldsymbol{B}$ being the partial differential operator as described in Equation (1.43) and $\boldsymbol{c}$ the matrix of material constants in Equation (1.44).

With $\boldsymbol{F}_b(\boldsymbol{x}, t)$ and $\boldsymbol{F}_s(\boldsymbol{x}, t)$ denoting the vectors of body and surface forces, respectively, we have:

$$T = \frac{1}{2}\int_{\Omega} \rho \dot{\boldsymbol{q}}^T \boldsymbol{U}^T \boldsymbol{U} \dot{\boldsymbol{q}} dV = \frac{1}{2}\dot{\boldsymbol{q}}^T \boldsymbol{M} \dot{\boldsymbol{q}} \quad \text{and}$$

$$V = \frac{1}{2}\int_{\Omega} \boldsymbol{q}^T \boldsymbol{U}^T \boldsymbol{B}^T \boldsymbol{c}\, \boldsymbol{B} \boldsymbol{U} \boldsymbol{q} dV - \int_{\Omega} \boldsymbol{q}^T \boldsymbol{U}^T \boldsymbol{F}_b dV - \int_{\Omega A} \boldsymbol{q}^T \boldsymbol{U}^T \boldsymbol{F}_s\, dA$$

$$= \frac{1}{2}\boldsymbol{q}^T \boldsymbol{K} \boldsymbol{q} - \boldsymbol{q}^T (\boldsymbol{F}_b + \boldsymbol{F}_s) \tag{3.2}$$

Here,

$$\boldsymbol{M} = \int_{\Omega} \rho \boldsymbol{U}^T \boldsymbol{U}\, dV, \quad \boldsymbol{K} = \int_{\Omega} \boldsymbol{U}^T \boldsymbol{B}^T \boldsymbol{c} \boldsymbol{B} \boldsymbol{U}\, dV \quad \text{and}$$

$$\boldsymbol{F}_1 = \int_{\Omega} \boldsymbol{U}^T \boldsymbol{F}_b dV \quad \text{and} \quad \boldsymbol{F}_2 = \int_{\Omega_A} \boldsymbol{U}^T \boldsymbol{F}_s\, dA \tag{3.3}$$

$\boldsymbol{M}$ and $\boldsymbol{K}$ represent the mass and stiffness matrices corresponding to the discretization and their size is dependent on the number of terms N in the assumed displacement field. The energy functional $I(\cdot)$ now expressed in terms of the unknown generalized coordinates $\boldsymbol{q}(t)$ and their derivatives is now extremised using the variational principle $\delta \int I(\boldsymbol{q}, \dot{\boldsymbol{q}}, t)dt = 0$. Following the procedure involving variational calculus and as described in Section 1.1, we obtain the following equations of motion for a linear system in the form of ordinary differential equations (ODEs) in $\boldsymbol{q}(t)$. With $\boldsymbol{F}(t) = \boldsymbol{F}_1(t) + \boldsymbol{F}_2(t)$, the ODEs take the form:

$$\boldsymbol{M}\ddot{\boldsymbol{q}}(t) + \boldsymbol{K}\boldsymbol{q}(t) = \boldsymbol{F}(t) \tag{3.4}$$

The above matrix-vector differential equations (DEs) constitute a multidegree-of-freedom (MDOF) system. $\boldsymbol{F}(t)$ is the vector of external forcing functions. The coordinate functions $\boldsymbol{q}(t)$, once determined enable us to evaluate the response variables $\boldsymbol{u}(\boldsymbol{x}, t)$ and any other derived quantities as well. If the stress–strain relationship is nonlinear, Equation (3.4) is no longer linear and leads to nonlinear ODEs in $\boldsymbol{q}(t)$. In essence, the Rayleigh–Ritz approximation transforms the continuous system into a discrete system with finite degrees of freedom represented by the coupled DEs in Equation (3.4) – linear or nonlinear.

The advantage of the Rayleigh–Ritz method is that solution of ODEs in Equation (3.4) is easier than that of the strong form of governing PDEs of motion in Equation (1.35). Also, the approximate solution in the assumed from in Equation (3.1) approaches the true solution in the limit as $N \to \infty$. Whilst in actual computation, this limit may not be realisable, accuracy of a solution with a finite number of terms, N often suffices an analysis or a design requirement. However, to obtain acceptable solutions from Equation (3.4), the need for the assumed trial functions $U(\boldsymbol{x})$ satisfying the essential boundary conditions over the specified domain $\partial\Omega$ is difficult to be complied with. Here, one notices the disadvantage of the method for practical application to a complex structure in which case it is obviously a complication in devising the admissible functions.

The following example on a one-dimensional beam (cantilever) illustrates the application of the method.

Example 3.1: Application of Rayleigh–Ritz method for a beam under flexure
To illustrate the application of Rayleigh–Ritz method to the beam under flexure, let us specifically consider a freely vibrating uniform cantilever beam as shown in Figure 3.1. The PDE governing the beam vibratory motion (see Equation (2.16)) with no external force is:

$$\frac{\partial^4 w}{\partial x^4} = -\frac{m}{EI}\frac{\partial^2 w}{\partial t^2} \tag{3.5}$$

Note that $w = 0$ and $\frac{\partial w}{\partial x} = 0$ are the essential (geometric) boundary conditions, and $EI\frac{\partial^2 w}{\partial x^2} = 0$ and $\frac{\partial}{\partial x}\left(EI\frac{\partial^2 w}{\partial x^2}\right) = 0$ are the natural (force) boundary conditions. Following the procedure enumerated above to obtain the Rayleigh–Ritz solution, let us discretize the energy integral $\boldsymbol{I}(\cdot) = \int_{t_1}^{t_2} (T - V)\,\boldsymbol{dt}$ by assuming an approximate solution in the form $\hat{w}(x,t) = \sum_{j=1}^{N} W_j(x)q_j(t)$, where the trial functions $W_j(x)$ are so chosen as to satisfy the essential boundary conditions. $q_j(t)$, $j = 1, 2, .. N$ are the unknown generalized coordinates. The trial functions are selected to form the complete set of monomials: $=\{1, x, x^2, x^3, \ldots\ldots\}$. If we limit our attention to $N = 2$, it requires for the cantilever beam:

$$\hat{w}(x,t) = x^2 q_1(t) + x^3 q_2(t) \tag{3.6}$$

Obviously our choice for the two trial functions is given by $W_1(x) = x^2$ and $W_2(x) = x^3$, which satisfy the two essential boundary conditions. Introducing nondimensional parameter $X = x/l$ and using the above trial functions, we obtain a 2 dof discrete system with system matrices evaluated as given by Equations (3.2) and (3.3):

$$\boldsymbol{M} = \rho\int_0^1 W^T W A\, dX = m\begin{bmatrix} \frac{1}{5} & \frac{1}{6} \\ \frac{1}{6} & \frac{1}{7} \end{bmatrix}$$

$$\boldsymbol{K} = E\int_{\Omega} z^2 \left(\frac{\partial^2 W}{\partial x^2}\right)^T \left(\frac{\partial^2 W}{\partial x^2}\right) dA\,dX$$

$$= EI\int_0^1 \left(\frac{\partial^2 W}{\partial x^2}\right)^T \left(\frac{\partial^2 W}{\partial x^2}\right) dX = EI\begin{bmatrix} 4 & 6 \\ 6 & 12 \end{bmatrix} \tag{3.7}$$

Here, $m = \rho A$ with ρ and A being the mass density per unit volume and area of cross section, respectively. Table 3.1 shows the matrices derived with $N = 2$, 3 and 4 with

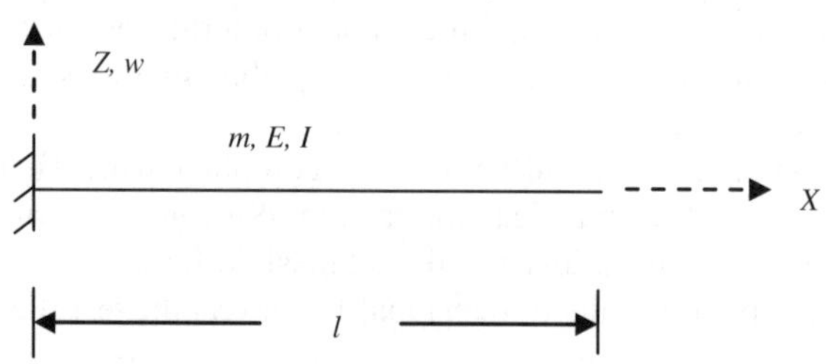

Figure 3.1 Application of Rayleigh–Ritz method to a beam under flexure (cantilever)

Table 3.1 Mass and stiffness matrices from Rayleigh–Ritz approximation

Number of trial functions, N	2	3	4
$w(x,t)$	$x^2q_1(t)+x^3q_2(t)$	$x^2q_1(t)+x^3q_2(t)+x^4q_3(t)$	$x^2q_1(t)+x^3q_2(t)+x^4q_3(t)+x^5q_4(t)$
Mass matrix	$m\begin{bmatrix}\frac{1}{5} & \frac{1}{6}\\ \frac{1}{6} & \frac{1}{7}\end{bmatrix}$	$m\begin{bmatrix}\frac{1}{5} & \frac{1}{6} & \frac{1}{7}\\ \frac{1}{6} & \frac{1}{7} & \frac{1}{8}\\ \frac{1}{7} & \frac{1}{8} & \frac{1}{9}\end{bmatrix}$	$m\begin{bmatrix}\frac{1}{5} & \frac{1}{6} & \frac{1}{7} & \frac{1}{8}\\ \frac{1}{6} & \frac{1}{7} & \frac{1}{8} & \frac{1}{9}\\ \frac{1}{7} & \frac{1}{8} & \frac{1}{9} & \frac{1}{10}\\ \frac{1}{8} & \frac{1}{9} & \frac{1}{10} & \frac{1}{11}\end{bmatrix}$
Stiffness matrix	$EI\begin{bmatrix}4 & 6\\ 6 & 12\end{bmatrix}$	$EI\begin{bmatrix}4 & 6 & 8\\ 6 & 12 & 18\\ 8 & 18 & \frac{144}{5}\end{bmatrix}$	$EI\begin{bmatrix}4 & 6 & 8 & 10\\ 6 & 12 & 18 & 24\\ 8 & 18 & \frac{144}{5} & 40\\ 10 & 24 & 40 & \frac{400}{7}\end{bmatrix}$

appropriate trial functions. With this definition of the system matrices, the Rayleigh–Ritz approximation results in the following equations of motion:

$$\boldsymbol{M}\ddot{\boldsymbol{q}}(t) + \boldsymbol{K}\boldsymbol{q}(t) = \boldsymbol{0} \tag{3.8}$$

The homogeneous linear ODEs (Equation (3.8)) above admits nontrivial solution expressible in terms of the Fourier basis:

$$\boldsymbol{q}(t) = \boldsymbol{Q}\exp(i\omega t) \tag{3.9}$$

Here, $i = \sqrt{-1} \cdot \omega$ is known as the eigenvalue of the discrete system in Equation (3.8) and $\boldsymbol{Q}$ is the associated eigenvector analogous to the eigenfunction to a self-adjoint boundary value problem (Chapter 2). The assumed harmonic solution above, if substituted in Equation (3.8) leads to the generalized eigenvalue problem:

$$\boldsymbol{KQ} = \omega^2\boldsymbol{MQ} \tag{3.10}$$

Note that the eigenvalues are the roots of the characteristic equation $\left|\boldsymbol{KQ} - \omega^2\boldsymbol{MQ}\right| = 0$. The eigensolution of the above generalized eigenvalue problem and its application in obtaining both the free and forced vibration solution of MDOF systems are detailed in Chapter 5. Moreover, numerical methods for the eigenextraction are considered in Chapter 7. The eigenvalue problem yields N approximate eigenvalues each of which remains an upper bound (Appendix C) to the corresponding exact eigenvalue of the continuous system. In other words, if ω_i^e, $i = 1, 2, \ldots$ correspond to the exact eigenvalues, then

$$\omega_i^e \le \omega_i^c, \; i = 1, 2, .., N \tag{3.11}$$

Table 3.2 Eigenvalue result from the Rayleigh–Ritz approximation with increasing N

Number of trial functions (N)	2	3	4	Analytical result from Equation (2.129)
Eigenvalue parameter, $\omega\sqrt{\frac{mL^4}{EI}}$	3.532732	3.517076	3.516021	3.5156250
	34.806893	22.23336	22.157855	22.0345538
		118.14443	63.347839	61.6972142
			281.590329	120.901916

where ω_i^c stand for the eigenvalues computed from the Rayleigh–Ritz approximation. This result follows from the eigenvalue separation property (Appendix C) that states that the eigenvalues of a symmetric matrix A of dimension $N-1$ are bracketed by those of the matrix of dimension N with a row and column added to A. In this connection, we note that the mass and stiffness matrices (see Table 3.1) obtained from the Rayleigh–Ritz approximation with $N-1$ number of trial functions are fully contained in those obtained with N number of trial functions. Thus, the eigenvalue separation property ensures:

$$\ldots\ldots \geq \omega_i^{N-2} \geq \omega_i^{N-1} \geq \omega_i^{N} \geq \ldots\ldots\ldots \geq \omega_i^{exact}, \; i = 1, 2\ldots. \tag{3.12}$$

Table 3.2 shows the eigenvalues via Rayleigh–Ritz approximation with increasing N as well as the exact ones. The determinant search method (finding the roots of $\left|\boldsymbol{KQ} - \omega^2\boldsymbol{MQ}\right| = 0$) is used to obtain the eigenvalues for the system of matrices listed in this table. It can be observed that the eigenvalues from the Rayleigh–Ritz approximation tend to converge to the corresponding exact values of the continuous system with increasing N and thus each eigenvalue remains an upper bound.

3.2.1 Rayleigh's Principle

Rayleigh–Ritz method is in fact an extension of Rayleigh's principle due to Lord Rayleigh (1945) that provides an estimate of the first fundamental frequency of a continuous system. Given the importance of the first fundamental frequency of a continuous structural system in obtaining its response, it is always of advantage to be able to readily obtain its estimate and Rayleigh's principle just provides such an approximation in the form of Rayleigh's quotient:

$$\omega^2 = \frac{\text{Maximum potential energy}}{\text{Maximum kinetic energy}} \tag{3.13}$$

Equation (3.13) is derived based on conservation of energy in a nondissipative vibrating system that amounts to equating the maximum kinetic and potential energies at a system frequency ω. Hence, for a freely vibrating body, with an assumed vibrating shape in the form of harmonic motion with $\boldsymbol{u}(\boldsymbol{x}, t) = \boldsymbol{U}(\boldsymbol{x})\, exp(i\omega t)$, the energy expressions in Equation (3.2) take the following form:

$$T = \left(\frac{1}{2}\omega^2 \int_\Omega \rho \boldsymbol{U}^T \boldsymbol{U} dV\right) \exp(2i\omega t) \quad \text{and} \quad V = \left(\frac{1}{2}\int_\Omega \boldsymbol{U}^T\boldsymbol{B}^T\boldsymbol{c}\,\boldsymbol{B}\boldsymbol{U}\,\boldsymbol{dV}\right)\exp(2i\omega t) \tag{3.14}$$

From Equation (3.13), it follows that

$$\omega^2 = \frac{\int_\Omega \boldsymbol{U}^T \boldsymbol{B}^T \boldsymbol{c}\,\boldsymbol{B}\boldsymbol{U}\,dV}{\int_\Omega \rho \boldsymbol{U}^T \boldsymbol{U} dV} \tag{3.15}$$

As an illustration, let us consider again the cantilever beam in Figure 3.1. In this one-dimensional case, if we assume the trial function in the form $U(X) = X^2$ with $X = x/l$, we can obtain the estimate for the first eigenvalue from Rayleigh's principle as $\omega^2 = \frac{EI\int_0^1 \left(\frac{\partial^2 U}{\partial x^2}\right)^2 dX}{ml^4 \int_0^1 U^2 dX}$ giving rise to $\omega\sqrt{\frac{ml^4}{EI}} = 4.4721$ as compared to the exact value of 3.515625 (see Table 3.2). Further elaboration on Rayleigh's quotient is provided in Chapter 5.

3.3 Weighted Residuals Method

The weighted residuals method (Finlayson and Scriven, 1966) is more general than the Rayleigh–Ritz method in obtaining an approximate solution in that it can be directly applied to the governing equations of motion – linear or nonlinear and initial value or boundary value or eigenvalue problems. It is preferred to the Rayleigh–Ritz method where an energy functional may not exist (e.g. for governing PDEs where the operators involving only the spatial variables are nonelliptic, nonconservative systems) or in cases where an energy functional may not be readily available. Let the system equations of motion be described by:

$$\boldsymbol{Lu} - \boldsymbol{f} = \boldsymbol{0} \text{ in } \Omega \subset \mathbb{R}^n$$

with boundary conditions:

$$\mathcal{B}_i \boldsymbol{u} = g(t), i = 1, ..p, \text{ on } \partial\Omega \subset \mathbb{R}^{n-1}$$

and initial conditions:

$$\boldsymbol{u}(\boldsymbol{x}, 0) = \boldsymbol{h}(\boldsymbol{x}), \quad \boldsymbol{x} \in \Omega \tag{3.16}$$

where $\boldsymbol{L}$ and $\mathcal{B}$ are differential operators and $\boldsymbol{f}(\boldsymbol{x}, t)$ is the forcing function. For example, in the case of a cantilever beam under flexure $\boldsymbol{L} = EI\frac{\partial^4}{\partial x^4} + m\frac{\partial^2}{\partial t^2}$ and $p = 4$. $\mathcal{B}_1 = 1.0$ and $\mathcal{B}_2 = \frac{\partial}{\partial x}$ corresponding to the essential boundary conditions $u(0, t) = 0$ and $\frac{\partial u(0,t)}{\partial x} = 0$ and $\mathcal{B}_3 = \text{EI}\frac{\partial^2}{\partial x^2}$ and $\mathcal{B}_4 = \frac{\partial}{\partial x}\left(\text{EI}\frac{\partial^2}{\partial x^2}\right)$ corresponding to the natural boundary conditions $EI\frac{\partial^2 u(l,t)}{\partial x^2} = 0$ and $\frac{\partial}{\partial x}\left(EI\frac{\partial^2 u(l,t)}{\partial x^2}\right) = 0$. If the exact solution for the displacement field $u(\boldsymbol{x}, t)$ is substituted into Equation (3.16), the right-hand sides of both the system DE and the boundary equations must be identically zero everywhere in the respective domains of interest. On the other hand, if we consider a 1D case and an approximate solution $\tilde{u}(\mathbf{x}, t) = \sum_{j=1}^N U_j^T(\boldsymbol{x})q_j(t)$ is assumed as in Equation (3.1) and substituted in Equation (3.16), the right-hand sides will no longer be zero and they will instead be given by the residual $R(\cdot)$ as follows:

$$R(\tilde{u}) = \boldsymbol{L}(u) - f \neq \boldsymbol{0} \tag{3.17}$$

Here $U_j(\boldsymbol{x}),\ j = 1, 2, ..N$ are known as trial functions. The basic idea in weighted residuals method is to render the residual minimum, in some sense, by using a set of

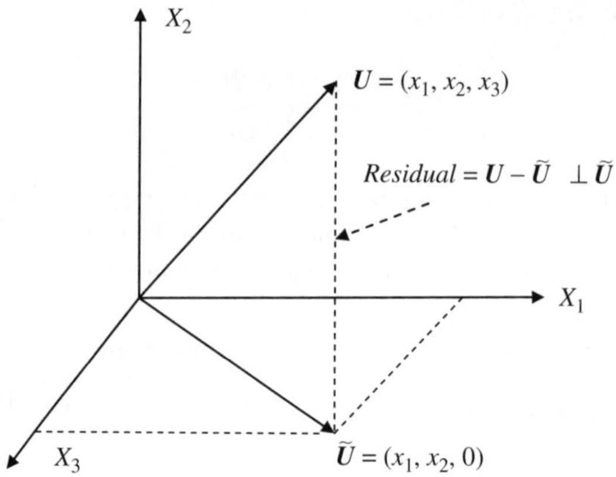

Figure 3.2 Orthogonal projection and minimum residual norm

orthogonality conditions. To this end, we prescribe a set of weighting functions $\boldsymbol{w}(\boldsymbol{x}) \in \mathbb{R}^N$ and the weighted integrals of the residual are set to zero, that is:

$$<w_j, R> = 0, \quad j = 1.2, \ldots N. \tag{3.18}$$

where $<w, v> = \int_\Omega w(x)v(x)dx$ stands for the inner product of two functions in Ω. The trial function vector $\boldsymbol{U}(\boldsymbol{x})$ are chosen from a set of comparison functions (Appendix C) that are functions that are sufficiently differentiable and satisfy both the essential and natural boundary conditions (see Chapter 1 for definition of these boundary conditions). Equation (3.18) amounts to an orthogonal projection of the residual onto a reduced (and finite-) dimensional linear (vector) space spanned by the elements of $\boldsymbol{w}(\boldsymbol{x})$. This principle is derived from the classic projection theorem that states that 'given a Hilbert space $\mathcal{H}$ and a closed nonempty subspace G of $\mathcal{H}$ and with $u \in \mathcal{H}$, then there exists a $v \in G$ such that the norm $\|u - v\| \leq \|u - z\|$ for all $z \in G$'. Since, in practice, the trial and weighting functions belong to a finite-dimensional vector space, the weighted residual method seeks to find the best approximant within this space in the sense as envisaged through the projection theorem. This can be intuitively understood from the fact that a position vector $\boldsymbol{U}$ in $\mathbb{R}^3$ space is represented by $\tilde{\boldsymbol{U}}$ in the reduced $\mathbb{R}^2$ space with minimum error when the residual $\boldsymbol{U} - \tilde{\boldsymbol{U}}$ is orthogonal to X_1 and X_2, as illustrated in Figure 3.2.

Now, the choice of weighting functions $\boldsymbol{w}(\boldsymbol{x})$ leads to the so-called variants of the weighted residual method – the Galerkin method, the collocation method, the least squares method and the subdomain method. In any case, Equation (3.18) leads to a set of ODEs in the case of the initial value problem and a set of algebraic equations for steady-state vibration problems. For an eigenvalue problem of the type $\boldsymbol{L}(u) = \lambda u$, the method leads to a discrete eigenvalue problem as in Equation (3.10).

3.3.1 Galerkin Method

Galerkin's method of weighted residuals is the most popular one and chooses the weighting functions, $\boldsymbol{w}(\boldsymbol{x})$ from the basis trial functions $\boldsymbol{U}(x)$ itself. Thus, Equation (3.18) takes the form:

$$<U_j, R> = 0 \Rightarrow \int_{\Omega} U_j(x) \left\{ \mathbf{L} \left(\sum_{j=1}^{N} U_j^T(\mathbf{x}) q_j(t) \right) - f(\mathbf{x}, t) \right\} d\mathbf{x} = 0, \; j = 1, 2, \ldots . N$$

$$\sum_{j=1}^{N} \int_{\Omega} U_j(x) \left\{ \mathbf{L} \left(U_j^T(\mathbf{x}) q_j(t) \right) - f(x,t) \right\} d\mathbf{x} = 0, \; j = 1, 2, \ldots . N \tag{3.19}$$

Example 3.2: Eigenvalue problem for a beam under flexure by the Galerkin method
As an example, let us consider free vibration of the uniform cantilever beam in Figure 3.1. The first task is to approximate the solution as $\hat{u}(x,t) = \sum_{j=1}^{N} U_j(x) q_j(t)$ with the trial functions $U_j(x)$, $j = 1, 2 \ldots, N$ that satisfy both the essential and natural boundary conditions. To find an approximate one-parameter solution with $N = 1$, let us start by choosing

$$\frac{\partial^3 U_1(x)}{\partial x^3} = \cos \frac{\pi x}{2l} \tag{3.20}$$

such that $U_1(x)$ satisfies the natural boundary condition $\frac{\partial^3 u(l,t)}{\partial x^3} = 0$. We can now construct $U_1(x)$ by integrating Equation (3.20) and finding the integration constants via imposition of the remaining two essential boundary conditions $u(0,t) = 0$ and $\frac{\partial u(0,t)}{\partial x} = 0$ and the one natural boundary condition $\frac{\partial^2 u(l,t)}{\partial x^2} = 0$. This results in:

$$U_1(x) = \frac{4l^2}{\pi^2} x - \frac{l}{\pi} x^2 - \frac{8l^3}{\pi^3} \sin \frac{\pi x}{2l} \tag{3.21}$$

For a free vibration problem, $q_j(t) = Q_j \exp(i\omega t)$ with $i = \sqrt{-1}$. The one parameter approximate function $\tilde{u}_1(x,t)$ is now given by

$$\tilde{u}_1(x,t) = \left(\frac{4l^2}{\pi^2} x - \frac{l}{\pi} x^2 - \frac{8l^3}{\pi^3} \sin \frac{\pi x}{2l} \right) Q_1 \exp(i\omega t) \tag{3.22}$$

Applying the Galerkin method, Equation (3.19) gives:

$$Q_1 \left[\left(\frac{6}{\pi^2} - \frac{16}{\pi^3} \right) - \omega \sqrt{\frac{ml^4}{EI}} \left(\frac{1}{5\pi^2} - \frac{2}{\pi^3} + \frac{16}{3\pi^4} + \frac{160}{\pi^6} - \frac{512}{\pi^7} \right) \right] = 0 \tag{3.23}$$

Let $\mu = \omega \sqrt{\frac{ml^4}{EI}}$. For a nontrivial solution ($Q_1 \neq 0$), we get $\mu = 3.5193$ that is close to the exact result of $\mu_1 = 3.5156250$ (Table 3.2). Improvement in the estimate of the eigenvalue can be obtained with $N = 2$. A two-parameter approximate function is assumed to be the form:

$$\tilde{u}(x,t) = \left(U_1(x) Q_1 + U_2(x) Q_2 \right) \exp(i\omega t) \tag{3.24}$$

By choosing $\frac{\partial^3 U_2(x)}{\partial x^3} = \cos \frac{3\pi x}{2l}$ and following the same steps as in the case of the one-parameter solution, we obtain:

$$U_2(x) = \left(\frac{4l^2}{9\pi^2} x + \frac{l}{3\pi} x^2 - \frac{8l^3}{27\pi^3} \sin \frac{3\pi x}{2l} \right) \tag{3.25}$$

Thus, with $N = 2$ and using the two parameter approximation for $\hat{u}(x, t)$, Equation (3.19) takes the form:

$$\int_{\Omega} U_k(x) \left\{ \boldsymbol{L}\left(\sum_{j=1}^{2} U_j^T(x) Q_j exp(i\omega t) \right) \right\} dx = 0, \ k = 1, 2 \tag{3.26}$$

The integral in Equation (3.26) results in the following two linear simultaneous equations in Q_1 and Q_2:

$$Q_1 \left[\left(\frac{6}{\pi^2} - \frac{16}{\pi^3} \right) - \mu \left(\frac{1}{5\pi^2} - \frac{2}{\pi^3} + \frac{16}{3\pi^4} + \frac{160}{\pi^6} - \frac{512}{\pi^7} \right) \right]$$
$$+ Q_2 \left[\left(-\frac{4}{\pi^2} + \frac{16}{9\pi^3} \right) - \mu \left(-\frac{1}{15\pi^2} + \frac{2}{9\pi^3} + \frac{16}{27\pi^4} - \frac{5248}{243\pi^6} + \frac{10496}{243\pi^7} \right) \right] = 0$$

and

$$Q_1 \left[\left(-\frac{4}{\pi^2} + \frac{16}{9\pi^3} \right) - \mu \left(-\frac{1}{15\pi^2} + \frac{2}{9\pi^3} + \frac{16}{17\pi^4} - \frac{5248}{243\pi^6} + \frac{10496}{243\pi^7} \right) \right]$$
$$+ Q_2 \left[\left(\frac{6}{9\pi^2} + \frac{16}{27\pi^3} \right) - \mu \left(\frac{1}{45\pi^2} + \frac{2}{27\pi^3} + \frac{16}{243\pi^4} + \frac{160}{729\pi^6} + \frac{512}{2187\pi^7} \right) \right] = 0 \tag{3.27}$$

Equation (3.27) corresponds to the generalized eigenvalue problem (Equation (3.10)) with $\boldsymbol{Q} = (Q_1 \ Q_2)^T$ and

$$\boldsymbol{K} = \begin{bmatrix} \left(\frac{6}{\pi^2} - \frac{16}{\pi^3} \right) & \left(-\frac{4}{\pi^2} + \frac{16}{9\pi^3} \right) \\ \left(-\frac{4}{\pi^2} + \frac{16}{9\pi^3} \right) & \left(\frac{6}{9\pi^2} + \frac{16}{27\pi^3} \right) \end{bmatrix}$$

and

$$\boldsymbol{M} = \begin{bmatrix} \left(\frac{1}{5\pi^2} - \frac{2}{\pi^3} + \frac{16}{3\pi^4} + \frac{160}{\pi^6} - \frac{512}{\pi^7} \right) & \left(-\frac{1}{15\pi^2} + \frac{2}{9\pi^3} + \frac{16}{27\pi^4} - \frac{5248}{243\pi^6} + \frac{10496}{243\pi^7} \right) \\ \left(-\frac{1}{15\pi^2} + \frac{2}{9\pi^3} + \frac{16}{27\pi^4} - \frac{5248}{243\pi^6} + \frac{10496}{243\pi^7} \right) & \left(\frac{1}{45\pi^2} + \frac{2}{27\pi^3} + \frac{16}{243\pi^4} + \frac{160}{729\pi^6} + \frac{512}{2187\pi^7} \right) \end{bmatrix} \tag{3.28}$$

Equation (3.28) is a symmetric eigenvalue problem and a nontrivial solution for $\boldsymbol{Q}$ gives the estimate for the first two eigenvalues as $\mu = \omega\sqrt{\frac{ml^4}{EI}} = 3.5160$ and 22.269 (see Table 3.2 for comparison with the result by the Rayleigh–Ritz method). Note that in the Galerkin method also the matrices $\boldsymbol{K}$ and $\boldsymbol{M}$ are symmetric like in the Rayleigh–Ritz method.

Example 3.3: Free vibration solution of a sagging cable by the Galerkin method
Figure 3.3 shows an extensible cable fixed at both ends with mass per unit length m, horizontal length l and area of cross section A. Let u and v be the dynamic displacements (i.e. the time-varying parts of the total displacements) (Figure 3.3a) over and above the static equilibrium and $T(s)$ and $\tau(s,t)$ denote the static and dynamic components of cable tension. Considering the equilibrium of any cable segment (Figure 3.3b), the planar oscillations of the sagging cable in free vibration are described by the following equations of motion (Irvine and Caughey, 1974):

$$\frac{\partial}{\partial s}\left[(T+\tau)\left\{\frac{\partial}{\partial s}(x+u)\right\}\right] = m\ddot{u}$$
$$\frac{\partial}{\partial s}\left[(T+\tau)\left\{\frac{\partial}{\partial s}(y+v)\right\}\right] = m\ddot{v} - mg \tag{3.29a,b}$$

Here, $\frac{d}{ds}\left(T\frac{dy}{ds}\right) = -mg$ represents the equation for the static equilibrium of the cable. Let H be the horizontal component of the cable tension given by $T\frac{dx}{ds}$. If $d\bar{s}$ and ds are the infinitesimal line segments in the deformed and undeformed configurations of the cable segment, the cable strain ε is given by:

$$\varepsilon = \frac{d\bar{s} - ds}{ds} \tag{3.30}$$

From the cable geometry, we have:

$$d\bar{s}^2 = (dx+du)^2 + (dy+dv)^2, \; ds^2 = (dx)^2 + (dy)^2 \tag{3.31}$$

The time-dependent incremental tension $\tau(s,t)$ is related to the cable strain ε by:

$$\varepsilon = \frac{\tau}{EA} \tag{3.32}$$

From Equations (3.30)–(3.32), we have (by retaining only the linear terms):

$$\tau = EA\left(\frac{\partial u}{\partial s} + \frac{dy}{ds}\frac{\partial v}{\partial s}\right) \tag{3.33}$$

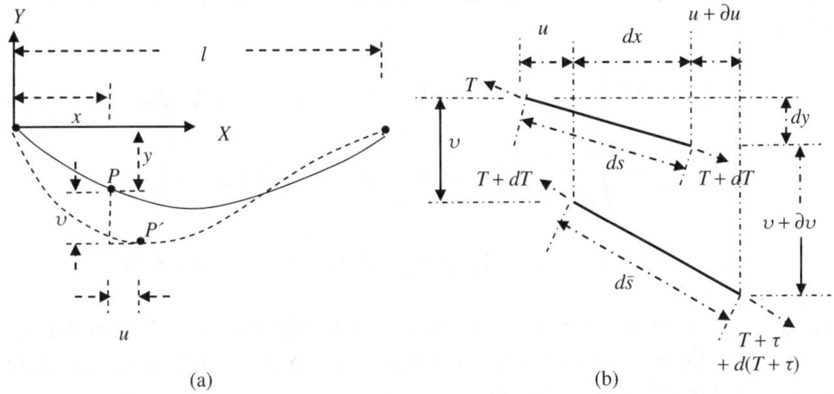

Figure 3.3 (a) Sagging cable in dynamic equilibrium; P – a point on an undeformed position and P' – on a deformed position (b) cable segment in equilibrium

For a low sag cable, $ds \approx dx$, $T \approx H$ and $\frac{dy}{ds} \approx \frac{dy}{dx}$. With this assumption, we also have the static profile given by $H\frac{d^2y}{dx^2} = -mg$ that yields $y(x) = \frac{\beta l}{2}\left(\frac{x}{l} - \frac{x^2}{l^2}\right)$. Here, $\beta = \frac{mgl}{H}$. Since the cable sag at midspan ($x = l/2$) is $\frac{\beta l}{8}$, the parameter β characterises the cable sag to span ratio. The assumption of low cable sag reduces Equation (3.29) to:

$$\boldsymbol{L}_1 u + \boldsymbol{L}_2 v = \frac{m}{H}\ddot{u}$$

$$\boldsymbol{L}_2 u + \boldsymbol{L}_3 v = \frac{m}{H}\ddot{v} \tag{3.34a,b}$$

Here, $\boldsymbol{L}_1, \boldsymbol{L}_2$ and $\boldsymbol{L}_3$ are, respectively, the differential operators $\frac{\partial}{\partial x}\left[P_1\frac{\partial}{\partial x}\right]$, $\frac{\partial}{\partial x}\left[P_2\frac{\partial}{\partial x}\right]$ and $\frac{\partial}{\partial x}\left[P_3\frac{\partial}{\partial x}\right]$ with $P_1 = 1 + \frac{EA}{H}$, $P_2 = \frac{EA}{H}\frac{dy}{dx}$ and $P_3 = 1 + \frac{EA}{H}\left(\frac{dy}{dx}\right)^2$. We note here that $\frac{EA}{H}$ represents the cable extensibility. For the free vibration solution, we use the Galerkin method by assuming approximate solutions for $u(x,t)$ and $v(x,t)$ in the form:

$$\hat{u}(x,t) = \left(\sum_{j=1}^{N} A_j \sin j\pi\frac{x}{l}\right)\exp(i\omega t) \quad \text{and} \quad \hat{v}(x,t) = \sum_{j=1}^{N} B_j \sin j\pi\frac{x}{l}\exp(i\omega t) \tag{3.35}$$

In the assumed series solutions above, $\sin j\pi\frac{x}{l}$ are the trial functions, which are indeed the eigenfunctions of a taut string (equivalent to a cable with negligible sag). Substituting Equation (3.35) in Equation (3.34) and following the Galerkin method, we obtain the eigenvalue problem as Iyengar and Rao (1988):

$$\sum_{j=1}^{N} A_i\left(I_{jk} - \mu^2\delta_{jk}\right) + B_j J_{jk} = 0$$

$$\sum_{j=1}^{N} A_i\left(J_{jk} - \mu^2\delta_{jk}\right) + B_i K_{jk} = 0 \quad k = 1, 2, \ldots \tag{3.36}$$

Here, δ_{jk} is the Kronecker delta defined as $\delta_{jk} = 1$ for $j = k$ and zero otherwise. $\mu = \sqrt{\frac{m}{H}}\frac{l}{\pi}\omega$ is an eigenvalue parameter. With $X = \frac{x}{l}$, I_{jk}, J_{jk} and K_{jk} in Equation (3.36) are given by:

$$I_{jk} = 2j\int_0^1 \left(j\pi P_1 \sin j\pi X - P_1' \cos j\pi X\right)\sin k\pi X dX$$

$$J_{jk} = 2j\int_0^1 \left(j\pi P_2 \sin j\pi X - P_2' \cos j\pi X\right)\sin k\pi X dX$$

$$K_{jk} = 2j\int_0^1 \left(j\pi P_3 \sin j\pi X - P_3' \cos j\pi X\right)\sin k\pi X dX \tag{3.37}$$

Here, primes denote differentiations with respect to X. Note that the differential operators $\boldsymbol{L}_1, \boldsymbol{L}_2$ and $\boldsymbol{L}_3$ in Equation (3.34) are self-adjoint (Chapter 2) and hence the eigenvalue problem in Equation (3.36) is symmetric.

If we denote $\frac{EA}{H}$ by γ, the eigensolution is obtained with $N = 4$ terms in the series expansion in Equation (3.35). The eigenvalue problem takes the standard form

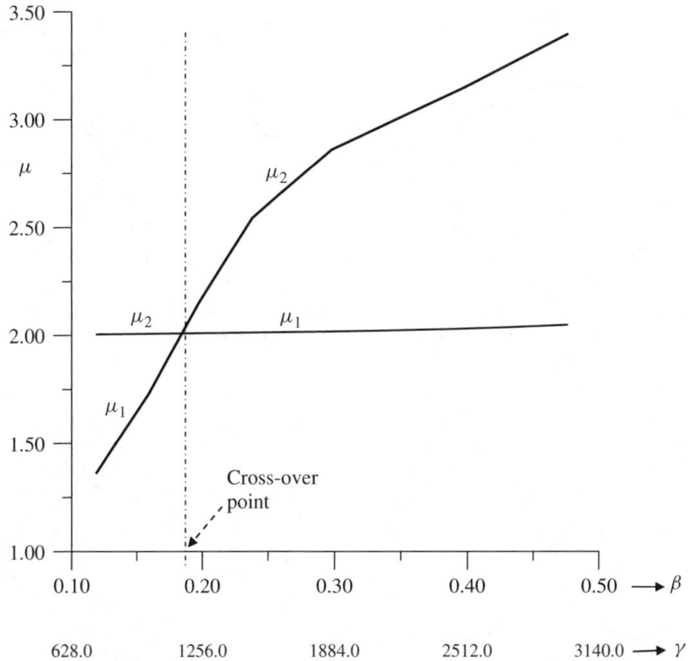

Figure 3.4 Free vibration solution of a sagging cable by the Galerkin method; $m = 2$ kg/m, $l = 400$ m and $EA = 4.929 \times 10^7$; mode crossover of frequencies (of vertical oscillation) with variation of the cable parameters, $\beta = \frac{mgl}{H}$ and $\gamma = \frac{EA}{H}$

$\left[\boldsymbol{A} - \mu^2 \boldsymbol{I}\right] \boldsymbol{\phi} = \boldsymbol{0}$ and is 8-dimensional. The eigenvalues or the natural frequencies of the cable are obtained in terms of the eigenvalue parameter μ by the Jacobi method (see Chapter 7 for a description of the method). It is the characteristic of a sagging cable to exhibit frequency crossover (with respect to the vertical oscillation) in that the usual fundamental symmetric mode crosses the antisymmetric mode as the cable sag increases (Irvine and Caughey, 1974; Shih and Tadjbakhsh, 1984). Variation of the first two frequencies are shown in Figure 3.4 that refers to a long cable with the following dimensions: length, $l = 400$ m, mass per unit length, $m = 2$ kg/m and $EA = 4.929 \times 10^7$. In the figure, the parameters β, γ are, respectively, varied in the intervals [0.1, 0.5] and [628–3140]. This corresponds to a reduction of H from 78480 to 15700 N. The graph corresponding to the symmetric first vertical mode of vibration crosses the antisymmetric mode at $(\beta, \gamma) \cong (0.185, 1194)$. The mode shapes of these two modes are shown in Figure 3.5 for two cases of cable parameters near the crossover.

In Figure 3.5, snapshots of different mode shape profiles are shown at different time instants within one period of oscillation.

3.3.2 Collocation Method

Referring to the different variants of the weighted residuals method, the collocation method employs forcing of the residual to zero at specific points in the domain. In this respect, the weighting function $w_i(\boldsymbol{x})$ in Equation (3.18) is taken to be $\delta(\boldsymbol{x} - \boldsymbol{x}_i)$, $i = 1, 2 \ldots, N$.

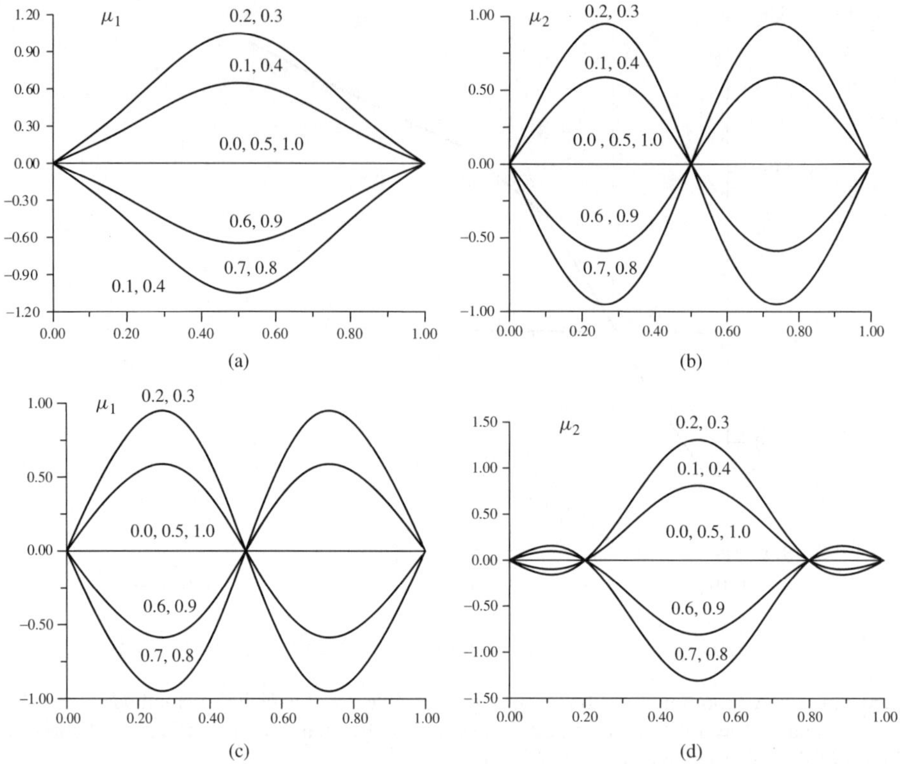

Figure 3.5 Free vibration solution of a sagging cable by the Galerkin method and phenomenon of frequency crossover; mode shapes of the first two natural frequencies of the cable in vertical oscillation, (a,b) no crossover for $\beta = 0.183$, $\gamma = 1150$ with $\mu_{1,2} = 1.989, 2.01$ (c,d) crossover for $\beta = 0.238$, $\gamma = 1495$ with $\mu_{1,2} = 2.01, 2.54$

Hence Equation (3.18) takes the form:

$$\int_{\Omega} \delta(\boldsymbol{x} - \boldsymbol{x}_i) R(\tilde{u}) d\boldsymbol{x} = 0 \Rightarrow R(\tilde{u}_i) = 0, \quad i = 1, 2 \ldots N \quad \text{with} \quad \tilde{u}_i = \tilde{u}(\boldsymbol{x}_i, t) \tag{3.38}$$

Example 3.4: Eigenvalue problem for a beam under flexure by the collocation method

Consider the free vibration of the uniform cantilever beam (Figure 3.1) of Example 3.1. Starting with a one-parameter solution with $N = 1$ in, let us start with $U_1(x)$ as in Equation (3.21). Applying the collocation method and choosing $x_1 = l$ (Equation (3.38)) as the specific point at which the residual is forced to become zero, Equation (3.38) gives:

$$R(\tilde{u}_1(L, t)) = 0 \Rightarrow \left(EI \frac{\partial^4}{\partial x^4} + m \frac{\partial^2}{\partial t^2} \right) Q_1 \left(\frac{4l^3}{\pi^2} - \frac{l^3}{\pi} - \frac{8l^3}{\pi^3} \right) \exp(i\omega t) = 0$$

$$\Rightarrow Q_1 \left[\frac{m\omega^2 l^4}{EI} \left\{ \frac{4}{\pi^2} - \frac{1}{\pi} - \frac{8}{\pi^3} \right\} + \frac{\pi}{2} \right] = 0 \tag{3.39}$$

Thus, for a nontrivial solution of Q_1, we obtain the estimate for the fundamental frequency of vibration as $\omega_1\sqrt{\frac{ml^4}{EI}} = 3.0305$ as against 3.516015 (Table 3.2). Improvement in the estimate of the eigenvalue can be obtained with $N = 2$. In this case, the residue $R(\cdot)$ is forced to zero at two points say, $x = l/2$ and l. With the two trial functions $U_1(x)$ and $U_2(x)$ as in Equations (3.21) and (3.25), Equation (3.38) finally leads to the following two linear homogeneous equations in the form:

$$\left\{-\frac{\pi}{2} - \mu^2\left(\frac{4}{\pi^2} - \frac{1}{\pi} - \frac{8}{\pi^3}\right)\right\} Q_1 + \left\{\frac{3\pi}{2} - \mu^2\left(\frac{4}{9\pi^2} + \frac{1}{3\pi} - \frac{8}{27\pi^3}\right)\right\} Q_2 = 0$$

$$\left\{-\frac{\pi}{2\sqrt{2}} - \mu^2\left(\frac{2}{\pi^2} - \frac{1}{4\pi} - \frac{8}{\sqrt{2}\pi^3}\right)\right\} Q_1$$

$$+\left\{-\frac{3\pi}{2\sqrt{2}} - \mu^2\left(\frac{2}{9\pi^2} + \frac{1}{12\pi} - \frac{8}{27\sqrt{2}\pi^3}\right)\right\} Q_2 = 0 \tag{3.40}$$

As in the previous cases, the above equations can be recast into a generalized eigenvalue problem (Equation (3.10)) with $\boldsymbol{Q} = (Q_1 \quad Q_2)^T$ and:

$$K = \begin{bmatrix} -\frac{\pi}{2} & \frac{3\pi}{2} \\ -\frac{\pi}{2\sqrt{2}} & -\frac{3\pi}{2\sqrt{2}} \end{bmatrix}, \quad M = \begin{bmatrix} \left(\frac{4}{\pi^2} - \frac{1}{\pi} - \frac{8}{\pi^3}\right) & \left(\frac{4}{9\pi^2} + \frac{1}{3\pi} - \frac{8}{27\pi^3}\right) \\ \left(\frac{2}{\pi^2} - \frac{1}{4\pi} - \frac{8}{\sqrt{2}\pi^3}\right) & \left(\frac{2}{9\pi^2} + \frac{1}{12\pi} - \frac{8}{27\sqrt{2}\pi^3}\right) \end{bmatrix} \tag{3.41}$$

For a nontrivial solution, Equation (3.39) yields the estimate for the eigenvalue parameter $\mu\left(= \omega\sqrt{\frac{mL^4}{EI}}\right)$ of the cantilever beam as $\mu_1 = 3.344$ and $\mu_2 = 20.133$. Thus, with $N = 2$, the method yields better estimates for μ than with $N = 1$. It is to be noticed here that the collocation method leads to an eigenvalue problem with unsymmetric matrices for any N. This is unlike the situation in the Rayleigh–Ritz and Galerkin methods where the matrices always remain symmetric. This is a pointer to the fact that the estimate for μ in the collocation method may no longer remain an upper bound.

3.3.3 Subdomain Method

In the subdomain method, the integral of the weighted residual is forced to zero over a number of nonoverlapping subsections of the domain. Thus, in this method, the integral over the entire domain is divided into a number of subdomains Ω_j such that, $\cup\Omega_j = \Omega$ and enough to evaluate the unknown parameters with increasing accuracy. Thus, we have

$$\int_{\Omega_j} R(\tilde{u})d\boldsymbol{x} = 0, \quad j = 1, 2 \ldots .. N \tag{3.42}$$

In this version of weighted residuals method, the weighting function $w(\boldsymbol{x})$ can be identified (see Equation (3.18)) as unity if $\boldsymbol{x} \in \Omega_j$ and zero otherwise.

Example 3.5: Eigenvalue problem for a beam under flexure by the subdomain method
Consider again the beam in Figure 3.1 of Example 3.1. For $N = 1$, the residue $R(\cdot)$ is integrated over the entire beam length. With the choice of $U_1(x)$ the same as that in Equation (3.21), the integral in Equation (3.42) gives:

$$\mu^2 = \frac{1}{\left(\frac{1}{3\pi} - \frac{2}{\pi^2} + \frac{16}{\pi^4}\right)} \Rightarrow \mu = \omega\sqrt{\frac{ml^4}{EI}} = 3.843 \tag{3.43}$$

A two-parameter solution is obtained by using two trial functions $U_1(x)$ and $U_2(x)$ of Equations (3.21) and (3.25). In this case two subdomains are chosen for integration in Equation (3.42) – one from 0 *to* $l/2$ and the other from $l/2$ *to* l. The resulting generalized eigenvalue problem as given by Equation (3.10) has the matrices:

$$K = \begin{bmatrix} -1 + \dfrac{1}{\sqrt{2}} & -1 - \dfrac{1}{\sqrt{2}} \\ -\dfrac{1}{\sqrt{2}} & \dfrac{1}{\sqrt{2}} \end{bmatrix},$$

$$M = \begin{bmatrix} -\dfrac{1}{24\pi} + \dfrac{1}{2\pi^2} + \dfrac{16}{\pi^4}\left(-1 + \dfrac{1}{\sqrt{2}}\right) & \dfrac{1}{72\pi} + \dfrac{1}{18\pi^2} + \dfrac{16}{\pi^4}\left(-1 - \dfrac{1}{\sqrt{2}}\right) \\ -\dfrac{7}{24\pi} + \dfrac{3}{2\pi^2} + \dfrac{16}{\pi^4\sqrt{2}} & -\dfrac{7}{24\pi} + \dfrac{1}{6\pi^2} + \dfrac{16}{81\pi^4\sqrt{2}} \end{bmatrix} \tag{3.44}$$

The matrices obtained from the subdomain method are unsymmetric as in the case of the collocation method. The estimate of the two eigenvalues obtained from the above eigenvalue problem is given by $\mu_1 = 3.542$ and $\mu_2 = 27.217$.

3.3.4 Least Squares Method

In the least squares method, as its name indicates, least square minimization is employed on the functional involving square of the residual $R(\cdot)$ with respect to the unknown parameters contained in the approximation $\tilde{u}(\boldsymbol{x}, t)$. Thus, if $I(\cdot) = \int_\Omega R^2(\tilde{u}, f, q_j)d\boldsymbol{x}$, the minimization with respect to the unknown parameter q_i results in

$$\frac{\partial I}{\partial q_i} = 2\int_\Omega \frac{\partial R}{\partial q_j} R(\tilde{u}, f, q_j)d\boldsymbol{x} = \int_\Omega w_j(\boldsymbol{x})R(\tilde{u}, f, q_j)d\boldsymbol{x} = 0 \tag{3.45}$$

Equation (3.45) indicates that $2\frac{\partial R}{\partial q_j}$ acts as the weighting function $w_j(\boldsymbol{x})$ and the integral of the weighted residual is set to zero, typical of a weighted residual method. The method leads to a quadratic eigenvalue problem:

$$\boldsymbol{A}\mu^4 + \boldsymbol{B}\mu^2 + \boldsymbol{C} = \boldsymbol{0} \tag{3.46}$$

The method results in $\boldsymbol{A}$, $\boldsymbol{B}$ and $\boldsymbol{C}$ being symmetric matrices but, however, with increasing number of trial functions, solving the quadratic eigenvalue problem may be more tedious from a computational viewpoint.

Example 3.6: Eigenvalue problem for a beam under flexure by the least squares method
With the beam described in Example 3.1, the least square minimization in Equation (3.45) with a one-parameter approximation ($N = 1$), that is $\tilde{u}(x,t) = Q_1(x)\exp(j\omega t)$ and $Q_1(x)$ as that in Equation (3.21) gives $A = 7.1325$, $B = -176.7082$ and $C = 1186.0700$. Equation (3.46) is a quadratic in μ and the estimate for $\mu = \omega\sqrt{\frac{ml^4}{EI}} = 3.5196$. For a two-parameter solution with $Q_1(x)$ and $Q_2(x)$ same as those in Equations (3.21) and (3.25), the method yields:

$$A = \begin{bmatrix} 7.1325 & -6.1858 \\ -6.1858 & 5.4053 \end{bmatrix}, \quad B = \begin{bmatrix} -176.7082 & 149.5130 \\ 149.5130 & -166.6600 \end{bmatrix} \quad \text{and}$$

$$C = \begin{bmatrix} 1186.0700 & 0.0 \\ 0.0 & 10674.6000 \end{bmatrix} \tag{3.47}$$

The estimates for μ are 3.5161 and 22.2832. Here, the quadratic eigenvalue problem in Equation (3.46) is solved by using its characteristic equation.

3.4 Conclusions

The analytical or classical methods of solving the PDEs presented in the last and present chapters may be insightful and even stimulating, but they are far from being general enough for applications to structural systems of practical interest. Indeed such system could be of complicated shapes and might even involve components/subsystems with incompatible mathematical models (think of a shell structure connected to a column!). Nevertheless, the common feature of finite-dimensional functional discretizations in these methods imparts a nontrivial lesson in that it enables reducing an infinite-dimensional governing PDE to a finite dimensional and numerically tractable system of ODEs. Thus, the eigenfunctions (in the analytical method) and trial functions (of the classical method) could be regarded as 'approximation functions' (or 'shape functions') constructed over the problem domain Ω. This aspect is shared even by the powerful computational techniques such as the FEM, wherein continuous or piecewise continuous polynomials are used to obtain the approximation functions. Polynomial shape functions are specifically of obvious advantages in that they can be constructed rather easily with the help of the monomial basis functions $1, \boldsymbol{x}, \boldsymbol{x}^2, \boldsymbol{x}^3, \ldots .\boldsymbol{x}^n$ and also the integration process (yielding the system matrices) involving polynomials is computationally less taxing. The next chapter provides a glimpse of the elements of the FEM, adapted specifically for structural dynamic systems.

Exercises

1. Estimate the first two natural frequencies of a circular stepped beam (Figure 3.6) in longitudinal vibration by the Raleigh–Ritz and Galerkin methods. Assume that the ration of the cross-sectional areas is $\frac{A_1}{A_2} = 2$ and adopt the trial functions $a_1 \sin\frac{\pi x}{2l}$ and $a_2 \sin\frac{3\pi x}{2l}$.
2. Derive the Rayleigh quotient for the cantilever (Figure 3.7) with a weight $W = Mg$ attached at its centre and find its fundamental frequency. Assume the trial function

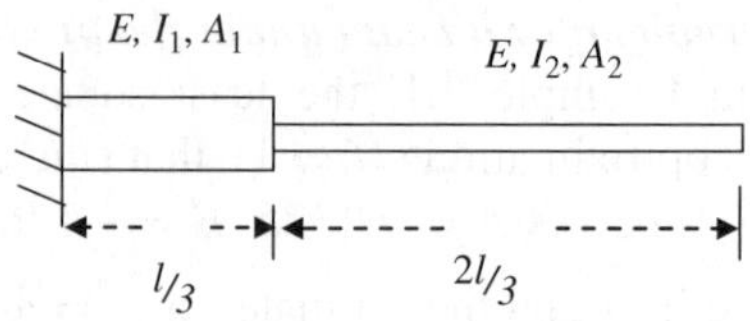

Figure 3.6 Stepped beam with $\frac{A_1}{A_2} = 2$

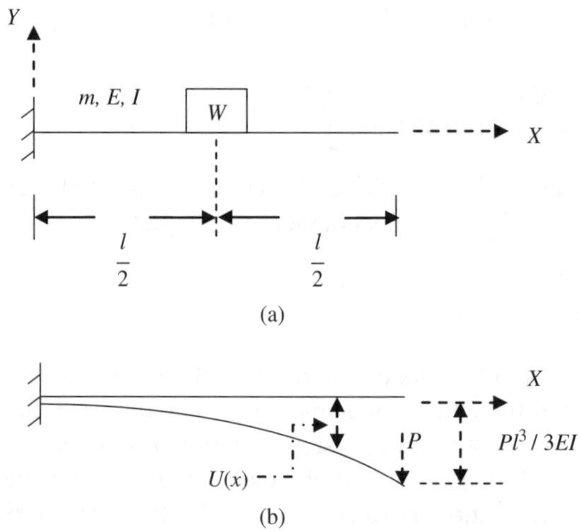

Figure 3.7 Application of the Rayleigh quotient; (a) cantilever with a weight at its centre and (b) deflected beam shape under a tip load P

$U(x) = \frac{Pl^3}{3EI}\frac{3x^2l - x^3}{2l^3}$, which actually is the deflected shape of the beam under a tip load P. (Hint. Kinetic energy T has an additional term due to the weight W in the form: $\frac{W}{2g}\omega^2\{U(x = \frac{l}{2})\}^2$.)

Notations

$\boldsymbol{A_1}, \boldsymbol{A_2}, \boldsymbol{A_3}$	Symmetric matrices (Equation (3.46))
$\mathcal{B}_i, i = 1, 2, 3, 4$	Boundary condition differential operator
$\boldsymbol{F}(t)$	Vector of external forcing functions
g	Acceleration due to gravity
H	Horizontal component of cable tension
$\boldsymbol{L}_1, \boldsymbol{L}_2, \boldsymbol{L}_3$	Differential operators (Equation (3.34))
N	Number of trial functions, size of the matrix
P_1, P_2, P_3	Functions, $1 + \frac{EA}{H}$, $\frac{EA}{H}\frac{dy}{dx}$, $1 + \frac{EA}{H}\left(\frac{dy}{dx}\right)^2$, respectively (Equation 3.37)
$\boldsymbol{Q}$	Eigenvector, $=(Q_1 \ Q_2)^T$
$R(\cdot)$	Residual (in weighted residuals method)

$T(s)$	Static component of cable tension
u(x,t)	Longitudinal displacement of a cable
$\boldsymbol{u}(\boldsymbol{x}, t)$	Displacement field variable ($= \{u_1, u_2, u_3\}^T$)
$\boldsymbol{U}(\boldsymbol{x})$	Assumed trial functions ($= (\boldsymbol{U_1}(\boldsymbol{x}), \boldsymbol{U_2}(\boldsymbol{x}), \boldsymbol{U_3}(\boldsymbol{x}))^T$)
$u(X, t)$	Longitudinal displacement of a cable
$v(X, t)$	Lateral displacement of a cable
$\boldsymbol{w}(\boldsymbol{x})$	Vector of weighting functions
$W_1(x)$, $W_2(x)$	Trial functions
$\hat{w}(\boldsymbol{x}, t)$	Approximate solution
X	Normalised coordinate, x/l
$y(x)$	Cable static profile $\left(= \frac{\beta l}{2}\left(\frac{x}{l} - \frac{x^2}{l^2}\right)\right)$
ω_i^e, ω_i^c	Exact and computed eigenvalues, respectively
β	Cable parameter $\left(= \frac{mgl}{H}\right)$
γ	Cable parameter $\left(= \frac{EA}{H}\right)$
$\tau(s, t)$	Dynamic component of cable tension

References

Finlayson, B.A. and Scriven, L.E. (1966) The method of weighted residuals-a review. *Applied Mechanics Reviews*, **19** (9), 735–748.

Irvine, H.M. and Caughey, T.K. (1974) The linear theory of free vibrations of a suspended cable. *Proceedings of the Royal Society of London*, **A341**, 299–315.

Iyengar, R.N.I. and Rao, G.V. (1988) Free vibration and parametric instability of a laterally loaded cable. *Journal of Sound and Vibration*, **127** (2), 231–243.

Lord Rayleigh, J.W.S. (1945) *The Theory of Sound*, 2nd edn, vols. **1** and **2**, Macmillan (Reprinted by Dover, New York).

Shih, B. and Tadjbakhsh, I.G. (1984) Small-amplitude vibrations of extensible cables. *Journal of Engineering Mechanics Division, ASCE*, **110** (4), 569–576.

Bibliography

Den Hartog, J.P. (1956) *Mechanical Vibrations*, 4th edn, McGraw-Hill, New York.

Finlayson, B.A. (1972) *The Method of Weighted Residuals and Variational Principles*, Academic Press, New York.

Meirovitch, L. (1967) *Methods of Analytical Dynamics*, Macmillan, New York.

Newland, D.E. (1989) *Mechanical Vibration Analysis and Computation*, Dover Publications, Inc., New York.

4

Finite Element Method and Structural Dynamics

4.1 Introduction

The Cauchy equations (Equations (1.35), (1.46) or (1.65)) of motion (partial differential equations (PDEs)) governing the dynamics of a structural system have been derived in Chapters 1 and 2. In differential operator form, the equations of motion are rewritten along with the boundary and initial conditions as:

$$\begin{aligned} \boldsymbol{L}\boldsymbol{u}(\boldsymbol{x},t) &= \boldsymbol{f}(\boldsymbol{x},t), \quad && \boldsymbol{x} \in \Omega \subset \mathbb{R}^n, t \in \mathbb{R}^+ \\ \boldsymbol{B}\boldsymbol{u}(\boldsymbol{x},t) &= \boldsymbol{g}(t), && \boldsymbol{x} \in \delta\boldsymbol{\Omega} \subset \mathbb{R}^{n-1}, t \in \mathbb{R}^+ \text{ and} \\ \boldsymbol{u}(\boldsymbol{x},0) &= \boldsymbol{h}(\boldsymbol{x}), && \boldsymbol{x} \in \Omega \end{aligned} \tag{4.1}$$

For example, if we consider the equation of a longitudinally vibrating rod of length l, acted upon by a distributed axial force $f(x,t)$, $x \in \Omega = [0,l] \in \mathbb{R}$, $\boldsymbol{L} = -\frac{\partial}{\partial x}\left(\alpha(x)\frac{\partial}{\partial x}\right) + m\frac{\partial^2}{\partial t^2}$. The rod is assumed to be fixed at the left end and free at right end. $B_1 = 1$ and $B_2 = \frac{\partial}{\partial x}$ corresponding to the essential (left end) and natural (right end) boundary conditions $u(0,t) = 0$ and $EA\frac{\partial u(l,t)}{\partial x} = 0$ respectively. Here $\alpha(x) = E(x)A(x)$ with E being the Young's modulus, A the cross-sectional area and m the mass per unit length of the rod. If for this vibrating rod, we have f, g and $h \in C(0,l)$ and $\alpha \in C^1(0,l)$, then the solution $u(x,t) \in C^2(0,l)$. These continuity or regularity requirements on α, f and g and hence on $u(x,t)$ may be too strong to be able to solve the equation of motion directly or otherwise. For instance, difficulties may arise if a nonuniformity is present in terms of a jump discontinuity in $\alpha(x)$ in which case $\alpha(x)$ is not even $C(0,l)$. Similar is the case when the external loading function, $f(x,t)$ is prescribed to be concentrated at a single point in $(0,l)$. Practical specification of such systems could therefore be complicated and it is thus desirable to have mathematical models with less stringent smoothness requirements which could then be incorporated into a suitable framework for obtaining approximate numerical solutions. Here, given the unavailability of $u(x,t)$ to start with, it seems convenient to work with a notion of error which is different from $\|u(x,t) - \hat{u}(x,t)\|$ where $\hat{u}(x,t)$ is

Elements of Structural Dynamics: A New Perspective, First Edition. Debasish Roy and G Visweswara Rao.

an approximate solution and $\|\cdot\|$ denotes any valid functional norm. Towards this, one may define the so called residual $R(u) = \boldsymbol{L}u(x,t) - f(x,t)$ and obtain $\hat{u}(x,t)$ in such a manner as to minimise, in a sense, $R(u)$ as in Galerkin's method (Chapter 3). One can then crosscheck if this approach indeed corresponds to the solution providing the extremum (minimum) for some possible norm (such as the norm provided by a strictly positive definite energy functional as in Rayleigh-Ritz method; see Chapter 3). Since these methods use certain integrals (to represent the energy functional or the weighted residual as the case may be) and not a point-wise collocation of the original differential equation, the solutions so obtained are broadly referred to as weak. In these methods, the solution $u(x,t)$ is approximated as a finite series involving known basis functions with unknown time-varying coefficients, valid uniformly over the domain Ω.

The finite element method (FEM), which provides a convenient generalisation, (Zienkiewicz and Cheung 1967, Zienkiewicz 1977, Hughes 1987, Szabo and Babuska, 1991, Bathe 1996), may be viewed as a piece-wise application of either Rayleigh-Ritz or Galerkin methods, given a discretization of the geometric domain Ω into a set of nonoverlapping elements, $\{\Omega_i\}, i = 1, 2, \ldots, N$ such that $\Omega_i \cap \Omega_j = \emptyset$, for $i \neq j$ and $\bigcap_{i=1}^{N} \Omega_i = \Omega$. With this semi-discretization (i.e. discretization of the geometric domain alone and not the time axis), the approximating solution $\hat{u}(x,t)$ is written as:

$$\hat{u}(\boldsymbol{x},t) = \sum_{j=1}^{N} q_j(t)\,\Phi_j(\boldsymbol{x}) \tag{4.2}$$

In the FEM, the basis functions $\{\Phi_j(\boldsymbol{x})\}$ are typically taken as piecewise polynomials over the elements. $q_j(t), j = 1, 2, \ldots N$ are known as the generalized coordinates.

More specifically, one utilises a special form of the basis set such that each element function of the basis is 'interpolating' as explained below. Consider Figure 4.1 where a 1D

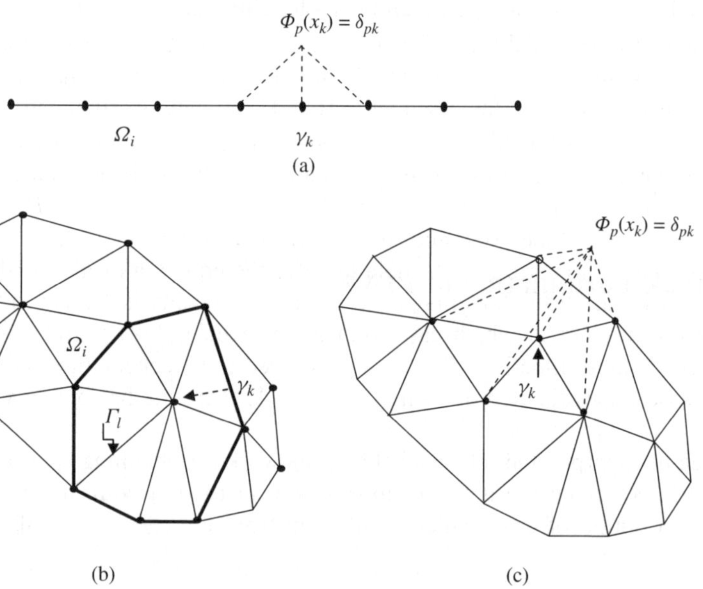

Figure 4.1 (a) 1D domain, (b) & (c) 2D domain. Semi-discretization- elements Ω_i, nodes γ_k and edges Γ_l, basis function (dashed line) $\Phi_p(\boldsymbol{x}_k) = \delta_{pk}$; $\boldsymbol{x}_k$ – coordinates of the node γ_k

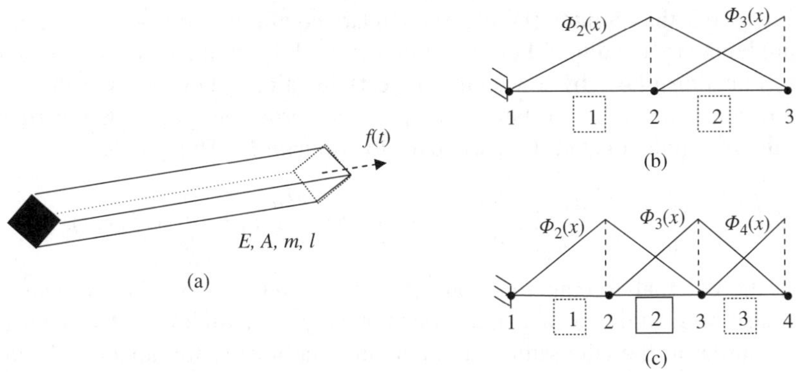

Figure 4.2 (a) Longitudinally vibrating rod; FE models with (b) two elements and (c) three elements

domain Ω is discretized through a set of line elements Ω_j, $j = 1, 2 \ldots, N_{el}$. The points of intersection of the elements give a set of nodes $\{\gamma_k\}$. In the same figure, a 2D domain Ω is shown which is discretized through a set of triangular elements Ω_j, $i = 1, 2 \ldots, N_{el}$. The set of nodes $\{\gamma_k\}$, $k = 1, 2, \ldots N$ appears as the intersection of the element edges Γ_l. The polynomial basis set $\{\Phi_j\}$ is called interpolating if for any node γ_k with coordinates $\boldsymbol{x}_k$, there exists only one basis function $\Phi_k(x)$ such that $\Phi_k(x_k) = 1$ and $\Phi_p(x_k) = 0, \forall p \neq k$. This yields the functional approximation with the special property:

$$u_k(t) := \hat{u}(x_k, t) = \sum_{j=1}^{N} q_j(t)\Phi_j(x_k) = q_k(t) \tag{4.3}$$

Thus, thanks to the FEM based functional discretization with interpolating bases, the generalized coordinates $q_k(t)$ are immediately identifiable with the physically meaningful (displacement) coordinates $u_k(t)$. Also, it is significant that the basis function Φ_k is supported over the subdomain bounded by the darkened edges and that $\Phi_k(x_k \in \Gamma_l) = 0$ for all Γ_l not passing through the node γ_k (see Figure 4.1b as an illustration; proof left to the reader as an exercise). From now onwards, following the convention, the interpolating functions $\Phi_j(x)$ are referred to as shape functions.

4.2 Weak Formulation of PDEs

The objective in deriving a weak formulation to a given PDE is to weaken the smoothness requirement of the functional approximants (describing either the response or the material/geometric properties) with respect to the space variables. Given this, it is also necessary to fix up appropriate function spaces to which these approximants shall belong (as it helps understand the structure and properties of errors). For ease of exposition, we consider the strong form PDE of a vibrating rod:

$$-\frac{\partial}{\partial x}\left(\alpha\frac{\partial u}{\partial x}\right) + m\frac{\partial^2 u}{\partial t^2} = f(x, t)$$

$$u(0, t) = 0, \quad \alpha\frac{\partial u(l, t)}{\partial x} = 0 \tag{4.4}$$

Here $u(0,t) = 0$ is the essential (Dirchlet) boundary condition and $\alpha\frac{\partial u(l,t)}{\partial x} = 0$, the natural (Neumann) boundary condition. For obtaining the weak form, the procedure is to multiply the PDE (Equation (4.4)) by a weight (or test) function $\psi(x)$ (see the discussion on weighted residual methods in Chapter 3) whose properties are yet to be ascertained and integrate the resulting equation by parts over the domain Ω. This yields:

$$-\psi\left(\alpha\frac{\partial u}{\partial x}\right)\Bigg|_0^l + \int_0^l \frac{d\psi}{dx}\left(\alpha\frac{\partial u}{\partial x}\right)dx + \int_0^l \psi m\frac{\partial^2 u}{\partial t^2}dx = \int_0^l \psi f(x,t)dx \tag{4.5}$$

$f(x,t)$ is the external forcing function distributed over the rod. The weight function $\psi(x)$ can also be regarded as a virtual (kinematically admissible) displacement imposed on $u(x,t)$ similar to the admissible variations in a variational formulation. Accordingly, $\psi(x)$ must vanish over the essential part of the boundary ($x = 0$ in the present case). Moreover, given that the Neumann boundary condition $\alpha\frac{\partial u(l,t)}{\partial x} = 0$, the first term in Equation (4.5) vanishes. Further, we also assume that the forcing function $f(x,t)$ can be written in a variable separable form as $P(t)F(x)$ which may be valid for most practical loading conditions on structural systems. Now, substituting the approximate solution (Equation (4.2)) for $u(x,t)$ in Equation (4.5) gives:

$$\sum_{j=1}^{N} q_j(t)\int_0^l \frac{d\psi}{dx}\left(\alpha\frac{d\Phi_j}{dx}\right)dx + \ddot{q}_j(t)\int_0^l \psi m\Phi_j dx = P(t)\int_0^l \psi F(x)dx$$

$$\Rightarrow \sum_{j=1}^{N}\left\{\mathcal{K}(\psi,\Phi_j)q_j + \mathcal{M}(\psi,\Phi_j)\ddot{q}_j\right\} = P\ell(\psi) \tag{4.6}$$

$\mathcal{K}(\psi,\Phi_j)$ and $\mathcal{M}(\psi,\Phi_j)$ are bilinear forms and $\ell(\psi)$, the linear form (Appendix D). Equation (4.6) is the weak formulation for the PDE Equation (4.4) and is in an integral form. Note that in case Neumann boundary conditions are inhomogeneous, they directly appear in the weak formulation following integration by parts. For instance, in the example of the vibrating rod, denote $\alpha(l)\frac{\partial u(l,t)}{\partial x} = \mu(t)$. Substitution of the Neumann boundary condition in Equation (4.5) along with $\psi\,(0) = 0$ yields:

$$\sum_{j=1}^{N} q_j(t)\left\{\int_0^l \frac{d\psi}{dx}\left(\alpha\frac{d\Phi_j}{dx}\right)dx\right\} + \ddot{q}_j(t)\int_0^l \psi m\Phi_j dx = P(t)\int_0^l \psi F(x)dx + \psi\,(l)\,\mu(t) \tag{4.7}$$

Comparing Equation (4.7) with Equation (4.6), we find that the bilinear form $K(\psi,\Phi_j)$ in Equation (4.7) acquires an additional term due to this nonzero Neumann boundary condition. Before we proceed further to discuss the above weak form, it is revealing to notice the equivalence of this formulation to those involving extremization of a (energy-like) functional and weighted residuals. In Chapter 3, Rayleigh-Ritz method is described in obtaining an approximate solution via minimization of the energy functional. The weak form in Equation (4.6) can also be interpreted as a minimization of a functional. Consider ψ as a variation over Φ_j and thus, with $\psi = \delta\Phi_j$, each term in Equation (4.6) takes the from:

$$\int_0^l \psi'\alpha\Phi_j' dx = \int_0^l \alpha\Phi_j' d(\delta\Phi_j') = \delta\int_0^l \alpha\Phi_j' d\Phi_j' = \delta\left(\frac{1}{2}\int_0^l \alpha\Phi_j'^2 dx\right)$$

$$\int_0^l m\psi\Phi_j dx = \int_0^l m\Phi_j\delta\Phi_j dx = \delta\int_0^l m\Phi_j^2 dx$$

$$\int_0^l F\psi dx = \delta\int_0^l F\Phi_j dx \tag{4.8}$$

Thus, the weak formulation is equivalent to finding Φ_j such that it satisfies the necessary condition $\delta I(\Phi_j) = 0$ for an extremum of a functional $I(\Phi_j)$ where:

$$I(\Phi_j) = \sum_j^N \left\{\frac{1}{2}q_j\int_0^l \alpha\Phi_j'^2 dx + \ddot{q}_j\int_0^l m\Phi_j^2 dx\right\} - P\int_0^l F\Phi_j dx \tag{4.9}$$

However, derivation of a convex, positive definite functional (Appendix D) as above may not be possible for all boundary value problems of interest in structural dynamics. Moreover, in such classical methods as the Rayleigh-Ritz, selection of $\Phi_j(x)$ (admissible functions over the whole domain Ω) is also a nontrivial task in that the choice is restricted to those functions that conform to the essential boundary conditions.

4.2.1 Well-Posedness of the Weak Form

To prove that the weak form is well posed, it suffices to show the bilinear and linear forms in Equation (4.6) are bounded (in the sense clarified below). To this end, let us start with the assumption that F, ψ and Φ_j, $j = 1, 2, .., N$ belong to a linear vector space $\boldsymbol{V}$. Since $\ell(\psi) = \int_0^l \psi F(x)dx$ is an inner product on $\boldsymbol{V}$, we have by Cauchy-Schwartz inequality (Appendix D):

$$|(F, \psi)| \leq \|F\|\,\|\psi\|, \forall\psi \in \boldsymbol{V} \tag{4.10}$$

If $\boldsymbol{V} \subset L_2(0, l)$, then F and ψ are square integrable, that is $\|F\|^2 = \int_0^l F^2 dx < \infty$ and $\|\psi\|^2 = \int_0^l \psi^2 dx < \infty$ and hence $\ell(\psi)$ is bounded. $\mathcal{K}(\psi, \Phi_j)$ and $\mathcal{M}(\psi, \Phi_j)$ are symmetric bilinear forms since $\mathcal{K}(\psi, \Phi_j) = \mathcal{K}(\Phi_j, \psi)$ and $\mathcal{M}(\psi, \Phi_j) = \mathcal{M}(\Phi_j, \psi)$. Also, $\mathcal{K}(\psi, \Phi_j)$ is positive semi-definite (Appendix D) since, for $\alpha(x) \geq 0, x \in (0, l)$:

$$\mathcal{K}(\psi, \psi) = \int_0^l \alpha\psi'^2 dx \geq 0 \tag{4.11}$$

The equality sign in Equation (4.11) arises for unrestrained systems and, under adequate restraints preventing rigid body motion, $\mathcal{K}(\psi, \Phi_j)$ is a positive definite bilinear form. Also $\mathcal{K}(\psi, \psi) = 0 \Longleftrightarrow \psi = 0$ (almost everywhere) and hence it is an inner product on V. Similarly $M(\psi, \Phi_i)$ is positive definite and an inner product since:

$$\mathcal{M}(\psi, \psi) = \int_0^l \psi^2 dx > 0 \text{ and } \mathcal{M}(\psi, \psi) = 0 \Longleftrightarrow \psi = 0 \text{ (a.e.)} \tag{4.12}$$

Hence, $\mathcal{K}(\psi, \Phi_j)$ and $\mathcal{M}(\psi, \Phi_j)$ satisfy the Cauchy-Schwartz inequality:

$$|\mathcal{K}(\psi, \Phi_j)| = \left|\int_0^l \psi'\left(\alpha\Phi'\right)dx\right| = \left|\left(\psi', \alpha\Phi_j'\right)\right| \leq \max_{x\in[0,l]} \alpha(x)\left\|\psi'\right\|\left\|\Phi_j'\right\|, \quad \forall\psi, \Phi_j \in \boldsymbol{V} \tag{4.13a}$$

and

$$\left|M(\psi, \Phi_j)\right| = \left|\int_0^l \psi m \Phi_j dx\right| = \left|\psi, m\Phi_j\right| \leq \max_{x\in[0,l]} m(x) \left\|\psi\right\| \left\|\Phi_j\right\|, \quad \forall \psi, \Phi_j \in \boldsymbol{V} \tag{4.13b}$$

From the above, we observe that if (i) $\alpha(x)$ and $m(x)$ are bounded in $(0, l)$ and (ii) $\|\psi'\|, \|\Phi_j'\| < \infty$, the two bilinear forms above are bounded. Thus, combining Equations (4.10) and (4.13), we find that for the weak form to be bounded, it is required that $\boldsymbol{V} \subset H^1(0, l)$, a Sobolev space which is a set of all functions in $L_2(0, l)$ with their first-order derivatives (where derivatives are defined in the weak sense, as explained in Appendix D) also in $L_2(0, l)$. These continuity requirements on f, α and Φ_j and in turn on $u(x, t)$ are less stringent compared to those demanded by the original PDE. From henceforth, we use 'test space' and 'solution space' for the spaces to which the weight function ψ and the shape function Φ_j respectively belong. For the example of vibrating rod, since ψ shall vanish at the boundary, the test space is given by $H_0^1(0, l) = \{v \in H^1(0, l) : v(0) = 0\}$. Given that $u(0, t) = 0$, the solution space of Φ_j also is $H_0^1(0, l)$. In case of nonzero essential boundary condition (for example if $u(0, t) = U_0$), we can assume that $u(x, t) = \bar{u}(x, t) + U_0$ such that $\bar{u}(0, t) = 0$ and both the test and solution spaces are again $H_0^1(0, l)$. Here, note that $H_0^1(0, l) \subset H^1(0, l)$.

Extension of the procedure to derive a corresponding weak form for 2D and 3D problems and identifying the appropriate functional spaces is straightforward. For example, consider the Poisson equation $-\Delta u + m\frac{\partial^2 u}{\partial t^2} = f(\boldsymbol{x}, t)$ corresponding to a vibrating membrane under transverse load $f(\boldsymbol{x}, t)$, where we have $\Delta u = \frac{\partial^2 u}{\partial x^2} + \frac{\partial^2 u}{\partial y^2}$ and Ω, a bounded domain in the plane $\mathbb{R}^2 = \boldsymbol{x} = (\boldsymbol{x}, \boldsymbol{y})$: $\boldsymbol{x}, \boldsymbol{y} \in \mathbb{R}$ with boundary Γ. Without loss of generality, if it is assumed that $u = 0$ on Γ, the essential boundary, we obtain the following weak form of equation similar to Equation (4.6). However, in this case, whilst the forms of $\mathcal{M}(\psi, \Phi_j)$ and $\ell(\psi)$ remain the same as in Equation (4.6), the bilinear form $\mathcal{K}(\psi, \Phi_j)$ is given by:

$$\mathcal{K}(\psi, \Phi_j) = \int_\Omega \left(\frac{d\psi}{dx}\frac{d\Phi_j}{dx} + \frac{d\psi}{dy}\frac{d\Phi_j}{dy}\right) d\boldsymbol{x} \tag{4.14}$$

In this case also, ψ and Φ_j shall belong to $H_0^1(\Omega)$ and $F(\boldsymbol{x})$ shall belong to $L_2(\Omega)$ to satisfy the regularity requirements which are again, of course, less stringent than those of the original PDE.

4.2.2 Uniqueness and Stability of Solution to Weak Form

Issues of uniqueness and stability (Brenner and Scott, 1994) of the solution to the weak form (henceforth referred to as the 'weak solution'), require reference to Riesz representation theorem. The theorem stipulates the sufficient conditions for the unique weak solution. Towards this, write the weak form generically as $\chi(v, u) = \Lambda(v)$ where $\chi(v, u) \left(= \int_\Omega v'\sigma u' dx\right)$ is a bilinear form and $\Lambda(v) \left(= \int_\Omega f v dx\right)$, a linear form. Comparison with the weak form in Equation (4.6) (which is typical of a vibrating structural system) shows that $\chi(v, u)$ is analogous to the sum $\sum_{j=1}^{N} \mathcal{K}(\psi, \Phi_j q_j(t))$. By linearity of bilinear form, we have $\sum_{j=1}^{N} \mathcal{K}(\psi, \Phi_j q_j(t)) = \mathcal{K}(\psi, \sum_{j=1}^{N} \Phi_j q_j(t)) = \hat{\mathcal{K}}(\psi, \hat{\Phi}), say.$ The second bilinear form $\mathcal{M}(\psi, \Phi_j)$ in Equation (4.6) being a coefficient of the inertia term can be treated as a pseudo forcing (for a given t) and transferred to the right-hand side of the equation. The semi-discretization also allows us to focus on the state of the system as a snapshot at, say, the current time t, and thus the additional forcing term combined with $\ell(\psi)$ can be assumed to be the new linear form $\Lambda(\psi)$. Thus, the weak form for a vibrating structure becomes:

$$\hat{K}(\psi, \hat{\Phi}) = \Lambda(\psi) \tag{4.15}$$

Riesz representation theorem states that the above weak form possesses a unique solution provided it satisfies three conditions:

$$\hat{\mathcal{K}}(\psi, \psi) \geq k\|\psi\|_V^2 \quad \forall\psi \in \boldsymbol{V}, \tag{4.16a}$$

$$|\hat{\mathcal{K}}(\psi, \hat{\Phi})| \leq C\|\psi\|_V \left\|\hat{\Phi}\right\|_V \quad \forall\psi, \hat{\Phi} \in \boldsymbol{V} \text{ and} \tag{4.16b}$$

$$\Lambda(\psi) \leq D\|\psi\|_V \quad \forall\psi \in \boldsymbol{V} \tag{4.16c}$$

Here k, C and D and are strictly positive constants. The first two conditions in Equations (4.16a and b) are related to the coercivity (ellipticity) and continuity of the $\hat{\mathcal{K}}(\psi, \hat{\Phi})$ respectively and the last one to the continuity of $\Lambda(\psi)$. Assuming that $\bar{\alpha} = \max_{x\in(0,l)} \alpha(x)$, we detail in what follows the verification of the above inequalities for a given weak form. In doing so, we restrict ourselves to $x \in R$ for expositional simplicity. Thus, consider the coercivity:

$$\hat{\mathcal{K}}(\psi, \psi) = \int_0^l \alpha(\psi')^2 dx \geq \bar{\alpha}\|\psi'\|^2 \tag{4.17}$$

Since $\psi \in H_0^1(0, l)$ we can write $\psi(x) = \int_0^x \psi'(x)dx = (1, \psi)$, which represents a bilinear form $(u, v) := \int_0^x uvdx$ on $H^1(0, x)$. We have from Cauchy-Schwartz inequality:

$$|\psi| = |(1, \psi)| \leq \left(\int_0^x 1^2 dx \int_0^x \left(\psi'(x)\right)^2 dx\right)^{1/2} = \sqrt{x}\left(\int_0^x \left(\psi'(x)\right)^2 dx\right)^{1/2}$$

$$\Rightarrow \|\psi\|^2 = \int_0^l |\psi|^2 dx \leq \int_0^l x\left(\int_0^x \left(\psi'(z)\right)^2 dz\right)dx$$

$$\leq \left(\int_0^l x\,dx\right)\left(\int_0^l \left(\psi'(z)\right)^2 dz\right) = \frac{1}{2}\|\psi'\|^2$$

$$\Rightarrow \|\psi'\|^2 \geq 2\|\psi\|^2 \tag{4.18}$$

In view of the above result (known as *Friedrich*'s inequality), Equation (4.17) leads to:

$$\hat{\mathcal{K}}(\psi, \psi) \geq \bar{\alpha}\left\|\psi'\right\|^2 = \bar{\alpha}\left(\frac{2}{3}\left\|\psi'\right\|^2 + \frac{1}{3}\left\|\psi'\right\|^2\right) \geq \frac{2}{3}\bar{\alpha}\left(\|\psi\|^2 + \left\|\psi'\right\|^2\right) = \frac{2}{3}\bar{\alpha}\left\|\psi\right\|_{H^1(0,l)}^2$$

$$\Rightarrow \hat{\mathcal{K}}(\psi, \psi) \geq \frac{2}{3}\bar{\alpha}\|\psi\|_{H^1(0,l)}^2 \tag{4.19}$$

Thus, the coercivity condition in Equation (4.16a) is satisfied with $k = 2\bar{\alpha}/3$. The continuity of $\hat{\mathcal{K}}(\psi, \hat{\Phi})$ in Equation (4.16b) follows from the Cauchy-Schwartz inequality:

$$\left|\hat{\mathcal{K}}(\psi, \hat{\Phi})\right| = \left|\int_0^l \alpha\psi'\hat{\Phi}'dx\right| = \left|\alpha\psi', \hat{\Phi}'\right| \leq \left\|\alpha\psi'\right\|\left\|\hat{\Phi}'\right\|$$

$$\leq \bar{\alpha}\left\|\psi'\right\|\left\|\hat{\Phi}'\right\| \leq \bar{\alpha}\|\psi\|_{H^1(0,l)}\left\|\hat{\Phi}\right\|_{H^1(0,l)} \tag{4.20}$$

The last step in the above equation is due to the fact that for any $u \in H^1(\Omega)$, $\|u\|^2_{H^1(\Omega)} = \|u\|^2 + \|u'\|^2$. The continuity of $\Lambda(\psi)$ also follows by steps similar to the above, that is:

$$|\Lambda(\psi)| = \left|\int_0^l F\psi dx\right| = |F, \psi| \leq \|F\| \, \|\psi\| \leq \|F\| \, \|\psi\|_{H^1(0,l)} \tag{4.21}$$

To prove uniqueness of the weak solution, we start with the hypothesis that two solutions $\hat{\Phi}_1$ and $\hat{\Phi}_2$ are possible to $\hat{\mathcal{K}}(\psi, \hat{\Phi}) = \Lambda(\psi)$. Then,

$$\mathcal{K}\left(\psi, \hat{\Phi}_1\right) = \Lambda(\psi) \text{ and } \mathcal{K}\left(\psi, \hat{\Phi}_2\right) = \Lambda(\psi) \tag{4.22}$$

Subtracting one from the other and using the linearity of the bilinear form, we obtain:

$$\left[\mathcal{K}\left(\psi, \hat{\Phi}_1\right) - \mathcal{K}\left(\psi, \hat{\Phi}_2\right)\right] = 0 \Rightarrow \mathcal{K}\left(\psi, \hat{\Phi}_1 - \hat{\Phi}_2\right) = 0, \quad \psi \in \boldsymbol{V} \tag{4.23}$$

Now, since $\boldsymbol{V}$ is a linear space, $\psi(x)$ can be $\hat{\Phi}_1 - \hat{\Phi}_2$ and Equation (4.23) gives:

$$\mathcal{K}\left(\hat{\Phi}_1 - \hat{\Phi}_2, \hat{\Phi}_1 - \hat{\Phi}_2\right) = 0 \Rightarrow \frac{2}{3}\bar{\alpha} \left\|\hat{\Phi}_1 - \hat{\Phi}_2\right\|^2_{H^1(0.l)} \leq 0 \quad \text{(from Equation (4.19))}$$

$$\Rightarrow \left\|\hat{\Phi}_1 - \hat{\Phi}_2\right\|_{H^1(0.l)} \leq 0 \tag{4.24}$$

Being a norm, $\left\|\hat{\Phi}_1 - \hat{\Phi}_2\right\|_{H^1(0.l)}$ is always greater than or equal to zero. Thus, the result in Equation (4.24) is a contradiction to the hypothesis and we must have that $\hat{\Phi}_1 = \hat{\Phi}_2$, thereby proving uniqueness.

The stability (upper bound to the norm) of the weak solution depends on the coercivity of the bilinear form and continuity of the linear form and is obtained as follows. With $\psi = \Phi_j$ for each j in the weak form of equation, we have:

$$\Rightarrow \frac{2}{3}\bar{\alpha}\|\psi\|^2_{H^1(0.l)} \leq \hat{\mathcal{K}}(\psi, \psi) = \Lambda(\psi) \leq \|F\| \, \|\psi\|_{H^1(0.l)}$$

$$\Rightarrow \psi_{H^1(0.l)} \leq \frac{3}{2\bar{\alpha}} \|F\| \tag{4.25}$$

In FEM, we seek for an approximate solution $\hat{u}(\boldsymbol{x}, t) = \sum_{j=1}^{N} q_j(t)\Phi_j(\boldsymbol{x})$ via the weak form $\sum_{j=1}^{N}\{\mathcal{K}(\psi, \Phi_j)q_j + \mathcal{M}(\psi, \Phi_j)\ddot{q}_j\} = P\ell(\psi)$. In the symmetric formulation (also called the Bubnov-Galerkin method), the weight function $\psi(x)$, being kinematically admissible and arbitrary otherwise, is taken to be Φ_j except when the node j is on the essential boundary. With this choice, the weak form reduces to a system of linear coupled ODEs which can be written in a matrix form:

$$\boldsymbol{M\ddot{q}} + \boldsymbol{Kq} = \boldsymbol{\alpha} P(t) \ \ \boldsymbol{\alpha} \in \mathbb{R}^N \tag{4.26}$$

Here $\boldsymbol{M}$ and $\boldsymbol{K}$ are the (symmetric) mass and stiffness matrices in $\mathbb{R}^{N\times N}$ that result from the bilinear forms $\mathcal{M}\left(\Phi_k, \Phi_j\right) = \int_0^l \Phi_k(m\Phi_j)dx$ and $\mathcal{K}(\Phi_k, \Phi_j) = \int_0^l \Phi'_k(\sigma\Phi'_j)dx$ respectively for $j, k = 1, 2, \ldots, N$. $\boldsymbol{q}(t)$ is the vector of unknown generalized coordinates at the nodes. $\boldsymbol{\alpha}$ is the vector of real constants generated by the linear form $\ell(\Phi_j) = \int_\Omega F(x)\Phi_j dx$ and it multiplies the time varying part $P(t)$ of the external loading. Note that, for a set of constrained nodes γ_k on the boundary Γ, $\Phi_j(x)$ may equivalently be selected as zero and evaluation of mass and stiffness matrix elements corresponding to the *dof*s of these nodes can be avoided.

4.2.3 Numerical Integration by Gauss Quadrature

Numerical implementation of the semi-discretization via the FEM requires evaluations of the domain integrals appearing in the weak form (i.e. those defining, for instance, the elements of the stiffness/mass matrices and the force vector; see Equation (4.6) or (4.7). Since both the test and trial functions in the FEM are typically Lagrange polynomials, these integrals over 1D spatial domains ($n = 1$) could be evaluated in closed form. However, for higher dimensional domains ($n \geq 2$) with involved mesh geometry, analytical evaluations of the integrals are often infeasible, which calls for an automated numerical route to computing them. Gauss quadrature provides for one of the eminent methods, wherein the integral is approximated via a weighted sum of the integrand evaluated at the so-called quadrature points. In order to elucidate further, consider the following 1D integral $\int_a^b \eta(x)dx$. If $\eta(x)$ is a continuous function over the closed interval $[a, b], b > a$ then, by Stone-Weierstrass theorem (Rudin, 1976), $\eta(x)$ can be uniformly approximated to any desired accuracy by polynomial functions. Using Lagrangian polynomials, for instance, such an approximation may take the form:

$$\eta(x) \approx P_\kappa(x) = \sum_{j=1}^{\kappa} H_j(x)\eta_j, \ \ \eta_j = \eta(x_j) \tag{4.27}$$

Here $H_j(x) = \frac{(x-x_1)(x-x_2)....(x-x_{j-1})(x-x_{j+1})....(x-x_\kappa)}{(x_j-x_1)(x_j-x_1)....(x_j-x_1)(x_j-x_1)....(x_j-x_\kappa)}$ is the κth order Lagrange polynomial function constructed over a partition $(a = x_1 < x_2 \cdots < x_\kappa = b)$ of $[a, b]$ and $\kappa \geq 1$ is an integer. Note that, over this partition, $H_j(x)$ is interpolating, that is $H_j(x) = \delta_{j\kappa}$, the Kronecker delta with $j, k \in [1, \kappa]$. Using this approximant $P_\kappa(x)$, we may thus write:

$$\int_a^b \eta(x)dx \approx \int_a^b P_\kappa(x)dx = \sum_{j=1}^{\kappa} \left(\int_a^b H_j(x)dx \right) \eta_j \tag{4.28}$$

Defining the set of weights $\{W_j = \int_a^b H_j(x)dx\}$, the original integral may thus be approximated through the following generic numerical quadrature form:

$$\int_a^b \eta(x)dx \approx \sum_{j=1}^{\kappa} W_j \eta_j \tag{4.29}$$

with higher accuracy of approximation for increasing κ. In deriving the above quadrature formula, the quadrature points were chosen a priori and the weights were derived subsequent to this choice. Consequently, if $(\kappa + 1)$ (the number of quadrature points) increases by 1, the formal order of the quadrature based approximation of the integral also improves similarly. In the Gauss quadrature, however, both the quadrature point locations and the associated weights are treated as unknown quantities. This enables a superior numerical accuracy as well as improved convergence of the quadrature scheme. More specifically, first assume that the interval $[a, b]$ is suitably scaled to $[-1, 1]$ through a change of variables so that the new integral takes the generic form $\int_{-1}^{+1} \eta(x)dx$. Now consider the following approximation of the integrand $\eta(x)$ via a $(2\kappa - 1)^{th}$ order polynomial:

$$\eta(x) \approx G_\kappa(x) = \sum_{j=1}^{\kappa} H_j(x)\eta_j + \upsilon(x) \sum_{j=1}^{\kappa} x^{j-1} a_j \tag{4.30}$$

Here a_k is a set of arbitrary coefficients and the polynomial $\upsilon(x)$ is given by:

$$\upsilon(x) = (x - x_1)(x - x_2)\ldots(x - x_\kappa) \tag{4.31}$$

Also note that the first summand on the RHS of Equation (4.30) is nothing but $P_\kappa(x)$, as in Equation (4.26). Now, at the sampling points $\{x_j | j \in [1, \kappa]\}$, we have $\upsilon(x) = 0$. Hence:

$$\eta(x_j) = G_\kappa(x_j), \ \ j \in [1, \kappa] \tag{4.32}$$

In Gauss quadrature, the quadrature point locations $\{x_j | j \in [1, \kappa]\}$ satisfy the following identity:

$$\int_{-1}^{1} \upsilon(x) x^{j-1} dx = 0 \tag{4.33}$$

Using the above, one may write:

$$\begin{aligned}\int_{-1}^{1} \eta(x)dx \approx \int_{-1}^{1} G_\kappa(x)dx &= \int_{-1}^{1} \sum_{j=1}^{\kappa} H_j(x)\eta_j dx + \sum_{j=1}^{\kappa} a_j \int_{-1}^{1} \upsilon(x) x^{j-1} dx \\ &= \int_{-1}^{1} \sum_{j=1}^{\kappa} H_j(x)\eta_j dx = \sum_{j=1}^{\kappa} W_j \eta_j \end{aligned} \tag{4.34}$$

This implies that the original integral is approximated as one of $(2\kappa - 1)^{th}$ order polynomial interpolating function constructed based on just κ sample points.

Now, consider a few simple specific cases. For $\kappa = 1$, we have:

$$\int_{-1}^{1} (x - x_1)x^{1-1} dx = 2x_1 = 0 \Rightarrow x_1 = 0 \tag{4.35a}$$

The weight W_1 at the quadrature point $x_1 = 0$ is given by:

$$W_1 = \int_{-1}^{1} H_1(x)dx = \int_{-1}^{1} 1dx = 2 \tag{4.35b}$$

Similarly, for $\kappa = 2$, we obtain the following equations for the quadrature points:

$$\int_{-1}^{1} (x - x_1)(x - x_2)\, x^{1-1} dx = \frac{2}{3} + 2x_1x_2 = 0$$

$$\text{and} \quad \int_{-1}^{1} (x - x_1)(x - x_2)\, x^{2-1} dx = \frac{2}{3}(x_1 + x_2) = 0 \tag{4.36a,b}$$

Solving for the roots of the last two algebraic equations yields the quadrature points as:

$$x_1 = -\sqrt{\frac{1}{3}}, \quad x_2 = \sqrt{\frac{1}{3}} \tag{4.37}$$

Integrating the associated Lagrange polynomials provides the pair of weights as:

$$W_1 = \int_{-1}^{1} H_1(x)dx = \int_{-1}^{1} \frac{x - \sqrt{\frac{1}{3}}}{-\sqrt{\frac{1}{3}} - \sqrt{\frac{1}{3}}} dx = 1$$

$$W_2 = \int_{-1}^{1} H_2(x)dx = \int_{-1}^{1} \frac{x + \sqrt{\frac{1}{3}}}{\sqrt{\frac{1}{3}} + \sqrt{\frac{1}{3}}} dx = \mathbf{1} \tag{4.38a,b}$$

Similar derivations for $\kappa = 3$, provide the quadrature points and weights as:

$$\begin{aligned} x_1 &= -\sqrt{\frac{3}{5}}, \quad x_2 = 0, \quad x_3 = \sqrt{\frac{3}{5}}, \\ W_1 &= W_3 = \frac{5}{9}, \quad W_2 = \frac{8}{9} \end{aligned} \tag{4.39a,b}$$

A similar extension of the same logic to the canonical 2D integral $\int_{-1}^{1} \int_{-1}^{1} \eta\,(x, y)\, dxdy$ gives:

$$\int_{-1}^{1} \int_{-1}^{1} \eta\,(x, y)\, dxdy = \sum_{(i,j)\in[1,\kappa]\times[1,\kappa]} W_i W_j \eta\left(x_i, y_j\right) \tag{4.40}$$

In deriving the above Gauss quadrature rule, the 1D formula may be used twice, first holding y fixed whilst applying the 1D formula over variable x and then doing just the other way round. Indeed, the 3D Gauss quadrature formula may also be derived to be in a precisely similar form as in Equation (4.40):

$$\int_{-1}^{1} \int_{-1}^{1} \int_{-1}^{1} \eta\,(x, y, z)\, dxdydz \approx \sum_{(i,j,k)\in[1,\kappa]\times[1,\kappa]\times[1,\kappa]} W_i W_j W_k \eta\left(x_i, y_j, z_k\right) \tag{4.41}$$

Such formulae for Gauss quadrature with different κ and over different spatial dimensions, along with the quadrature point locations and weights arranged in tabular forms, are available in Abramowitz and Stegun (1972).

Example 4.1: Axially vibrating rod: free vibration solution using FEM
Consider an axially vibrating uniform rod (Figure 4.2) of length l, area of cross section A and mass density m. Let E be the Young's modulus of elasticity (also assumed uniformly constant). In order to illustrate the semi-discretization, two simple cases are shown in Figure 4.2b,c). Here each element has two nodes with only 1 *dof* (the axial displacement).

- **Case 1.** The semi-discretization involves two elements $= 2$ and 3 nodes. Since the essential boundary is at node 1, we may choose the global shape function, $\Phi_1(x) = 0$, $x \in [0, l]$. $\Phi_2(x)$ and $\Phi_3(x)$ are shown in Figure 4.3.
- **Case 2.** The semi-discretization in this case involves three elements $= 3$ and 4 nodes. Since in this case also, the essential boundary is at node 1,the global shape function at this node. $\Phi_1(x) = 0, x \in [0, l]$, $\Phi_2(x)$, $\Phi_3(x)$ and $\Phi_4(x)$ are shown in Figure 4.4.

Determination of $\boldsymbol{M}$ and $\boldsymbol{K}$ matrices is the next step. With known global shape functions, (Figures 4.3 and 4.4) the matrices are evaluated for the two cases of semi-discretization and given in Table 4.1. In free vibration, we seek nontrivial solutions (Chapters 2 and 3) of the homogeneous equation:

$$\boldsymbol{M}\ddot{\boldsymbol{q}} + \boldsymbol{K}\boldsymbol{q} = 0 \tag{4.42}$$

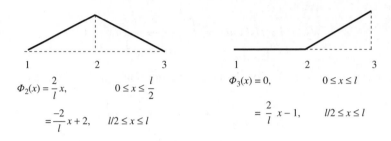

$$\Phi_2(x) = \frac{2}{l}x, \qquad 0 \le x \le \frac{l}{2}$$
$$= \frac{-2}{l}x + 2, \qquad l/2 \le x \le l$$

$$\Phi_3(x) = 0, \qquad 0 \le x \le l$$
$$= \frac{2}{l}x - 1, \qquad l/2 \le x \le l$$

Figure 4.3 Global shape functions $\Phi_2(x)$ and $\Phi_3(x)$ for Example 4.1 with two elements

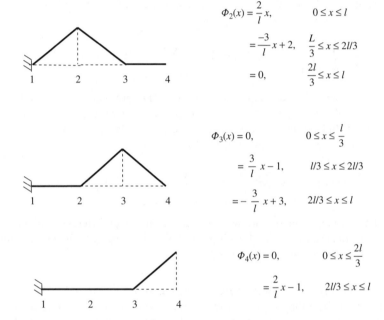

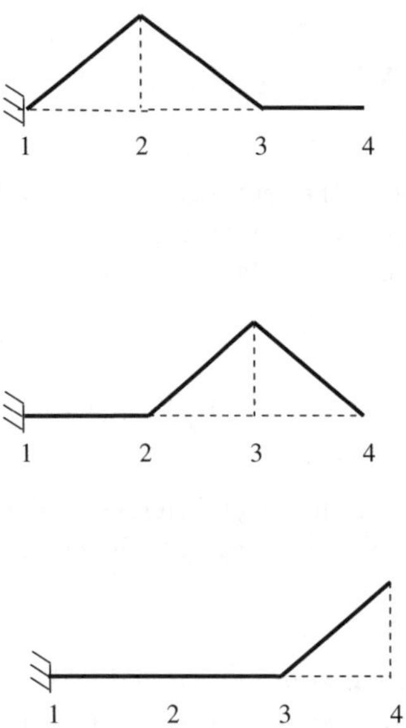

$$\Phi_2(x) = \frac{2}{l}x, \qquad 0 \le x \le l$$
$$= \frac{-3}{l}x + 2, \qquad \frac{L}{3} \le x \le 2l/3$$
$$= 0, \qquad \frac{2l}{3} \le x \le l$$

$$\Phi_3(x) = 0, \qquad 0 \le x \le \frac{l}{3}$$
$$= \frac{3}{l}x - 1, \qquad l/3 \le x \le 2l/3$$
$$= -\frac{3}{l}x + 3, \qquad 2l/3 \le x \le l$$

$$\Phi_4(x) = 0, \qquad 0 \le x \le \frac{2l}{3}$$
$$= \frac{2}{l}x - 1, \qquad 2l/3 \le x \le l$$

Figure 4.4 Global shape functions $\Phi_2(x)$, $\Phi_3(x)$ and $\Phi_4(x)$ for the Example 4.1 with three elements

Table 4.1 Mass and stiffness matrices for longitudinally vibrating rod in Example 4.1 for the two cases of discretization

Case 1. Semi-discretization with 2 elements	Case 2. Semi-discretization with 3 elements
$\boldsymbol{M} = ml\begin{bmatrix} \frac{1}{3} & \frac{1}{12} \\ \frac{1}{12} & \frac{1}{6} \end{bmatrix}$	$\boldsymbol{M} = ml\begin{bmatrix} \frac{2}{9} & \frac{1}{18} & 0 \\ \frac{1}{18} & \frac{2}{9} & \frac{1}{18} \\ 0 & \frac{1}{18} & \frac{1}{9} \end{bmatrix}$
$\boldsymbol{K} = \frac{EA}{l}\begin{bmatrix} 4 & -2 \\ -2 & 2 \end{bmatrix}$	$\boldsymbol{K} = \frac{EA}{l}\begin{bmatrix} 6 & -3 & 0 \\ -3 & 6 & -3 \\ 0 & -3 & 3 \end{bmatrix}$

Table 4.2 Eigenvalue result by FEM for the longitudinally vibrating rod in Example 4.1

Eigenvalue parameter	FEM with two elements	FEM with three elements	Exact solution (Clough and Penzien, 1982)
$\lambda_i = \omega_i \sqrt{\dfrac{ml^2}{EA}}$	1.611	1.588	1.5708
	5.629	5.196	4.712
	–	9.437	7.854

From the associated eigenvalue problem (Chapter 3), we obtain the eigenvalues (natural frequencies, $\omega_i, i = 1, 2..$ of the system) in terms of the eigenvalue parameter, $\lambda_i = \omega_i l \sqrt{\frac{m}{EA}}$. These eigenvalues are shown in Table 4.2. The exact solution (Clough and Penzien, 1982) of the vibrating rod as a continuous system is also provided in the table for comparison.

4.3 Element-Wise Representation of the Weak Form and the FEM

In the FEM, the finite dimensional space $\boldsymbol{V}$ (containing ψ, Φ_j) typically consists of piecewise polynomial functions with the set of global shape functions $\{\Phi_j\}_{j=1}^N$ as the basis. For any interior node, the shape function is compactly supported (Appendix D) within Ω. This leads to sparsity and bandedness in $\boldsymbol{K}$ and $\boldsymbol{M}$ matrices whose elements are zero whenever the intersection of Φ_i and Φ_j is empty (see the tri-diagonal $\boldsymbol{K}$ and $\boldsymbol{M}$ matrices of case 2 in Table 4.1). Sparse and banded matrices are more convenient to handle from both storage and computational points of view. An added computational advantage within the FEM is the element-wise restriction of the global shape functions (henceforth referred to as element shape functions) which in turn enables an element-wise splitting of the weak form. Specifically, denoting by $\Omega^e \subset \Omega$ the subdomain corresponding to the element e, the domain integration $\int_\Omega d\Omega$ (in the weak form) can be replaced by $\sum_e \int_{\Omega^e} d\Omega^e$. Thus, the bilinear forms $\mathcal{K}(\Phi_k, \Phi_j)$ and $\mathcal{M}(\Phi_k, \Phi_j)$ and the linear form $l(\Phi_k)$ can now be written as a sum of integrals over each element sub-domain. If the integrand associated with $\mathcal{K}(\Phi_k, \Phi_j)$ in Equation (4.6) is denoted as $W(\Phi_k, \Phi_j)$, then we have:

$$\begin{aligned} \mathcal{K}(\Phi_k, \Phi_j) &= \int_\Omega W(\Phi_k, \Phi_j) d\Omega \\ &= \sum_e \int_{\Omega^e} W(\Phi_k^e, \Phi_j^e)\Omega^e \\ &= \sum_e \mathcal{K}^e(\Phi_k^e, \Phi_j^e) \end{aligned} \tag{4.43}$$

Here $\mathcal{K}^e(\Phi_k^e, \Phi_j^e)$ is the bilinear form at the element level. $\mathcal{M}^e(\Phi_k^e, \Phi_j^e)$ is a similar bilinear form at the element level for $\mathcal{M}(\Phi_k, \Phi_j)$ and $l^e(\Phi_k^e)$, the element-level linear form for $l(\Phi_k)$. The system (global) matrices and force vector are obtained via an 'assembly' in that equations for such nodal *dof* s, that are common for more than one element, must be appropriately superimposed. Whilst the global shape functions directly yield the system matrices $\boldsymbol{M}$ and $\boldsymbol{K}$ bypassing the assembly procedure, the element-based operations enables the computational scheme to be strictly modular and hence more easily implementable for complex structural systems. The following example illustrates generation of system matrices utilising the elemental shape functions that are linear.

Example 4.2: Axially vibrating rod: generation of system matrices using element shape functions

Consider the finite element (FE) model of the axially vibrating rod (constrained at the left end against axial movement) with four nodes and three elements (Figure 4.2c) of uniform size $h = l/3$. The three elements are shown in Figure 4.5a along with a typical 1D element with two nodes per element ($n_e = 2$) and 1 *dof* per node ($n_d = 1$). Denote $P_k(\Omega_e)$ to be polynomials of degree $\leq k$. The element-wise shape functions Φ_j^e and Φ_k^e (Figure 4.5b) are, in particular, chosen to be linear polynomials $P_1(\Omega_e) = a + bx$ with $ab \in \mathbb{R}$ such that Φ_j^e and $\Phi_k^e \in H^1(\Omega)$. Thus, we assume that $\Phi_j^e = a_1 + b_1 x$ and $\Phi_k^e = a_2 + b_2 x$, where the real constants a_i and b_i, $i = 1, 2$ are obtained from the delta property of the shape functions defined by:

$$\Phi_j^e(x_j) = 1,\ \Phi_j^e(x_k) = 0 \quad \text{and} \quad \Phi_k^e(x_k) = 1,\ \Phi_k^e(x_j) = 0 \tag{4.44}$$

Here x_j and x_k are the coordinates of the jth and k^{th} nodes respectively (Figure 4.5b). The above delta property of the shape functions is derived from the fact that the FE approximate solution shall equal $u_j(t)(:= u(x_j, t))$ and $u_k(t)$ (see Equation (4.3)). Using Equation (4.44), we obtain a_i and b_i, $i = 1, 2$ from:

$$\begin{bmatrix} 1 & x_j \\ 1 & x_k \end{bmatrix} \begin{bmatrix} a_1 & a_2 \\ b_1 & b_2 \end{bmatrix} = \begin{bmatrix} 1 & 0 \\ 0 & 1 \end{bmatrix} \Rightarrow \begin{bmatrix} a_1 & a_2 \\ b_1 & b_2 \end{bmatrix} = \begin{bmatrix} 1 & x_j \\ 1 & x_k \end{bmatrix}^{-1} \tag{4.45}$$

Since $\det\begin{bmatrix} 1 & x_j \\ 1 & x_k \end{bmatrix} = x_k - x_j = h \neq 0$, the inverse in Equation (4.45) exists. Now, with Φ_k^e, Φ_j^e thus known, we obtain from the weak form (at the element level) the element

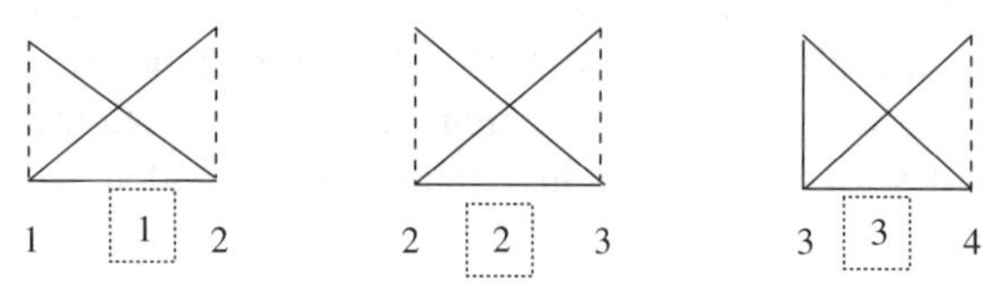

(a)

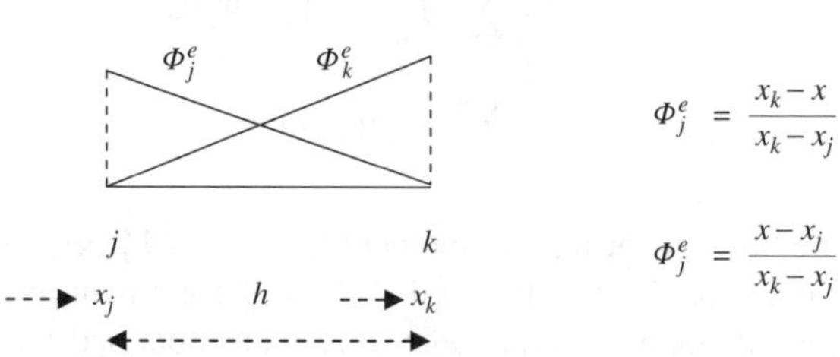

(b)

Figure 4.5 Axially vibrating rod (Example 4.2); (a) element shape functions for the three elements of uniform size and (b) typical two-noded finite element and shape functions, $\Phi_j^e(x)$ and $\Phi_k^e(x)$

matrices $\boldsymbol{M}^e(\Phi_k^e, \Phi_j^e)$ and $\boldsymbol{K}^e(\Phi_k^e, \Phi_j^e)$ as:

$$\boldsymbol{M}^e(\Phi_k^e, \Phi_j^e) = m\int_{\Omega_e} \Phi_k^e \Phi_j^e dx \quad \text{and} \quad \boldsymbol{K}^e(\Phi_k^e, \Phi_j^e) = EA\int_{\Omega_e} \Phi_k^{e\prime} \Phi_j^{e\prime} dx \tag{4.46}$$

Here each element domain is given by $\Omega_e = \left(x_j, x_k\right) \subset (0, l)$.

Following the evaluation of the element-domain integrals appearing in Equation (4.46), we get:

$$\boldsymbol{M}^e(\Phi_k^e, \Phi_j^e) = \frac{mh}{3}\begin{bmatrix} 1 & \frac{1}{2} \\ \frac{1}{2} & 1 \end{bmatrix} \quad \text{and} \quad \boldsymbol{K}^e(\Phi_k^e, \Phi_j^e) = \frac{EA}{h}\begin{bmatrix} 1 & -1 \\ -1 & 1 \end{bmatrix} \tag{4.47}$$

Assembling the element matrices appropriately, we obtain the global matrices $\boldsymbol{M}$ and $\boldsymbol{K}$ as:

$$\boldsymbol{M} = \frac{mh}{3}\begin{bmatrix} 1 & \frac{1}{2} & 0 & 0 \\ \frac{1}{2} & 2 & \frac{1}{2} & 0 \\ 0 & \frac{1}{2} & 2 & \frac{1}{2} \\ 0 & 0 & \frac{1}{2} & 1 \end{bmatrix} \quad \text{and} \quad \boldsymbol{K} = \frac{EA}{h}\begin{bmatrix} 1 & -1 & 0 & 0 \\ -1 & 2 & -1 & 0 \\ 0 & -1 & 2 & -1 \\ 0 & 0 & -1 & 1 \end{bmatrix} \tag{4.48}$$

Since at the clamped left end of the rod, the displacement at the associated node $u_k(t)$ is zero (i.e. no longer an unknown), the first row and column (highlighted in Equation (4.48) can be eliminated from the above matrices by way of imposing this essential boundary condition. Now, with $h = l/3$, it can be observed that the 3×3 matrices appearing in the rightmost corners of the matrices above are identical to those obtained using global shape functions (Table 4.1, case 2).

4.4 Application of the FEM to 2D Problems

For 2D problems, FEM typically accomplishes the discretization of Ω using triangular or quadrilateral elements.

4.4.1 Membrane Vibrations and FEM

Consider a vibrating membrane (Figure 4.6) under a transverse load density $f(\boldsymbol{x}, t)$ so that the governing PDE of motion takes the form:

$$-\Delta u + m\frac{\partial^2 u}{\partial t^2} = f(\boldsymbol{x}, t), u = g \text{ on } \Gamma_g \text{ and } (\nabla u \cdot \vec{n})\vec{n} = h \text{ on } \Gamma_h \text{ so that } \Gamma_g \cup \Gamma_h = \partial\Omega \tag{4.49}$$

Here $\boldsymbol{x} = (x_1, x_2)$ and, towards a fairly general treatment of this specific problem, both Dirichlet (Γ_g) and Neumann (Γ_h) boundaries are presently included (Figure 4.6). h is the flux magnitude along $\vec{n}$, the unit normal to Γ_h at x. We denote the set of triangular elements (Figure 4.1) as $T_e, e = 1, 2, \ldots N_{el}$. Following the familiar steps to derive the weak form and recalling the associated function spaces, we seek to find an approximate

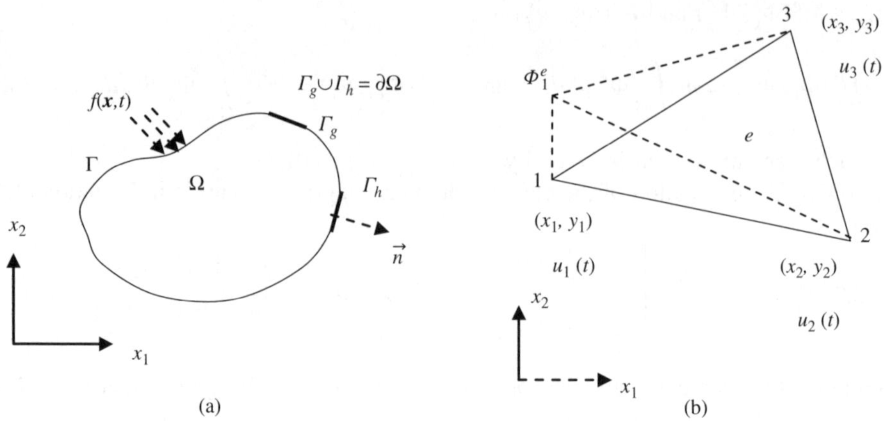

Figure 4.6 Transverse vibrations of a membrane; (a) 2D domain and its essential (Γ_g) and natural (Γ_h) boundaries and (b) a triangular element and a typical linear (element) shape function, Φ_j^e at node 1 with nodal *dof*s $u_j(t)$, $j = 1, 2$ and 3

solution $\hat{u}(\boldsymbol{x}, t)$ via the weak form split element-wise as:

$$\sum_{j=1}^{N_{el}}\left(q_j(t)\left\{\int_{T_j}(\nabla\psi_j)^{\mathrm{T}}(\nabla\Phi_j)\,d\boldsymbol{x}\right\}+\ddot{q}_j(t)\int_{T_j}\psi_j m\Phi_j d\boldsymbol{x}\right)$$
$$=\sum_{j=1}^{N_{el}}\left(P(t)\int_{T_j}\psi_j F(\boldsymbol{x})d\boldsymbol{x}+\int_{T_j\cap\Gamma_h}\psi_j h d\boldsymbol{x}\right) \tag{4.50}$$

Here $\nabla\psi_j = \left(\frac{\partial\psi_j}{\partial x_1}, \frac{\partial\psi_j}{\partial x_2}\right)^T$ and $\nabla\Phi_j = \left(\frac{\partial\Phi_j}{\partial x_1}, \frac{\partial\Phi_j}{\partial x_2}\right)^T$. In writing the above equation, it is assumed that $u = \bar{u} + u_g$ with $u_g = g$ on Γ_g, such that $\bar{u} = 0$ on Γ_g. Since u_g is known, it suffices to approximate $\bar{u}$ only so that we may consider both $\psi_j, \Phi_j \in H_0^1(\Omega)$. The element level approximation is represented by $\hat{u}^e(\boldsymbol{x}, t) = \sum_{j=1}^{3}\Phi_j^e(\boldsymbol{x})\,u_j(t)$ for any generic point $\boldsymbol{x}$ in T_e and Φ_j^e, $j = 1, 2, 3$, the element shape functions, are the restrictions of the global shape functions Φ_j on element e. Since the global shape function $\Phi_j \in H_0^1(\Omega)$, one may choose Φ_j^e to be linear function of the form $a_j + b_j x + c_j y$, with a_j, b_j and $c_j \in \mathbb{R}$. Note that the number of *dof*s of an element is equal to that of unknown real constants in Φ_j^e.

By virtue of their delta property, the linear shape functions satisfy the equation $\boldsymbol{A}_e\boldsymbol{a}_e = \mathbf{I}$, where $\mathbf{I}$ is the 3×3 identity matrix and the matrices $\boldsymbol{A}$ and $\boldsymbol{a}$ are given by:

$$\boldsymbol{A}_e = \begin{bmatrix} 1 & x_1 & y_1 \\ 1 & x_2 & y_2 \\ 1 & x_3 & y_3 \end{bmatrix}, \quad \boldsymbol{a}_e = \begin{bmatrix} a_1 & a_2 & a_3 \\ b_1 & b_2 & b_3 \\ c_1 & c_2 & c_3 \end{bmatrix} \tag{4.51}$$

Knowing the local coordinates $\boldsymbol{x}_k$, $k = 1, 2, 3$ of the three nodes of T_e, we obtain these coefficients by the inversion of $\boldsymbol{A}$. Note that since det $\boldsymbol{A} = 2 \times$ area (T_e) for each e, the nonsingularity of $\boldsymbol{A}_e$ is ensured. Figure 4.6b shows a typical shape function Φ_1^e at node 1 of the element. Corresponding to each element T_e with the unknown *dof*s given by $u_j(t)$, $j = 1, 2$ and 3, we have the following equations for the weak form:

$$\sum_{e=1}^{N_{el}} \left(\boldsymbol{M}^e(\Phi_k^e, \Phi_j^e) \begin{Bmatrix} \ddot{u}_1 \\ \ddot{u}_2 \\ \ddot{u}_3 \end{Bmatrix}^e + \boldsymbol{K}^e(\Phi_k^e, \Phi_j^e) \begin{Bmatrix} u_1 \\ u_2 \\ u_3 \end{Bmatrix}^e \right)$$

$$= \sum_{e=1}^{N_{el}} \left(P(t) l \left(\Phi_k^e \right) + \int_{T_j \cap \Gamma_h} \Phi_k^e h d\boldsymbol{x} \right) \tag{4.52}$$

Here explicit time-dependence of the functions g and h, which is presently avoided for convenience, may be readily incorporated. The matrices $\boldsymbol{M}^e(\Phi_k^e, \Phi_j^e)$ and $\boldsymbol{K}^e(\Phi_k^e, \Phi_j^e)$ are obtainable from Equation (4.50). Once the element matrices are available, the global matrices are obtained by a proper assembly keeping in view the nodes common to more than one element. Since $u = g$ is known on the boundary Γ_g, the equations corresponding to the nodes on Γ_g may be excluded from the finally assembled equations. In addition, appropriate changes need to be effected in the remaining equations in that the contributions due to these boundary nodes may be transferred to the right-hand side of these equations.

4.4.2 Plane (2D) Elasticity Problems – Plane Stress and Plane Strain

Plane elasticity problems arise from simplification of (dimensional descent from) 3D elasticity theory, based on the observation that either the kinematic variables (strains) or their stress equivalents are zero or of insignificant variations in the third dimension. Thus, when the thickness t of a 3D solid in, say, the z-direction is small compared to its geometrical spread in the other two (say x and y) dimensions, we may adopt the plane stress model (Figure 4.7a) provided the external forces act only in the $x - y$ plane (here the normal and shear stresses in the z-direction are nearly zero). If, on the other hand, the geometric spread along z is too large compared to the dimensions in x and y, it is a 2D plane strain problem for an arbitrarily chosen cross section sufficiently away from the z-boundaries (Figure 4.7b) if the external forces are z-invariant. For these 2D models, Cauchy's equations derived from say, Newton's force balance (Equation (1.46)) or linear momentum balance (Equation (1.65)) reduce to:

$$\frac{\partial \sigma_{xx}}{\partial x} + \frac{\partial \sigma_{yx}}{\partial y} + f_x = m\ddot{u}$$
$$\frac{\partial \sigma_{xy}}{\partial x} + \frac{\partial \sigma_{yy}}{\partial y} + f_y = m\ddot{v} \tag{4.53}$$

Here m denotes the mass density per unit surface area. $u(x, y, t)$ and $v(x, y, t)$ are respectively the planar displacements along x and y directions. The strain components (presently written as a vector for notational simplicity and making use of the symmetry of the associated tensor) at a point is given (Chapter 1) in terms of the two displacement fields as:

$$\varepsilon = \begin{Bmatrix} \varepsilon_{xx} \\ \varepsilon_{yy} \\ \varepsilon_{xy} \end{Bmatrix} = \boldsymbol{B} \begin{Bmatrix} u \\ v \end{Bmatrix} \quad \text{with} \quad \boldsymbol{B} = \begin{bmatrix} \dfrac{\partial}{\partial x} & 0 \\ 0 & \dfrac{\partial}{\partial y} \\ \dfrac{\partial}{\partial y} & \dfrac{\partial}{\partial x} \end{bmatrix} \tag{4.54}$$

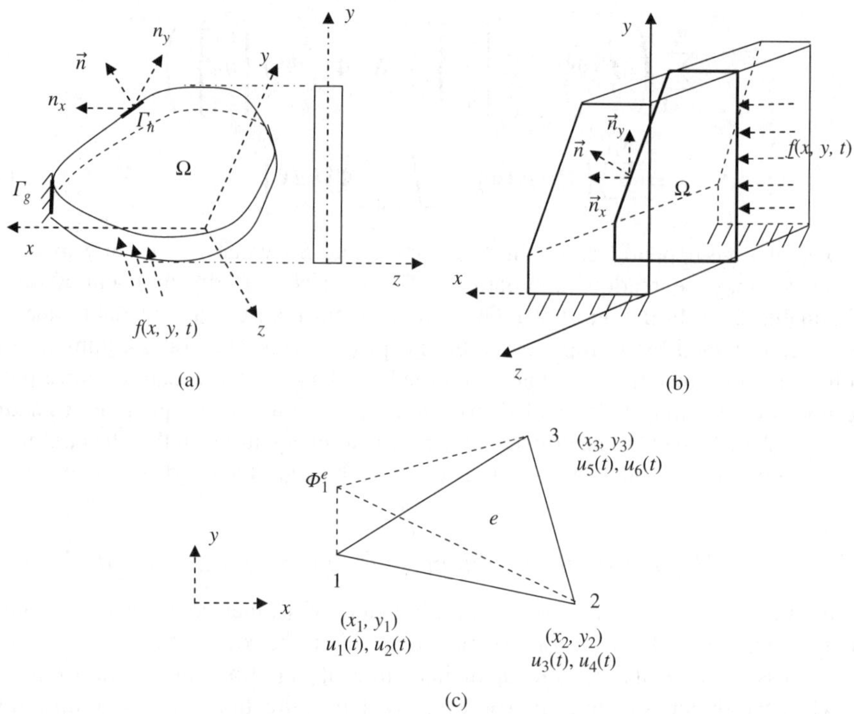

Figure 4.7 Plane elasticity; (a) plane stress problem, (b) plane strain problem and (c) a typical element shape function for a triangular element with 2 *dof*s/node – $u_{2j-1}(t)$ and $u_{2j}(t)$, $j = 1, 2$ and 3

Table 4.3 Stress-strain constitutive relationship and the matrix of material constants, $\boldsymbol{c}$

Plane stress	Plane strain
$\boldsymbol{\sigma} = \boldsymbol{c}\boldsymbol{\varepsilon}$	$\boldsymbol{\sigma} = \boldsymbol{c}\boldsymbol{\varepsilon}$
$\boldsymbol{c} = \dfrac{E}{1-\nu^2}\begin{bmatrix} 1 & \nu & 0 \\ \nu & 1 & 0 \\ 0 & 0 & \dfrac{1-\nu}{2} \end{bmatrix}$	$\boldsymbol{c} = \dfrac{E(1-\nu)}{(1+\nu)(1-2\nu)} \times \begin{bmatrix} 1 & \dfrac{\nu}{(1-\nu)} & 0 \\ \dfrac{\nu}{(1-\nu)} & 1 & 0 \\ 0 & 0 & \dfrac{1-2\nu}{2(1-\nu)} \end{bmatrix}$
E: Youngs modulus	
ν: Poisson's ratio	

The associated stress 'vector' is given by $\boldsymbol{\sigma} = (\sigma_{xx}, \sigma_{yy}, \sigma_{xy})^T$ and the stress-strain constitutive relationship may be written (in the matrix-vector form) as $\boldsymbol{\sigma} = \boldsymbol{c}\boldsymbol{\varepsilon}$. Here, $\boldsymbol{c}$ is the matrix of material constants (appropriately reduced from its fourth order tensor counterpart), defined by the three rows $(c_{11}, c_{12}, 0)$, $(c_{12}, c_{22}, 0)$ and $(0, 0, c_{66})$, as given in Table 4.3. Substituting Equation (4.54) in $\boldsymbol{\sigma} = \boldsymbol{c}\boldsymbol{\varepsilon}$, and the result in Equation (4.53), we obtain the equations of motion (Reddy, 1984) as:

$$\frac{\partial}{\partial x}\left(c_{11}\frac{\partial u}{\partial x}+c_{12}\frac{\partial v}{\partial y}\right)+c_{66}\frac{\partial}{\partial y}\left(\frac{\partial u}{\partial y}+\frac{\partial v}{\partial x}\right)+m\ddot{u}=f_x$$

$$c_{66}\frac{\partial}{\partial x}\left(\frac{\partial u}{\partial y}+\frac{\partial v}{\partial x}\right)+\frac{\partial}{\partial y}\left(c_{12}\frac{\partial u}{\partial x}+c_{22}\frac{\partial v}{\partial y}\right)+m\ddot{v}=f_y \quad (4.55)$$

f_x and f_y are the external inplane force densities in x and y directions respectively. Let the boundary conditions – essential and natural – be specified as:

$$u=u_g(x,y),\quad v=v_g(x,y)\quad \text{on } \Gamma_g$$

$$\sigma_{xx}n_x+\sigma_{xy}n_y=b_u\quad \text{on } \Gamma_h$$

$$\sigma_{xy}n_x+\sigma_{yy}n_y=b_v\quad \text{on } \Gamma_h \quad (4.56\text{a–c})$$

$\vec{n}$ is the unit normal to Γ at $\boldsymbol{x}$ in Γ_h and $n_x := |\vec{n}_x|$, and so on. The natural boundary conditions are rewritten in terms of u and v and their derivatives as:

$$\left(c_{11}\frac{\partial u}{\partial x}+c_{12}\frac{\partial v}{\partial y}\right)n_x+c_{66}\left(\frac{\partial u}{\partial y}+\frac{\partial v}{\partial x}\right)n_y=b_u\quad \text{on } \Gamma_h$$

$$c_{66}\left(\frac{\partial u}{\partial y}+\frac{\partial v}{\partial x}\right)n_x+\left(c_{12}\frac{\partial u}{\partial x}+c_{22}\frac{\partial v}{\partial y}\right)n_y=b_v\quad \text{on } \Gamma_h \quad (4.57\text{d,e})$$

We obtain the weak form of the two equations 4.39 by employing two weight functions $\psi_u(x,y)$ and $\psi_v(x,y)$. For FEM domain-discretization, we may mesh Ω into triangular (Figure 4.7c) and/or quadrilateral elements with 2 dofs (nodal u_j and v_j) per node. The two displacement fields are approximated by the interpolants expressed element-wise as:

$$\hat{u}(x,y,t)=\sum_{j=1}^{N_e}\Phi^e_{uj}(x,y)u_j(t),\ \hat{v}(x,y,t)=\sum_{j=1}^{N_e}\Phi^e_{vj}(x,y)v_j(t) \quad (4.58)$$

Using the same set of shape functions for the two interpolants (i.e. $\Phi^e_{uj}\equiv\Phi^e_{vj}$ so that the suffix u or v is dropped henceforth), the weak form finally reduces to the following matrix-vector equations:

$$\sum_{j=1}^{N_{el}}[\boldsymbol{M}^e_{uu}(\Phi^e_k,\Phi^e_j)\ddot{u}_j(t)+\boldsymbol{M}^e_{uv}(\Phi^e_k,\Phi^e_j)\ddot{v}_j(t)$$
$$+\boldsymbol{K}^e_{uu}(\Phi^e_k,\Phi^e_j)u_j(t)+\boldsymbol{K}^e_{uv}(\Phi^e_k,\Phi^e_j)v_j(t)]$$
$$=P_u(t)l_u\left(\Phi^e_k\right)+\int_{T_j\cap\Gamma_h}\Phi^e_k b_u\,dxdy$$

$$\sum_{j=1}^{N_{el}}[\boldsymbol{M}^e_{vu}(\Phi^e_k,\Phi^e_j)\ddot{u}_j(t)+\boldsymbol{M}^e_{vv}(\Phi^e_k,\Phi^e_j)\ddot{v}_j(t)$$
$$+\boldsymbol{K}^e_{vu}(\Phi^e_k,\Phi^e_j)u_j(t)+\boldsymbol{K}^e_{vv}(\Phi^e_k,\Phi^e_j)v_j(t)]$$
$$=P_v(t)l_v\left(\Phi^e_k\right)+\int_{T_j\cap\Gamma_h}\Phi^e_k b_v\,dxdy \quad (4.59\text{a,b})$$

The mass and stiffness matrices are given presently by:

$$M^e_{uu}(\Phi^e_k, \Phi^e_j) = \int_{T_j} \Phi^e_k m \Phi^e_j dxdy,$$

$$M^e_{uv}(\Phi^e_k, \Phi^e_j) = M^e_{vu}(\Phi^e_k, \Phi^e_j) = \int_{T_j} \Phi^e_k m \Phi^e_{vj} dxdy,$$

$$M^e_{vv}(\Phi^e_k, \Phi^e_j) = \int_{T_j} \Phi^e_k m \Phi^e_j dxdy \tag{4.60a–c}$$

$$K^e_{uu}(\Phi^e_k, \Phi^e_j) = \int_{T_j} \left(c_{11}\Phi^e_{k,x}\Phi^e_{j,x} + c_{66}\Phi^e_{k,y}\Phi^e_{j,y}\right) dxdy$$

$$K^e_{uv}(\Phi^e_k, \Phi^e_j) = K^e_{vu}(\Phi^e_k, \Phi^e_j) = \int_{T_j} \left(c_{11}\Phi^e_{k,x}\Phi^e_{,y} + c_{66}\Phi^e_{k,y} m \Phi^e_{j,x}\right) dxdy$$

$$K^e_{vv}(\Phi^e_k, \Phi^e_j) = \int_{T_j} \left(c_{66}\Phi^e_{k,x}\Phi^e_{j,x} + c_{22}\Phi^e_{k,y}\Phi^e_{j,y}\right) dxdy \tag{4.61a–c}$$

The right-hand side forces $l_u\left(\Phi^e_k\right)$ and $l_v\left(\Phi^e_k\right)$ in Equation (4.59) are given by:

$$l_u\left(\Phi^e_k\right) = \int_{T_j} \Phi^e_k F(x, y) dxdy, \quad l_v\left(\Phi^e_k\right) = \int_{T_j} \Phi^e_k F(x, y) dxdy \tag{4.62}$$

Here $\Phi^e_{k,x}$ and $\Phi^e_{k,y}$ stand for $\frac{\partial \Phi^e_k}{\partial x}$ and $\frac{d\partial \Phi^e_k}{\partial y}$ respectively. As with other cases, the current weak form also demands less stringent continuity requirements in that the element shape functions and their first-order derivatives need only to be $L^2(\Omega)$. Thus, the element shape functions can even be linear. In case of triangular elements, the element shape functions can be determined following the same procedure as in the membrane problem. However, since the number of dofs per node is presently two, the element matrices are of size 6×6. Shape functions for a quadrilateral element suitable for the plane elasticity problem are described in a later section.

4.5 Higher Order Polynomial Basis Functions

For beam and plate vibrations, the PDEs of motion (Chapter 2) contain fourth order derivatives so that the respective weak forms contain second-order derivatives. Thus, the element shape functions are to be chosen from polynomials, $P_k(\Omega), k > 1$.

4.5.1 Beam Vibrations and FEM

Equation (2.16) (Chapter 2) gives the PDE of motion for a uniform beam of length l as $EI\frac{\partial^4 u}{\partial x^4} + m\frac{\partial^2 u}{\partial t^2} = f(x, t)$. E is the Young's modulus, I the area moment of inertia of the beam cross section and m the mass per unit length of the beam. The essential and natural boundary conditions are given by Equations (2.13) and (2.14) of the same Chapter. In general, the weak form corresponding to the beam equation is:

$$\sum_{j=1}^{N}\left\{EI\int_0^l \frac{d^2v}{dx^2}\frac{d^2\Phi_j}{dx^2}dx\right\}q_j+\left\{m\int_0^l v\Phi_j dx\right\}\ddot{q}_j$$
$$=\int_0^l vf(x,t)dx-\left(EI\frac{\partial^2 u}{\partial x^2}\frac{dv}{dx}\right)_0^l+\left(EI\frac{\partial^3 u}{\partial x^3}v\right)_0^l \tag{4.63}$$

The weak form involves second derivatives (against those of fourth order in the strong form) for both the weighting and shape functions, v and Φ_j respectively. Further, vanishing of the weighting function at the essential boundaries requires that the test space, **V** must be a subset of $H_0^2=\{v\in H^2(0,l):v(0)=v'(0)=0\}$. Here, the essential boundaries correspond to a cantilever beam. If these boundary conditions are zero, Φ_j also $\in H_0^2(0,l)$. If we look at the domain terms (specifically the first one) of the weak form in Equation (4.63), we see that an element shape function needs at least to be a quadratic polynomial. On the other hand, the natural boundary conditions on the forces and the moments involve third and second-order derivatives of $u(x,t)$ respectively (see the boundary terms in Equation (4.63). This requires a cubic polynomial for the element shape function, $\Phi_j^e(x)$ so that the shear forces at the natural boundaries are nonzero. Thus, we may take $\Phi_j^e(x)=a_j+b_jx+c_jx^2+d_jx^3$ with $j=1,2$ and a_j,b_j,c_j and $d_j\in\mathbb{R}$. Since the polynomial has four real constants as unknowns, each beam element must have 4 *dof*s (i.e. 2 *dof*s per node, one displacement and the other slope) to enable a unique determination of these constants. These shape functions have to satisfy the essential boundary conditions at the ends of each element in the form (see Equation (4.3)):

$$\sum_{j=1}^{4}u_j(t)\Phi_j^e(x_k)=u_k(t)\text{ for the two deflections }(k=1,2)\text{ and}$$

$$\sum_{j=1}^{4}u_j(t)\frac{d\Phi_j^e(x_k)}{dx}=\theta_k(t),\text{ for the two slopes }(k=1,2) \tag{4.64}$$

The above equations exhibit the delta property of the shape functions in that $\Phi_j^e(x_k)=\delta_{jk}$ and $\frac{d\Phi_j^e(x_k)}{dx}=\delta_{jk}$, $j=1,2,3,4$ and $k=1,2$ from which we obtain the following simultaneous equations:

$$\begin{bmatrix}1 & x_1 & x_1^2 & x_1^3\\ 0 & 1 & 2x_1 & 3x_1^2\\ 1 & x_2 & x_2^2 & x_2^3\\ 0 & 1 & 2x_2 & 3x_2^2\end{bmatrix}\begin{bmatrix}a_1 & a_2 & a_3 & a_4\\ b_1 & b_2 & b_3 & b_4\\ c_1 & c_2 & c_3 & c_4\\ d_1 & d_2 & d_3 & d_4\end{bmatrix}=\begin{bmatrix}1&0&0&0\\0&1&0&0\\0&0&1&0\\0&0&0&1\end{bmatrix} \tag{4.65}$$

If $x_1=0$ and $x_2=h$ are taken as the local coordinates of the two ends of the element (Figure 4.8), we can conveniently evaluate the unknown constants and have the element shape functions (Figure 4.8b) as:

$$\Phi_1^e(x)=1-3\left(\frac{x}{h}\right)^2+2\left(\frac{x}{h}\right)^3,\quad \Phi_2^e(x)=x\left\{1-\left(\frac{x}{h}\right)\right\}^2$$
$$\Phi_3^e(x)=3\left(\frac{x}{h}\right)^2-2\left(\frac{x}{h}\right)^3,\quad \Phi_4^e(x)=x\left\{\left(\frac{x}{h}\right)^2-\left(\frac{x}{h}\right)\right\} \tag{4.66}$$

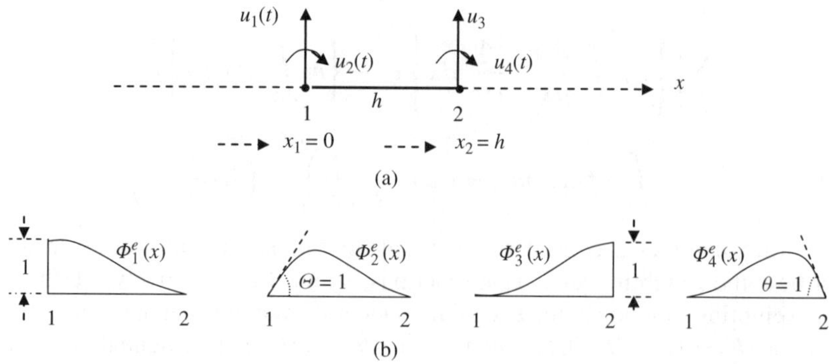

Figure 4.8 Beam vibrations; (a) two noded element with 4 *dof*s and (b) shape functions, $\Phi_j^e(x)$, $j = 1, 2, 3$ and 4

Utilising these element shape functions, we obtain the element matrices and the force vector as:

$$\boldsymbol{K}^e(\Phi_k^e, \Phi_j^e) = EI\int_0^h \frac{d^2\Phi_k^e}{dx^2}\frac{d^2\Phi_j^e}{dx^2}dx,\ \boldsymbol{M}^e(\Phi_k^e, \Phi_j^e) = m\int_0^h \Phi_k^e\Phi_j^e dx,$$

$$f^e(\Phi_k^e) = \int_0^h \Phi_k^e f(x,t)dx \tag{4.67a–c}$$

The element matrices are of size 4×4 and $\left(u_1(t), \theta_1(t) = u_2(t), u_3(t), \theta_2(t) = u_4(t)\right)^T$ constitutes the vector of unknowns at the element level.

4.5.2 Plate Vibrations and FEM

The transverse vibration (Chapter 2) of an isotropic uniform rectangular plate with domain Ω, is governed by the biharmonic equation $D\Delta(\Delta \mathrm{w}) + \mathrm{m}\frac{\partial^2 \mathrm{w}}{\partial t^2} = f(x, y, t)$ where $\Delta = \frac{\partial^2}{\partial x^2} + \frac{\partial^2}{\partial y^2}$ and $\mathrm{D} = \frac{Eh^3}{12(1-\nu^2)}$. E is the Young/s modulus, ν the Poisson's ratio, h the thickness of the plate and m the mass per unit surface area. We obtain the weak form for plate vibrations as:

$$\sum_{j=1}^{N}\left\{D\int_\Omega \Delta v \Delta\Phi_j dxdy\right\}q_j + \left\{m\int_\Omega v\Phi_j dxdy\right\}\ddot{q}_j = \int_\Omega vf(x,y,t)dxdy$$
$$-\int_0^b \left|\left(\frac{\partial^2 w}{\partial x^2} + \nu\frac{\partial^2 w}{\partial y^2}\right)\left(\frac{\partial v}{\partial x}\right)\right|_0^a dy - \int_0^a \left|\left(\nu\frac{\partial^2 w}{\partial x^2} + \frac{\partial^2 w}{\partial y^2}\right)\left(\frac{\partial v}{\partial y}\right)\right|_0^b dx$$
$$-\int_0^b \left|\frac{\partial\left(\frac{\partial^2 w}{\partial x^2} + \nu\frac{\partial^2 w}{\partial y^2}\right)}{\partial x}v\right|_0^a dy - \int_0^a \left|\frac{\partial\left(\nu\frac{\partial^2 w}{\partial x^2} + \frac{\partial^2 w}{\partial y^2}\right)}{\partial y}v\right|_0^b dx \tag{4.68}$$

Here $w(x,t) = \sum_{j=1}^{N}\phi_j(x)q_j(t)$ where $x = (x, y)^T$ and $q_j(t) = w(\hat{x}_j, t)$. $\hat{x}_j = (x_j, y_j)^T$ are the coordinates of jth node. For a rectangular plate element (Figure 4.9), the possible

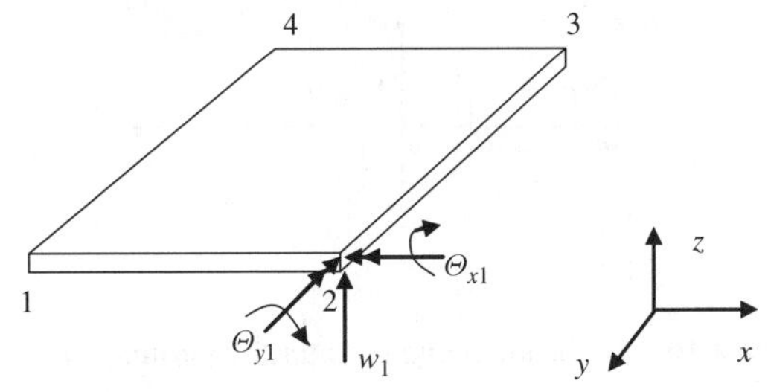

Figure 4.9 Plate element with 3 *dof*s per node

boundary conditions – both essential and natural – on the four edges are given by Equation (2.50). By arguments similar to the case of the beam equation, in the present case too, we need the element shape function Φ_j to be a cubic polynomial (even though the weak form may contain only second-order derivatives in w) so that the shear forces at the natural boundaries are nonzero. A cubic polynomial in x and y for Φ_j is given by $a_{1j} + a_{2j}x + a_{3j}y + a_{4j}x^2 + a_{5j}xy + a_{6j}y^2 + a_{7j}x^3 + a_{8j}x^2y + a_{9j}xy^2 + a_{10}y^3$ and has 10 real unknown constants. Now consider the element to be 4-noded with each vertex acting as a node. Then we must have the slopes θ_x and θ_y also as *dof*s in addition to the displacement *dof* at each node. This also ensures partial continuity of slopes across elements. Now, with these 12 *dof*s per element, Φ_j need to be a fourth degree polynomial but restricted to 12 constants.

Thus, Φ_j is generally taken as (Zienkiewicz and Taylor, 1989):

$$\Phi_j(x) = a_1 + a_2x + a_3y + a_4x^2 + a_5xy + a_6y^2 + a_7x^3 + a_8x^2y + a_9xy^2 + a_{10}y^3 + a_{11}x^3y + a_{12}xy^3 \tag{4.69}$$

The constants $a_i, i = 1, 2, \ldots, 12$ can be found in terms of the nodal coordinates, for each of $\Phi_j(x), j = 1, 2.., 12$ by the usual procedure described in the previous cases. The details of descretization of the boundary integrals in equation 4.68 and further simplification of the weak form leading to the standard MDOF equations are left as an exercise to the reader.

4.6 Some Computational Issues in FEM

FEM being a numerical method is computationally intensive. Its effectiveness in analysing complicated systems depends on the adoption of efficient procedural steps during the semi-discretization and the generation of shape functions. In this context, one useful feature of the FEM is the utilisation of normalised coordinates (also known as natural coordinates). Specifically, use of natural coordinates enables restricting the element shapes to within a few canonical variants. Thus, whilst discretising 2D elasticity problems using quadrilateral elements of general shapes, natural coordinates allow the element-based formulation to be entirely through square elements based on appropriate coordinate mapping. Details of this and other numerical artifices that allow for an easier computer implementation are available in numerous text books on FEM, a few of which are listed in the bibliography. The brief discussion that summarises a few of such important numerical features is included for a quick reference.

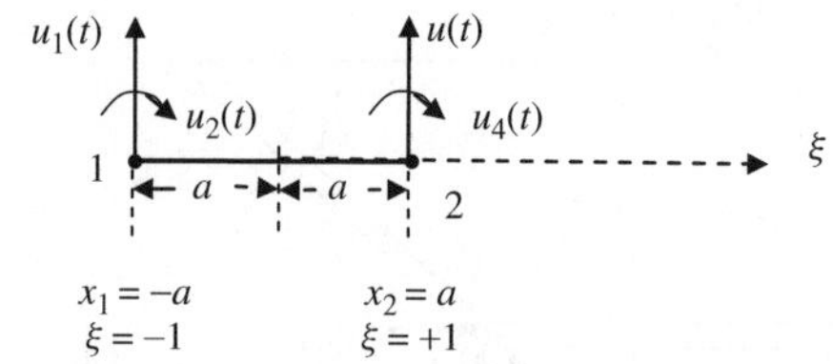

Figure 4.10 1D beam element; natural coordinate, $\xi = x/a$

4.6.1 Element Shape Functions in Natural Coordinates

4.6.1.1 Beam Elements

Consider a 1D beam element (Figure 4.10) with origin located at the midpoint and the element domain as $(-a, +a)$. The natural coordinate is given by $\xi = x/a$ and the shape function written as:

$$\Phi_j^e(x) = a_j + b_j\xi + c_j\xi^2 + d_j\xi^3, \; j = 1, 2, 3, 4 \tag{4.70}$$

Following the same procedure as in Equations (4.64) and (4.65), we can obtain the four shape functions in terms of ξ as:

$$\Phi_1^e(\xi) = \frac{1}{4}\left(2 - 3\xi + \xi^3\right), \; \Phi_2^e(\xi) = \frac{a}{4}(1 - \xi - \xi^2 + \xi^3)$$

$$\Phi_3^e(\xi) = \frac{1}{4}(2 + 3\xi - \xi^3), \; \Phi_4^e(\xi) = \frac{a}{4}(-1 - \xi + \xi^2 + \xi^3) \tag{4.71a–d}$$

Note that, in further evaluation of element matrices, the derivatives in natural and original coordinates are related via the chain rule, viz., $\frac{d\Phi_j^e(x)}{dx} = \frac{d\Phi_j^e(\xi)}{d\xi}\frac{d\xi}{dx} = \frac{1}{a}\frac{d\Phi_j^e(\xi)}{d\xi}$, $\frac{d^2\Phi_j^e(x)}{dx^2} = \frac{1}{a^2}\frac{d^2\Phi_j^e(\xi)}{d\xi^2}$, and so on.

4.6.1.2 Quadrilateral Elements

Suppose the semi-discretization of the domain for the plane elasticity problem uses quadrilateral elements (Figure 4.11), each element now carries 8 dofs with 2 dof per node, in terms of the inplane displacement field variables u and v. As described earlier, the element shape functions $\Phi_{uj}^e(x, y)$ and $\Phi_{vj}^e(x, y)$ corresponding to the two variables can be identical. As such, we have u(x, y, t) and v(x, y, t) at any location (x, y) within an element as $\begin{Bmatrix} u \\ v \end{Bmatrix} = Nu$ where:

$$\boldsymbol{N} = \begin{bmatrix} \Phi_1^e & 0 & \Phi_2^e & 0 & \Phi_3^e & 0 & \Phi_4^e & 0 \\ 0 & \Phi_1^e & 0 & \Phi_2^e & 0 & \Phi_3^e & 0 & \Phi_4^e \end{bmatrix} \text{ and}$$

$$\boldsymbol{u} = \left(u_1(t)\; u_2(t)\; u_3(t)\; u_4(t)\; u_5(t)\; u_6(t)\; u_7(t)\; u_8(t)\right)^T \tag{4.72a,b}$$

With origin at the centre of the element in the natural coordinate system, one can construct the shape functions in terms of ξ and η first and relate them later to the local (physical) coordinates x and y by appropriate mapping. The weak form of the equation demands

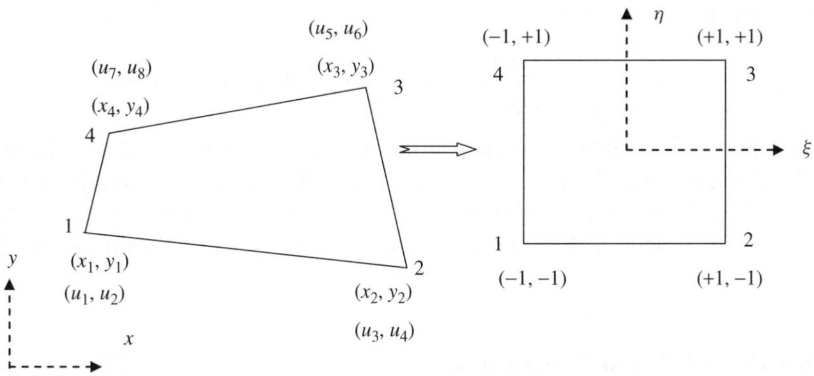

Figure 4.11 Plane elasticity problem with quadrilateral element mapped into a square one in natural coordinates

only a linear polynomial to represent each element shape function Φ_j^e since it belongs to $H^1(\Omega)$. However, since we need for each element, four distinct shape functions, Φ_j^e, $j = 1, 2, 3$ and 4, a quadratic polynomial limited to four real constants is required and one choice for each Φ_j^e is $a_j + b_j\xi + c_j\eta + d_j\xi\eta$, $j = 1, 2, 3$ and 4. These functions can be systematically generated by following the procedure described earlier. However, due to the simplicity of the square element in the natural coordinate system, we can directly fix the shape functions as:

$$\Phi_1^e = \frac{1}{4}(1-\xi)(1-\eta), \quad \Phi_2^e = \frac{1}{4}(1+\xi)(1-\eta)$$
$$\Phi_3^e = \frac{1}{4}(1+\xi)(1+\eta), \quad \Phi_4^e = \frac{1}{4}(1-\xi)(1+\eta) \tag{4.73a,b}$$

Similar to the equations $\begin{Bmatrix} u \\ v \end{Bmatrix} = \boldsymbol{Nu}$ relating the nodal displacements to the field displacement variables, one can interpolate the physical coordinates (x, y) of any point within the element e in terms of ξ and η by using the same shape function matrix $N(\xi, \eta)$. It follows that:

$$\begin{Bmatrix} x(\xi, \eta) \\ y(\xi, \eta) \end{Bmatrix}^e = \begin{Bmatrix} \sum_{j=1}^4 \Phi_j^e(\xi, \eta)x_j \\ \sum_{j=1}^4 \Phi_j^e(\xi, \eta)\, y_j \end{Bmatrix} \tag{4.74}$$

(x_j, y_j), $j = 1, 2, 3$ and 4 are the nodal coordinates of the quadrilateral element in physical coordinate system. Note that the four edges of the quadrilateral element in the physical coordinate system are mapped into the four edges of the square in the natural coordinate system. This can be verified by substituting, for example $\xi = 1$ (corresponding to the edge 2–3 of the square element) in Equation (4.74):

$$x = \frac{1}{2}(1-\eta)x_2 + \frac{1}{2}(1+\eta)x_3 = \frac{1}{2}(x_2 + x_3) + \frac{1}{2}\eta(x_3 - x_2)$$
$$y = \frac{1}{2}(1-\eta)y_2 + \frac{1}{2}(1+\eta)y_3 = \frac{1}{2}(y_2 + y_3) + \frac{1}{2}\eta(y_3 - y_2) \tag{4.75a,b}$$

The above equation leads to:

$$y = \frac{(x_3 - x_2)}{(y_3 - y_2)}\left\{x - \frac{1}{2}(x_2 + x_3)\right\} + \frac{1}{2}(y_2 + y_3) \tag{4.76}$$

Equation (4.59) is a straight line between the nodes 2 and 3 in the physical coordinate system. This implies that the edge 2–3 in the physical coordinate system is mapped into the edge 2–3 in the natural coordinate system. Similar is the case with the other edges and thus the full domain of the element is mapped into a square domain by the mapping in Equation (4.74).

4.7 FEM and Error Estimates

FEM yields an approximate solution to the weak form (Equation (4.15)) via a semi-discretization of Ω and the displacement interpolation between the nodal points. If $u \in \boldsymbol{V}$ is the weak solution such that:

$$\hat{\mathcal{K}}(u, v) = \Lambda(v) \quad \forall v \in \boldsymbol{V} \tag{4.77}$$

then, in FEM we seek $u_h \in V_h \subset \boldsymbol{V}$ such that:

$$\hat{\mathcal{K}}(u_h, v) = \Lambda(v) \quad \forall v \in V_h \tag{4.78}$$

It is of interest to know how close u_h is to u in some sense and in this regard one can obtain (Strang and Fix, 1973, Brenner and Scott, 1994) both *a priori* and *a posteriori* estimates on $\|u - u_h\|$. A brief discussion on a-*priori error estimate* follows. We recall that $V_h = \text{span}(\Phi_j(x))$ where $\Phi_j(x)$, $j = 1, 2.., N$ are the piece-wise linear global basis functions such that $u_h(x, t) = \sum_{j=1}^{N} \Phi_j(x) u_j(t)$ with $u_j(t)$ being the physical coordinate at each jth node.

4.7.1 A-Priori Error Estimate

Note that V_h represents a function space corresponding to a mesh size h and we have a sequence of such spaces with decreasing h. Now, with $v \in V_h(\subset \boldsymbol{V})$ and subtracting Equation (4.78) from Equation (4.77), we have:

$$\hat{\mathcal{K}}(u, v_h) - \hat{\mathcal{K}}(u_h, v_h) = 0 \Rightarrow \hat{\mathcal{K}}(e_h, v_h) = 0 \text{ with } e_h = u - u_h \text{ and } \forall v_h \in V_h \tag{4.79}$$

The above result is known as Galerkin orthogonality. We define the energy norm $\|v\|_E := \hat{K}(v, v)$, $v \in V_h$, which is in fact twice the strain energy in a properly supported structural system undergoing displacement v and is greater than zero for all $v \neq 0$. Using the result in Equation (4.79) and the linearity of a bilinear form, we have:

$$\begin{aligned}
\hat{\mathcal{K}}(u, u) &= \hat{\mathcal{K}}(u_h + e_h, u_h + e_h) = \hat{\mathcal{K}}(u_h, u_h) + 2\hat{\mathcal{K}}(e_h, u_h) + \hat{\mathcal{K}}(e_h, e_h) \\
&= \hat{\mathcal{K}}(u_h, u_h) + \hat{\mathcal{K}}(e_h, e_h) \\
&\Rightarrow \hat{\mathcal{K}}(u_h, u_h) \leq \hat{\mathcal{K}}(u, u) \text{ since } \hat{\mathcal{K}}(e_h, e_h) > 0
\end{aligned} \tag{4.80}$$

The above result implies that the strain energy corresponding to a FE solution u_h is always smaller than or equal to that of an exact solution u. This also means that the

FE-discretized system is 'stiffer' than the original one. Further, it can be shown that the FE solution corresponds to the minimum energy norm, that is $\hat{\mathcal{K}}(e_h, e_h)$ is a minimum. This is evident from the following:

$$\begin{aligned}\hat{\mathcal{K}}(e_h + w_h, e_h + w_h) &= \hat{\mathcal{K}}(e_h, e_h) + \hat{\mathcal{K}}(w_h, w_h) \text{ for any } w_h \in V_h \\ &\Rightarrow \hat{\mathcal{K}}(e_h, e_h) \leq \hat{\mathcal{K}}(e_h + w_h, e_h + w_h) \text{ since } \hat{\mathcal{K}}(w_h, w_h) > 0\end{aligned} \tag{4.81}$$

With $w_h = u_h - v_h$, Equation (4.81) becomes:

$$\hat{\mathcal{K}}(e_h, e_h) \leq \hat{\mathcal{K}}(u - v_h, u - v_h) \quad \forall v_h \in V_h \tag{4.82}$$

The above result is known as Cea's lemma. In FEM, $v_h = \pi u \in V_h$ where π is the nodal interpolation operator, that is πu is an interpolant of u. Thus, to apply Cea's lemma and get an estimate on e_h, a bound for the interpolation error $u - \pi u$ is needed. By use of Taylor's theorem, this bound is given in a general form (Appendix D):

$$\|(u - \pi u)^{(n)}\|_{L_2(a,b)} \leq C h^{m-n} \|u^{\{m\}}\|_{L_2(a,b)} \tag{4.83}$$

In the above general formula, $m - 1$ is the degree of polynomial interpolant πu. C is a constant independent of u and h and $u^{\{k\}}$ the kth derivative of u. In case, πu is a linear polynomial (for instance, refer to the weak form of the axially vibrating rod with $\hat{\mathcal{K}}(u, u) = \int_0^l \alpha(u')^2 dx$ in Equation (4.11), we have $m = 2$ in the above general formula. Also for $m = 2$ and $n = 1$, one can show that $C = \sqrt{3}$ (Appendix D). Thus, we have from Equation (4.82):

$$\begin{aligned}\hat{\mathcal{K}}(u - v_h, u - v_h) &= \|u - \pi u\|_E^2 \\ &= \int_0^l \alpha((u - \pi u)')^2 dx \\ &= \sum_{i=1}^{N} \int_{x_{i-1}}^{x_i} \alpha((u - \pi u)')^2 dx \\ &\leq \sum_{i=1}^{N} \bar{\alpha} \|(u - \pi u)'\|_{L_2(x_{i-1}, x_i)}^2 \text{ by Cauchy-Schwartz inequality} \\ &\qquad \text{and } \bar{\alpha} = \max_{x \in (0,l)} \alpha(x) \\ &\leq \sum_{i=1}^{N} \bar{\alpha} \frac{h^2}{3} \|u''\|_{L_2(x_{i-1}, x_i)}^2 \text{ by Equation (4.83)} \\ &= \bar{\alpha} \frac{h^2}{3} \|u''\|^2\end{aligned} \tag{4.84}$$

Thus, from Equation (4.82), $\hat{\mathcal{K}}(e_h, e_h) = \|u - v_h\|_E^2 \leq \|u - \pi u\|_E^2 \leq \bar{\alpha} \frac{h^2}{3} \|u''\|^2$ and we have the *a priori* error estimate via energy norm:

$$\|u - u_h\|_E \leq C h \|u''\| \text{ with } C = \sqrt{\frac{\bar{\alpha}}{3}} \tag{4.85}$$

Thus, $\|u - u_h\|_E = O(h)$ and as $h \to 0$ (adopting finer mesh), the FE solution converges to the exact solution. However, this error estimate is only a mean square error and not a pointwise error. Moreover, as the exact solution u is not known, this estimate is of little use in guiding the optimal mesh-refinement needed to achieve a prescribed level of numerical accuracy.

4.8 Conclusions

With the limited aim of introducing the elements of the FEM, this chapter excludes a discourse on the method's versatility across a broad range of problems and the algorithmic aspects towards efficient computer implementation. Given a spatially continuous dynamic system model, the semi-discretization via FEM leads to a set of coupled ODEs in the form of matrix vector equations of motion (see Equation (4.26)), constituting what are referred to as multi-degree-of-freedom (MDOF) system. A logical way forward would therefore be to arrive at a methodology for solving these MDOF systems and thus obtain the time histories of nodal displacements and finally those of the displacement fields $\hat{u}(\boldsymbol{x}, t)$ or their derivatives (e.g. stress or strain components of design interest). This briefly describes the line of discussion in the chapters to follow, wherein we start with a typical MDOF system and describe methods to integrate the associated ODEs whilst emphasising some of the underlying mathematical concepts wherever applicable.

Exercises

1. Suppose $u(x) = (1 - |x|), x \in (-\infty, \infty)$. The function is shown in Figure 4.12. Find the weak derivative of $u(x)$. (Hint: use integration by parts).
2. For the one-dimensional boundary value problem:

$$-\frac{d^2u}{dt^2} + u = f, \quad 0 \neq f(t) \in L^2(0, 1) \tag{4.86}$$

with boundary conditions, $u(t = 0) = u(t = 1) = 0$,
(a) write the weak form and
(b) show that the bilinear form is coercive

3. Consider the 1D boundary value problem:

$$-\frac{d^2u}{dt^2} + \alpha u = f, \quad 0 \neq f(t) \in L^2(0, 1) \tag{4.87}$$

(a) write the weak form and show that the bilinear form is not necessarily coercive for $\alpha \gg 1$

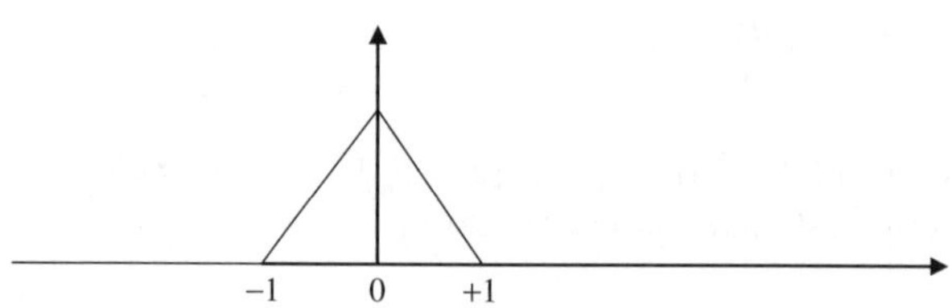

Figure 4.12 Continuous function u(x)

4. Consider the dynamic response of a bar subjected to a harmonic load at mid span (i.e. at $x = \frac{l}{2}$) with the form of governing equation given by:

$$-\frac{EA}{m}\frac{\partial^2 u}{\partial x^2} + \frac{\partial^2 u}{\partial t^2} = \sin\lambda t\delta\left(x - \frac{l}{2}\right) \tag{4.88}$$

l is the length of the bar, A the area of cross-section, m the mass per unit length and E Young's modulus. The boundary conditions which are strictly Neumann are given by $\frac{\partial u(0,t)}{\partial x} = \frac{\partial u(l,t)}{\partial x} = 0$. Explain why a unique solution is not possible for the boundary value problem.

Notations

a, a_1, a_2, a_3, a_4	real constants
$a_{ij}, i = 1, 2, .., 10$	coefficients (real constants) in a cubic (interpolating) polynomial
$\boldsymbol{a_e}$	matrix (3 × 3) of real constants
b, b_1, b_2, b_3, b_4	real constants
$\boldsymbol{B}_i$	partial differential operators
c_1, c_2, c_3, c_4, C	real constants
d_1, d_2, d_3, d_4, D	real constants
e_h	function $(=(u-v_h) \in V_h)$
f_x, f_y	external forcing functions in x and y directions
$F(x)$	function of spatial variable
g	real constant
$\boldsymbol{g}(t)$	boundary condition (possibly a function of time)
$G_\kappa(x)$	polynomial of $(2\kappa - 1)^{th}$ order
h	real constant
$h(x)$	initial condition (possibly a function of the spatial variable)
$H_j(x)$	Lagrange polynomial function
$H^m(0,l)$, $H_0^m(a,b)$	Sobolev spaces
$\boldsymbol{K^e}$	element level stiffness matrix
$\mathcal{K}(\psi, \Phi_j)$	bilinear form
$\hat{\mathcal{K}}(u, v)$	bilinear form
$\mathcal{K}^e(\Phi_k^e, \Phi_j^e)$	bilinear form at the element level
$\ell(\psi)$	linear form
$l^e(\Phi_k^e)$	linear form at the element level
m	integer
$\mathcal{M}(\psi, \Phi_j)$	bilinear form
$\boldsymbol{M^e}$	element level mass matrix
$\mathcal{M}^e(\Phi_k^e, \Phi_j^e)$	bilinear form at the element level
n	integer
$\vec{n}$	unit normal on Γ_h
$\vec{n}_x, \vec{n}_y$	components of $\vec{n}$ (in x and y directions)
$\boldsymbol{N}$	shape function matrix
N_{el}	number of elements in a FE model
n_d	number of *dof* per node

N_d	number of nodes per element
$P(t)$	time varying part of the external forcing function
$P_\kappa(x)$	approximant (Equation (4.27))
$P_k(\Omega_e)$	polynomial of degree $\leq k$
T_e	triangular element
u	weak solution $\in V$
u_g	real constant
u_h	function ($\in V_h$) – FE solution with mesh size h
$u_k(t)$	displacement coordinates
$u^{\{k\}}$	k^{th} derivative of u
U_0	real constant
$u(x, y, t)$	displacement field variable
v	function $\in V$
$v(x)$	polynomial in x
v_h	function ($\in V_h$)
$v(x, y, t)$	displacement field variable
V_h	function space ($\subset V$) corresponding to a mesh size h
w_h	function ($\in V_h$)
W_j	weights in Gauss quadrature
x_1, x_2, x_3, x_4	nodal x coordinates
x_j	quadrature points
y_1, y_2, y_3, y_4	nodal y coordinates
$\bar{\alpha}$	constant ($= \max_{x\in(0,l)} \alpha(x)$)
$\alpha(x)$	parameter $= E(x)I(x)$ in problem of vibrating rod
$\{\gamma_k\}$	set of nodes
$\delta\Phi_j$	variation over Φ_j
$\eta(x)$	function in x
$\eta(x, y)$	function in x and y
$\eta(x, y, z)$	function in x, y and z
η	local coordinate axis
Θ_{xj}, Θ_{yj}	rotations at j^{th} node about x and y axes
$\mu(t)$	time function (Equation (4.7))
ξ	local coordinate axis
π	nodal interpolation operator
$\chi(v, u)$	bilinear form
$\psi(x)$	test (weight) function
$\psi_u(x, y)$ and $\psi_v(x, y)$	weight functions – plane stress/strain problem
Γ_g, Γ_h	Dirichlet and Neumann boundaries
Γ_l	element edges
$\Lambda(v)$	linear form
$\Phi_j(\boldsymbol{x})$	basis functions, interpolating (shape) functions
Φ_j^e	element-wise shape function
Ω_e	element domain
$\{\Omega_j\}$	set of elements

References

Abramowitz, M. and Stegun, A. (eds) (1972) *Handbook of Mathematical Functions (with Formulae, Graphs and Mathematical Tables)*, 10th printing with corrections, National Bureau of Standards, Washington, DC (also New York, Dover, 1965).

Bathe, K.J. (1996) *Finite Element Procedures*, Prentice-Hall International, Inc., Englewood Cliffs, NJ.

Brenner, S.C. and Scott, L.R. (1994) *The Mathematical Theory of Finite Element Method*, Springer-Verlag, New York.

Clough, R.W. and Penzien, J. (1982) *Dynamics of Structures*, McGraw Hill, Singapore.

Hughes, T.J.R. (1987) *The Finite Element Method*, Prentice-Hall International, Inc., Englewood Cliffs, NJ.

Noor, A.K. (1991) Bibliography of books and monographs on finite element technology. *Applied Mechanics Reviews*, **44** (6), 307 –317.

Reddy, J.N. (1984) *Energy and Variational Methods in Applied Mechanics*, John Wiley & Sons, Inc., New York.

Rudin, W. (1976) *Principles of Mathematical Analysis*, 3 edn, McGraw-Hill, New York.

Strang, G. and Fix, G.F. (1973) *An Analysis of the Finite Element Method*, Prentice-Hall International, Inc., Englewood Cliffs, NJ.

Szabo, B. and Babuska, I. (1991) *Introduction to Finite Element Analysis*, John Wiley & Sons, Inc., New York.

Zienkiewicz, O.C. (1977) *The Finite Element Method*, 3rd edn, McGraw-Hill Book Co., London.

Zienkiewicz, O.C. and Cheung, Y.K. (1967) *The Finite Element Method in Structural and Continuum Mechanics*, McGraw-Hill Book Co., London.

Zienkiewicz, O.C. and Taylor, R.L. (1989) *The Finite Element Method*, Basic Formulations and Linear Problems, Vol. **1**, 4th edn, McGraw-Hill Book Co., London.

Bibliography

Cook, R.D., Malkus, D.S. and Piesha, M.E. (1989) *Concepts and Applications of Finite Element Analysis*, 3rd edn, John Wiley & Sons, Inc., New York.

Noor, A.K. (1991) Bibliography of Books and Monographs on Finite Element Technology, *Applied Mechanics Reviews*, 44(6), pp 307–317.

Rao, S.S. (2005) *The Finite Element Method in Engineering*, 4th edn, Elsevier Inc., USA.

5

MDOF Systems and Eigenvalue Problems

5.1 Introduction

From the solution methods presented in Chapters 3 and 4, it is apparent that for most engineering systems, whose governing equations of motion are in the form of partial differential equation (PDE), both classical element method and finite element method (FEM) essentially supply us with semi-discretization strategies leading to simpler mathematical models in the form of ordinary differential equations (ODEs). Given the level of complexity in most engineering systems of practical interest, reduction to ODEs (linear or nonlinear) is definitely a welcome relief in that the mathematical description of the evolving dynamics is well captured within a vastly simplified model and that there exists a vast repertoire of well-known schemes for solving these ODEs with far less computation overhead than would be typically needed for the parent PDEs (Simmons and Krantz, 1925; Kreyszig, 1999; Bathe and Wilson, 1976). Recall that the semi-discretized set of coupled ODEs that we obtain in Equation (3.4) or (4.26) represents an MDOF system. In fact, as will be seen in the next section, such an MDOF system may sometimes be ingeniously arrived at from a given continuous system by directly idealising the latter through a discrete system of lumped masses and stiffnesses, thereby avoiding a more involved semi-discretization strategy altogether. In these equations the vector-valued dependent variable $\boldsymbol{x}(t) : \mathbb{R}$ to $\mathbb{R}^N$ is the unknown to be determined. Thus, $\boldsymbol{x}(t)$ represents the N-dimensional vector of generalized coordinates in the case of classical methods or the vector of nodal *degrees of freedom* in the FEM (see Equations (3.4) and (4.25)). Once $\boldsymbol{x}(t)$ is obtained for a discretized forcing vector $\boldsymbol{f}(t) : \mathbb{R}$ to $\mathbb{R}^N$ and substituted in the assumed series solution in Equation (3.1) or (4.2), we approximately obtain the dynamic response function $\boldsymbol{u}(\boldsymbol{x}, t)$ of the structural system.

By far, the most general route to evaluate $\boldsymbol{x}(t)$ for MDOF systems, linear or otherwise, is through one of the many available direct integration schemes (Bathe, 1996; Bathe and Wilson, 1976). Indeed, for most nonlinear as well as time-varying linear systems, direct integration methods are often the only available recourse for numerical evaluation of $\boldsymbol{x}(t)$ and we will be discussing them in Chapter 8. Presently, however, our focus is on linear time-invariant MDOF systems wherein a simplified analytical approach, based

Elements of Structural Dynamics: A New Perspective, First Edition. Debasish Roy and G Visweswara Rao.

on eigenvalue (spectral) decomposition, yields the solution $\boldsymbol{x}(t)$. Since the principle of superposition is valid for such systems, the eigenanalysis enables decomposing the equations of motion into a set of n uncoupled second-order ODEs (each representing an single degree of freedom (SDOF) system) and then finding the system response as a weighted sum of the individual responses of the SDOF systems. Such an approach, if applicable, provides insightful expressions that aid in the understanding of the system dynamical behaviour and hence can be contrasted with a purely numerical route (e.g. one provided by a direct integration scheme), wherein a recursive application of the scheme hinders an understanding of the long-time or global behaviour (i.e. the response over the entire phase space (space consisting of all possible values of position and momentum variables) of the system. This aspect will be borne out more clearly in Chapter 8. Motivated by this, we proceed to outline the basics of the eigenanalysis applied to linear time-invariant MDOF systems in this chapter.

Following the semi-discretization, the coupled ODEs representing a linear, time-invariant MDOF system with N degree of freedom are given by the following matrix-vector equations:

$$\boldsymbol{M}\ddot{\boldsymbol{x}} + \boldsymbol{K}\,\boldsymbol{x} = \boldsymbol{f}(t) \tag{5.1}$$

$\boldsymbol{M}$ and $\boldsymbol{K}$ are $N \times N$ matrices which are generally symmetric, especially if the parent PDEs of motion are self-adjoint (see Chapter 2 for details on self-adjoint property). $\boldsymbol{x} := (x_1, x_2, \ldots x_N)^T$ is the displacement vector. (Non-symmetric matrices could occur in many cases of non-selfadjoint systems where, among others, effects of moving load, rotations, etc. are to be considered.) It can be noted that the coupled second-order ODEs in Equation (5.1) pertain to a conservative systems with no energy dissipation terms and thus are referred to as an undamped system.

5.2 Discrete Systems through a Lumped Parameter Approach

As noted above, in some practical cases of interest, instead of adopting classical methods or the FEM, one can derive with less effort, the semi-discretized dynamic equations of motion by an appropriate idealisation via a lumped parameter model and thus capture the response characteristics with reasonable accuracy. Such (MDOF) systems can be derived by a direct application of Newton's second law of motion or by using Lagrange's equations of motion (Chapter 1). As illustrations, let us consider a few structural systems and derive the equations of motion by this lumped parameter approach.

Example 5.1: An axially loaded bar (Craig, 1981) and coupled second-order linear ODEs

Figure 5.1 shows a cantilever bar of uniform cross section and E, where $m = \rho Al$ is the total mass with ρ being the mass density; A, the area of cross section and l, the length. Out of the apparently many ways of arriving at the lumped model, one way would be, say, to consider three lumped masses, one ($= m/2$) at the centre and the other two (each equal to $m/4$) at the tips. Note that the mass at the fixity contributes no inertial force and hence the idealisation in Figure 5.1 semi-discretizes the axially loaded bar into essentially a 2 dof model. Specifically, the *degrees of freedom* are the axial displacements $x_1(t)$ and $x_2(t)$ at each of the two mass points at the centre and the right tip as shown in the figure. The axial stiffnesses k_1 and k_2 of each of the two equal halves of the rod are the same (i.e. $k_1 = k_2 = k$). k is approximated by finding the force required to produce

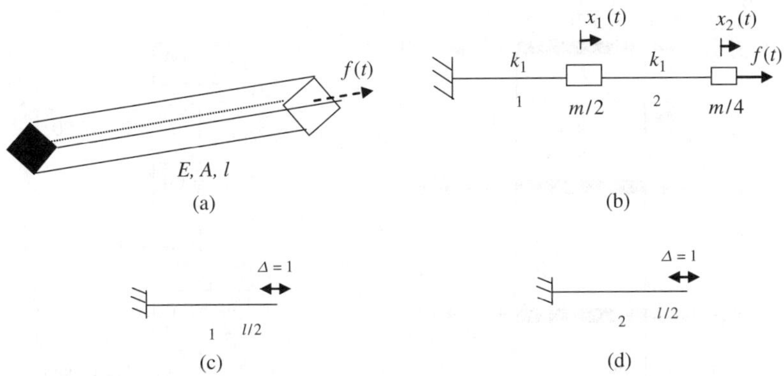

Figure 5.1 (a) Axially loaded bar (b) idealisation as a 2 dof system and (c,d) stiffness determination

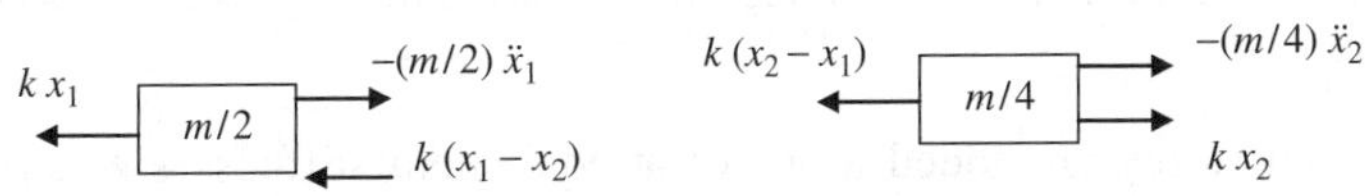

Figure 5.2 Axially loaded bar; Force equilibrium at the two mass points

unit elongation at one end whilst constraining the other end as shown in Figure 5.1c,d. Thus, $k = 2EA/l$.

The force equilibrium at the two mass points is shown in Figure 5.2 which readily yields the following equations of motion:

$$(m/2)\ddot{x}_1 + kx_1 + k(x_1 - x_2) = 0$$
$$(m/4)\ddot{x}_2 + kx_2 + k(x_2 - x_1) = f(t) \tag{5.2a,b}$$

In matrix-vector form, Equation (5.2) can be written similar to Equation (5.1) as:

$$\begin{bmatrix} m/2 & 0 \\ 0 & m/4 \end{bmatrix} \begin{Bmatrix} \ddot{x}_1 \\ \ddot{x}_2 \end{Bmatrix} + \begin{bmatrix} 2k & -k \\ -k & k \end{bmatrix} \begin{Bmatrix} x_1 \\ x_2 \end{Bmatrix} = \begin{Bmatrix} 0 \\ f(t) \end{Bmatrix} \tag{5.3}$$

At this stage, the system has been considered as undamped.

Example 5.2: A shear frame and coupled second-order linear ODEs

Another illustration (Clough and Penzien, 1982) of a continuous system being idealised directly as a semi-discrete system is the case of a shear frame shown in Figure 5.3. This model assumes significance, for instance, in the dynamics of a multi-storied reinforced concrete building under base excitation (e.g. during an earthquake event). Here, the *degrees of freedom* refer to the lateral displacement coordinates $x(t) j = 1, 2, 3$ at each of the three storey levels of the frame.

The idealisation in Figure 5.3a results in a 3 dof degree system for the shear frame; the *degrees of freedom* presently being $x_j(t)$, $j = 1, 2, 3$. Since the storey floors are assumed to be rigid (i.e. far stiffer than the columns connecting them), they are modelled by lumped masses. Each pair of column stiffnesses k_j, $j = 1, 2,$ and 3 on either side of the shear

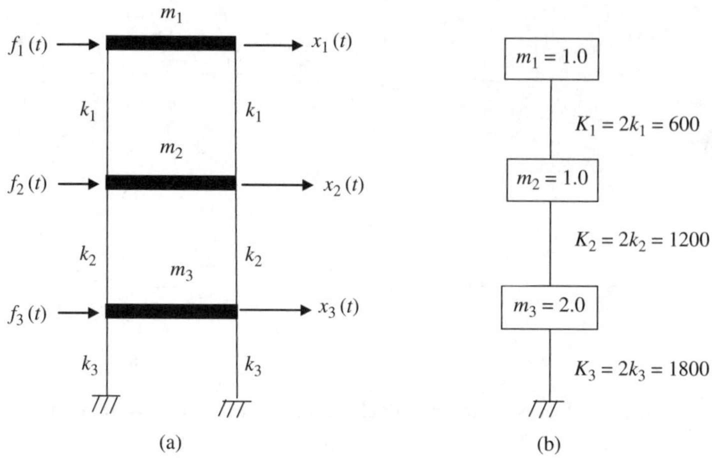

Figure 5.3 (a) Shear frame and (b) lumped mass and stiffness model of shear frame

frame below each storey are added to arrive at equivalent stiffnesses K_j, $j = 1, 2$ and 3 (Figure 5.3b). The numerical values associated with these column stiffnesses are given in the same figure. The semi-discrete equations of motion are finally given by:

$$\begin{bmatrix} m_1 & 0 & 0 \\ 0 & m_2 & 0 \\ 0 & 0 & m_3 \end{bmatrix} \begin{Bmatrix} \ddot{x}_1 \\ \ddot{x}_2 \\ \ddot{x}_3 \end{Bmatrix} + \begin{bmatrix} K_1 & -K_1 & 0 \\ -K_1 & K_1 + K_2 & -K_2 \\ 0 & -K_2 & K_2 + K_2 \end{bmatrix} \begin{Bmatrix} x_1 \\ x_2 \\ x_3 \end{Bmatrix} = \begin{Bmatrix} f_1(t) \\ f_2(t) \\ f_3(t \end{Bmatrix} \tag{5.4}$$

The equations of motion of the shear frame in Equation (5.4) represent again an undamped MDOF system with number of *degrees of freedom* $N = 3$.

5.2.1 Positive Definite and Semi-Definite Systems

The mass and stiffness matrices for the above two examples are both symmetric. Further the stiffness matrix is positive definite. Here, a word about positive definiteness or otherwise of a matrix may be in order. Let $\boldsymbol{B}$ be an $N \times N$ square symmetric matrix (with real entries). $\boldsymbol{B}$ is said to be positive definite if the quadratic scalar function $\boldsymbol{x}^T \boldsymbol{B} \boldsymbol{x} > 0, \forall \boldsymbol{x} > \boldsymbol{0}$ and $\boldsymbol{x}^T \boldsymbol{B} \boldsymbol{x} = 0$ only for $\boldsymbol{x} = \boldsymbol{0}$. $\boldsymbol{B}$ is said to be positive semi-definite if $\boldsymbol{x}^T \boldsymbol{B} x \geq 0, \forall \boldsymbol{x} > \boldsymbol{0}$. An $N \times N$ positive definite matrix has rank N (full rank) and is thus nonsingular. It is easy to verify that the two stiffness matrices in the above two examples are nonsingular and thus their rank is equal to N, the order of the matrices. To further verify the positive definiteness, let us form the quadratic function $\boldsymbol{x}^T \boldsymbol{B}\, x$, for example for the stiffness matrix in Example 5.1. This gives us:

$$\begin{pmatrix} x_1 & x_2 \end{pmatrix} \begin{bmatrix} 2k_1 & -k_1 \\ -k_1 & k_1 \end{bmatrix} \begin{pmatrix} x_1 \\ x_2 \end{pmatrix} = 2k_1 {x_1}^2 - 2k_1 x_1 x_2 + k_1 {x_2}^2 = k_1 \left\{ {x_1}^2 + \left(x_1 - x_2\right)^2 \right\} \tag{5.5}$$

which shows the positive definiteness of the stiffness matrix. In structural dynamics it is common to have systems with nonpositive definite stiffness matrices. For example, systems having no constraints during vibration such as an aircraft structure in flight or

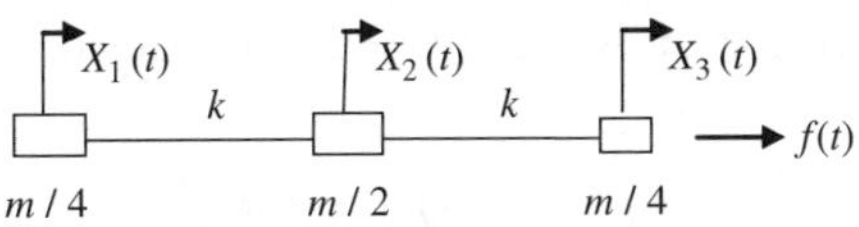

Figure 5.4 Free-free rod

a ship in floating condition, are examples of those having positive semi-definite stiffness matrices. As an illustration, consider the axially vibrating rod in Example 5.1 and impose free-free boundary conditions as shown in Figure 5.4. With the balance mass $m/4$ attached to the left end of the rod, the system has now 3 *dof*s.

Example 5.3: Axially vibrating rod with free-free boundary conditions – positive semi-definite system

The governing equations for the (semi-discrete) free-free rod are given by:

$$\begin{bmatrix} m/4 & 0 & 0 \\ 0 & m/2 & 0 \\ 0 & 0 & m/4 \end{bmatrix} \begin{Bmatrix} \ddot{x}_1 \\ \ddot{x}_2 \\ \ddot{x}_3 \end{Bmatrix} + \begin{bmatrix} k_1 & -k_1 & 0 \\ -k_1 & 2k_1 & -k_1 \\ 0 & -k_1 & k_1 \end{bmatrix} \begin{Bmatrix} x_1 \\ x_2 \\ x_2 \end{Bmatrix} = \begin{Bmatrix} 0 \\ 0 \\ f(t) \end{Bmatrix} \tag{5.6}$$

For this example, the quadratic form can be expanded to yield:

$$x^T Bx = \begin{pmatrix} x_1 & x_2 & x_3 \end{pmatrix} \begin{bmatrix} k_1 & -k_1 & 0 \\ -k_1 & 2k_1 & -k_1 \\ 0 & -k_1 & k_1 \end{bmatrix} \begin{pmatrix} x_1 \\ x_2 \\ x_2 \end{pmatrix} = k_1 \left\{ (x_1 - x_2)^2 + (x_2 - x_3)^2 \right\} \tag{5.7}$$

This function is identically zero for nonzero $x_j(t) j = 1, 2, 3$ with $x_1 = x_2 = x_3$, which renders the system positive semi-definite. More specifically, note that the 3×3 stiffness matrix is singular with rank = 2 (left as an exercise).

We will once more be coming across the notion of positive definiteness or otherwise of a system whilst discussing about the eigenvalue problem of symmetric matrices later in this chapter.

5.3 Coupled Linear ODEs and the Linear Differential Operator

The coupled, linear, second-order ODEs in Equation (5.1) representing an MDOF system describe, in general, the dynamics of either discrete systems or that of semi-discrete forms of continuous systems. It is now our interest to obtain solutions of these ODEs under different forcing functions and prescribed initial conditions. In the context of structural dynamics, one would often intuitively expect the solutions to be 'oscillatory' or periodic. This in turn leads us to the need to know the conditions under which such solutions are possible. Towards this, first recast the second-order ODEs in Equation (5.1) into a set of $2N$ first-order ODEs:

$$\dot{\boldsymbol{y}} = \boldsymbol{A}\boldsymbol{y} + \boldsymbol{g}(t) \tag{5.8}$$

The $2N$-dimensional state-space vector $\boldsymbol{y}(\mathrm{t}) = \left\{x_1, x_2, x_3, \ldots, x_n, \dot{x}_1, \dot{x}_2, \dot{x}_3 \ldots\ldots, \dot{x}_n\right\}^T$ includes both displacement and velocities as the unknown generalized coordinates or nodal variables. $\boldsymbol{g}(t) = \{\boldsymbol{0}, \boldsymbol{f}(t)\}^T$ is the forcing vector with '**0**' denoting the N

dimensional zero (null) vector. $\boldsymbol{A}$ is the $2N \times 2N$ dimensional (time-invariant) system matrix given by:

$$\boldsymbol{A} = \begin{bmatrix} \boldsymbol{O} & \boldsymbol{I} \\ -\boldsymbol{M}^{-1}\boldsymbol{K} & \boldsymbol{O} \end{bmatrix} \tag{5.9}$$

Here, $\boldsymbol{I}$ is the $N \times N$ identity matrix and $\boldsymbol{O}$ the $N \times N$ null matrix. The $N \times N$ matrix $\boldsymbol{M}^{-1}\boldsymbol{K}$ may be unsymmetric even if $\boldsymbol{M}$ and $\boldsymbol{K}$ are symmetric. We can look for a unique solution of Equation (5.8) provided the initial condition vector $\boldsymbol{y}(t = 0) := \boldsymbol{y}_0$ (contains both the initial displacement and velocity vectors) is specified. Indeed, a similar specification of the initial conditions is needed even if one attempts directly solving the second-order ODEs in Equation (5.1). If the forcing vector $\boldsymbol{g}(t) = \boldsymbol{0}$, the first-order ODEs are called autonomous or homogeneous (otherwise nonautonomous or nonhomogeneous).

Equivalently, one may write the coupled first-order ODEs $\dot{\boldsymbol{y}} = \boldsymbol{A}\boldsymbol{y}$ into a single linear homogeneous ODE of order $n = 2N$ with constant coefficients:

$$y^{(n)} + a_{n-1}y^{(n-1)} + \cdots\cdots + a_1 y^{(1)} + a_0 y = 0 \tag{5.10}$$

Here, $a_j \in \mathbb{R}$, $j = 0, 1, 2 \ldots n-1$. With $D = d/dt$, the associated linear differential operator is given by:

$$P(D) = D^n + a_{n-1}D^{n-1} + \ldots\ldots + a_1 D + a_0 \tag{5.11}$$

We know from the fundamental theorem of algebra that the auxiliary polynomial $P(m) = m^n + a_{n-1}m^{n-1} + \cdots\cdots + a_1 m + a_0$ factors into n linear factors as

$$P(m) = \prod_{j=1}^{n} \left(m - c_j\right) \tag{5.12}$$

In Equation (5.12), $c_j \in \mathbb{C}$, $j = 1, 2.....n$ (in general complex constants – distinct or otherwise). For any complex number c_j, the null space or Kernel (Appendix E) of the differential operator $(D - c_j)$ has $\left\{e^{c_j t}\right\}$ as a basis. Hence the general solution to the Equation (5.10) can be written as the following linear combination of elements in the basis set $\left\{e^{c_j t}\right\}$ whose elements are linearly independent (Appendix E) :

$$y(t) = \sum_{j=1}^{n} d_j e^{c_j t} \tag{5.13}$$

Here, $d_j \,(\in \mathbb{C})$, $j = 1, 2..n$ is a vector of constants that depend on the initial conditions.

5.4 Coupled Linear ODEs and Eigensolution

Albeit in a slightly different form, we exploit the fundamental feature of linear, time-invariant differential operators, embodied in Equation (5.11), in solving the coupled ODEs (Equation (5.8)). In the process, we make use of elementary linear algebra, the subject dealing with linear transformations (Appendix E), in a vector (linear) space $\boldsymbol{V}$. This reformulation has considerable practical significance as it enables dealing more easily with large dimensional systems. For purposes of clarity, note that the set of all solutions $\{y(t)\}$ to $\dot{y} = Ay$ belongs to a vector space V and the matrix $A : V \to V$ is the linear transformation of interest. Assume a solution to the homogeneous system $\dot{y} = Ay$ in the form:

$$y(t) = e^{\lambda t} v;\ v \in V \tag{5.14}$$

Substituting the above in $\dot{y} = Ay$, we obtain:

$$A\,v = \lambda v \Rightarrow [A - \lambda I]\,v = 0 \tag{5.15}$$

Equation (5.15) is known as the standard eigenvalue problem. The nontrivial solutions v of this equation form the null space of $A - \lambda I$. A nontrivial solution for the homogeneous Equation (5.15) is possible if and only if the coefficient matrix $A - \lambda I$ is singular. Such a condition of singularity is equivalent to the vanishing of the determinant, that is $\det(A - \lambda I) = 0$. Expansion of the determinant results in an algebraic polynomial equation (called the characteristic equation) in λ of degree $n = 2N$:

$$f_A(\lambda) = \lambda_n + \alpha_1 \lambda_{n-1} + \cdots + \alpha_n = 0 \tag{5.16}$$

$f_A(\lambda)$ is called the characteristic polynomial and $a_j \in \mathbb{R}$, $j = 1, 2 \ldots n$. The eigenvalues of A are the roots of the characteristic equation $f_A(\lambda) = 0$. Upon factorisation, we may write:

$$f_A(\lambda) = \prod_{j=1}^{n} (\lambda - \lambda_j) \tag{5.17}$$

The solution for $\dot{y} = Ay$ is now of the form:

$$y(t) = \sum_{j=1}^{n} c_j v_j e^{\lambda_j t} \tag{5.18}$$

$c_j \in \mathbb{C}$ are constants (in general complex), to be determined from the initial conditions known a priori. The vector $v_j \in v$ is an eigenvector of A corresponding to the eigenvalue λ_j if and only if $v_j \in N\left(A - \lambda_j I\right)$ where $N(\cdot)$ is the null space (Appendix E) of the linear transformation A. Thus, one may define the eigenspace E_{λ_j} corresponding to λ_j as:

$$E_\lambda = \left\{v_j \in V, A v_j = \lambda_j v_j\right\} = N\left(A - \lambda_j I\right) \tag{5.19}$$

and the eigenspace of the square matrix A is given by $\bigcup_j E_{\lambda_j}$. Note that, if v is an eigenvector corresponding to an eigenvalue λ, then cv is also an eigenvector for any scalar constant c. This is clear from Equation (5.15) since $Av = \lambda v \Rightarrow A(cv) = \lambda(cv)$.

The facts that the forms of the characteristic polynomials, $f_A(\lambda)$ and $P(m)$ are identical and that the present reformulation directly deals with a system of first-order (state space) equations bring into focus the advantage of the latter approach, especially for large dimensional dynamical systems, recast in the state space form. Indeed, given the naturally arising system of second-order ODEs for most structural dynamical systems, one could directly define an appropriate eigenvalue problem for such systems without taking recourse to the state space form. This is, in fact, the route commonly adopted in structural dynamics, as elaborated shortly afterwards.

Before we discuss further on the properties of eigenvalues and eigenvectors and their implications in obtaining the system response, it may be illuminating to consider eigensolutions for a few simple structural systems.

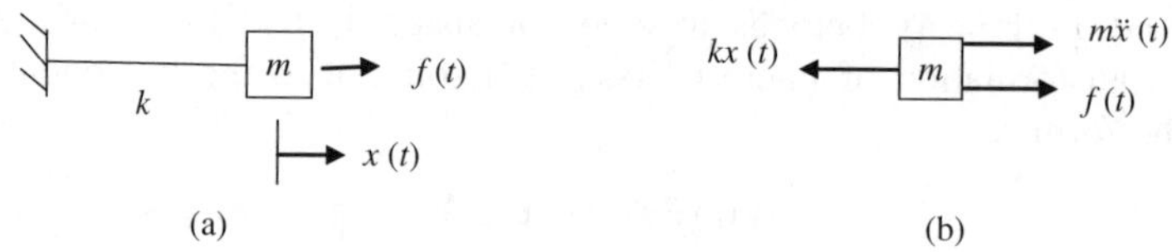

Figure 5.5 (a) SDOF oscillator and (b) free body diagram

Example 5.4: An SDOF oscillator

Consider the case of SDOF oscillator shown in Figure 5.5. The equation of motion for the undamped SDOF oscillator is given by the second-order ODE:

$$m\ddot{x}(t) + kx(t) = f(t) \tag{5.20}$$

The state space form corresponding to Equation (5.20) is now obtained as:

$$\begin{Bmatrix} \dot{y}_1(t) \\ \dot{y}_2(t) \end{Bmatrix} = \begin{bmatrix} 0 & 1 \\ -\dfrac{k}{m} & 0 \end{bmatrix} \begin{Bmatrix} y_1(t) \\ y_2(t) \end{Bmatrix} + \begin{Bmatrix} 0 \\ f(t) \end{Bmatrix} \tag{5.21}$$

In Equation (5.21), $y_1(t) = x(t)$ and $y_2(t) = \dot{x}(t) = \dot{y}_1(t)$ Thus, $\boldsymbol{A} = \begin{bmatrix} 0 & 1 \\ -\frac{k}{m} & 0 \end{bmatrix}$ and the eigensolution is obtained by equating to zero the determinant of $[\boldsymbol{A} - \lambda \boldsymbol{I}]$. Hence the characteristic equation is given by:

$$\det \left| \begin{bmatrix} 0 & 1 \\ -\dfrac{k}{m} & 0 \end{bmatrix} - \lambda \begin{bmatrix} 1 & 0 \\ 0 & 1 \end{bmatrix} \right| = 0 \Rightarrow \lambda^2 + \frac{k}{m} = 0 \Rightarrow \lambda = \pm i\omega \quad \text{with} \quad \omega = \sqrt{\frac{k}{m}} \tag{5.22}$$

The eigenvalues are $\lambda = \pm i\omega$ which are purely imaginary. Here, $i = \sqrt{-1}$. The null space for the eigenvalue $\lambda_1 = i\omega$ (henceforth called the first eigenvalue) is given by the span of the so-called first eigenvector $\mathbf{y}^{(1)} \in \boldsymbol{N}(\boldsymbol{A} - \lambda_1 \boldsymbol{I})$ (see Appendix E for the definition of the 'span' of a set of vectors). Thus, we seek nonzero solutions of:

$$\begin{aligned} &\left[\begin{bmatrix} 0 & 1 \\ -\dfrac{k}{m} & 0 \end{bmatrix} - \lambda_1 \begin{bmatrix} 1 & 0 \\ 0 & 1 \end{bmatrix} \right] \begin{Bmatrix} y_1 \\ y_2 \end{Bmatrix}^{(1)} = \begin{pmatrix} 0 \\ 0 \end{pmatrix} \Rightarrow \begin{bmatrix} -i\omega & 1 \\ -\dfrac{k}{m} & -i\omega \end{bmatrix} \begin{Bmatrix} y_1 \\ y_2 \end{Bmatrix}^{(1)} = \begin{pmatrix} 0 \\ 0 \end{pmatrix} \\ &\Rightarrow -i\omega y_1^{(1)} + y_2^{(1)} = 0 \\ &\quad -\frac{k}{m} y_1^{(1)} - i\omega y_2^{(1)} = 0 \end{aligned} \tag{5.23}$$

The 2×2 coefficient matrix in Equation (5.23) is singular and of rank 1. Utilising any one of the two equations and assuming $y_1 = 1$, we obtain the solution as:

$$\begin{Bmatrix} y_1 \\ y_2 \end{Bmatrix}^{(1)} = \begin{Bmatrix} 1 \\ i\omega \end{Bmatrix} \tag{5.24}$$

Since the two eigenvalues are complex conjugates, the second eigenvector (corresponding to λ_2) is the complex conjugate of $y^{(1)} = \begin{Bmatrix} 1 \\ i\omega \end{Bmatrix}$, that is $y^{(2)} = \begin{Bmatrix} y_1 \\ y_2 \end{Bmatrix}^{(2)} = \begin{Bmatrix} 1 \\ -i\omega \end{Bmatrix}$. Thus, the eigenspace $\boldsymbol{Q}$ of $\boldsymbol{A}$ is given by span $\begin{bmatrix} 1 & 1 \\ i\omega & -i\omega \end{bmatrix}$. The homogeneous solution to the oscillator is now written as:

$$\begin{aligned} y_1(t) &= c_1 e^{i\omega t} + c_2 e^{-i\omega t} \\ y_2(t) &= c_1 i\omega e^{i\omega t} - c_2 i\omega e^{-i\omega t} \end{aligned} \tag{5.25}$$

Note that $c_j (\in \mathbb{C})$, $j = 1, 2$ are constants to be determined from the initial conditions and are complex conjugates. Observe that, in Equation (5.25), $y_2(t)$ (being a velocity *degree of freedom*) is the derivative of $y_1(t)$. Further, these *degrees of freedom*, being response quantities of a physical system, must be real. In fact the solution in Equation (5.25), if expanded using Euler's formula, yields:

$$\begin{aligned} y_1(t) &= \left(\frac{c_1 + c_2}{2}\right) \cos \omega t + \left(\frac{c_1 - c_2}{2i}\right) \sin \omega t \\ y_2(t) &= -\left(\frac{c_1 + c_2}{2}\right) \omega \sin \omega t + \left(\frac{c_1 - c_2}{2i}\right) \omega \cos \omega t \end{aligned} \tag{5.26}$$

Example 5.5: Axially vibrating rod of Figure 5.1

Now consider the ODEs in Equation (5.3), corresponding to the lumped parameter idealisation of the axially loaded rod in Figure 5.1. The state space form for the governing equations of motion is:

$$\begin{Bmatrix} \dot{y}_1 \\ \dot{y}_2 \\ \dot{y}_3 \\ \dot{y}_4 \end{Bmatrix} = \begin{bmatrix} 0 & 0 & 1 & 0 \\ 0 & 0 & 0 & 1 \\ -4\dfrac{K_1}{m} & 2\dfrac{K_1}{m} & 0 & 0 \\ 4\dfrac{K_1}{m} & 4\dfrac{K_1}{m} & 0 & 0 \end{bmatrix} \begin{Bmatrix} y_1 \\ y_2 \\ y_3 \\ y_4 \end{Bmatrix} + \begin{Bmatrix} 0 \\ 0 \\ 0 \\ 4\dfrac{f(t)}{m} \end{Bmatrix} \tag{5.27}$$

where $y_1 := x_1$, $y_2 := x_2$, $y_3 := \dot{x}_1$, $y_4 := \dot{x}_2$. As in the last example, the present 4×4 system matrix $\boldsymbol{A}$ is unsymmetric even if the mass and stiffness matrices are symmetric. The characteristic equation is given by:

$$\det \left\| \begin{bmatrix} 0 & 0 & 1 & 0 \\ 0 & 0 & 0 & 1 \\ -4\dfrac{K_1}{m} & 2\dfrac{K_1}{m} & 0 & 0 \\ 4\dfrac{K_1}{m} & -4\dfrac{K_1}{m} & 0 & 0 \end{bmatrix} - \lambda \begin{bmatrix} 1 & 0 & 0 & 0 \\ 0 & 1 & 0 & 0 \\ 0 & 0 & 1 & 0 \\ 0 & 0 & 0 & 1 \end{bmatrix} \right\| = 0 \tag{5.28}$$

Expanding the above determinant and substituting $2EA/l$ for K_1 and ρAl for m, we obtain the following characteristic equation for λ:

$$\lambda^4 + 16\frac{E}{\rho}\lambda^2 + 32\frac{E^2}{\rho^2} = 0 \tag{5.29}$$

The four roots of the characteristic equation are given by $\lambda = \pm i2.343\sqrt{\frac{E}{\rho}}, \pm i13.657\sqrt{\frac{E}{\rho}}$, all of which are purely imaginary. With $\boldsymbol{y} = (y_1, y_2, y_3, y_4)^T$, the null spaces for the two pairs of imaginary eigenvalues are given by the span of eigenvectors $\boldsymbol{y} \in \boldsymbol{N}(\boldsymbol{A} - \lambda\boldsymbol{I})$. Thus, we seek nonzero solutions of:

$$\left[\begin{bmatrix} 0 & 0 & 1 & 0 \\ 0 & 0 & 0 & 1 \\ -4\frac{K_1}{m} & 2\frac{K_1}{m} & 0 & 0 \\ 4\frac{K_1}{m} & -4\frac{K_1}{m} & 0 & 0 \end{bmatrix} - \lambda \begin{bmatrix} 1 & 0 & 0 & 0 \\ 0 & 1 & 0 & 0 \\ 0 & 0 & 1 & 0 \\ 0 & 0 & 0 & 1 \end{bmatrix}\right] \begin{Bmatrix} y_1 \\ y_2 \\ y_3 \\ y_4 \end{Bmatrix} = \boldsymbol{0}$$

$$\Rightarrow \begin{bmatrix} -\lambda & 0 & 1 & 0 \\ 0 & -\lambda & 0 & 1 \\ -4\frac{K_1}{m} & 2\frac{K_1}{m} & -\lambda & 0 \\ 4\frac{K_1}{m} & -4\frac{K_1}{m} & 0 & -\lambda \end{bmatrix} \begin{Bmatrix} y_1 \\ y_2 \\ y_3 \\ y_4 \end{Bmatrix} = \boldsymbol{0} \tag{5.30}$$

Solving Equation (5.30) for each of the four eigenvectors and arranging them column-wise in the so-called eigenvector matrix $\boldsymbol{\phi}$, we get:

$$\boldsymbol{\phi} = \begin{bmatrix} 0.707 & 0.707 & -0.707 & -0.707 \\ 1.0 & 1.0 & 1.0 & 1.0 \\ 0.707\lambda_1 & 0.707\lambda_2 & -0.707\lambda_3 & -0.707\lambda_4 \\ \lambda_1 & \lambda_2 & \lambda_3 & \lambda_4 \end{bmatrix} \tag{5.31}$$

First two rows of $\boldsymbol{\phi}$ correspond to the two displacement coordinates y_1 and y_2 of the structural system and the last two rows to the velocity coordinates.

Example 5.6: Shear frame in Example 5.2

Considering Equation (5.4) corresponding to a shear frame, we have the following state space form:

$$\begin{Bmatrix} \dot{y}_1 \\ \dot{y}_2 \\ \dot{y}_3 \\ \dot{y}_4 \\ \dot{y}_5 \\ \dot{y}_6 \end{Bmatrix} = \begin{bmatrix} 0 & 0 & 0 & 1 & 0 & 0 \\ 0 & 0 & 0 & 0 & 1 & 0 \\ 0 & 0 & 0 & 0 & 0 & 1 \\ -\frac{K_1}{m_1} & \frac{K_1}{m_1} & 0 & 0 & 0 & 0 \\ \frac{K_1}{m_2} & -\frac{K_1+K_2}{m_3} & \frac{K_2}{m_2} & 0 & 0 & 0 \\ 0 & \frac{K_2}{m_3} & -\frac{K_2+K_3}{m_3} & 0 & 0 & 0 \end{bmatrix} \begin{Bmatrix} y_1 \\ y_2 \\ y_3 \\ y_4 \\ y_5 \\ y_6 \end{Bmatrix} + \begin{Bmatrix} 0 \\ 0 \\ 0 \\ \frac{f_1(t)}{m_1} \\ \frac{f_2(t)}{m_2} \\ \frac{f_3(t)}{m_3} \end{Bmatrix} \tag{5.32}$$

The state vector is $\{y(t)\} = \{y_1, y_2, y_3, y_4, y_5, y_6\}^T$ with $y_1 := x_1, y_2 := x_2, y_3 := x_3, y_4 := \dot{x}_1, y_5 := \dot{x}_2, y_6 := \dot{x}_3$ and the system matrix $\boldsymbol{A}$, of size 6×6, is again observed to be unsymmetric. Substituting the numerical values for the lumped stiffness and mass parameters, the characteristic equation for this example is given by:

$$\det\left|\begin{bmatrix} 0 & 0 & 0 & 1 & 0 & 0 \\ 0 & 0 & 0 & 0 & 1 & 0 \\ 0 & 0 & 0 & 0 & 0 & 1 \\ -600 & 600 & 0 & 0 & 0 & 0 \\ 400 & -1200 & 800 & 0 & 0 & 0 \\ 0 & 600 & -1500 & 0 & 0 & 0 \end{bmatrix} - \lambda \begin{bmatrix} 1 & 0 & 0 & 0 & 0 & 0 \\ 0 & 1 & 0 & 0 & 0 & 0 \\ 0 & 0 & 1 & 0 & 0 & 0 \\ 0 & 0 & 0 & 1 & 0 & 0 \\ 0 & 0 & 0 & 0 & 1 & 0 \\ 0 & 0 & 0 & 0 & 0 & 1 \end{bmatrix}\right| = 0 \quad (5.33)$$

Equation (5.33) simplifies to $\lambda^6 + 3300\,\lambda^4 + 270\,x\,10^4\lambda^2 + 432\,x\,10^6 = 0$ and is cubic in λ^2 and its roots are given by $\lambda^2 = -210$, -966 and -2124. Accordingly, the eigenvalues are $\lambda = \pm 14.5\,i$, $\pm 31.1\,i$ and $\pm\,46.1\,i$ all of which are again purely imaginary. The cubic equation is presently solved by the bisection method. First, the three intervals bracketing the three roots are found by evaluating the characteristic polynomial with respect to different λ^2. The three intervals thus obtained are $(-200, -250)$, $(-950, -1000)$ and $(-2100, -2150)$. For each of these intervals the bisection method is iteratively applied to approach the three roots within a reasonable tolerance. Figure 5.6 shows the graph of the characteristic polynomial versus $\lambda^2 \in (-4000, -100)$. A typical result is shown in Figure 5.7 in obtaining ${\lambda_1}^2 = -210.0$. In Figure 5.7, the two numerical values in small brackets stand for ${\lambda_1}^2$ and the corresponding value of the characteristic polynomial.

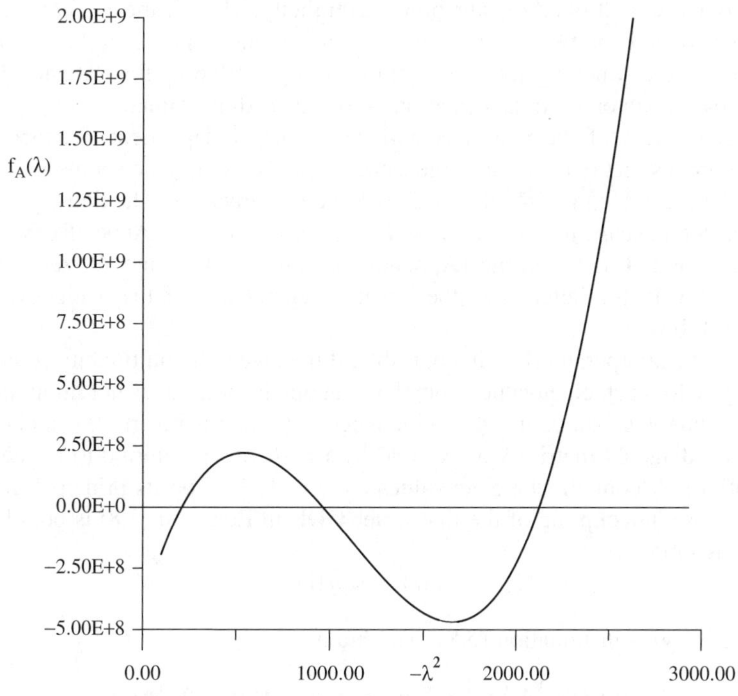

Figure 5.6 Graph of the characteristic polynomial corresponding to Equation (5.33)

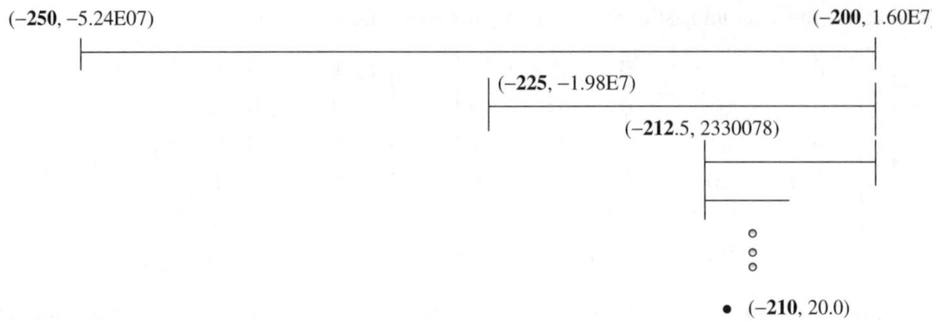

Figure 5.7 Bisection method, applied over the first interval $(-200, -250)$, to capture the first eigenvalue $= -210.0$ (numerical values in each of the brackets stand for $\lambda_1{}^2$ and the corresponding value of the characteristic polynomial)

5.5 First Order Equations and Uncoupling

In general, all the eigenvalues may not be distinct, that is the eigenvalue λ_j may appear m_j times (we say λ_j has an algebraic multiplicity m_j). Then $(\lambda - \lambda_j)^{m_j}$ is an elementary divisor of $f_A(\lambda)$. On the other hand, geometric multiplicity is the dimension of the eigenspace E_j and it is given by the maximum number of linearly independent vectors (Appendix E) in the associated null space. Clearly, when the eigenvalues are not distinct, we have geometric multiplicity $\leq$ algebraic multiplicity. Thus, if the algebraic multiplicity of every eigenvalue is unity, all the eigenvalues are distinct. In case all the eigenvalues are distinct, there is a distinct eigenvector v_j for each eigenvalue λ_j, that satisfies $\boldsymbol{A}\boldsymbol{v}_j = \lambda_j \boldsymbol{v}_j$.

For a linear operator or transformation $\boldsymbol{A}$ on an n-dimensional vector space $\boldsymbol{V}$, the element eigenvectors of the matrix $\boldsymbol{\phi}$ will be linearly independent if either (i) all the eigenvalues are distinct (so that each eigenspace E_j is 1D) or (ii) the geometric multiplicity is equal to the algebraic multiplicity for all repeated eigenvalues. The linearly independent set of eigenvectors spans the vector space $\boldsymbol{V}$ and forms its basis (Appendix E). Moreover, the transformation $\boldsymbol{A}$ is invertible (Appendix E) only if there are no zero eigenvalues in the spectrum. In the latter case, the linear independence of the n eigenvectors of $\boldsymbol{\phi}$ immediately follows.

Finally, the linear operator $\boldsymbol{A}$ is diagonisable, if the algebraic multiplicity equals geometric multiplicity for each eigenvalue. Note that diagonalisability does not imply invertibility of $\boldsymbol{A}$. If $\boldsymbol{A}$ is diagonisable, matrix $\boldsymbol{\phi}$ of eigenvectors (which is invertible) can be made use of to obtain a diagonal matrix $\boldsymbol{\Lambda}$ as $\boldsymbol{\Lambda} = \boldsymbol{\phi}^{-1}\boldsymbol{A}\boldsymbol{\phi} = diag[\lambda]$ (referred to as the similarity transformation). Λ contains the eigenvalues λ_j, $j = 1, 2\ldots$ on its principal diagonal.

In such a case, uncoupling of the first-order ODE in Equation (5.8) is possible. To this end, let us assume:

$$\boldsymbol{y}(t) = \boldsymbol{\phi}\, \boldsymbol{q}(t) \tag{5.34}$$

Substituting for $\boldsymbol{y}(t)$ in Equation (5.8), we obtain:

$$\dot{\boldsymbol{q}} = \boldsymbol{\phi}^{-1}\boldsymbol{A}\boldsymbol{\phi}\boldsymbol{q} + \boldsymbol{\phi}^{-1}\boldsymbol{g}(t) \Rightarrow \dot{\boldsymbol{q}} = \boldsymbol{D}\, q + \boldsymbol{\phi}^{-1}\boldsymbol{g}(t) \tag{5.35}$$

Equation (5.35) represents a set of n uncoupled first-order ODEs. These ODEs can be solved for $\boldsymbol{q}(t)$ either through the differential operator approach or the Laplace transform method. Following this, $\boldsymbol{y}(t)$ may be obtained from Equation (5.34). We take Example 5.6 to illustrate the diagonisation of the operator $\boldsymbol{A}$.

Example 5.7: Uncoupling of the system equations for the shear frame in Examples 5.2 and 5.6

For this case, the first-order ODEs are given in Equation (5.32). The system matrix $\boldsymbol{A}$ is given by:

$$\boldsymbol{A} = \begin{bmatrix} 0 & 0 & 0 & 1 & 0 & 0 \\ 0 & 0 & 0 & 0 & 1 & 0 \\ 0 & 0 & 0 & 0 & 0 & 1 \\ -\dfrac{K_1}{m_1} & \frac{K_1}{m_1} & 0 & 0 & 0 & 0 \\ \dfrac{K_1}{m_2} & -\dfrac{K_1+K_2}{m_3} & \dfrac{K_2}{m_2} & 0 & 0 & 0 \\ 0 & \dfrac{K_2}{m_3} & -\dfrac{K_2+K_3}{m_3} & 0 & 0 & 0 \end{bmatrix} \tag{5.36}$$

With substitution of the numerical values for the stiffness and mass parameters K_1, K_2, m_1, m_2 and m_3, the eigenvalues are found to be $\lambda = \pm 14.5\,i, \pm 31.1\,i$ *and* $\pm$ 46.1 i. For each of the six distinct eigenvalues, we find the associated eigenspaces (i.e. the six distinct eigenvectors) and thus form the eigenvector matrix as:

$$\boldsymbol{\phi} = \begin{bmatrix} 1.0 & 1.0 & 1.0 & 1.0 & 1.0 & 1.0 \\ 0.644 & 0.644 & -0.601 & -0.601 & -2.57 & -2.57 \\ 0.30 & 0.30 & -0.676 & -0.676 & 2.47 & 2.47 \\ \lambda_1 & \lambda_2 & \lambda_3 & \lambda_4 & \lambda_5 & \lambda_6 \\ 0.644\,\lambda_1 & 0.644\,\lambda_2 & -0.601\,\lambda_3 & -0.601\,\lambda_4 & -2.57\,\lambda_5 & -2.57\,\lambda_6 \\ 0.30\,\lambda_1 & 0.30\,\lambda_2 & -0.676\,\lambda_3 & -0.676\,\lambda_4 & 2.47\,\lambda_5 & 2.47\,\lambda_6 \end{bmatrix} \tag{5.37}$$

It can be verified that $\boldsymbol{\phi}^{-1}\boldsymbol{A}\boldsymbol{\phi} = \boldsymbol{\Lambda}$.

5.6 First Order versus Second Order ODE and Eigensolutions

The equations of motion (Equation (5.1)) with $\boldsymbol{f}(t) = \boldsymbol{0}$ constitute a conservative system in that the energy is conserved over any finite interval $[0, T]$ of interest. If we premultiply both sides of the Equation (5.1) by $\dot{\boldsymbol{x}}^T$, we obtain:

$$\dot{\boldsymbol{x}}^T\boldsymbol{M}\ddot{\boldsymbol{x}} + \dot{\boldsymbol{x}}^T\boldsymbol{K}\,\boldsymbol{x} = \dot{\boldsymbol{x}}^T\boldsymbol{f} \tag{5.38}$$

Equation (5.38) can be rewritten as

$$\frac{d}{dt}\left(\frac{1}{2}\dot{\boldsymbol{x}}\boldsymbol{M}\dot{\boldsymbol{x}}\right) + \frac{d}{dt}\left(\frac{1}{2}\boldsymbol{x}^T\,\boldsymbol{K}\,\boldsymbol{x}\right) = \dot{\boldsymbol{x}}^T\boldsymbol{f} \tag{5.39}$$

The two terms within the parentheses in Equation (5.39) are recognised as the kinetic energy (KE) and the potential energy (PE) of the structural system respectively (in the order of their appearance) and hence in the absence of external forces, we can rewrite Equation (5.39) as:

$$\frac{d}{dt}(KE + PE) = 0 \tag{5.40}$$

This shows that for a conservative and unforced system, the total energy $= KE + PE$ is time-invariant and hence conserved. The equations considered so far (Equations (5.1) and (5.8)) pertain to undamped structural systems. The eigenvalue problem in Equation (5.15) refers to the state-space representation via the coupled first-order ODEs (Equation (5.8)) of a conservative system. Through this formalism, we have established that the eigenvalues are purely imaginary (and hence the solutions oscillatory) for such conservative systems. We also observe that the matrix $\boldsymbol{A}$ is always of the form in Equation (5.9) and the eigenvalue problem $[\boldsymbol{A} - \lambda \boldsymbol{I}]\,\boldsymbol{v} = \boldsymbol{0}$ can also be stated as

$$\begin{bmatrix} -\lambda & \boldsymbol{I} \\ -\boldsymbol{M}^{-1}\boldsymbol{K} & -\lambda \end{bmatrix} \boldsymbol{v} = \boldsymbol{0} \tag{5.41}$$

The characteristic equation $\det(\boldsymbol{A} - \lambda \boldsymbol{I}) = 0$ takes the form:

$$\left\| \begin{bmatrix} -\lambda & \boldsymbol{I} \\ -\boldsymbol{M}^{-1}\boldsymbol{K} & -\lambda \end{bmatrix} \right\| = 0 \Rightarrow \left| \left(\boldsymbol{M}^{-1}\boldsymbol{K} + \lambda^2 \boldsymbol{I} \right) \right| = 0 \tag{5.42}$$

Since $\boldsymbol{y}(t) = \left\{ \boldsymbol{x}^T(t), \dot{\boldsymbol{x}}^T(t) \right\}^T$ and $\boldsymbol{y}(t) = e^{\lambda t}\boldsymbol{v}$, we can identify the $2N \times 1$ eigenvector, $\boldsymbol{v} = \begin{Bmatrix} \boldsymbol{u} \\ \lambda \boldsymbol{u} \end{Bmatrix}$, where $\boldsymbol{u}$ denotes the $N \times 1$ vector of displacement coordinates $\boldsymbol{x}(t)$ and $\lambda \boldsymbol{u}$, the vector of velocity coordinates $\dot{\boldsymbol{x}}(t)$. These two solution vectors can be expressed as:

$$\boldsymbol{x}(t) = e^{\lambda t}\boldsymbol{u} \;\text{ and }\; \dot{\boldsymbol{x}}(t) = \lambda e^{\lambda t}\boldsymbol{u} \tag{5.43}$$

Hence from Equation (5.42), it is possible to write an equivalent eigenvalue problem within a reduced N-dimensional vector space in terms of the eigenvalue parameter λ^2 (not λ) as:

$$\left[\boldsymbol{M}^{-1}\boldsymbol{K} + \lambda^2 \boldsymbol{I} \right] \boldsymbol{u} = \boldsymbol{0} \Rightarrow \boldsymbol{M}^{-1}\boldsymbol{K}\,\boldsymbol{u} = \omega^2 \boldsymbol{I}\,\boldsymbol{u} \text{ with } \omega^2 = -\lambda^2 \tag{5.44}$$

For a conservative structural system, one can then directly solve the above (reduced) eigenvalue problem, in preference to the state space approach that results in a higher sized eigenvalue problem.

As noted before, the matrix $\boldsymbol{M}^{-1}\boldsymbol{K}$ in Equation (5.44) may be unsymmetric, even if $\boldsymbol{M}$ and $\boldsymbol{K}$ are symmetric. However, the eigenvalue problem above (Equation (5.44)) may be readily converted to a symmetric form by rewriting it as:

$$\boldsymbol{K}\,\boldsymbol{u} - \omega^2 \boldsymbol{M}\,\boldsymbol{u} = \boldsymbol{0} \Rightarrow \boldsymbol{K}\,\boldsymbol{u} = \omega^2 \boldsymbol{M}\,\boldsymbol{u} \tag{5.45}$$

This is known as the generalized eigenvalue problem. Indeed, it is possible to reduce this form into the canonical form of the standard eigenvalue problem $[\boldsymbol{B} - \lambda \boldsymbol{I}]\,\boldsymbol{u} = \boldsymbol{0}$, whilst maintaining the symmetry of $\boldsymbol{B}$, by Cholesky decomposition of either the stiffness or the mass matrix. Decomposing, for instance, the mass matrix (assumed to be positive definite) as $\boldsymbol{M} = \boldsymbol{L}\,\boldsymbol{L}^T$ (where $\boldsymbol{L}$ is a nonsingular lower triangular matrix), substituting $\boldsymbol{L}\,\boldsymbol{L}^T$ for $\boldsymbol{M}$ in Equation (5.44) and finally writing the response vector $\boldsymbol{u} = \boldsymbol{L}^{-T}\boldsymbol{q}$, we get:

$$\mathrm{K}\,\boldsymbol{u} - \omega^2 \boldsymbol{L}\,\boldsymbol{L}^T \boldsymbol{u} = 0 \Rightarrow \boldsymbol{L}^{-1}\boldsymbol{K}\boldsymbol{L}^{-T}\boldsymbol{q} = \omega^2 \boldsymbol{q} \Rightarrow \boldsymbol{B}\,\boldsymbol{q} = \omega^2 \boldsymbol{q} \tag{5.46}$$

The matrix $\boldsymbol{B} = \boldsymbol{L}^{-1}\boldsymbol{K}\boldsymbol{L}^{-T}$ is indeed symmetric. Equation (5.46) is now a symmetric eigenvalue problem yielding the same eigenvalues as those of Equation (5.44). Moreover, the eigenvectors $\boldsymbol{u}$ are linearly related to $\boldsymbol{q}$ via $\boldsymbol{u} = \boldsymbol{L}^{-\mathrm{T}}\boldsymbol{q}$.

It is known that for a symmetric eigenvalue problem, all the eigenvalues ω^2 are real and the eigenvectors are real and orthogonal (Appendix E). The orthogonal property is given by

$$\boldsymbol{q}_j{}^T\boldsymbol{q}_k = \delta_{jk} \quad \text{and} \quad \boldsymbol{q}_j{}^T\boldsymbol{B}\boldsymbol{q}_k = \omega_j{}^2\delta_{jk} \tag{5.47}$$

Thus, the eigenvalue problem in Equation (5.45) gives positive real eigenvalues in terms of ω^2. Since $\lambda^2 = -\omega^2$, the eigenvalues λ of the generalized eigenvalue problem in Equation (5.45) or of the standard form in Equation (5.46) are pure imaginary which is quite consistent with our earlier observation on a conservative system. Accordingly, whilst adopting the symmetric eigenvalue problem (Equation (5.45)), the solution to the Equation (5.1) must be taken as $\boldsymbol{x}(\mathrm{t}) = \mathrm{e}^{\mathrm{i}\lambda \mathrm{t}}\boldsymbol{u}$. This eventually ensures that the response is finally obtained in real quantities (as it ought to be) for the physical system.

In view of the transformation $\boldsymbol{u} = \boldsymbol{L}^{-\mathrm{T}}\boldsymbol{q}$, the orthogonal property in Equation (5.47) can be rewritten in terms of $\boldsymbol{u}$ and it is of the form:

$$\begin{aligned} &\boldsymbol{u}_j{}^T\boldsymbol{L}\boldsymbol{L}^T\boldsymbol{u}_k = \delta_{jk} \Rightarrow \boldsymbol{u}_j^T\boldsymbol{M}\boldsymbol{u}_k = \delta_{jk} \text{ and} \\ &\boldsymbol{u}_j^T\boldsymbol{L}\boldsymbol{L}^{-1}\boldsymbol{K}\,\boldsymbol{L}^{-T}\boldsymbol{L}^T\boldsymbol{u}_k = \omega^2\delta_{jk} \Rightarrow \boldsymbol{u}_j{}^T\boldsymbol{K}\boldsymbol{u}_k = \omega_j{}^2\delta_{jk} \end{aligned} \tag{5.48}$$

If now we take the eigenvector matrix as $\boldsymbol{\phi} = \left[\boldsymbol{u}_1, \boldsymbol{u}_2 \ldots\ldots.., \boldsymbol{u}_N\right]$, then the orthogonal properties in Equation (5.48) can be expressed as:

$$\boldsymbol{\phi}^T\boldsymbol{M}\boldsymbol{\phi} = \boldsymbol{I} \text{ and } \boldsymbol{\phi}^T\boldsymbol{K}\boldsymbol{\phi} = [\boldsymbol{\Lambda}] = diag\left[\omega^2\right] \tag{5.49}$$

Equation (5.49) indicates that the eigenvector matrix $\boldsymbol{\phi}$ is orthogonal to both the mass and stiffness matrices.

5.7 MDOF Systems and Modal Dynamics

With the help of the orthogonal property of $\boldsymbol{\phi}$ in Equation (5.49), we decouple the equations of motion (5.1). To this end, we use the transformation:

$$\boldsymbol{x}(t) = \boldsymbol{\phi}\boldsymbol{q}(t) = q_1(t)\boldsymbol{\phi}_1 + q_1(t)\boldsymbol{\phi}_2 + \cdots\cdots\cdots\cdots + q_N(t)\boldsymbol{\phi}_N \tag{5.50}$$

$\boldsymbol{q}(t)$ is an N-dimensional vector of unknown functions of time called generalized coordinates or modal coordinates. Substitution in Equation (5.1) yields:

$$\boldsymbol{M}\boldsymbol{\phi}\ddot{\boldsymbol{q}} + \boldsymbol{K}\boldsymbol{\phi}\boldsymbol{q} = \boldsymbol{f}(\mathrm{t}) \tag{5.51}$$

Premultiplying both sides by $\boldsymbol{\phi}^T$ and utilising the orthogonality of $\boldsymbol{\phi}$, Equation (5.51) takes the form:

$$\ddot{q}(\boldsymbol{t}) + \boldsymbol{\Lambda}\boldsymbol{q}(\boldsymbol{t}) = \boldsymbol{\phi}^T\boldsymbol{f}(t) \tag{5.52}$$

Equation (5.52) is a set of uncoupled ODEs in the elements of $\boldsymbol{q}\ (\boldsymbol{t})$. Each of these modal equations represents an SDOF oscillator, which can be solved far more easily than the parent coupled system. Solution of each ODE requires information on initial conditions

$\boldsymbol{q}_0 = \boldsymbol{q}(0)$ and $\dot{\boldsymbol{q}}_0 = \dot{\boldsymbol{q}}(0)$. These initial conditions in modal space are derivable from the specified initial condition vectors $\boldsymbol{x}_0 = \boldsymbol{x}(0)$ and $\dot{\boldsymbol{x}}_0 = \dot{\boldsymbol{x}}(0)$ as:

$$\boldsymbol{q}_0 = \boldsymbol{\phi}^T \boldsymbol{M}\, \boldsymbol{x}_0 \text{ and } \dot{\boldsymbol{q}}_0 = \boldsymbol{\phi}^T \boldsymbol{M}\, \dot{\boldsymbol{x}}_0 \tag{5.53}$$

Thus, for conservative systems characterised by symmetric mass and stiffness matrices, the coupled ODEs can be decoupled by a similarity transformation using the eigenvector matrix $\boldsymbol{\phi}$. The matrix $\boldsymbol{\phi}$ is also known as the modal matrix with each column $\boldsymbol{\phi}_j$ representing an eigenvector or a vibration mode corresponding to each eigenvalue λ_j. The set $\{\lambda_j\}$ is more popularly referred to as the set of natural frequencies. The system response $\boldsymbol{x}(t)$ in Equation (5.50) is a superposition of modal vectors multiplied by the generalized coordinate vector $\boldsymbol{q}(t)$. This is the so-called 'modal superposition technique' in structural dynamics which is based on the concept that any vector $\boldsymbol{x}$ in $\boldsymbol{v}$ is a linear combination of the basis vectors.

5.7.1 SDOF Oscillator and Modal Solution

The equation of motion for an undamped SDOF oscillator is given in Equation (5.20) and in standard form it is written as:

$$\ddot{q}(t) + \omega_n{}^2 q(t) = \mathrm{f}(t) \tag{5.54}$$

ω_n is the undamped natural frequency of the SDOF oscillator. In transform approach, we obtain the solution in s-domain in the form:

$$\begin{aligned} \left(s^2 + \omega_n{}^2\right) Q(s) &= s q_0 + \dot{q}_0 + F(s) \\ \Rightarrow Q(s) &= \frac{s q_0 + \dot{q}_0}{s^2 + \omega_n{}^2} + \frac{F(s)}{s^2 + \omega_n{}^2} \end{aligned} \tag{5.55a,b}$$

q_0 and $\dot{q}_0$ are the initial conditions. $Q(s)$ and $F(s)$ are the Laplace transforms of $q(t)$ and $f(t)$. In Equation (5.55b), the first term on right-hand side is the contribution to system response from initial conditions, the so-called complementary solution. The second term represents the particular solution due to the external forcing function $f(t)$. Given the initial conditions, the complementary solution is obtained as the inverse transform of $\frac{sq_0+\dot{q}_0}{s^2+\omega_n{}^2}$ and given by:

$$q^{(c)}(t) = q_0 \cos \omega_n t + \frac{\dot{q}_0}{\omega_n} \sin \omega_n t \tag{5.56}$$

With $H(s) = \frac{1}{s^2+\omega_n{}^2}$ the particular solution is obtained as:

$$q^{(p)}(t) = \mathcal{L}^{-1} H(s) F(s) = H(s)^* F(s) \tag{5.57}$$

'*' stands for convolution integral given by:

$$\begin{aligned} q^{(p)}(t) &= \int_0^t h\,(t-\tau)\,\mathrm{f}(\tau) d\tau \text{ or} \\ &= \int_0^t h(\tau) \mathrm{f}(t-\tau) d\tau \end{aligned} \tag{5.58}$$

Here, $h(t) = \mathcal{L}^{-1}H(s)$. Notice that if $F(s) = 1$, $q^{(p)}(t) = h(t)$. Since the Laplace transform of an impulse function $I(t)$ is unity, $h(t)$ is known as the impulse response function of the oscillator. It is given by:

$$h(t) = \mathcal{L}^{-1}\left(\frac{1}{s^2 + \omega_n{}^2}\right) = \frac{\sin \omega_{\mathrm{n}} \mathrm{t}}{\omega_{\mathrm{n}}} \tag{5.59}$$

For linear time invariant, stable and causal systems (which is the subject matter of this chapter), note that $\int_{-\infty}^{\infty} |h(\tau)|\, d\tau < \infty$ (stability condition) and $h(\tau) = 0$ for $\tau < 0$ (causality condition). Further, the time invariant nature is manifest in the property that if an impulse $\delta(t)$ at $t = 0$ causes the response $h(t)$, the impulse $\delta(t - \tau)$ at $t = \tau$ causes the response $h(t - \tau)$. By treating each modal equation in (Equation (5.52)) as an SDOF oscillator, the modal response vector $\boldsymbol{q}(t)$ is obtained as $\boldsymbol{q}(t) = \boldsymbol{q}^{(c)}(t) + \boldsymbol{q}^{(p)}(t)$ from which the response of the MDOF system (Equation (5.1)) can be determined from Equation (5.50).

Example 5.8: Shear frame in Example 5.2: Response under harmonic excitation

Referring to Figure 5.3, the coupled second-order ODEs in terms of the stiffness and mass matrices are given by Equation (5.4). With the numerical values given in the figure, the generalized eigenvalue problem (Equation (5.45)) is given by:

$$\left[\begin{bmatrix} 600 & -600 & 0 \\ -600 & 1800 & -1200 \\ 0 & -1200 & 3000 \end{bmatrix} - \omega^2 \begin{bmatrix} 1.0 & 0 & 0 \\ 0 & 1.5 & 0 \\ 0 & 0 & 2 \end{bmatrix}\right] \boldsymbol{u} = \boldsymbol{0} \tag{5.60}$$

It is easy to formulate the characteristic equation from the above equation as:

$$\left|\begin{bmatrix} 600 - \omega^2 & -600 & 0 \\ -600 & 1800 - 1.5\omega^2 & -1200 \\ 0 & -1200 & 3000 - 2\omega^2 \end{bmatrix}\right| = 0$$

$$\Rightarrow \omega^6 - 3300\omega^4 + 270 \times 10^4\omega^2 - 432 \times 10^6 = 0 \tag{5.61}$$

This is similar to the one obtained with Equation (5.33) but now yields positive real eigenvalues. Thus, we get the eigenvalues or the natural frequencies of the system as 14.5, 31.1 and 46.1 rad/s. The corresponding eigenvector/modeshape for each ω are obtained as the solution from the homogeneous equations:

$$\begin{bmatrix} 600 - \omega^2 & -600 & 0 \\ -600 & 1800 - 1.5\omega^2 & -1200 \\ 0 & -1200 & 3000 - 2\omega^2 \end{bmatrix} \boldsymbol{u} = \boldsymbol{0} \tag{5.62}$$

The eigenvectors are $u_1 = \{1.0 \quad 0.644 \quad 0.3\}^T$ for $\omega_1 = 14.5$ rad/s, $u_2 = \{1.0 \quad -0.601 \quad -0.676\}^T$ for $\omega_2 = 31.1$ rad/s and $u_3 = \{1.0 \quad -2.57 \quad 2.47\}^T$ for $\omega_3 = 46.1$ rad/s. The vectors are so normalised that the first elements are unity. Thus, the modal matrix for this example is given by:

$$\boldsymbol{\phi} = \begin{bmatrix} 1.0 & 1.0 & 1.0 \\ 0.644 & -0.601 & -2.57 \\ 0.300 & -0.676 & 2.47 \end{bmatrix} \tag{5.63}$$

The modal matrix is orthogonal to mass and stiffness matrices. For example:

$$\boldsymbol{\phi}^T \boldsymbol{M} \boldsymbol{\phi} = \begin{bmatrix} 1.803 & 0 & 0 \\ 0 & 2.456 & 0 \\ 0 & 0 & 23.10 \end{bmatrix} \text{ and } \boldsymbol{\phi}^T \boldsymbol{K} \boldsymbol{\phi} = \begin{bmatrix} 380.04 & 0 & 0 \\ 0 & 2367.23 & 0 \\ 0 & 0 & 49110.48 \end{bmatrix} \tag{5.64a,b}$$

The numerical quantities on the diagonal of $\boldsymbol{\phi}^T \boldsymbol{M} \boldsymbol{\phi}$ above are the modal masses m_j, $j = 1, 2$ and 3 and the values on the diagonal of $\boldsymbol{\phi}^T \boldsymbol{K} \boldsymbol{\phi}$ are the corresponding modal stiffnesses k_j, $j = 1, 2$ and 3. Thus, the 3 dof discrete system is reduced to three independent SDOF oscillators (modal equations) represented by second-order DE (see Equation (5.52)):

$$m_j \ddot{q}(t) + k_j q(t) = \sum_{k=1}^{3} \phi_{jk} f_k(t) \tag{5.65}$$

Ratio of each pair of stiffness k_j and mass m_j gives the square of the undamped natural frequencies, $\omega_j{}^2$, $j = 1, 2$ and 3 of the three modes. We can orthonormalise the modal matrix such that $\boldsymbol{\phi}^T \boldsymbol{M} \boldsymbol{\phi} = \boldsymbol{K}$ and $\boldsymbol{\phi}^T \boldsymbol{K} \boldsymbol{\phi} = diag[\omega^2]$. To this end, we divide each column vector – that is each mode shape – by the square root of its modal mass. This gives us:

$$\boldsymbol{\phi} = \begin{bmatrix} 0.745 & 0.638 & 0.208 \\ 0.480 & -0.383 & -0.535 \\ 0.225 & -0.431 & 0.514 \end{bmatrix} \tag{5.66}$$

The mode shapes for the shear frame are shown in Figure 5.8 below.

Now it is possible to write the modal equations as in Equation (5.65) as:

$$\ddot{q}_j + \omega_j{}^2 q_j = \sum_{i=1}^{3} \phi_{ij} f_i(t), \; j = 1, 2 \text{ and } 3 \tag{5.67}$$

For the sake of illustration, let us take two cases of forcing functions – (i) step input and (ii) harmonic excitation – acting at the first mass $m_1 = 1.0$. Let the excitation vector

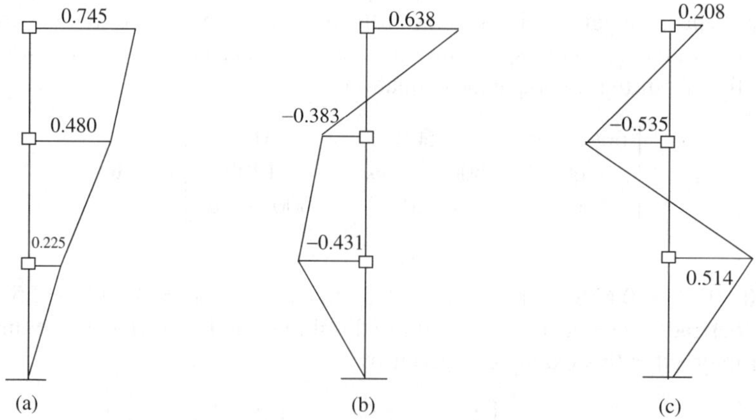

Figure 5.8 8 Mode shapes of shear frame; (a) $\omega = 14.52$ rad/s, (b) $\omega = 31.1$ rad/s and (c) 46.1 rad/s. (see Equation 5.66)

$f(t) = \{f_1(t), \quad 0.0, \quad 0.0\}^T$. The modal forces in Equation (5.67) are $\boldsymbol{\phi}^T f(t)$, presently given by $\boldsymbol{\phi}^T f(t) = \{0.745 f_1, 0.480 f_1, 0.225 f_1\}^T$.

Case (a) Step function input at the top mass point (Figure 5.9b).

The step load (suddenly applied load at $t = 0$) is defined as:

$$f_1(t) = F, \quad t \geq 0 \\ = 0, \quad \text{otherwise} \tag{5.68}$$

The modal equations are given by:

$$\ddot{q}_1 + {\omega_1}^2 q_1 = 0.745 f_1(t) \\ \ddot{q}_2 + {\omega_2}^2 q_2 = 0.480 f_1(t) \\ \ddot{q}_3 + {\omega_3}^2 q_3 = 0.225 f_1(t) \tag{5.69a–c}$$

We refer to Equation (5.54) and obtain solution to the above modal equations by transform approach. Noting that in the present case of suddenly applied load (Equation (5.68)), $F(s) = F/s$ the solution due to the suddenly applied load is obtained as:

$$Q(s) = \frac{sq_0 + \dot{q}_0}{s^2 + {\omega_n}^2} + \frac{F}{s\left(s^2 + {\omega_n}^2\right)} = \frac{sq_0 + \dot{q}_0}{s^2 + {\omega_n}^2} + \frac{F}{{\omega_n}^2}\left(\frac{1}{s} - \frac{s}{s^2 + {\omega_n}^2}\right) \tag{5.70}$$

q_0 and $\dot{q}_0$ are the initial conditions. Inverse transform gives the solution for $q(t)$ as:

$$q(t) = q_0 \cos \omega_n t + \frac{\dot{q}_0}{\omega_n} \sin \omega_n t + \frac{F}{{\omega_n}^2}\left(1 - \cos \omega_n t\right) \tag{5.71}$$

The last term in the above equation represents the particular solution and it can also be obtained from convolution integral in Equation 5.58. Suppose if we substitute the

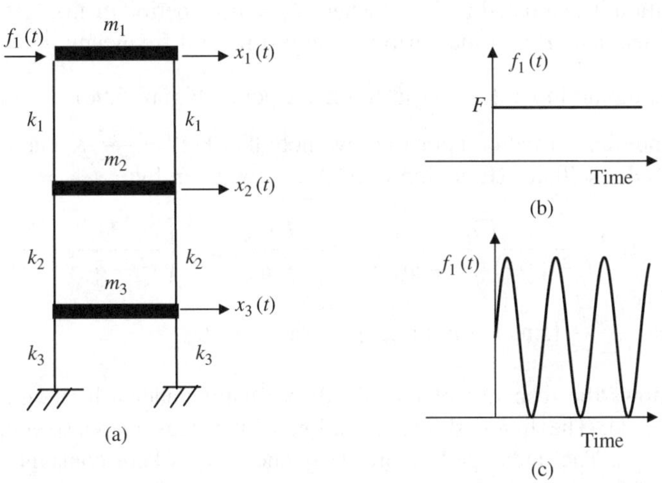

Figure 5.9 Shear frame with suddenly applied load at top; (a) lumped parameter model, (b) step input and (c) harmonic input $f_1(t) = F \sin \lambda t$

expressions of $h(t)$ (see Equation (5.59)) and $f_1(t)$ in the convolution integral and perform the integration, we get:

$$\begin{aligned} q^{(p)}(t) &= \frac{F}{\omega_n}\int_0^t \sin\omega_n(t-\tau)\,d\tau \\ &= \frac{F}{\omega_n}\left\{\sin\omega_n t\int_0^t \cos\omega_n\tau d\tau - \cos\omega_n t\int_0^t \sin\omega_n\tau d\tau\right\} \\ &= \frac{F}{{\omega_n}^2}(1-\cos\omega_n t) \end{aligned} \tag{5.72}$$

This solution is same as the one obtained by transform approach. Now in view of Equation (5.71), we can write the solution to Equation (5.69a–c) as:

$$\begin{aligned} q_1(t) &= \frac{F}{{\omega_1}^2}\left(1-\cos\omega_1 t\right) \\ q_2(t) &= \frac{F}{{\omega_2}^2}\left(1-\cos\omega_2 t\right) \\ q_3(t) &= \frac{F}{{\omega_3}^2}\left(1-\cos\omega_3 t\right) \end{aligned} \tag{5.73a–c}$$

In writing the solutions for $q_j(t)$, $j = 1, 2$ and 3, zero initial conditions are assumed for $\boldsymbol{x}(t)$ and $\dot{\boldsymbol{x}}(t)$ and hence in view of Equation (5.53), $\boldsymbol{q}_0$ and $\dot{\boldsymbol{q}}_0$ are both zero. The solution for the physical coordinates $\boldsymbol{x}(t)$ is now given by the modal summation in Equation (5.50) as:

$$x_i(t) = \sum_{j=1}^{3}\phi_{ij}q_j(t), \quad i = 1, 2 \text{ and } 3 \tag{5.74}$$

Figure 5.10 shows the response histories obtained from Equation (5.74) for the three mass points. The total duration of time is taken to be ~10 times T_1 and the time step at which the solution is evaluated is $T_3/10$ where T_1 = time period of first natural frequency $= \frac{2\pi}{\omega_1} = 0.433$ sec. and T_3 = time period of largest natural frequency $\frac{2\pi}{\omega_3} = 0.136$ sec.

Case (b) Harmonic input $F\sin\lambda t$ at top mass point (Figure 5.9c)

Following the transformation approach, we note that $F(s) = \frac{F\lambda}{s^2+\lambda^2}$. The particular solution for an SDOF oscillator (Equation (5.54)) is now given by:

$$\begin{aligned} q^{(p)}(t) &= \mathcal{L}^{-1}\frac{F\lambda}{\left(s^2+\lambda^2\right)\left(s^2+{\omega_n}^2\right)} = \frac{F\lambda}{\lambda^2-{\omega_n}^2}\mathcal{L}^{-1}\left\{\frac{1}{s^2+{\omega_n}^2}-\frac{1}{s^2+\lambda^2}\right\}. \\ &= \frac{F/{\omega_n}^2}{1-r^2}\left(\sin\lambda t - r\sin\omega_n t\right) \quad \text{with } r = \lambda/\omega_n \end{aligned} \tag{5.75}$$

Figure 5.11 shows the response of the SDOF oscillator under a harmonic input of unit amplitude ($F = 1$). The figure shows variation of the peak response amplitude vs the parameter $r = \lambda/\omega$. The undamped natural frequency, ω_n is kept constant at 14.52 rad/s, the frequency of first fundamental mode of the shear frame. The peak response amplitude (positive or negative) for each value of λ/ω is obtained from corresponding time history of $q^{(p)}(t)$ in Equation (5.75). The response graph in Figure 5.11, known as the frequency

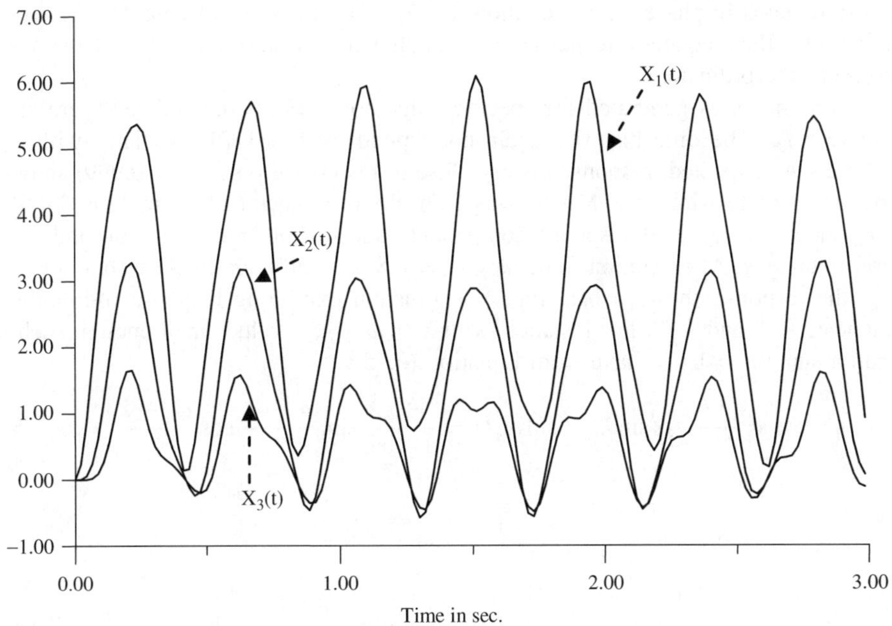

Figure 5.10 Shear frame under step input loading; displacement response at the three mass points

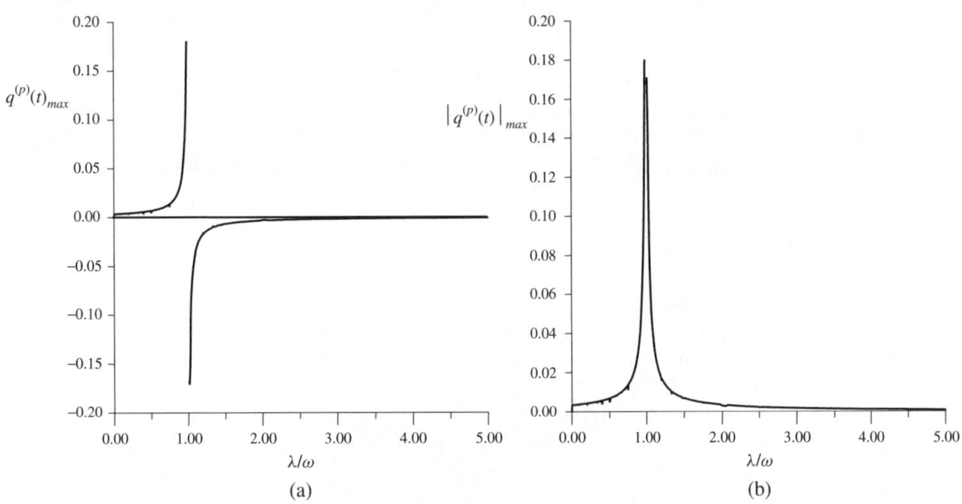

Figure 5.11 Response of an SDOF oscillator under harmonic input $F \sin \lambda t$ (with $F = 1$); (a) peak response $q^{(p)}(t)_{\max}$ vs $^{\lambda}/_{\omega}$ and (b) amplitude (absolute peak), $|q^{(p)}(t)|_{\max}$ vs $^{\lambda}/_{\omega}$

response curve, shows discontinuity at $\lambda = \omega$. As $\frac{\lambda}{\omega} \to 1^-$ (i.e. from the left), the solution approaches ∞ because of the term $(1 - r^2)$ in the denominator of $q^{(p)}(t)$, and if approached from right, it tends to $-\infty$. This is commonly known as a state of resonance where excitation frequency coincides with the natural frequency of the system. Note that at the point of resonance, the response phase angle changes from 0 to 180°. The response remains in phase with excitation till $\lambda/\omega < 1$ and is out of phase for $\lambda/\omega > 1$. In Figure 5.11b, the frequency response curve is shown with absolute value of the system response as the ordinate.

Few time histories generated for specific values of $r = 0.5, 0.8$ and 0.999 are shown in Figure 5.12. The time histories again correspond to the SDOF oscillator with $\omega_n =$ 14.52 rad/s As expected, response history close to resonance point ($r = 0.999$) shows an unbounded growth with time. Much away from the resonance (for example $r = 0.5$) the history shows presence of response components due to both the excitation and system natural frequency. At excitation frequency $\lambda(= 0.8\ \omega_n\ )$ sufficiently close (but not equal) to ω_n, the response shows a beat type phenomenon exhibiting response oscillations at frequencies $\frac{\lambda+\omega}{2}$ and $\frac{\lambda-\omega}{2}$. For instance, with $\lambda \approx \omega - \varepsilon$, a value small enough such that we can assume $r = 1$, we have from Equation (5.75):

$$q^{(p)}(t) = \frac{F/\omega_n{}^2}{1-r^2}\left(\sin\lambda t - \sin\omega_n t\right) = \frac{2F/\omega_n{}^2}{1-r^2}\sin\frac{\omega_n - \lambda}{2}\cos\frac{\omega_n + \lambda}{2} \tag{5.76}$$

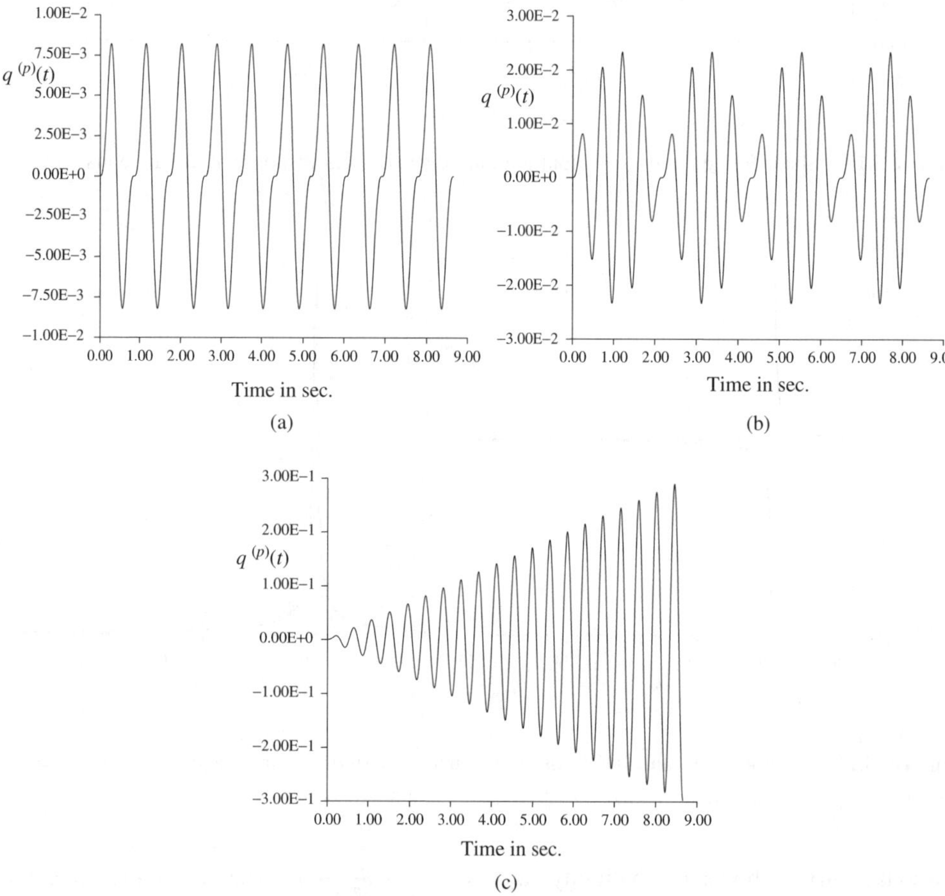

Figure 5.12 Time histories of $q^{(p)}(t)$ of an SDOF oscillator under harmonic input $F\sin\lambda t$ (with $F = 1$); (a) $\frac{\lambda}{\omega} = 0.5$ (bounded solution away from resonance), (b) $\frac{\lambda}{\omega} = 0.8$ (beat phenomenon near resonance) and (c) $\frac{\lambda}{\omega} = 0.999$ (unbounded solution at resonance)

Figure 5.12b shows the response history with a long period $\left(=\frac{2\pi}{\frac{\omega_n-\lambda}{2}}\right)$ component enveloping the frequency component with much short period $\left(=\frac{2\pi}{\frac{\omega_n+\lambda}{2}}\right)$.

In view of Equation (5.75), we can write the solution to all the three modal Equation (5.69a–c) as:

$$q_1(t) == \frac{F/\omega_1{}^2}{1-r_1{}^2}\left(\sin\lambda t - r_1 \sin\omega_1 t\right), q_2(t) == \frac{F/\omega_2{}^2}{1-r_2{}^2}\left(\sin\lambda t - r_2\sin\omega_2 t\right) \text{ and}$$

$$q_3(t) == \frac{F/\omega_3{}^2}{1-r_3{}^2}\left(\sin\lambda t - r_3\sin\omega_3 t\right) \tag{5.77a–c}$$

In the above equation, $r_1 = \lambda/\omega_1$, $r_2 = \lambda/\omega_2$and $r_3 = \lambda/\omega_3$. The solution for the physical coordinate vector $\boldsymbol{x}(t)$ is given by the modal summation in Equation (5.50). Note that at each of the resonance points viz., $\lambda = \omega_1$ or ω_3 or ω_3, the corresponding mode exhibits an unbounded growth (see Figure 5.12c) in its response. This gets reflected in the system responses as well. Presence of damping (internal or external) in a physical system can check this unbounded growth of amplitudes. The effect of damping on system behaviour is discussed in a later section of this Chapter. Figure 5.13 shows the shear frame (undamped) response histories under harmonic loading at each of the three mass points for the three cases of $\lambda/\omega_1 = 0.433$, $\lambda/\omega_2 = 1.01$ and $\lambda/\omega_3 = 1.09$. The first case refers to the excitation frequency being well away from the three system natural frequencies and as such results in a bounded solution for the system *dof* s (Figures 5.13*a, b* and *c*). The case with $\lambda/\omega_2 = 1.01$ is a case of resonance in the second mode that eventually drives the system to a state of excessive (unbounded) amplitudes as shown in the Figures 5.13*d, e* and *f*. The last case with $\lambda/\omega_3 = 1.09$ simulates a situation similar to the one in Figure 5.12b with λ near to the third system natural frequency causing a phenomenon of beats in all the system responses but bounded (see Figures 5.13*g, h* and *i*).

5.7.2 *Rayleigh Quotient*

One can extend the Rayleigh minimum principle introduced in Chapter 3 to MDOF systems also, by writing the quotient in the form:

$$\rho(\boldsymbol{v}) = \frac{\boldsymbol{v}^T \boldsymbol{K}\boldsymbol{v}}{\boldsymbol{v}^T \boldsymbol{M}\boldsymbol{v}} \tag{5.78}$$

Here, $\rho(\cdot)$ is the Rayleigh quotient in terms of an assumed mode shape vector $\boldsymbol{v}$ of the eigenvalue problem in Equation (5.45). We readily see that if $\boldsymbol{v}$ happens to be one of the eigenvector $\boldsymbol{\phi}_r$ of the MDOF system, the above Rayleigh quotient gives the corresponding eigenvalue λ_r (this also follows from the orthogonality relationship in Equation (5.49)). In general, the minimum of $\rho(\boldsymbol{v})$ obtained for all possible vectors $\boldsymbol{v}$ that satisfy (at least) the essential boundary conditions, yields an upper bound to λ_1. Thus, $\lambda_1{}^2 \leq \rho(\cdot) \leq \lambda_N{}^2$. The proof goes as follows.

If $\boldsymbol{v}$ is expressed as a linear combination of the eigenvectors (basis vectors) as:

$$\boldsymbol{v} = \sum_{j=1}^{N} c_j \boldsymbol{\phi}_j \tag{5.79}$$

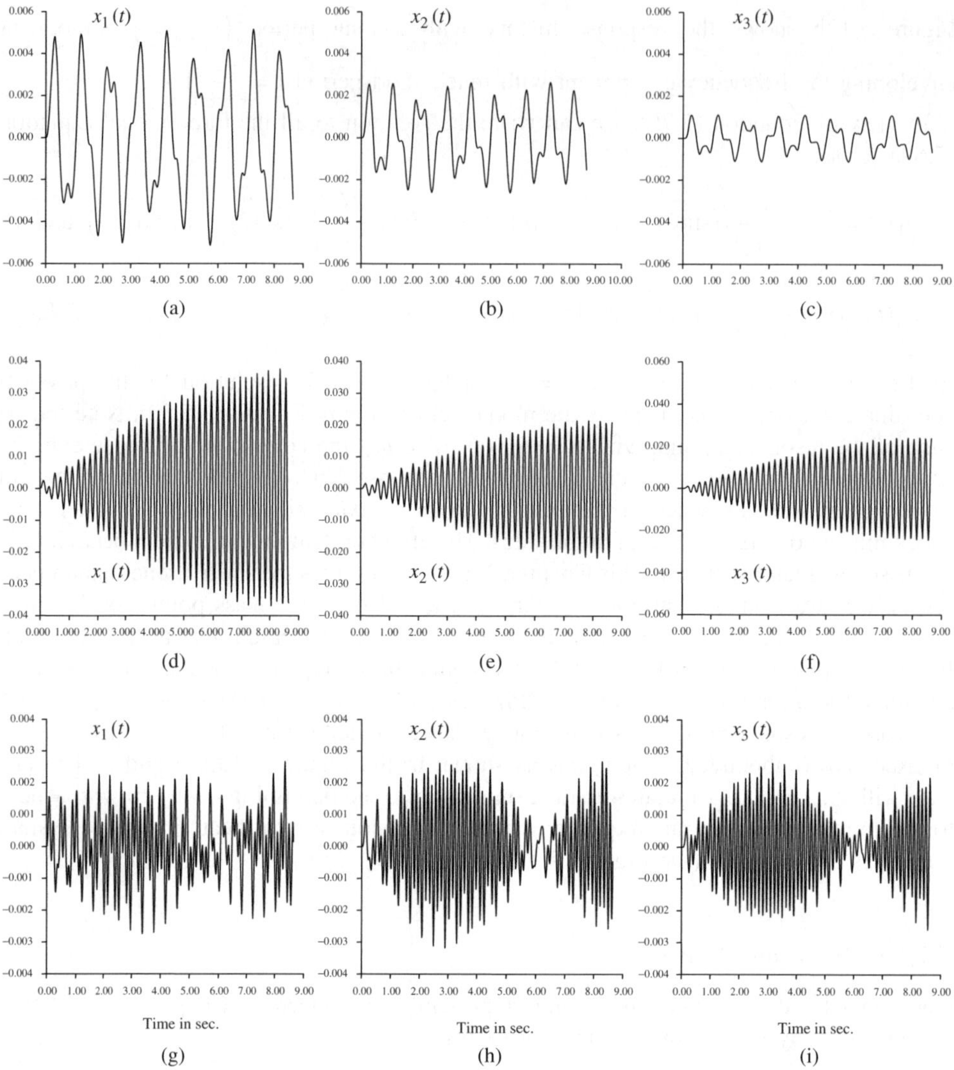

Figure 5.13 Shear frame under harmonic input loading $F\sin\lambda t$ (with $F = 1$); displacement response plots of $x_1(t)$, $x_2(t)$ and $x_3(t)$, (a – c) $\lambda/\omega_1 = 0.433$, (d – f) $\lambda/\omega_2 = 1.01$ and (g – i) $\lambda/\omega_3 = 1.09$

then, the Rayleigh quotient is given by:

$$\rho(\boldsymbol{v}) = \frac{\sum\limits_{j=1}^{N}\sum\limits_{k=1}^{N} c_j c_k \boldsymbol{\phi}_j{}^T \boldsymbol{K} \boldsymbol{\phi}_j}{\sum\limits_{j=1}^{N}\sum\limits_{k=1}^{N} c_j c_k \boldsymbol{\phi}_j{}^T \boldsymbol{M} \boldsymbol{\phi}_j} = \frac{\sum\limits_{j=1}^{N} c_j{}^2 \lambda_j{}^2}{\sum\limits_{j=1}^{N} c_j{}^2} \tag{5.80}$$

The orthogonality property of the eigenvectors $\{\boldsymbol{\phi}_j\}$ is utilised in getting the above expression for $\rho(\boldsymbol{v})$. It further follows that:

$$\rho(\boldsymbol{v}) = {\lambda_1}^2 \frac{\sum\limits_{j=1}^{N} {c_j}^2 \left(\frac{\lambda_j}{\lambda_1}\right)^2}{\sum\limits_{j=1}^{N} {c_j}^2} \quad \text{or} \quad \rho(\boldsymbol{v}) = {\lambda_N}^2 \frac{\sum\limits_{j=1}^{N} {c_j}^2 \left(\frac{\lambda_j}{\lambda_N}\right)^2}{\sum\limits_{j=1}^{N} {c_j}^2} \tag{5.81}$$

From the above two expressions of $\rho(\boldsymbol{v})$, we find that ${\lambda_1}^2 \le \rho(\cdot) \le {\lambda_N}^2$.

Example 5.9: Rayleigh quotient and the fundamental frequency of an MDOF system
Consider the 3-dof system of the shear frame in Figure 5.3 and the corresponding eigenvalue problem (5.60). Assuming $= (1, 1, 1)^T$, we have the Rayleigh quotient $\rho(\cdot) = 330$ and the associated fundamental frequency $= \sqrt{330} = 18.17 \frac{rad}{sec}$ as against the actual value for $\lambda_1 = 14.52 \frac{rad}{sec}$.

It is also informative that if $\boldsymbol{v}$ is an approximation to $\boldsymbol{\phi}_1$ to an order of $\varepsilon > 0$, that is $\boldsymbol{v} = \boldsymbol{\phi}_1 + \varepsilon \boldsymbol{u}$, then the Rayleigh quotient gives an approximation to λ_1 to an order of ε^2. To prove this, we consider $\rho\left(\boldsymbol{\phi}_1 + \varepsilon \boldsymbol{u}\right)$ and simplify using $\boldsymbol{u} = \sum_{j=1}^{N} c_j \boldsymbol{\phi}_j$ as:

$$\rho\left(\boldsymbol{\phi}_1 + \boldsymbol{\varepsilon u}\right) = \frac{\left(\boldsymbol{\phi}_1 + \boldsymbol{\varepsilon u}\right)^T \boldsymbol{K} \left(\boldsymbol{\phi}_1 + \boldsymbol{\varepsilon u}\right)}{\left(\boldsymbol{\phi}_1 + \boldsymbol{\varepsilon u}\right)^T \boldsymbol{M} \left(\boldsymbol{\phi}_1 + \boldsymbol{\varepsilon u}\right)} = \frac{{\lambda_1}^2 + \varepsilon^2 \sum\limits_{j=1}^{N} {c_j}^2 {\lambda_j}^2}{1 + \varepsilon^2 \sum\limits_{j=1}^{N} {c_j}^2}$$

$$= \left({\lambda_1}^2 + \varepsilon^2 \sum_{j=1}^{N} {c_j}^2 {\lambda_j}^2\right) \left(1 + \varepsilon^2 \sum_{j=1}^{N} {c_j}^2\right)^{-1}$$

$$= \left({\lambda_1}^2 + \varepsilon^2 \sum_{j=1}^{N} {c_j}^2 {\lambda_j}^2\right) \left(1 - \varepsilon^2 \sum_{j=1}^{N} {c_j}^2 + \varepsilon^4 \left(\sum_{j=1}^{N} {c_j}^2\right)^2 - \cdots\cdots\right)$$

$$= {\lambda_1}^2 + \varepsilon^2 \left(\sum_{j=1}^{N} {c_j}^2 {\lambda_j}^2 - {\lambda_1}^2 \sum_{j=1}^{N} {c_j}^2\right) + \text{higher order terms in } \varepsilon^2 \tag{5.82}$$

The above equation shows that the approximation to λ_1 is of order ε^2.

5.7.3 Rayleigh–Ritz Method for MDOF Systems

Using the ideas underlying the Rayleigh quotient, it is also possible to obtain approximations to a set of lowest modes of an MDOF system. This approach is referred to as the Rayleigh–Ritz method as applicable to discrete systems. To this end, we use the vector $\boldsymbol{v}$, which is expressed as a linear combination of the trial vectors $\{\bar{\phi}_j\}$, $j = 1, 2..r \ll N$, in the form:

$$\boldsymbol{v} = \sum_{j=1}^{N} c_j \bar{\phi}_j \tag{5.83}$$

The vectors $\bar{\phi}_j \in \boldsymbol{V}_r$, a subspace of the solution space $\boldsymbol{V}$ spanned by the eigenvectors of the original eigenvalue problem in Equation (5.45). We choose these vectors to be linearly independent. Substituting Equation (5.83) in Equation (5.78), we get the Rayleigh quotient as a function of the unknown coefficients, c_j (also known as the generalized coordinates) in the following form:

$$\rho(\boldsymbol{v}) = \frac{\sum_{j=1}^{r}\sum_{k=1}^{r} c_j c_k \bar{\phi}_j{}^T \boldsymbol{K} \bar{\phi}_k}{\sum_{j=1}^{r}\sum_{k=1}^{r} c_j c_k \bar{\phi}_j{}^T \boldsymbol{M} \bar{\phi}_k} = \frac{\sum_{j=1}^{r}\sum_{k=1}^{r} c_j c_k \bar{\boldsymbol{K}}_{jk}}{\sum_{j=1}^{r}\sum_{k=1}^{r} c_j c_k \bar{\boldsymbol{M}}_{jk}} = \mu^2, \text{ say} \tag{5.84}$$

$\bar{\boldsymbol{K}}_{jk}$ and $\bar{\boldsymbol{M}}_{jk}$ are given by $\bar{\boldsymbol{\phi}}_j{}^T \boldsymbol{K} \bar{\boldsymbol{\phi}}_k$ and $\bar{\boldsymbol{\phi}}_j{}^T \boldsymbol{M} \bar{\boldsymbol{\phi}}_k$ respectively. We now invoke Rayleigh's minimum principle and impose the necessary condition $\frac{\partial \rho(\boldsymbol{v})}{\partial c_j} = 0,\ j = 1, 2, ..r$ so as to minimise $\rho(\boldsymbol{v})$ with respect to the unknown coefficients c_j. This yields:

$$\sum_{j=1}^{r} c_j \left[\bar{\boldsymbol{K}}_{jk} - \mu^2 \bar{\boldsymbol{M}}_{jk}\right] = 0 \Rightarrow \left[\bar{\boldsymbol{K}} - \mu^2 \bar{\boldsymbol{M}}\right]\boldsymbol{c} = \boldsymbol{0} \tag{5.85}$$

Since $r \ll N$ in general, the Equation (5.85) represents a reduced eigenvalue problem of size $r \times r$. The eigenvalues $\{\mu_j\}$ give approximations to the r lowest eigenvalues λ_j, $j = 1, 2, ..r$ of the original eigenvalue problem. The eigenvectors $\{\boldsymbol{c}_j\}$ of the reduced eigenvalue problem can be utilised as approximations to the original eigenvectors $\boldsymbol{\phi}_j$, $j = 1, 2, ..r$ from Equation (5.83).

Example 5.10: Eigensolution to the shear frame in Figure 5.3 by the Rayleigh–Ritz method

For the shear frame, assume (with $r = 2$):

$$\boldsymbol{v} = c_1 \begin{pmatrix} 1 \\ 0.5 \\ 0.2 \end{pmatrix} + c_1 \begin{pmatrix} 1 \\ -0.5 \\ -0.2 \end{pmatrix} \tag{5.86}$$

The two Ritz vectors $(1, 0.5, 0.2)^T$ and $(1, -0.5, 0.2)^T$ are chosen to be linearly independent (but otherwise arbitrary). The reduced eigenvalue problem (Equation (5.85)) is obtained as:

$$\begin{bmatrix} 330 & 270 \\ 270 & 2010 \end{bmatrix} - \mu^2 \begin{bmatrix} 1.455 & 0.705 \\ 0.705 & 1l455 \end{bmatrix} = \boldsymbol{0} \tag{5.87}$$

The approximate eigenvalues for the shear frame are obtained as the roots of the characteristic polynomial associated with the above eigenvalue problem and they are 14.88 and 33.34 rad/s. These are indeed good approximations and close to $\lambda_{1,2} = 14.52$ and 31.1 rad/s, the true values. The corresponding approximations to the first two eigenvectors are as $(0.84, 0.55, 0.22)^T$ and $(0.74, -0.62, -0.25)^T$ respectively.

5.8 Damped MDOF Systems

Our discussion so far focused on undamped systems, wherein we implicitly assume that energy dissipation via damping is small enough to be ignored. This assumption has led to the response amplitudes getting unbounded at resonance. Moreover, such systems are also characterised by the persistence of terms that depend on the specified (nonzero) initial conditions (Figure 5.12). However, it is common knowledge that damping is inherently present in physical systems, wherein vibrations tend to cease after a whilst without external forcing. Different sources of energy dissipation can be traced (Lord Rayleigh, 1945; Clough and Penzien, 1982; Newland, 1989; Berger, 2002) in structural systems. These sources include, for instance, structural joints, presence of viscoelastic materials, resilient elements, internal friction in material, and so on. As such a typical vibrating system experiences energy dissipation with the result that the transient motion (i.e. the set of terms that depends explicitly on the initial conditions) eventually tends to vanish as t $\rightarrow \infty$. Precise description of these energy dissipation mechanisms in the form of appropriate mathematical models could be quite formidable (Adhikari, 2000). However, thanks to its simple mathematical description, the most common model for damping is the linear viscous (linearly proportional to velocity) damping. Damping forces are nonconservative as is the case of external forces acting on a system. If Lagrange's equations (Equation (1.14)) are to be modified to include viscous damping effects, a dissipation energy function may be included for the damping model in the form $E_D = \frac{1}{2}\sum_j^N \sum_k^N c_{jk}\ \dot{x}_j \dot{x}_k$. The velocities $\dot{x}_j$ and $\dot{x}_k$ represent generalized or nodal velocities of a semi-discretized system. $c_{ij}(= c_{ji})$ are the viscous damping coefficients and form the elements of a symmetric damping matrix $\boldsymbol{C}$. The Lagrange's equations of motion can now be written as:

$$\frac{d}{dt}\left(\frac{\partial L}{\partial \dot{x}_k}\right) - \frac{\partial L}{\partial x_k} + \frac{\partial E_D}{\partial \dot{x}_k} = Q_k \tag{5.88}$$

The viscous damping force associated with each velocity coordinate is given by:

$$F_{Dj} = -\frac{\partial E_D}{\partial \dot{x}_j} = -\sum_i^N c_{ij}\dot{x}_i \tag{5.89}$$

With these viscous damping forces included in the semi-discrete model, the system equations of motion become:

$$\sum_i^N m_{ij}\ddot{x}_j + c_{ij}\dot{x}_j + k_{ij}x_j = F_j,\ \ j = 1, 2 \ldots .N$$

$$\Rightarrow \boldsymbol{M}\ddot{\boldsymbol{x}} + \boldsymbol{C}\dot{\boldsymbol{x}} + \boldsymbol{K}\ \boldsymbol{x} = \boldsymbol{F} \text{ (in matrix form)s} \tag{5.90}$$

5.8.1 *Damped System and Quadratic Eigenvalue Problem*

We note that, with the damping matrix $\boldsymbol{C}$ present in the system, the coupled ODEs (5.90) with $\boldsymbol{F} = 0$ results in a quadratic eigenvalue problem of the form:

$$\left[\boldsymbol{K} + \lambda\boldsymbol{C} + \lambda^2\boldsymbol{M}\right]\boldsymbol{u} = 0 \tag{5.91}$$

The above equation is obtained by assuming the free vibration solution as $\boldsymbol{x}(t) = e^{\lambda t}\boldsymbol{u}$. For large dynamical systems, solution of the quadratic eigenvalue problem (5.91) could become quite involved. However, for a damped SDOF oscillator (Figure 5.14) governed by the equation:

$$m\ddot{x} + c\dot{x} + kx = f(t) \tag{5.92}$$

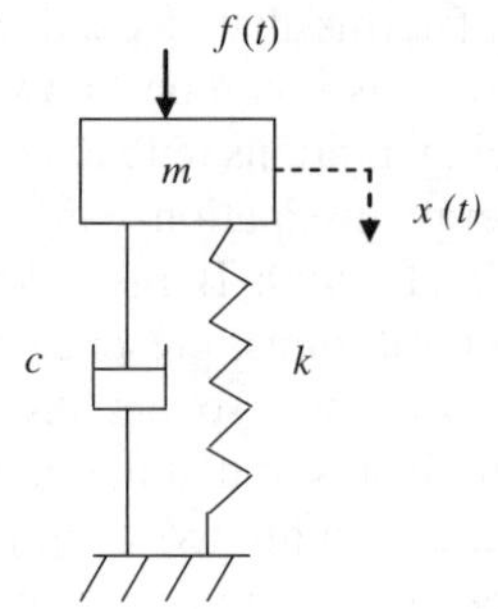

Figure 5.14 Damped SDOF oscillator (Equation (5.92))

the quadratic eigenvalue problem $\left[k + \lambda c + \lambda^2 m\right] u = 0$ gives complex eigenvalues (natural frequencies) as:

$$\lambda == -\frac{c}{2m} \pm i\sqrt{\left(\frac{k}{m}\right)^2 - \left(\frac{c}{2m}\right)^1} = \omega_n\left(-\xi \pm i\sqrt{1-\xi^2}\right) = -\xi\omega_n \pm i\omega_d \quad (5.93)$$

In the above equation $\xi = \frac{c}{2\sqrt{km}}$ is the damping ratio. $\omega_d = \omega_n\sqrt{1-\xi^2}$ is the damped natural frequency (with $\omega_n = \sqrt{\frac{k}{m}}$ being the natural frequency of an undamped SDOF system). Equation (5.92) can be written in the following canonical form:

$$\ddot{x} + 2\xi\omega_n\dot{x} + {\omega_n}^2 x = \frac{F(t)}{m} \quad (5.94)$$

The natural frequencies are complex conjugates provided that $c < 2\sqrt{km}$ or $\xi < 1$ and, in such a case, the oscillator is said to be under damped. Under this condition, the complementary solution or the so-called free vibration solution of the damped SDOF oscillator is given by:

$$x(t) = A_1 e^{(-\xi\omega_n + i\omega_d)t} + A_2 e^{(-\xi\omega_n - i\omega_d)t} \quad (5.95)$$

The arbitrary constants A_1 and A_2 are complex conjugates to be determined from the initial conditions. With $A_1 = a + ib$, $x(t)$ finally takes the form:

$$x(t) = e^{-\xi\omega_n t}\left(a\cos\omega_d t + b\sin\omega_d t\right) \quad (5.96)$$

The above equation validates the earlier observation that, in the presence of damping, the transient solution gets bounded (even at resonance) and eventually damped out. In fact, the free vibration solution $x(t)$ in Equation (5.96) follows a damped 'periodic' solution oscillating with period $= {}^{2\pi}/_{\omega_d}$ and exponentially decays to zero as $t \to \infty$.

5.8.2 Damped System and Unsymmetric Eigenvalue Problem

With increasing system size ($N \gg 1$), it is difficult to solve the damped MDOF system in Equation (5.90) by directly attempting to deal with the quadratic eigenvalue problem.

Alternatively one can reduce Equation (5.90) to a system of coupled first-order ODEs as in Equation (5.8) using a state space formulation. This results in a $2N\,x\,2N$ eigenvalue problem of the form similar to the one in Equation (5.9):

$$\begin{bmatrix} \boldsymbol{O} & \boldsymbol{I} \\ -\boldsymbol{M}^{-1}\boldsymbol{K} & -\boldsymbol{M}^{-1}\boldsymbol{C} \end{bmatrix} \boldsymbol{u} = \lambda \boldsymbol{u} \tag{5.97}$$

The system matrix in Equation (5.97) is unsymmetric, even if the matrices $\boldsymbol{M}$, $\boldsymbol{K}$ and $\boldsymbol{C}$ are symmetric. Once again, as with the quadratic eigenvalue problem, solving an unsymmetric eigenvalue problem is also nontrivial. This subject matter is dealt with in Chapter 7 in detail.

5.8.3 Proportional Damping and Uncoupling MDOF Systems

The above difficulties involved in solving viscously damped structural systems may be overcome in some special cases. One case where the decoupling procedure described for conservative systems is applicable to damped systems also is when the damping matrix $\boldsymbol{C}$ is expressible as a linear combination of the system stiffness and mass matrices (case of proportional damping) as:

$$\boldsymbol{C} = \alpha \boldsymbol{K} + \beta \boldsymbol{M} \tag{5.98}$$

α and β are scalar constants. Substituting in Equation (5.90), we get:

$$\boldsymbol{M}\ddot{\boldsymbol{x}} + (\alpha \boldsymbol{K} + \beta \boldsymbol{M})\,\dot{\boldsymbol{x}} + \boldsymbol{K}\,\boldsymbol{x} = \boldsymbol{F} \tag{5.99}$$

Recalling the orthogonality properties of the modal matrix, $\boldsymbol{\phi}$ of the conservative system, we use the same transformation $\boldsymbol{x} = \boldsymbol{\phi}\boldsymbol{q}$ as in Equation (5.50) and we get:

$$\begin{aligned} &\boldsymbol{\phi}^T\boldsymbol{M}\boldsymbol{\phi}\ddot{\boldsymbol{q}} + \boldsymbol{\phi}^T(\alpha\boldsymbol{K} + \beta\boldsymbol{M})\,\boldsymbol{\phi}\dot{\boldsymbol{q}} + \boldsymbol{\phi}^T\boldsymbol{K}\boldsymbol{\phi}\boldsymbol{x} = \boldsymbol{\phi}^T\boldsymbol{F} \\ &\Rightarrow \boldsymbol{I}\ddot{\boldsymbol{q}} + (\boldsymbol{\alpha}\boldsymbol{\Lambda} + \boldsymbol{\beta}\boldsymbol{I})\,\dot{\boldsymbol{q}} + \boldsymbol{\Lambda}\boldsymbol{q} = \boldsymbol{\phi}^T\boldsymbol{F} \end{aligned} \tag{5.100}$$

Equation (5.100) is a set of uncoupled second-order ODEs in the modal coordinates q_j, $j = 1, 2....N$. $\boldsymbol{\phi}^T\boldsymbol{F} = \{f_j(t)\}$ is the vector of modal forcing functions. Expressing $\boldsymbol{\alpha\Lambda} + \boldsymbol{\beta I} = 2\,diag[\boldsymbol{\xi\omega}]$ with $\boldsymbol{\omega}$ representing the set of system eigenvalues, $(\omega_1, \omega_2, \ldots.., \omega_N)$, $\boldsymbol{\xi} = (\xi_1, \xi_2, \ldots.., \xi_N)$, the set of modal damping ratios and $\boldsymbol{\xi\omega} = (\xi_1\omega_1, \xi_2\omega_2, \ldots.., \xi_N\omega_N)$, we write the modal equations as:

$$\ddot{q}_j + 2\xi_j\omega_j\dot{q}_j + {\omega_j}^2 q_j = f_j(t),\ j = 1, 2, \ldots\ldots., N \tag{5.101}$$

Each of the modal equations in Equation (5.101) has the same form as the equation of the damped SDOF system in Equation (5.94). It remains now how to specify the mass and stiffness proportional constants α and β. From $\boldsymbol{\alpha\Lambda} + \boldsymbol{\beta}I = 2diag[\boldsymbol{\xi\omega}]$, the modal damping ratio ξ_j is related to system natural frequencies ω_j as:

$$\xi_j = \frac{\alpha{\omega_j}^2 + \beta}{2\omega_j} = \frac{\alpha\omega_j}{2} + \frac{\beta}{2\omega_j} \tag{5.102}$$

By specifying ξ_j for any two modes, we obtain two simultaneous equations in the unknown constants α and β. Once these are evaluated, modal damping ratios can be evaluated for the rest of the modes from Equation (5.102).

Example 5.11: Proportional damping and modal damping ratios

The natural frequencies for the three modes of the shear frame are obtained in Example 5.8 as 14.5 , 31.1 and 46.1 rad/s. Suppose that the damping ratios ξ_1 and ξ_3 are given to be 0.05 corresponding to the first and third modes of the shear frame, we use Equation (5.102) to have

$$\begin{pmatrix} \xi_1 \\ \xi_3 \end{pmatrix} = \begin{pmatrix} 0.05 \\ 0.05 \end{pmatrix} = \begin{pmatrix} \dfrac{14.5\alpha}{2} + \dfrac{\beta}{29.0} \\ \dfrac{46.1\alpha}{2} + \dfrac{\beta}{92.2} \end{pmatrix} \tag{5.103}$$

From the two simultaneous equations in Equation (5.103), we get $\alpha = 0.00165$ and $\beta = 1.1$. With these two constants known, the damping ratio for the second mode is obtained as $\xi_2 = \frac{\alpha\omega_2}{2} + \frac{\beta}{2\omega_2} = 0.0433$.

With the proportionality constants α and β known, the damping matrix $\boldsymbol{C}$ is realised in terms of the mass and stiffness matrices as

$$\boldsymbol{C} = \alpha\boldsymbol{K} + \beta\boldsymbol{M} = \begin{bmatrix} 2.09 & -0.99 & 0.0 \\ -0.99 & 4.62 & -1.98 \\ 0.0 & -1.98 & 7.15 \end{bmatrix}$$

5.8.4 Damped Systems and Impulse Response

The complementary solution due to initial conditions for each of modal Equation (5.101) is given by the Equation (5.96) and represents the transient response. The particular solution $q^p(t)$ is obtained by either Laplace transform approach or the convolution integral. In the former case, we have:

$$q^p(t) = \mathcal{L}^{-1}H(s)F(s) \tag{5.104}$$

$H(s)$ is the system transform function and $F(s)$ the Laplace transform of the input function $f(t)$. The inverse transform is given by the convolution integral:

$$q^p(t) = \int_0^t h(t-\tau)\, f(\tau)d\tau \tag{5.105}$$

$h(t)$ is the impulse response function of the damped SDOF oscillator and is given by the response of the oscillator to an impulse excitation. Hence as observed earlier:

$$h(t) = \mathcal{L}^{-1}H(s) = \mathcal{L}^{-1}\frac{1}{(s^2 + 2\xi\omega s + \omega^2)} = \mathcal{L}^{-1}\frac{1}{(s+\xi\omega)^2 + \omega^2(1-\xi^2)} \tag{5.106}$$

With the notation $\omega_d = \omega\sqrt{(1-\xi^2)}$, the impulse response function for a damped oscillator is obtained from the above equation as:

$$h(t) = \frac{e^{-\xi\omega t}}{\omega_d}\sin\omega_d t \tag{5.107}$$

5.8.5 Response under General Loading

Note that $H(s)$ is known as the system transfer function. It is the Laplace transform of the system impulse response function $h(t)$. In fact any general input $f(t)$ is expressible in terms of delayed impulse functions in the discretized form:

$$f(t) = \sum_k f\left(t_k\right) I(t - t_k) \tag{5.108}$$

The above approximate representation of $f(t)$ is shown in Figure 5.15. At any time t_k, the forcing amplitude $f\left(t_k\right)$ represents a delayed impulse $I(t - t_k)$ of intensity $f\left(t_k\right) dt$ acting at time t_k defined by:

$$\begin{aligned} I\left(t - t_k\right) &= f\left(t_k\right) dt, \quad t = t_k \\ &= 0, \quad otherwise \end{aligned} \tag{5.109}$$

Noting that $\mathcal{L}\left(I\left(t - t_k\right)\right) = e^{-st_k} f\left(t_k\right) dt$, the differential response contribution due to this delayed impulse is given by:

$$h_k(t) = f\left(t_k\right)\left\{\mathcal{L}^{-1} e^{-st_k} H(s)\right\} dt = f\left(t_k\right) h\left(t - t_k\right) dt \tag{5.110}$$

The response (particular solution) due to the input $f(t)$ is given by the summation of these differential responses:

$$q^{(p)}(t) = \sum_k f\left(t_k\right) h\left(t - t_k\right) dt \tag{5.111}$$

In the limit as $dt \to 0$, the above summation attains the form of the convolution integral in Equation (5.105).

5.8.6 Response under Harmonic Input

Under harmonic input $f(t) = F \sin \lambda t$, the particular solution if obtained by transform approach, is given by:

$$q^{(p)}(t) = \mathcal{L}^{-1} \frac{F\lambda}{\left(s^2 + 2\xi\omega s + \omega^2\right)\left(s^2 + \lambda^2\right)} \tag{5.112}$$

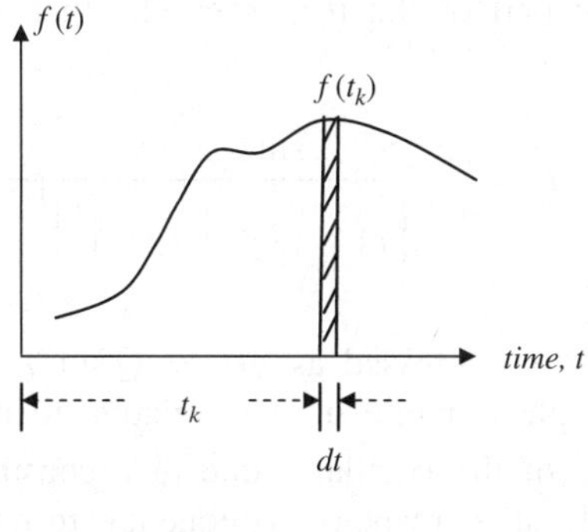

Figure 5.15 Discrete representation of $f(t)$

By partial fractions expansion, we get:

$$\frac{\lambda}{\left(s^2+2\xi\omega s+\omega^2\right)\left(s^2+\lambda^2\right)}=\frac{a_1+a_2 s}{s^2+\lambda^2}+\frac{a_3+a_4 s}{s^2+2\xi\omega s+\omega^2}$$

$$\text{with } a_1=\frac{\lambda\left(1-r^2\right)}{\omega^2\left\{\left(1-r^2\right)^2+4\xi^2r^2\right\}},$$

$$-a_2=a_4=\frac{2\xi r}{\omega^2\left\{\left(1-r^2\right)^2+4\xi^2r^2\right\}}\text{ and } a_3=-\frac{\lambda\left\{\left(1-r^2\right)-4\xi^2\right\}}{\omega^2\left\{\left(1-r^2\right)^2+4\xi^2r^2\right\}} \tag{5.113}$$

Here, $r=\frac{\lambda}{\omega}$. In view of the above partial fraction expansion, the inverse transform in Equation (5.112) gives:

$$\begin{aligned}q^{(p)}(t)&={}^{F}/_{\omega^2}\frac{\sin\left(\lambda t-\theta\right)}{D(\lambda)}\\&\quad+F\left[e^{-\xi\omega t}\left(a_4\cos\omega_d t+\left(a_3-\xi\omega a_4\right)\sin\omega_d t\right)\right]\end{aligned} \tag{5.114}$$

Here, $D(\lambda)=\left[\left(1-r^2\right)^2+\left(2\xi r\right)^2\right]^{1/2}$ and $\theta(\lambda)=\tan^{-1}\frac{2\xi r}{1-r^2}$.

The total solution is obtained by combining the above particular solution and the complementary solution in Equation (5.96). With the assumption of zero initial conditions, the unknown constants a and b in the complementary solution are obtained as:

$$a=0\quad\text{and}\quad b=0 \tag{5.115}$$

With the above constants included in the solution, the total response of a damped SDOF oscillator under a harmonic input excitation takes the form:

$$q(t)=\frac{F}{\omega^2}\frac{\sin\left(\lambda t-\theta\right)}{D(\lambda)}+\left[\frac{Fe^{-\xi\omega t}}{\omega^2D^2(\lambda)}\left(2\xi r\cos\omega_d t-\frac{r\left\{\left(1-r^2\right)-2\xi^2\right\}}{\left(1-\xi^2\right)^{1/2}}\sin\omega_d t\right)\right] \tag{5.116}$$

We notice from the above equation that the second term decays to zero as time, $t\to\infty$ and represents the transient part of the response. The first term gives the response in steady state as:

$$q(t)=\frac{F}{\omega^2}\frac{\sin\left(\lambda t-\theta\right)}{\left[\left(1-r^2\right)^2+\left(2\xi r\right)^2\right]^{1/2}} \tag{5.117}$$

If the steady state response is expressed as $q(t)=Q\sin\left(\lambda t-\theta\right)$, Q is the amplitude of oscillation and θ is the phase angle of $q(t)$ relative to the excitation. Treating $\frac{F}{\omega^2}$ as a sort of static response of the oscillator due to a constant force of magnitude F, the amplitude of the steady state response (frequency response) can be expressed in nondimensional form as:

$$\frac{Q(\lambda)}{F/_{\omega^2}} = \mathcal{M}(\lambda) = \frac{1}{\left[\left(1-r^2\right)^2 + (2\xi r)^2\right]^{1/2}} \tag{5.118a}$$

Note that for an SDOF oscillator in a standard form as in Equation (5.94), the frequency response is given by:

$$\frac{X(\lambda)}{F/_{m\omega^2}} = \frac{1}{\left[\left(1-r^2\right)^2 + (2\xi r)^2\right]^{1/2}} \tag{5.118b}$$

where it is apparent that $F/_{m\omega^2} = F/k$ is indeed the static response. In the nondimensional form, $\mathcal{M}(\lambda)$ is known as the magnification factor and is plotted in Figure 5.16a as a function of the frequency ratio, $r = \frac{\lambda}{\omega}$. The phase angle θ is also a function of the excitation frequency and its variation is shown in Figure 5.16b.

From Equation (5.118) and also from Figure 5.16a, we observe that at resonance ($r = 1$), the peak response is no longer infinity (as in the case of an undamped system) but is restricted to a finite value given by

$$\mathcal{M}(\lambda = \omega) = \frac{1}{2\xi} \tag{5.119}$$

From the force diagrams in Figure 5.17, we observe that for $r\left(=\frac{\lambda}{\omega}\right) \ll 1$, the steady state response lags the excitation by a small angle ($<90^\circ$) and the excitation force is opposed mostly by the spring force. At resonance ($r = 1 \Rightarrow \lambda = \omega$) the phase angle $\theta = 90^\circ$ and excitation force balances the damping force whilst the inertia and spring forces cancel out each other. For $r\left(=\frac{\lambda}{\omega}\right) \gg 1$, the phase angle $\theta > 90^\circ$ and the excitation force mostly opposes the inertia force.

5.8.7 Complex Frequency Response

From Equation (5.117), we note that in steady state, an SDOF system oscillates at the frequency of the driving harmonic force. This is always true for linear time invariant systems. Thus, steady state response under harmonic input can also be obtained for a SDOF oscillator (Equation (5.101)) by directly assuming:

$$q(t) = Q\exp(i\lambda t) \tag{5.120}$$

Substituting the above in Equation (5.101) we get:

$$\left[\left(\omega^2-\lambda^2\right) + i2\xi\omega\lambda\right]Q = F \Rightarrow Q(i\lambda) = \frac{F}{\left(\omega^2-\lambda^2\right) + i2\xi\omega\lambda} \Rightarrow \frac{F/_{\omega^2}}{\left(1-r^2\right) + i2\xi r} \tag{5.121}$$

$Q(i\lambda)$ is the complex frequency response function and we note that the magnitude $\frac{|Q(i\lambda)|}{F/_{\omega^2}}$ is given by the same expression as in Equation (5.118). For MDOF system Equation 5.121 can be used to obtain the system complex frequency response at any *degree of freedom* by modal superposition. Utilising Equation (5.50), we thus have:

$$\boldsymbol{X}(i\lambda) = \boldsymbol{\Phi}\boldsymbol{Q}(i\lambda) \tag{5.122}$$

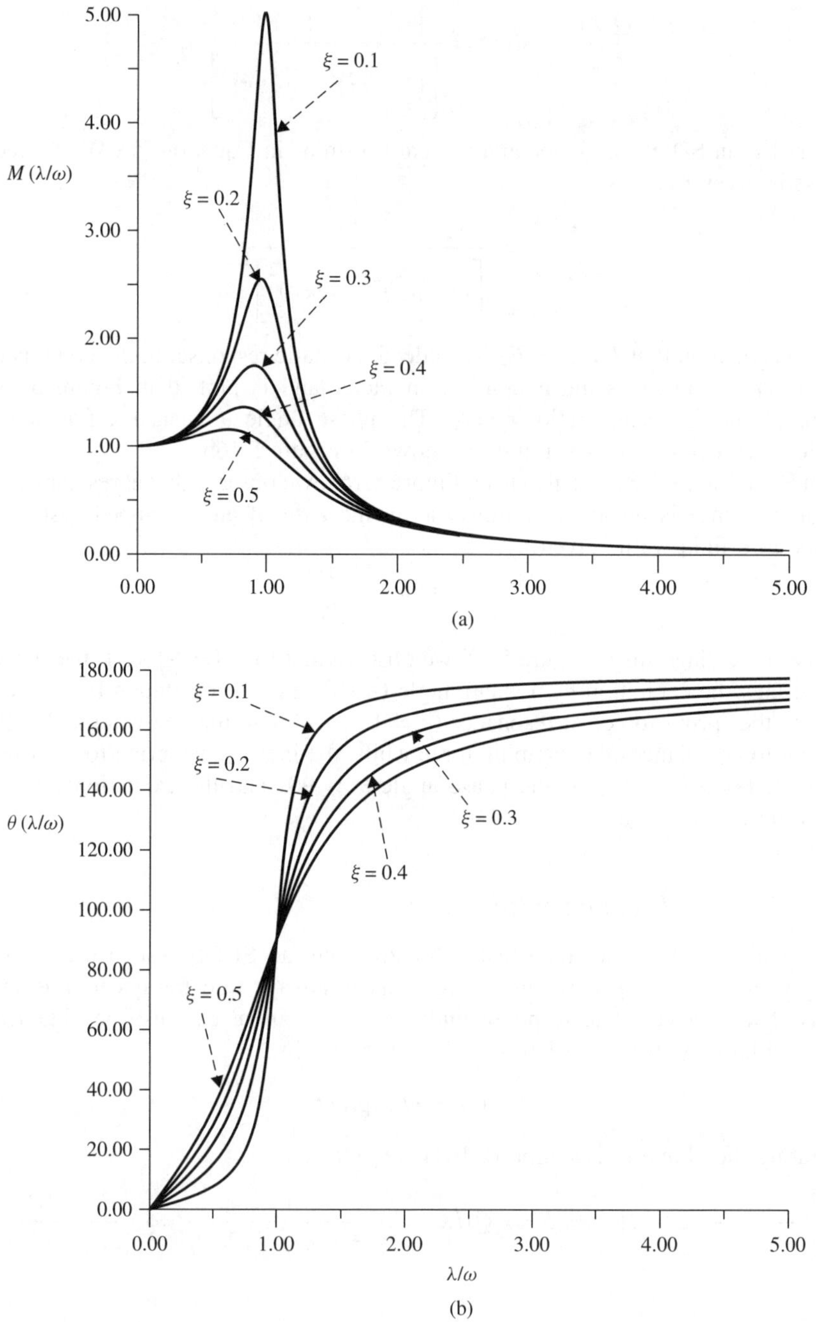

Figure 5.16 Frequency response of a damped SDOF oscillator; (a) magnification factor, $\mathcal{M}(\lambda/\omega)$ and (b) phase angle, $\theta(\lambda/\omega)$

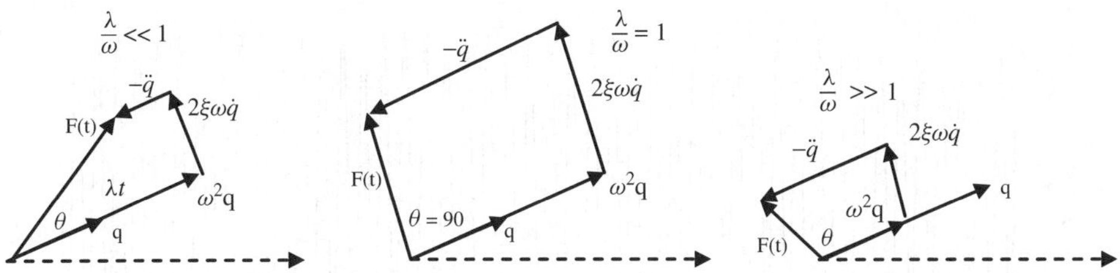

Figure 5.17 Force equilibrium diagrams

Example 5.12: Shear frame and response under harmonic excitation

In Example 5.8 the mass and stiffness matrices are specified and associated generalized eigenvalue problem is solved for the eigenvalues/natural frequencies and eigenvectors. The orthonormalised modal matrix $\boldsymbol{\phi}$ is given by Equation (5.66). To evaluate the system response under a harmonic input $F \sin \lambda t$, we first obtain the responses of the three modes using the damping ratios given in Example (5.11). Each modal response (complementary solution + particular solution) at a specified excitation frequency λ is evaluated using Equation (5.116). The frequency ratio is so chosen that the first mode is at resonance, that is $r_1 = \frac{\lambda}{\omega_1} = 1.0$. At this excitation frequency λ, $r_2 = \frac{\lambda}{\omega_2} = 0.466$ and $r_3 = \frac{\lambda}{\omega_3} = 0.315$. By superposition (Equation (5.50)) of these modal responses, the response time histories of the three mass points of the shear frame are obtained and shown in Figure 5.18.

Steady state responses (frequency responses $|Q_j(i\lambda)|$, $j = 1, 2$ and 3 in Equation (5.121)) of the three modes are shown in Figure 5.19. The peak amplitudes of the three modes in the figure correspond to $\frac{1}{2\xi_j}$, $j = 1, 2$ and 3 (see Equation (5.119)).

The complex frequency responses $X_j(i\lambda)$, $j = 1, 2$, and 3 are obtained at the three mass points by modal superposition using the Equation (5.122). Thus:

$$\begin{Bmatrix} X_1(i\lambda) \\ X_2(i\lambda) \\ X_3(i\lambda) \end{Bmatrix} = \begin{bmatrix} 0.745 & 0.638 & 0.208 \\ 0.480 & -0.383 & -0.535 \\ 0.225 & -0.431 & 0.514 \end{bmatrix} \begin{Bmatrix} Q_1(i\lambda) \\ Q_2(i\lambda) \\ Q_3(i\lambda) \end{Bmatrix} \tag{5.123}$$

$Q_j(i\lambda)$, j = 1,2 and 3 are evaluated using Equation (5.121). The amplitude and phase plots for the 3 *dof* of the shear frame are given in Figure 5.20.

5.8.8 Force Transmissibility

From Equation (5.92), it is recognised that the sum $c\dot{x} + kx$ constitutes the force exerted by the SDOF system on its base. Knowing that the system steady state response is $x(t) = X \sin(\lambda t - \theta)$ (see Equation (5.118)) under harmonic excitation $F \sin \lambda t$, the transmitted force is given by:

$$\begin{aligned} c\dot{x} + kx &= c\lambda X \cos(\lambda t - \theta) + kX \sin(\lambda t - \theta) \\ &\Rightarrow |c\dot{x} + kx| = X\left(c^2\lambda^2 + k^2\right)^{1/2} \\ &= X\left(m^2 (2\xi\omega)^2 \lambda^2 + \omega^4\right)^{1/2} \quad (\text{since } c/m = 2\xi\omega \text{ and } k/m = \omega^2) \\ &= Xm\omega^2 \left\{1 + (2\xi r)^2\right\}^{1/2} \quad (\text{with } r = \lambda/\omega) \end{aligned} \tag{5.124}$$

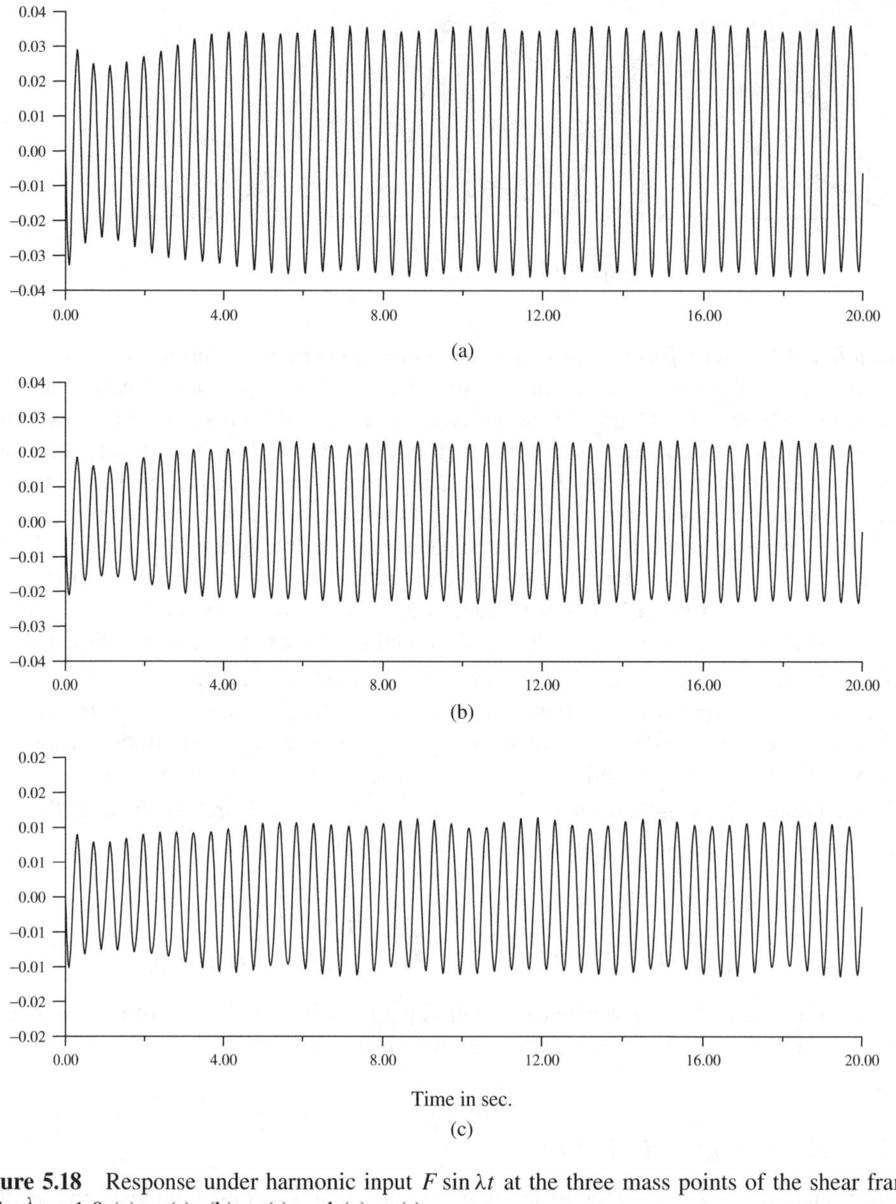

Figure 5.18 Response under harmonic input $F \sin \lambda t$ at the three mass points of the shear frame with $\frac{\lambda}{\omega_1} = 1.0$ (a) $x_1(t)$, (b) $x_2(t)$ and (c) $x_3(t)$

If we define force transmissibility (TR) as the ratio of the amplitude of transmitted force to that of the external harmonic input, we have:

$$TR = \frac{Xm\omega^2 \left\{1 + (2\xi r)^2\right\}^{1/2}}{F} = \frac{X}{F/m\omega^2} \left\{1 + (2\xi r)^2\right\}^{1/2}$$

$$\Rightarrow TR = \frac{\left\{1 + (2\xi r)^2\right\}^{1/2}}{\left[\left(1 - r^2\right)^2 + (2\xi r)^2\right]^{1/2}} \text{ (from Equation (5.118b))} \quad (5.125)$$

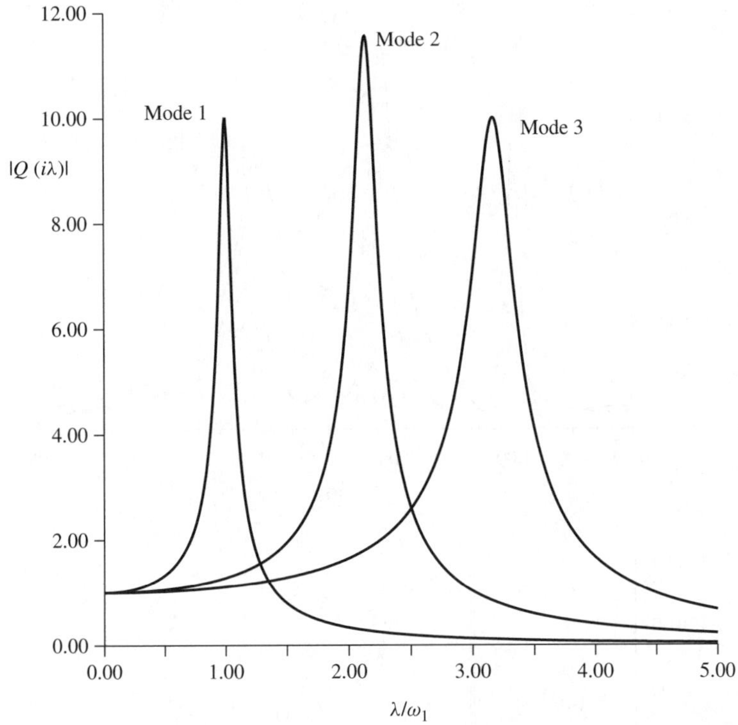

Figure 5.19 Steady state frequency responses of the three modes of shear frame under harmonic input; peak values of the curves: $\frac{1}{2\xi_1} = 10.0$, $\frac{1}{2\xi_2} = 11.55$, $\frac{1}{2\xi_3} = 10.0$

Figure 5.21 shows typical curves of the force transmissibility vs. frequency ratio for different values of ξ. Modelling a structural system as an SDOF oscillator may be an extreme idealisation. Yet, knowledge of the transmitted forces via this simplified model may help in providing basic guidelines for the design of isolation systems to protect any sensitive equipments mounted on the base of the structure (e.g. a turbo-generator). A typical observation from the Figure 5.21 is that all the curves intersect at the point where $r = \sqrt{2}$ beyond which the isolation is ineffective with further increase in damping.

5.8.9 System Response and Measurement of Damping

To measure damping in a system experimentally, for instance in terms of the parameter ξ of the viscous damping model, methods commonly adopted are logarithmic decrement method or half power method.

5.8.9.1 Logarithmic Decrement Method

This method uses the decay rate of a free vibration record. Free vibration solution of a system oscillating in any one of its modes is given by the Equation (5.96). With initial conditions $q(0) = q_0$ and $\dot{q}(0) = \dot{q}_0$, it is rewritten as

$$q(t) = e^{-\xi \omega t} A \sin\left(\omega_d t - \theta\right) \tag{5.126}$$

Here, $A = \sqrt{{q_0}^2 + \left(\frac{\dot{q}_0 + \xi \omega q_0}{\omega_d}\right)^2}$ and $\theta = \tan^{-1} \frac{\omega_d q_0}{\dot{q}_0 + \xi \omega q_0}$

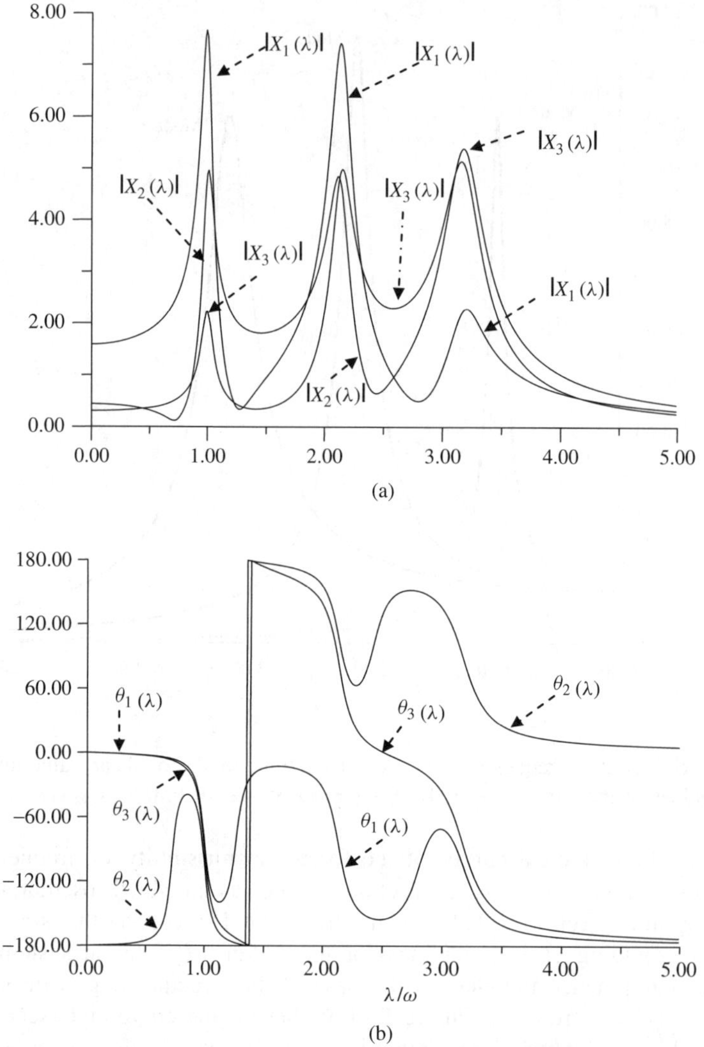

Figure 5.20 Frequency response at the three mass points of shear frame; (a) amplitude plots and (b) phase plots

Logarithmic decrement, δ represents the decay rate of the free vibration and is defined as

$$\delta = Ln\frac{q_1}{q_2} \tag{5.127}$$

Figure 5.22 shows such a free vibration record of $q(t)$. q_1 and q_2 are the successive response amplitudes, one period $t_d(=\frac{2\pi}{\omega_d})$ apart. Utilising Equation (5.126), we have

$$\delta = Ln\frac{e^{-\xi\omega t}\sin\left(\omega_d t - \theta\right)}{e^{-\xi\omega(t+t_d)}\sin\left(\omega_d\left(t+t_d\right)-\theta\right)} \Rightarrow \delta = Ln\ e^{\xi\omega t_d} \Rightarrow \delta = \xi\omega t_d \tag{5.128}$$

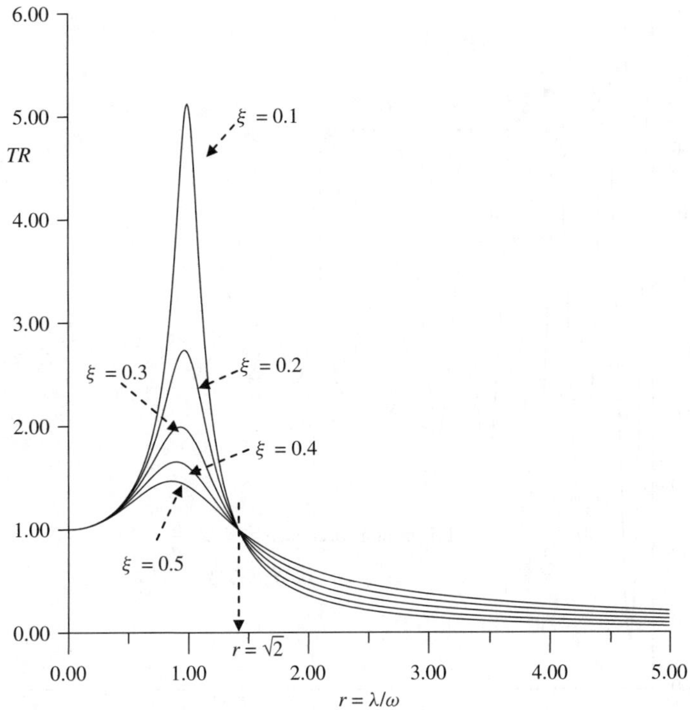

Figure 5.21 Force transmissibility versus Frequency ratio, $r = \lambda/\omega$

Ln denotes the natural logarithm. Substituting for $t_d = \frac{2\pi}{\omega_d} = \frac{2\pi}{\omega(1-\xi^2)}$, the logarithmic decrement takes the form:

$$\delta = \frac{2\pi\xi}{\left(1-\xi^2\right)} \tag{5.129}$$

Thus, for lightly damped systems with $\xi \ll 1$,

$$\delta \cong 2\pi\xi \Rightarrow \xi = {}^{\delta}\!/_{2\pi} \tag{5.130}$$

In case the amplitude ratio is measured with respect to amplitudes which are N periods apart (for convenience of measurement), Equation (5.128) takes the form

$$\delta = Ln\frac{q_1}{q_N} \Rightarrow \delta = \xi\omega N t_d \Rightarrow \xi = \frac{\delta}{\omega N t_d} \tag{5.131}$$

For small damping values, Equation (5.131) approximates to $\xi = \frac{Ln\frac{q_1}{q_N}}{2\pi N}$. Impulse response is also a free vibration solution under nonzero velocity initial condition (Equation (5.107)). Free vibration records generated by impulse or impact hammers are normally used for finding the logarithmic decrement and thereby estimating ξ. In this method, it is tacitly assumed that any desired mode can be excited in a system to find the corresponding damping ratio. In this context, the method has a limitation in that creating the appropriate

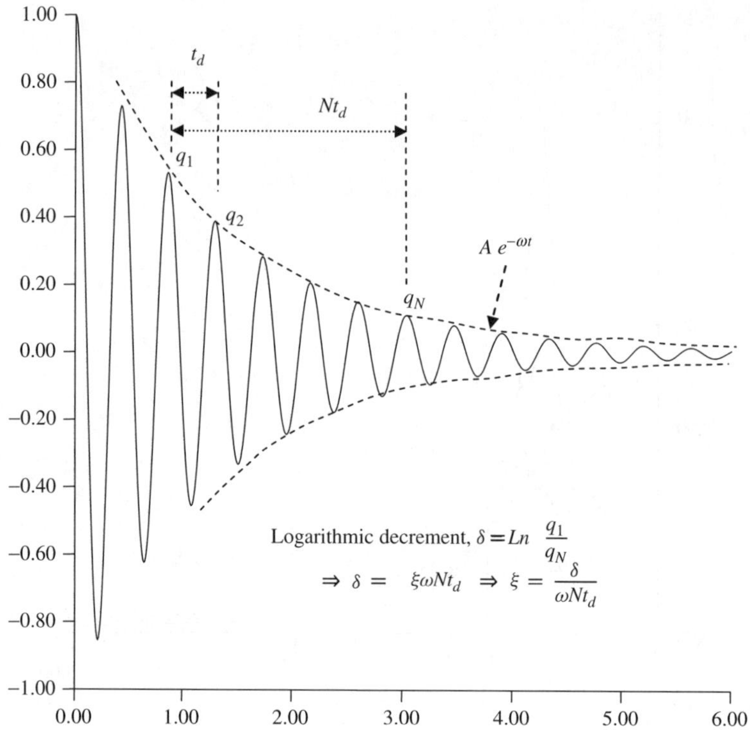

Figure 5.22 Free vibration solution of an SDOF oscillator and logarithmic decrement

initial conditions and making the system to vibrate in that particular mode may indeed be a formidable task.

5.8.9.2 Half Power Method

Referring to Equation (5.119) and Figure 5.16a, we recognise that the normalised amplitude (with reference to static displacement) at resonance is related to $^1/_{2\xi}$. This relationship may be utilised to obtain the damping ratio ξ. However, it requires finding the reference static displacement F/ω^2. Also it is required to capture the resonance point with sufficient accuracy. Particularly for lightly damped systems this requirement is difficult to be met with and hence use of this relationship is rarely recommended to find ξ. In the half power method, two points on the frequency response graph are located corresponding to $\mathcal{M}(\lambda) = \frac{1}{\sqrt{2}}\left(\frac{1}{2\xi}\right)$ (Equation (5.118)) and the corresponding frequencies λ_1 and λ_2 are identified. The word 'half power' is to be understood with an analogy to an electrical circuit where the word 'power' is related to square of the electric current and if power is halved, the magnitude of the current is reduced to $1/\sqrt{2}$ times the original value. In the present case, Figure 5.23 shows the so-called half power points at which the square of the normalised amplitude is $\frac{1}{2}\left(\frac{1}{2\lambda}\right)^2$. From Equation (5.119), we have at resonance

$$\{\mathcal{M}\,(r = 1)\}^2 = \left(\frac{1}{2\xi}\right)^2 \tag{5.132}$$

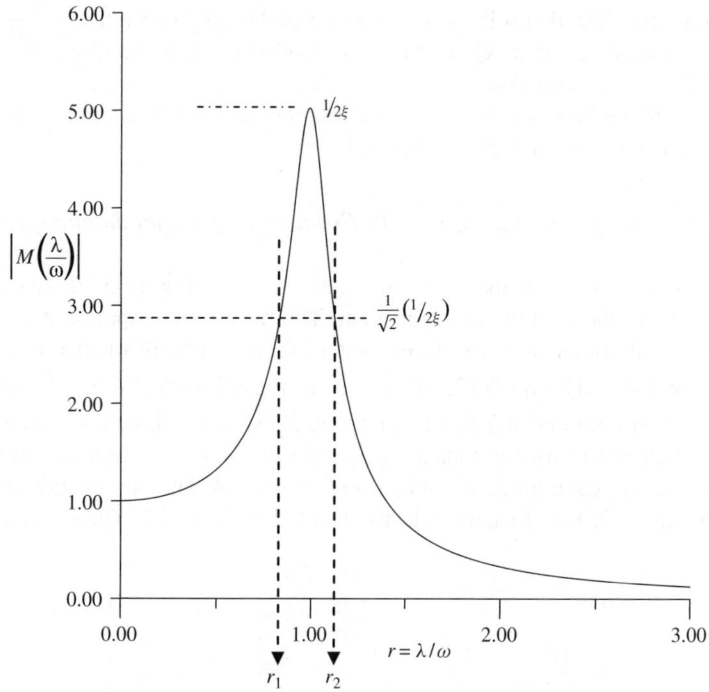

Figure 5.23 Half power method. Estimation of damping ratio

We need to find the value of $r = \frac{\lambda}{\omega}$ for which $\{M(r)\}^2 = \frac{1}{2}\left(\frac{1}{2\xi}\right)^2$. This leads to the equation:

$$\frac{1}{\left(1 - r^2\right)^2 + (2\xi r)^2} = \frac{1}{2}\left(\frac{1}{2\xi}\right)^2 \tag{5.133}$$

Above equation simplifies to:

$$r^4 - 2\left(1 - 2\xi^2\right) r^2 + \left(1 - 8\xi^2\right) = 0 \tag{5.134}$$

From Equation (5.134), the possible two roots for r^2 are obtained as:

$$r^2 = \left(1 - 2\xi^2\right) \pm 2\xi\left(1 - \xi^2\right) \tag{5.135}$$

For cases of $\xi \ll 1$, we have the half power frequencies as:

$$r^2 = 1 \pm 2\xi \Rightarrow \left(\frac{\lambda_1}{\omega}\right)^2 = 1 - 2\xi \text{ and } \left(\frac{\lambda_2}{\omega}\right)^2 = 1 + 2\xi \tag{5.136}$$

From the above equation, we obtain:

$$4\xi = \frac{{\lambda_2}^2 - {\lambda_1}^2}{\omega^2} = \frac{\left(\lambda_2 + \lambda_1\right)\left(\lambda_2 - \lambda_1\right)}{\omega^2} \cong \frac{2\left(\lambda_2 - \lambda_1\right)}{\omega} \Rightarrow \xi = \frac{\left(r_2 - r_1\right)}{2} \tag{5.137}$$

With the knowledge of r_1 and r_2, we estimate the damping ratio ξ from Equation (5.137). The frequency interval $\left(\lambda_2 - \lambda_1\right)$ around the resonance point is called the bandwidth of the modal equation (SDOF oscillator) and is given by 2ξ. The ratio $\frac{r}{(r_2-r_1)} = \frac{\lambda}{\lambda_2-\lambda_1}$ is known as the quality factor or Q–factor. The bandwidth and the Q-factor measure the sharpness of the resonance curve.

The half power method is effective in estimating the modal damping ratios provided the natural frequencies are not closely spaced.

Example 5.13: Shear frame Example 5.10: Estimation of modal damping ratios by half power method

Frequency response curves of the shear frame are given in Figure 5.20. Assume that the response curves are obtained by an experiment. The frequency response at each *degree of freedom* has contributions from the three system modes. The resonance points in terms of r $\left(= \frac{\lambda}{\omega_1}\right)$ are 1.0, 2.12 and 3.12. For instance, the response, $X_1\left(\frac{\lambda}{\omega}\right)$ at the top mass point shows peak amplitudes of 7.6744, 7.416 and 2.281 at the three resonance points. The corresponding half power magnitudes are respectively 5.4266, 5.244 and 1.6131. The half power frequencies for each mode are shown in Figure 5.24. The band width, the estimated damping ratio and Q factor of each mode are detailed in Table 5.1. The estimated damping

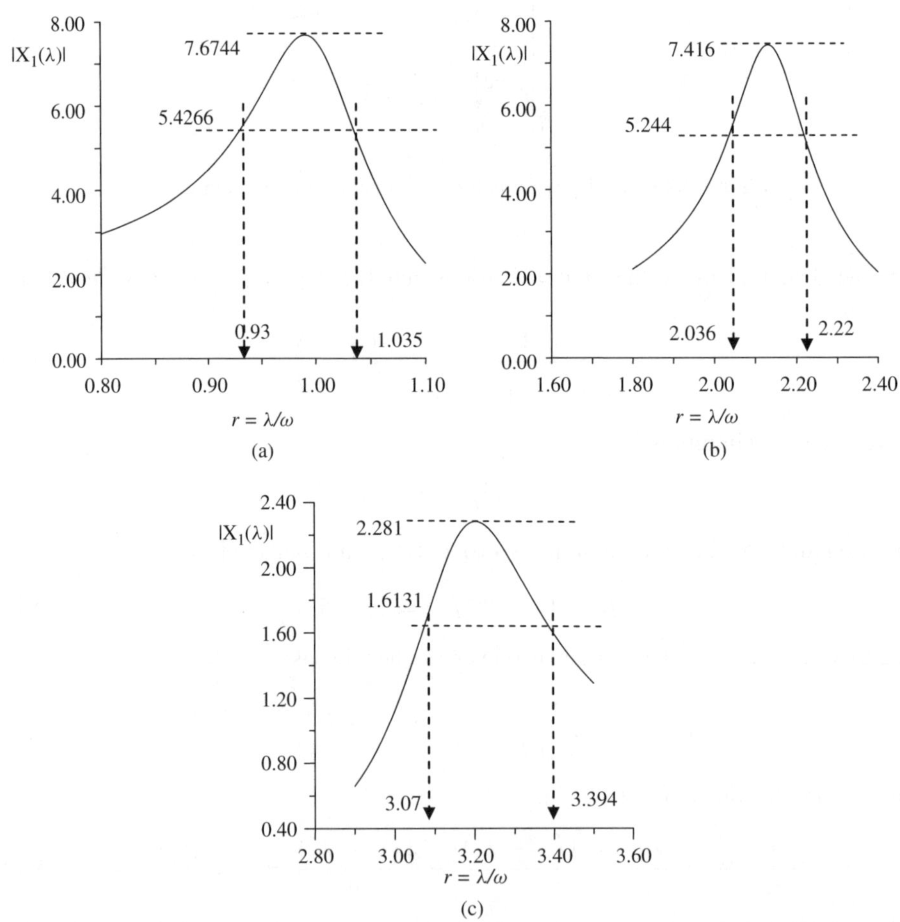

Figure 5.24 Shear frame: estimation of modal damping ratios by half-power method; (a) first mode, (b) second mode and (c) third mode of vibration

Table 5.1 Shear frame and half power method of finding estimates for ξ, bandwidth and Q-factor of each mode

Mode number	Bandwidth (rad/s)	Q-factor	Estimated ξ
1	1.52	9.545	0.0524
2	2.67	11.531	0.0432
3	4.68	9.662	0.0505

ratios fairly match with the values actually used to arrive at the frequency curves (see Example 5.11). This is the result from the frequency response obtained with a resolution of 0.01 in terms of frequency ratio r.

5.9 Conclusions

In this chapter, we have undertaken the study of discrete (MDOF) systems and their solutions by modal dynamics. The exposition, limited to linear time-invariant (LTI) systems, uses the familiar notions of n- dimensional linear differential operators, their characteristic values and series solutions in terms of the so-called complementary functions to finally define a generalized eigenvalue problem. Analogous to the eigenfunctions for a linear continuous system (Chapter 2), the eigenvectors of the eigenvalue problem for the discrete system span the solution vector space. Traversing through the matter presented in the preceding chapters leading up to the present one, we notice the vital role an SDOF oscillator plays as a building block in constructing the free or forced vibration solution via modal summation, irrespective of whether the system is continuous or discrete. As such, much attention is devoted in this chapter to the solution procedures for an SDOF oscillator and their features in both time and frequency domains. The modal summation approach is followed in the next chapter too, wherein solutions of MDOF systems under support excitations form the subject of study. The exclusive nature of this particular case of excitation arises from the need to simulate the familiar case of ground motion due to the incidence of earthquakes and from their importance in seismic risk management.

Exercises

1. A semi-definite system (unconstrained) is shown in Figure 5.25. Write the equations of motion for this system and also derive the frequency equation. Find the natural frequencies and mode shapes.

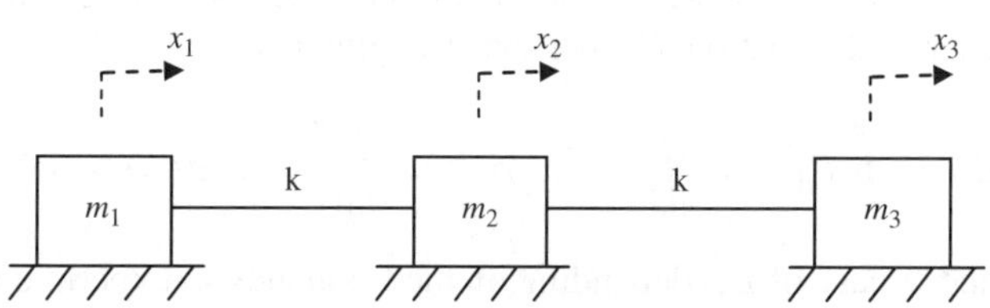

Figure 5.25 An unconstrained 3-*dof* system

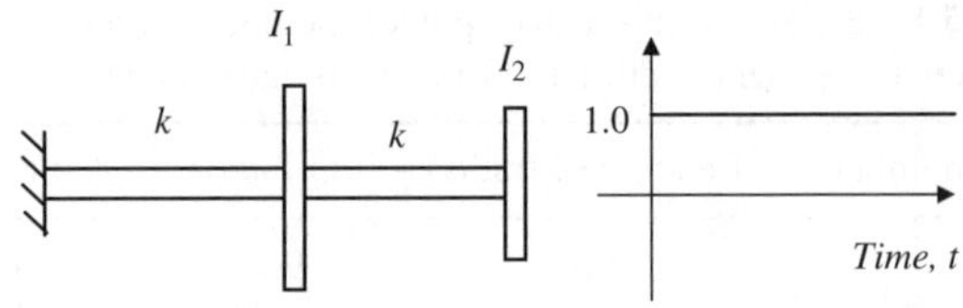

Figure 5.26 (a) A 2-*dof* torsional system (b) unit step input

2. A torsional system with 2 dofs is shown in Figure 5.26. Write the equations of motion and find the eigensolution. Using the modal superposition technique, determine the response at the two disk locations due to a unit step torque applied at the right disk at time $t = 0$. The following data are given $I_1 = 3\,\text{kg}\,\text{m}^2$, $I_2 = 1.5\,\text{kg}\,\text{m}^2$, $k = 200\,\text{Nm/rad}$. Also assume a uniform modal (viscous) damping ratio of 0.05 (for each mode) and zero initial conditions.
3. Use the Rayleigh–Ritz method to solve for the natural frequencies of the 2-dof torsional system in Exercise 2 above. Assume appropriate trial functions satisfying the essential boundary conditions.
4. For an *N dof* system, a hysteretic damping model is assumed (see the notes below on hysteretic damping) and the modal decomposition approach results in the following 2nd order ODE for each of the modes:

$$\ddot{q}_j + \omega_j{}^2 (1 + i\gamma)\, q_j = f_j(t)\,, i = \sqrt{-1} \tag{a}$$

The reader would be familiar with the notations q_j, ω_j and $f_j(t)$, $j = 1, 2, \ldots N$. The parameter γ is known as the hysteretic damping factor or structural damping factor. Using modal superposition, derive the steady state solution of the MDOF system under harmonic input (assume $f_j(t) = A_j \sin \lambda t$).

Notes on hysteretic damping:

Structural materials are known to exhibit a damping mechanism different from that of the simple viscous damping model (Section 5.7). The latter is an idealization of an otherwise complex phenomenon and is adopted for structural applications mostly because of the mathematical ease it offers in response computations.

Under cyclic stress, the internal damping in all materials exhibit a hysteretic behaviour as shown in Figure 5.27 and the energy dissipation E_D per cycle is given by the area within the hysteresis loop in the figure. We have already learnt that, in the case of viscous damping, the damping force is proportional to velocity, i.e., $F_d = c\dot{x}$ where c is the viscous damping coefficient. Further, under harmonic loading, $x(t) = X \sin(\lambda t - \varphi)$ and $\dot{x}(t) = \lambda X \cos(\lambda t - \varphi)$. The hysteresis loop for the viscous damping model is thus an ellipse and the energy dissipation E_V per cycle is given by:

$$E_V = \oint F_d dx = \int_0^{2\pi/\lambda} (c\dot{x})\dot{x}dt = \int_0^{2\pi/\lambda} c\dot{x}^2 dt = \pi c \lambda X^2 \tag{b}$$

Contrary to the above case of E_V depending on the frequency λ, experimental observations suggest that the energy dissipation per cycle due to the internal damping is independent

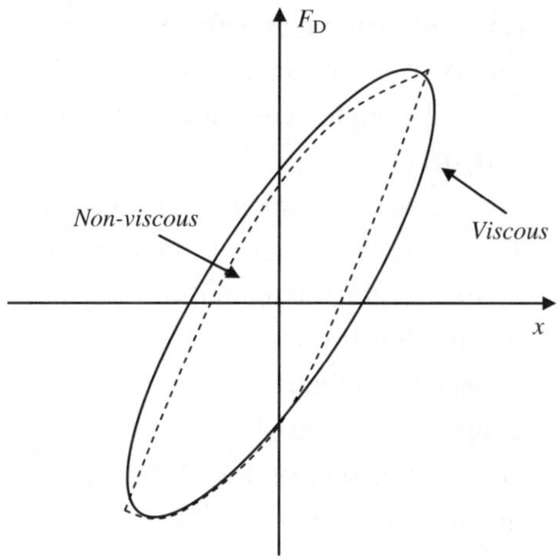

Figure 5.27 Hysteresis loop; viscous and non-viscous type of damping; F_D denotes the damping force and x, the displacement

of λ. In the hysteretic damping model, the energy dissipation per cycle E_D is taken as αX^2, where α is a real constant. Now, one can define an equivalent viscous damping coefficient c_{eq} by equating E_D to E_V (Equation b) as:

$$\pi c_{eq}\lambda X^2 = \alpha X^2 \Rightarrow c_{eq} = \frac{\alpha}{\pi\lambda} \tag{c}$$

Substituting c_{eq} for c in Equation 5.92 of an SDOF oscillator, we obtain the equation of motion as:

$$m\ddot{x} + \frac{\alpha}{\pi\lambda}\dot{x} + kx = f(t)$$

$$\Rightarrow m\ddot{x} + k\left(\frac{i\alpha}{\pi k} + 1\right)x = f(t) \quad \text{(assuming } x(t) = Xe^{i\lambda t} \text{ under harmonic input)}$$

$$\Rightarrow m\ddot{x} + k(1 + i\gamma)x = f(t) \tag{d}$$

In the above equation, $\gamma = \frac{\alpha}{\pi k}$ is known as the hysteretic damping (or structural damping) factor and $k(1 + i\gamma)$ is termed as the complex stiffness. Equation (a) is of a precisely similar form as equation (d), the former being in the modal form in terms of $q_j(t)$, $j = 1, 2, \ldots N$

Notations

a	Real constant
A_1, A_2	Complex conjugates
$\boldsymbol{A}$	$2N \times 2N$ (time-invariant) system matrix
b	Real constant
$\boldsymbol{B}$	Matrix ($= \boldsymbol{L}^{-1}\boldsymbol{K}\boldsymbol{L}^{-T}$-equation 5.46)

c	Constant, damping coefficient
$\boldsymbol{c}$	Matrix of eigenvectors (Equation (5.85))
c_j, d_j	General complex constants ($\in \mathbb{C},\ j = 1, 2 \ldots\ldots n$)
D	$= d/dt$
E_λ	Eigen space corresponding to the eigenvalue λ
$f(t)$	Time function (generally de noting forcing function)
$f_A(\lambda)$	Characteristic polynomial of the matrix A
$f_1(t), f_2(t), f_3(t)$	Forcing functions
F	Constant amplitude
$F(s)$	Laplace transform of $f(t)$
$\boldsymbol{g}(t)$	Forcing function vector ($2N \times 1$)
$I(t)$	Impulse function
$\boldsymbol{I}$	$N \times N$ identity matrix
$\mathcal{M}(\lambda)$	Magnification factor – function of excitation frequency λ
$\boldsymbol{x}(\boldsymbol{t})$	Displacement vector ($= (x_1, x_2, \ldots x_N)^T$)
$\boldsymbol{y}(t)$	$2N$-dimensional state-space vector $\left(\{x_1, x_2, x_3, \ldots, x_n, \dot{x}_1, \dot{x}_2, \dot{x}_3 \ldots\ldots, \dot{x}_n\}^{\mathrm{T}}\right)$
$y_1(t), y_2(t), y_3(t), y_4(t), y_5(t), y_6(t)$	Functions of time
k_j, K_j	Stiffnesses
$[\bar{\boldsymbol{K}}_{jk}]$	Stiffness matrix
$\boldsymbol{L}$	Non-singular lower triangular matrix
m	Roots of a polynomial
m_j	Masses, algebraic multiplicity
$[\bar{\boldsymbol{M}}_{jk}]$	Mass matrix
N	Integer number
N	Number of system *dof* s
$\boldsymbol{N}(\cdot)$	Null space
$P(m)$	Auxiliary polynomial
$\boldsymbol{q}$	Vector quantity (Equation (5.46))
$\boldsymbol{q_0}, \dot{\boldsymbol{q}}_\mathbf{0}$	Initial conditions on $\boldsymbol{q}(t)$
$q_j(t) j = 1, 2, .., N$	Solution to modal equations
$\boldsymbol{q}^{(c)}(t), \boldsymbol{q}^{(p)}(t)$	Complementary and particular solutions of a second-order ODE respectively
Q	Amplitude of steady state response
$Q(s)$	Laplace transform of $q(t)$
$Q(i\lambda)$	Complex frequency response function of modal coordinates
r	Frequency ratio $\left(=\frac{\lambda}{\omega_n}\right)$
t_d	Time period of oscillation $\left(=\frac{2\pi}{\omega_d}\right)$
U	$N \times 1$ vector of displacement coordinates

v	Vector quantity, an assumed mode shape vector
$\boldsymbol{v}_j$	Eigenvector corresponding to the eigenvalue λ_j
V_n	Subspace of the solution space, V
$X_j(i\lambda)$	Complex frequency response function of system coordinates
A	Real constant
B	Real constant
Δ	Logarithmic decrement
ε	Non-zero infinitesimal value
θ	Phase angle of steady state solution
λ	Eigenvalue, forcing frequency
$\boldsymbol{\xi}$	Vector of damping ratios $\left(= \left(\xi_1, \xi_2, \ldots.., \xi_N\right)\right)$
$\boldsymbol{\omega}$	Vector of eigenvalues, $\left(= \omega_1, \omega_2, \ldots.., \omega_N\right)$
$\boldsymbol{\phi}$	Eigenvector matrix
$\overline{\phi}_j$	Vectors $\in V_n$

References

Adhikari, S. (2000) Damping models for structural vibration, PhD thesis Cambridge University, UK.

Bathe, E.L. (1996) *Finite Element Procedures*, Prentice-Hall, Englewood Cliffs, NJ.

Bathe, E.J. and Wilson, E.L. (1976) *Numerical Methods in Finite Element Analysis*, Prentice-Hall, Englewood Cliffs, NJ.

Berger, E.J. (2002) Friction modeling for dynamic system simulation. *Applied Mechanics Reviews*, **55** (6), 535–577.

Clough, R.W. and Penzien, J. (1982) *Dynamics of Structures*, McGraw-Hill, Inc., Singapore.

Craig, R.R. Jr. (1981) *Structural Dynamics- An Introduction to Computer Methods*, John Wiley & Sons, Inc., New York.

Kreyszig, E. (1999) *Advanced Engineering Mathematics*, 8th edn, John Wiley & Sons, Inc., Singapore.

Lord Rayleigh, J.W.S. (1945) *The Theory of Sound*, 2nd edn, vols. 1 and 2, Macmillan (reprinted by Dover, New York).

Newland, D.E. (1989) *Mechanical Vibration Analysis and Computation*, Dover Publications, Inc., New York.

Simmons, G.F. and Krantz, S.G. (1925) *Differential Equations-Theory, Technique and Practice*, McGraw-Hill, Inc. , New York.

Bibliography

On Damping in Materials

Bert, C.W. (1973) Material damping: an introductory review mathematical models, measure and experimental techniques. *Journal of Sound and Vibration*, **29** (2), 129–153.

Crandall, S.H. (1991) The hysteretic damping model in vibration theory. *Journal of Mechanical Engineering Science*, **205**, 23–28.

Hasselsman, T.K. (1972) A method of constructing a full modal damping matrix from experimental measurements. *AIAA Journal*, **10** (4), 526–527.

Lazan, B.J. (1968) *Damping of Materials and Members in Structural Mechanics*, Pergamon Press, Oxford.

On Modal Dynamics

Caughey, T.K. (1960) Classical normal modes in linear dynamical systems. *Transactions of ASME, Journal of Applied Mechanics*, **27**, 269–271.

Caughey, T.K. and O'Kelly, M.E.J. (1965) Classical normal modes in linear dynamical systems. *Transactions of ASME, Journal of Applied Mechanics*, **32**, 583–588.
Evins, D.J. (1984) *Modal Testing: Theory and Practice*, Research Studies Press, Taunton, UK.
Veletsos, A.S. and Ventura, C.E. (1986) Modal analysis of non-classically damped linear systems. *Earthquake Engineering and Structural Dynamics*, **14**, 217–243.

On Nonproportional Damping

Clough, R.W. and Mojtahedi, S. (1976) Earthquake response analysis considering non-proportional damping. *Earthquake Engineering and Structural Dynamics*, **4**, 489–496.
Hurty, W.C. and Rubinstein, M.F. (1964) *Dynamics of Structures*, Prentice Hall.

6

Structures under Support Excitations

6.1 Introduction

Analysis of structures subjected to base excitation is of vital interest, especially in view of their importance in power plants and nuclear installations. In this context, the need for seismic qualification of these structures, associated components and equipments is self-evident and in fact insisted upon globally in relevant codes of practice (Indian Standard, 1893; ASCE Standard 4-98, 2000). Studies on base-excited structures could be significant even with nonseismic inputs such as sea-wave motion on structural components in ships, induced vibrations on airborne electronic equipment, road/rail undulations for vehicles and even for equipments mounted on these vehicles. Structural analysis under these base excitations essentially requires specification of appropriate mathematical models for these loads. In most of these cases, measured data related to these loading phenomena is available. For example, accelerogram (acceleration time history) records of past earthquake events are provided in EERL (1980). Figure 6.1 shows the accelerogram records (acceleration in 'g' units versus time in seconds) of the three orthogonal components of the El Centro earthquake 1940, California. Certain codes of practice specify deterministic functions (such as half-sine, triangle, etc.) that describe profile variations corresponding to either rail/roads or aircraft flight conditions. These test functions are actually meant for use in experimental investigations to assess the integrity and functional reliability of an equipment. However, analytical studies simulating the specified test procedures are also meaningful and are in fact preferred prior to and during a prototype test to help in the initial stages of design as well as to finally arrive at an acceptable design.

Whilst prescribing models for the base motions in the time domain is apparently more straightforward, it is also common practice to specify such motions in the frequency domain. For instance, International Electrotechnical Commission (1968) recommends harmonic base excitation to qualify electric traction equipment for simulation of shunting shocks and mechanical vibration (Figure 6.2). Despite the simplicity of deterministic functions used to model base motions, it is known that the nature of such support excitations is often inherently random and is best represented by models derived from theory of

Elements of Structural Dynamics: A New Perspective, First Edition. Debasish Roy and G Visweswara Rao.

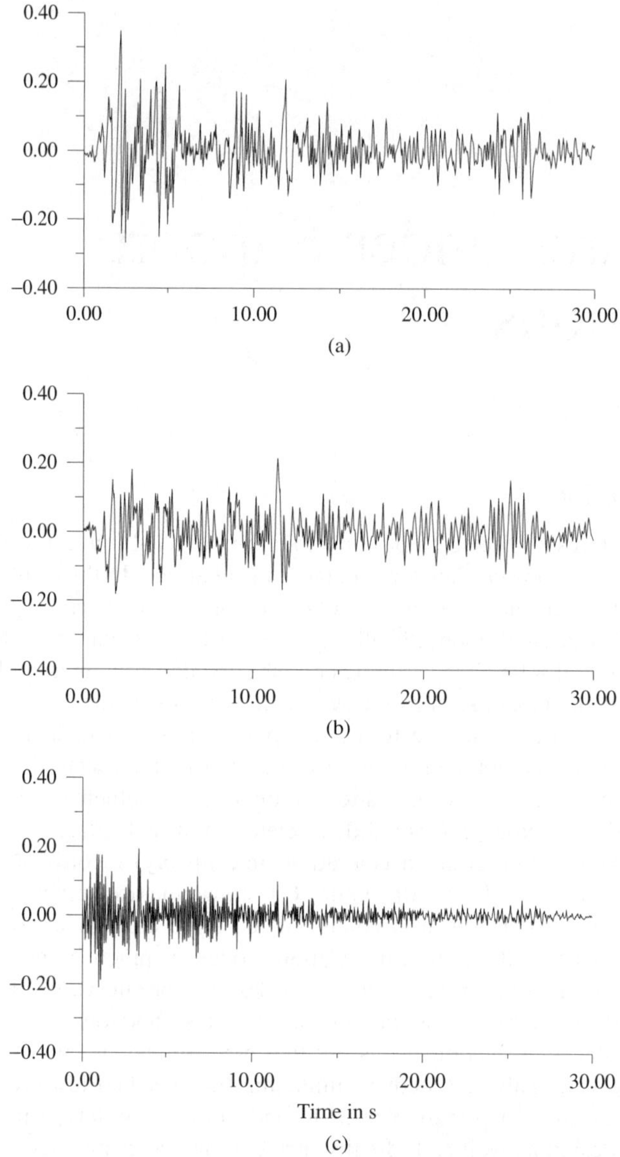

Figure 6.1 El Centro 1940 earthquake acceleration records; (a) S00E (b) S90W and (c) vertical components

stochastic processes. Discussion on the stochastic nature of external loads and elementary methods for the evaluation of structural response are considered in Chapter 9.

In the present chapter, however, deterministic procedures to arrive at the response of linear systems under support excitation are enumerated with emphasis on the mode superposition technique. Related aspects such as usefulness of response spectrum methods and their significance vis-à-vis a more involved numerical integration of the dynamic

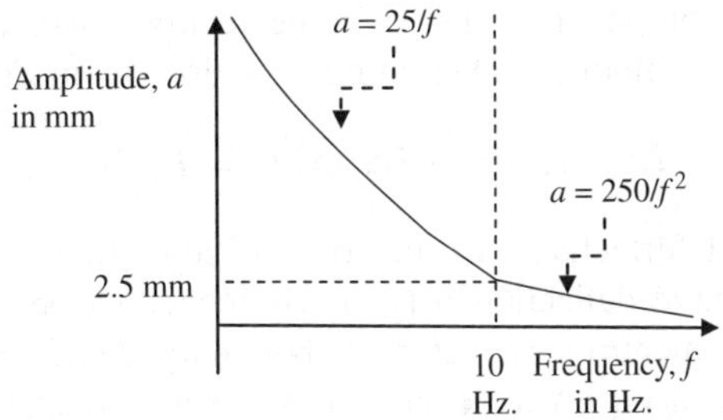

Figure 6.2 Base excitation frequency spectrum recommended by International Electrotechnical Commission (1968) for electric traction equipment. Modified from IEC 77, 1968

system equations are also highlighted. The presentation in this chapter first considers the ground excitation to be uniform in that the support accelerations corresponding to a specified degree of freedom have identical time histories, irrespective of the support locations. This may be contrasted with the case of multisupport excitations, wherein these time histories are 'asynchronous', that is they are identical or otherwise modulo appropriate phase lags. A brief exposition on the structural response under multi-support excitations is presented in the last section of the chapter.

6.2 Continuous Systems and Base Excitations

Continuous systems subjected to base excitations (and with no other external loads) are described by the following partial differential equation (PDE):

$$\boldsymbol{L}\boldsymbol{u}(\boldsymbol{x},t) = 0 \text{ in } \Omega \subset \mathbb{R}^n \tag{6.1}$$

$\boldsymbol{L}(.)$ is a linear partial differential operator. As the name suggests, the 'base' motion is typically applied at the system degrees of freedom (translational and/or rotational) that are located at the domain boundary. Accordingly, the base motion often appears in the form of the following nonhomogeneous boundary condition:

$$\boldsymbol{B}\boldsymbol{u}\left(\boldsymbol{x}_k,t\right) = h_k(t), k = 1, 2..n_g \tag{6.2}$$

n_g is the number of relevant system degrees of freedom on the boundary at which the base motion is applied. $\boldsymbol{h}(t) = \left(h_1(t), h_2(t), \ldots h_{n_g}(t)\right)^T$ is the vector of base motions – translational or rotational. Note that since the base motion is assumed to be uniform, the maximum number of distinct time-history components in $\boldsymbol{h}(t)$ is ≤ 6. $\boldsymbol{B}(.)$ is a differential operator and $\boldsymbol{x}_k$ the spatial coordinates of the excited locations on the boundary. Equations 6.1 and 6.2 constitute a PDE with inhomogeneous boundary conditions (see Section 2.8). To solve Equation 6.1 with the nonhomogeneous boundary condition, we express the displacement $\boldsymbol{u}(\boldsymbol{x},t)$ as a sum of a dynamic component, $\boldsymbol{u}_d(\boldsymbol{x},t)$, and a 'pseudo' static component, $\boldsymbol{u}_s(\boldsymbol{x},t)$ that could have been induced by the support motion if it were applied statically. Thus, the total displacement $\boldsymbol{u}(\boldsymbol{x},t)$ is given by:

$$\boldsymbol{u}(\boldsymbol{x},t) = \boldsymbol{u}_d(\boldsymbol{x},t) + \boldsymbol{u}_s(\boldsymbol{x},\mathrm{t}) \tag{6.3}$$

Note that the dynamic component $\boldsymbol{u}_d(\boldsymbol{x}, t)$ is the relative displacement of the structure with respect to the ground motion. Substituting Equation 6.3 in Equation 6.1, we obtain:

$$\boldsymbol{L}\boldsymbol{u}_d(\boldsymbol{x}, t) = -\boldsymbol{L}\boldsymbol{u}_s(\boldsymbol{x}, t) = \boldsymbol{P}_{eff}(t) \tag{6.4}$$

$\boldsymbol{P}_{eff}(t)$ is the so-called 'effective' loading on the continuous system due to the base motion. Knowing the static deflection patterns (influence shapes) caused by application of unit support displacement/rotation at each boundary degree of freedom, we evaluate the pseudostatic displacement $\boldsymbol{u}_s(\boldsymbol{x}, t)$ and hence the effective loading $\boldsymbol{P}_{eff}(t)$ (see steps in Equations 2.215–2.220). The PDE (Equation 6.4) in terms of the dynamic displacement component, $\boldsymbol{u}_d(\boldsymbol{x}, t)$, is now inhomogeneous with homogeneous boundary conditions:

$$\boldsymbol{B}\boldsymbol{u}_d\left(\boldsymbol{x}_k, t\right) = 0, k = 1, 2..n_g \tag{6.5}$$

The continuous system governed by the Equations 6.4 test and 6.5 is now solvable by any of the methods described earlier in Chapters 2 and 3. As an illustration, we take a 1D beam subjected to base motion as shown in Figure 6.3 below.

The equation of motion of the (unforced) uniform beam is given by the following PDE:

$$m\frac{\partial^2 y}{\partial t^2} + EI\frac{\partial^4 y}{\partial x^4} = 0 \tag{6.6}$$

where the linear partial differential operator is $\boldsymbol{L} = m\frac{\partial^2}{\partial t^2} + EI\frac{\partial^4}{\partial x^4}$. The parameters m, E and I, respectively, denote the mass per unit length, Young's modulus and moment of inertia of the beam cross section. l is the length of the beam. In this example, $n_g = 1$ and the corresponding degree of freedom is the beam translation at the fixed end in the Y-direction. The excitation at the base appears as the boundary condition:

$$y(0, t) = h(t) \tag{6.7}$$

To solve Equation 6.6 with the nonhomogeneous boundary condition in Equation 6.7, we express the beam displacement $y(x, t)$ as a sum of dynamic displacement component and a pseudostatic component, as explained earlier. In the present case the pseudostatic component is $y_s(x, t) = h(t)$. Thus, the total displacement $y(x, t)$ is given by:

$$y(x, t) = y_d(x, t) + h(t) \tag{6.8}$$

Substituting Equation 6.8 in Equation 6.6, we obtain:

$$m\frac{\partial^2 y_d}{\partial t^2} + EI\frac{\partial^4 y_d}{\partial x^4} = f_d(t) \tag{6.9}$$

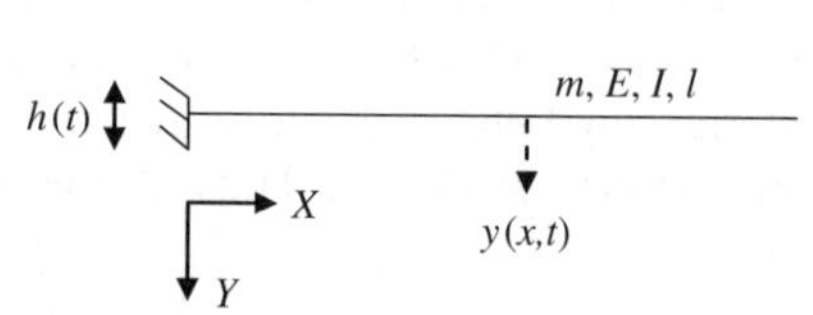

Figure 6.3 A uniform beam under base excitation

$f_d(t)$ is the effective reaction force on the beam at its base and is presently given by:

$$f_d(t) = -\boldsymbol{L} y_s(\boldsymbol{x}, t) = -m\ddot{h}(t) \tag{6.10}$$

The equation of motion (Equation 6.9) in $y_d(x, t)$ is now associated with homogeneous boundary condition $y_d(0, t) = 0$. The forced-vibration response $y_d(x, t)$ can be obtained as an eigenexpansion in terms of the beam eigenfunctions as:

$$y_d(x, t) = \sum_{k=1}^{N} \Phi_k(x) Y_k(t) \tag{6.11}$$

$\Phi_k(x) k = 1, 2, \ldots, N$ are the eigenfunctions of the beam and $Y_k(t)$ the time-dependent response of the kth mode. Substituting Equation 6.11 in Equation 6.9 we have:

$$\sum_{k=1}^{N} \left(m\Phi_k(x)\ddot{Y}_k(t) + EI \frac{\partial^4 \Phi_k}{\partial x^4} Y_k(t) \right) = f_d(t) \tag{6.12}$$

Since $\Phi_k(x)$ are orthogonal, we have:

$$m \int_0^l \Phi_k(x)\Phi_j(x)dx = M_j \delta_{jk},\ EI \int_0^l \Phi_k(x) \frac{\partial^4 \Phi_j}{\partial x^4} dx = K_j \delta_{jk} \tag{6.13}$$

M_j and K_j are known as the generalized mass and stiffness, respectively. Multiplying both sides of Equation 6.12 by $\Phi_j(x)$ and integrating over the beam length, l yields:

$$\sum_{k=1}^{N} \ddot{Y}_k(t) m \int_0^l \Phi_k(x)\Phi_j(x)dx + \sum_{j=1}^{N} Y_k(t) EI \int_0^l \Phi_j(x) \frac{\partial^4 \Phi_k}{\partial x^4} dx$$
$$= \int_0^l \Phi_j(x) f_d(t) dx \tag{6.14}$$

By using Equation 6.13, we obtain from Equation 6.14 a set of uncoupled modal equations as second-order ODEs:

$$\ddot{Y}_j(t) + \omega_j^2 Y_j(t) = p_j(t),\ j = 1, 2 \ldots, N \tag{6.15}$$

Here, $\omega_j^2 = \frac{K_j}{M_j}$ and $p_j(t)$ is the normalised generalized force associated with the jth mode and is given by:

$$p_j(t) = \frac{-m\ddot{h}(t)}{M_j} \int_0^l \Phi_j(x)dx = P_{fj}\ddot{h}(t) \tag{6.16}$$

where

$$P_{fj} = -\frac{m}{M_j} \int_0^l \Phi_j(x)dx \tag{6.17}$$

P_{fj} is known as the modal participation factor of mode j. The response of each mode is now obtained as the convolution integral:

$$Y_j(t) = \frac{1}{\omega_j} \int_0^t p_j(\tau) \sin \omega_j (t - \tau)\, d\tau \tag{6.18}$$

If the system is viscously damped, the above integral takes the form:

$$Y_j(t) = \frac{1}{\omega_{jD}} \int_0^t p_j(\tau) e^{-\xi_j \omega_j (t-\tau)} sin\omega_{jD}(t-\tau)\, d\tau \tag{6.19}$$

ξ_j is the viscous damping ratio and ω_{jD}, the damped natural frequency of the jth mode.

Example 6.1: A chimney subjected to seismic excitation

Figure 6.4 shows a chimney (Newland, 1989) of uniformly hollow circular cross section under base motion (length = 42 m, diameter = 2.25 m, wall thickness = 6.8 mm). We also adopt the following material data: Young's modulus $E = 2.1 \times 10^{11}$ N/m^2, $m = 500$ kg/m and the total weight of the chimney = 21 tonnes. The second moment of inertia of the cross section is given by $I = \frac{\pi}{64}\left\{D^4 - (\mathrm{D} - 2t)^4\right\}$. The eigenvalue parameter λ in the free-vibration solution (see Chapter 2) of the cantilever chimney satisfies the characteristic equation (Table 2.1): $\cos\lambda l \cosh\lambda l + 1 = 0$. With $\lambda^4 = \frac{m\omega^2}{EI}$, the first few modes correspond to the characteristic values $\lambda_1 l = 1.8751, \lambda_2 l = 4.6941$, $\lambda_3 l = 7.8548, \lambda_4 l = 10.996$, and so on. Accordingly, the first four natural frequencies of the chimney are: $\omega_1 = 7.1241$, $\omega_2 = 44.646$, $\omega_3 = 125.0128$ and $\omega_4 = 244.9930$ rad/s. The mode shapes of a cantilever beam are described in Table 2.1. From Equation 6.15, we have the modal equations (with damping included):

$$\ddot{Y}_j(t) + 2\xi_j\omega_j\dot{Y}_j(t) + \omega_j^2 Y_j(t) = p_j(t),\ j = 1, 2.\dots N \tag{6.20}$$

Four terms are considered in the eigenexpansion in Equation 6.11 and thus $N = 4$. Viscous damping is assumed for the system with a modal damping ratio $\xi_j = 0.05$, $j = 1, 2, 3$ and 4. The effective loading in terms of the generalized force $p_j(t)$ is given by $P_{fj}\ddot{\boldsymbol{h}}(t)$ as in Equation 6.16.

The modal participation factors, (P_{fj}) for the first four modes are 0.09232, –0.00952, 0.00039 and 0.00001, respectively. Note that the participation of the first mode is far more significant compared to the rest of the modes and it is expected that the transverse

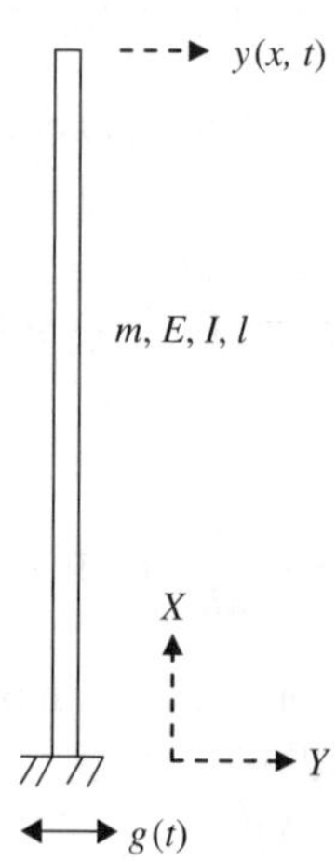

Figure 6.4 Chimney as a continuous system subjected to base excitation

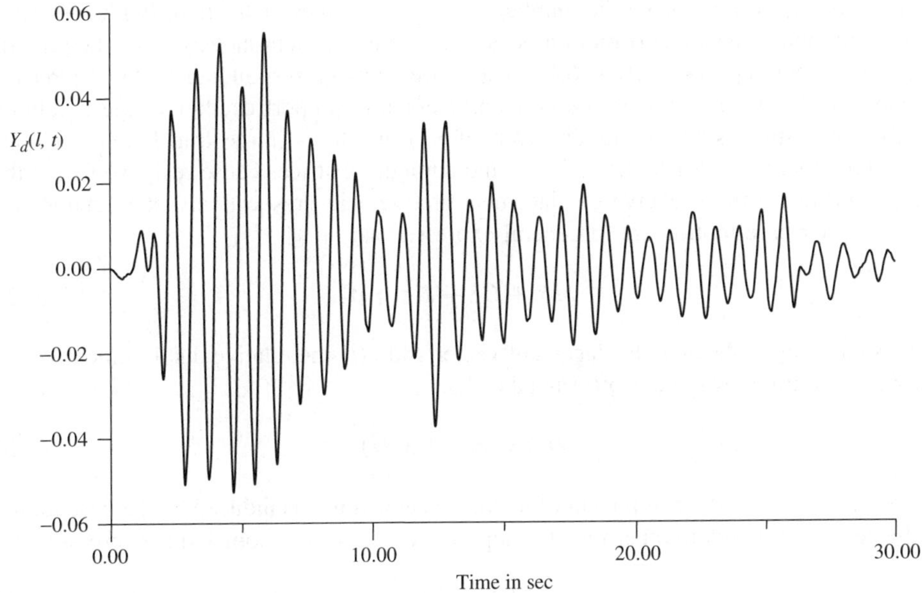

Figure 6.5 Dynamic response of the chimney relative to the base motion at its tip (free end)

deflection $Y_d(l, t)$ is dominated by the response of the first mode. If the El Centro earthquake component S00E (Figure 6.1a) is taken as the base acceleration $\ddot{h}(t)$, the modal responses can be obtained from the convolution integral (Equation 6.19). The acceleration time history data is of 30 s duration uniformly spaced at 0.02 s. To evaluate the modal response (Equation 6.19), the numerical procedure given in Appendix F (Clough and Penzien, 1982), is followed. Accordingly, the convolution integral in Equation 6.18 is written in the form:

$$Y_j(t) = A(t)\sin\omega_{jD}t - B(t)\cos\omega_{jD}t \tag{6.21}$$

Here, $A(\mathrm{t}) = \frac{1}{\omega_{jD}}\int_0^t p_j(\tau)e^{-\xi_j\omega_j(t-\tau)}\cos\omega_{jD}\tau\, d\tau$ and

$$B(\mathrm{t}) = \frac{1}{\omega_{jD}}\int_0^t p_j(\tau)e^{-\xi_j\omega_j(t-\tau)}\sin\omega_{jD}\tau\, d\tau \tag{6.22}$$

The above two integrals are evaluated by an incremental summation procedure using Simpson's rule (Kreyszig, 1999). The integration time step is taken as $T_4/10$, where T_4 is the time period corresponding to the 4th mode of vibration and equal to $\frac{2\pi}{\omega_4}(= 0.02505)$. With the modal responses obtained from Equation 6.21, the dynamic component of system response, $y_d(x, t)$, is evaluated by the modal summation in Equation 6.11. The time history of this component at the tip (free end) of the chimney is shown in Figure 6.5.

6.3 MDOF Systems under Support Excitation

Consider an multidegree-of-freedom (MDOF) system with $\boldsymbol{M}$, C and $\boldsymbol{K}$ being the mass, stiffness and damping matrices, respectively. Let N be the active degrees of freedom

(= total *dof* s – n_g), n_g being the number of support degrees of freedom (with prescribed support motion). The support motion is assumed to act 'synchronously' (see the remarks in the last paragraph of Section 6.1) in the directions corresponding to the degrees of freedom of the fixed nodes on the boundary. For this support excitation, the maximum number of distinct support motions (out of n_g) are six – three translational and the other three rotational. Since the stiffness and damping matrices cause resistive forces that are proportional to the relative displacement and velocity, respectively, the equations of motion under base motion are in general expressed as:

$$\boldsymbol{M}\ddot{\boldsymbol{x}} + \boldsymbol{C}\dot{\boldsymbol{z}} + \boldsymbol{K}\boldsymbol{z} = 0 \tag{6.23}$$

$\boldsymbol{x}(t)$ is the total or absolute displacement vector and $\boldsymbol{z}(t)$ the relative displacement vector. As with a continuous system, $\boldsymbol{x}(t)$ is taken as:

$$\boldsymbol{x}(t) = \boldsymbol{z}(t) + \boldsymbol{x}_s(t) \tag{6.24}$$

$\boldsymbol{x}_s(t)$ is the pseudostatic component of the displacement that is induced by the base motion at the degrees of freedom other than the support degrees of freedom and is expressible as:

$$\boldsymbol{x}_s(t) = \boldsymbol{T}\boldsymbol{g}(t) \tag{6.25}$$

$\boldsymbol{T}$ is a matrix of influence vectors, transforming the base motion to a pseudostatic displacement vector. $\boldsymbol{g}(t)$ is the vector of distinct support motions. Hence, $\boldsymbol{T}$ is at most of size $N \times 6$. Equation 6.23 can now be rewritten in terms of relative displacements alone as

$$\boldsymbol{M}\ddot{\boldsymbol{z}} + \boldsymbol{C}\dot{\boldsymbol{z}} + \boldsymbol{K}\boldsymbol{z} = -\boldsymbol{M}\boldsymbol{T}\ddot{\boldsymbol{g}}(t) \tag{6.26}$$

Examples 6.2 and 6.3 should further aid in illustrating the derivation of the above equations.

Example 6.2: Shear frame under base excitation

In the lumped-parameter model of the shear frame shown in Figure 6.6a, viscous dampers are present in between the mass points. The equilibrium of the three mass points is shown in Figure 6.6b.In this example, $n_g = 1$ and $g(t)$ acts in the transverse direction X. The force equilibrium of the mass points leads to the following the equations of motion:

$$\begin{bmatrix} m_1 & 0 & 0 \\ 0 & m_2 & 0 \\ 0 & 0 & m_3 \end{bmatrix} \begin{Bmatrix} \ddot{x}_1 \\ \ddot{x}_2 \\ \ddot{x}_3 \end{Bmatrix} + \begin{bmatrix} C_1 & -C_1 & 0 \\ -C_1 & C_2 + C_3 & -C_3 \\ 0 & -C_3 & C_3 \end{bmatrix} \begin{Bmatrix} \dot{x}_1 - \dot{g} \\ \dot{x}_2 - \dot{g} \\ \dot{x}_3 - \dot{g} \end{Bmatrix}$$
$$+ \begin{bmatrix} K_1 & -K_1 & 0 \\ -K_1 & K_2 + K_3 & -K_3 \\ 0 & -K_3 & K_3 \end{bmatrix} \begin{Bmatrix} x_1 - g \\ x_2 - g \\ x_3 - g \end{Bmatrix} = 0 \tag{6.27}$$

$\boldsymbol{x}(t) = \{x_1(t), x_2(t), x_3(t)\}^T$ and is the displacement vector. As with continuous systems, the displacement $\boldsymbol{x}(t)$ is expressed as the sum of dynamic component (displacement relative to $g(t)$) and the pseudostatic displacement, which is a rigid-body displacement

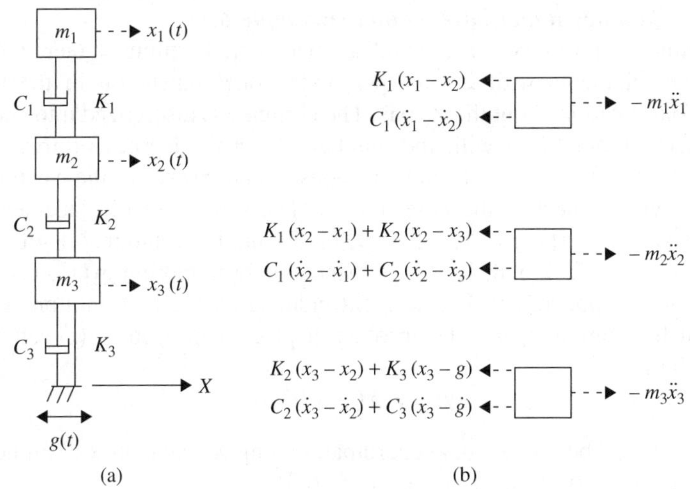

Figure 6.6 Shear frame under base motion; (a) 3 *dof* lumped-parameter system and (b) free-body diagrams and forces in equilibrium

$g(t)$ at all the three mass points. Thus, introducing relative displacements and velocities, respectively, defined by $z_i(t) = x_i(t) - g(t)$ and $\dot{z}_i(t) = \dot{x}_i(t) - \dot{g}(t)$, $i = 1, 2, 3$, Equation 6.27 reduces to:

$$\begin{bmatrix} m_1 & 0 & 0 \\ 0 & m_2 & 0 \\ 0 & 0 & m_3 \end{bmatrix} \begin{Bmatrix} \ddot{x}_1 \\ \ddot{x}_2 \\ \ddot{x}_3 \end{Bmatrix} + \begin{bmatrix} C_1 & -C_1 & 0 \\ -C_1 & C_2+C_3 & -C_3 \\ 0 & -C_3 & C_3 \end{bmatrix} \begin{Bmatrix} \dot{z}_1 \\ \dot{z}_2 \\ \dot{z}_3 \end{Bmatrix}$$

$$+ \begin{bmatrix} K_1 & -K_1 & 0 \\ -K_1 & K_2+K_3 & -K_3 \\ 0 & -K_3 & K_3 \end{bmatrix} \begin{Bmatrix} z_1 \\ z_2 \\ z_3 \end{Bmatrix} = 0 \tag{6.28}$$

Writing $\ddot{x}_i = \ddot{z}_i + \ddot{g}$, Equation 6.28 is recast in $z(t)$ alone as:

$$\begin{bmatrix} m_1 & 0 & 0 \\ 0 & m_2 & 0 \\ 0 & 0 & m_3 \end{bmatrix} \begin{Bmatrix} \ddot{z}_1 \\ \ddot{z}_2 \\ \ddot{z}_3 \end{Bmatrix} + \begin{bmatrix} C_1 & -C_1 & 0 \\ -C_1 & C_2+C_3 & -C_3 \\ 0 & -C_3 & C_3 \end{bmatrix} \begin{Bmatrix} \dot{z}_1 \\ \dot{z}_2 \\ \dot{z}_3 \end{Bmatrix}$$

$$+ \begin{bmatrix} K_1 & -K_1 & 0 \\ -K_1 & K_2+K_3 & -K_3 \\ 0 & -K_3 & K_3 \end{bmatrix} \begin{Bmatrix} z_1 \\ z_2 \\ z_3 \end{Bmatrix} = - \begin{Bmatrix} m_1 \\ m_2 \\ m_3 \end{Bmatrix} \ddot{g} = - \begin{bmatrix} m_1 & 0 & 0 \\ 0 & m_2 & 0 \\ 0 & 0 & m_3 \end{bmatrix} \begin{Bmatrix} 1 \\ 1 \\ 1 \end{Bmatrix} \ddot{g} \tag{6.29}$$

In matrix form, the above equation has the form:

$$\boldsymbol{M}\ddot{z} + \boldsymbol{C}\dot{z} + \boldsymbol{K}z = -\boldsymbol{M}\{\mathbf{1}\}\ddot{g} \tag{6.30}$$

Comparison with Equation 6.26 shows that $T = \{\mathbf{1}\}$ and it is a matrix with a single column of ones.

Example 6.3: Chimney under base motion (Example 6.1)

The steel chimney in Example 6.1 is semidiscretized here by finite element method (FEM) using 1D beam elements with 2 *dof*s per node – one translation in the transverse Y direction and one rotation about the Z-axis. The chimney is discretized into seven elements and eight nodes (Figure 6.7) with the number of active degrees of freedom $N = 14$. Let $\boldsymbol{y}(t)_{N\times 1}$ be the vector of absolute responses and $\boldsymbol{z}(t)_{N\times 1}$, the vector of relative displacements with respect to the base motion. The transverse displacement degrees of freedom are given by $y_j(t)$, $j = 1, 3, 5, 7, 9, 11, 13$ and the rotational degrees of freedom denoted by $y_j(t)$, $j = 2, 4, 6, 8, 10, 12, 14$. Since the base motion $g(t)$ acts only along Y, the pseudostatic components of displacement induced by the base motion appear only at the degrees of freedom along Y. The absolute displacement vector $\boldsymbol{y}(t)$ can be expressed in terms of $\boldsymbol{z}(t)$ as:

$$\boldsymbol{y}(t) = \boldsymbol{z}(t) + \boldsymbol{T}\boldsymbol{g}(t) \tag{6.31}$$

With $x_i, i = 1, 2, .., 7$ being the nodal coordinates along X, the matrix of influence vectors, $\boldsymbol{T} = \begin{bmatrix} 1 & 0 & 1 & 0 & 1 & 0 & 1 & 0 & 1 & 0 & 1 & 0 & 1 & 0 \\ x_1 & 1 & x_2 & 1 & x_3 & 1 & x_4 & 1 & x_5 & 1 & x_6 & 1 & x_7 & 1 \end{bmatrix}^T$. In obtaining the matrix $\boldsymbol{T}$, rigid body displacements of the structure are assumed to occur due to the translational and rotational components of the base excitation. $\boldsymbol{g}(t)$ is given by the vector $(g(t), 0)^T$. The equation of motion under base excitation is:

$$\boldsymbol{M}\ddot{\boldsymbol{y}} + \boldsymbol{C}\dot{\boldsymbol{z}} + \boldsymbol{K}\boldsymbol{z} = \boldsymbol{0} \tag{6.32}$$

By Equation 6.31, the Equation 6.32 corresponds to the general form in Equation 6.26.

The modal superposition technique (Chapter 5) can be employed to solve the linear coupled ordinary differential equation (ODE) (Equation 6.26). The free-vibration (undamped) solution is characterised by the eigenvalue problem:

$$\left[\boldsymbol{K} - \omega^2\boldsymbol{M}\right]\boldsymbol{\Phi} = \boldsymbol{0} \tag{6.33}$$

Using the mode shape matrix $\boldsymbol{\Phi}$, the damping matrix $\boldsymbol{C}$ in Equation 6.32 can be diagonalised by assuming proportional damping (Chapter 5) and by premultiplying

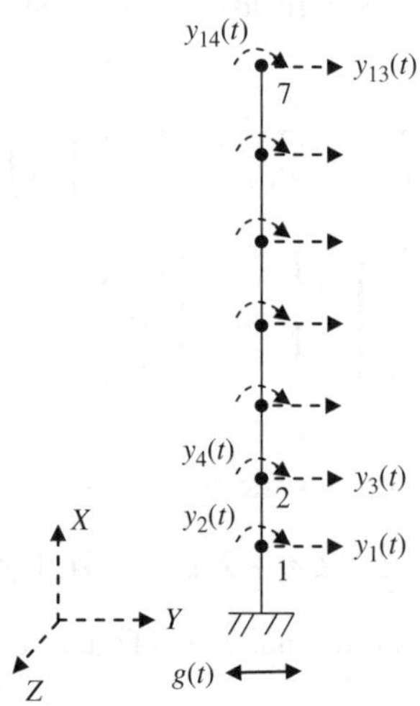

Figure 6.7 Finite element model of chimney of Example 6.1 under base motion

Equation 6.26 by $\mathbf{\Phi}^T$. Alternatively, if modal viscous damping ratios ξ_j, $j = 1, 2...N$ are considered, the modal equations are given by:

$$\ddot{q}_j + 2\xi_j\omega_j\dot{q}_j + \omega_j^2 q_j = p_j(t), \; j = 1, 2 \ldots, N \tag{6.34}$$

The effective loading vector $\boldsymbol{p}(t) = \left(p_1(t), p_2(t), \ldots, p_N(t)\right)^T$ is given by:

$$\boldsymbol{p}(t) = -\mathbf{\Phi}^T \boldsymbol{M}\ \boldsymbol{T}\ddot{\boldsymbol{g}}(t) \tag{6.35}$$

$-\mathbf{\Phi}^T \boldsymbol{M}\ \boldsymbol{T}$ is known as the participation factor matrix. Once the relative displacement $z(t)$ is obtained (see Example 6.4) by solving Equation 6.34 and modal summation, the absolute response $\boldsymbol{y}(t)$ is derived using the relation in Equation 6.24.

Example 6.4: Response of the chimney in Figure 6.7

The support excitation $g(t)$ is taken to be the El Centro 1940 S00E component (Figure 6.1a) acting only in the Y direction. The eigenvalues/natural frequencies of the chimney obtained from the eigenvalue problem in Equation 6.33 are listed in Table 6.1.The first few modeshapes of the chimney are shown in Figure 6.8. Figure 6.9 shows the response $y(t)$ at the chimney tip under the base excitation. It is known (ASCE Standard 4-98, 2000) that the dominant frequencies in a seismic signal lie within the range of 0–33 Hz and it is evident from Table 6.2 that only the first few modes ($\leq$4) are likely to significantly

Table 6.1 Eigensolution for chimney as a MDOF system (Figure 6.7)

Mode number	Natural frequency (rad/s)	Natural frequency (Hz)	Mode number	Natural frequency (rad/s)	Natural frequency (Hz)
1	7.086	1.128	8	1131.710	180.117
2	44.196	7.034	9	1445.989	230.136
3	122.888	19.560	10	1839.032	292.691
4	238.731	38.000	11	2312.099	367.982
5	391.129	62.250	12	2861.300	455.350
6	578.991	92.150	13	3424.282	544.991
7	793.288	126.300	14	3917.832	623.542

Table 6.2 Chimney under base excitation. Modal participation factors

Mode number	ω_j (Hz)	Transverse *degree of freedom* (Y- direction)	Rotational *degree of freedom* (about Z-axis)	Mode number	ω_j (Hz)	Transverse degree of freedom (Y-direction)	Rotational *degree of freedom* (about Z direction)
1	1.128	1.134E+2	−3.464E+3	8	180.1	9.744E+0	−1.930E+1
2	7.034	−6.279E+1	5.540E+2	9	230.1	−9.330E+0	1.718E+1
3	19.560	3.667E+1	−1.987E+2	10	292.7	−8.087E+0	1.408E+1
4	38.000	2.607E+1	−1.020E+2	11	368.0	6.788E+0	−1.137E+1
5	62.250	−2.011E+1	6.212E+1	12	455.4	5.317E+0	−8.685E+0
6	92.150	−1.622E+1	4.185E+1	13	545.0	−3.394E+0	5.462E+0
7	126.300	−1.276E+1	2.881E+1	14	623.5	−5.604E-1	8.946E-1

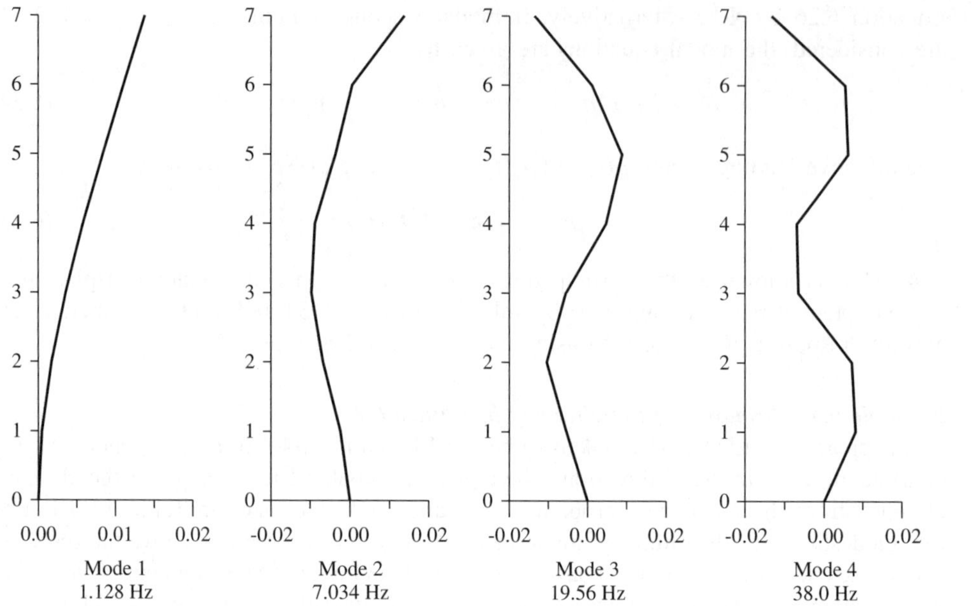

Figure 6.8 Mode shapes of the chimney, semidiscretized as an MDOF system

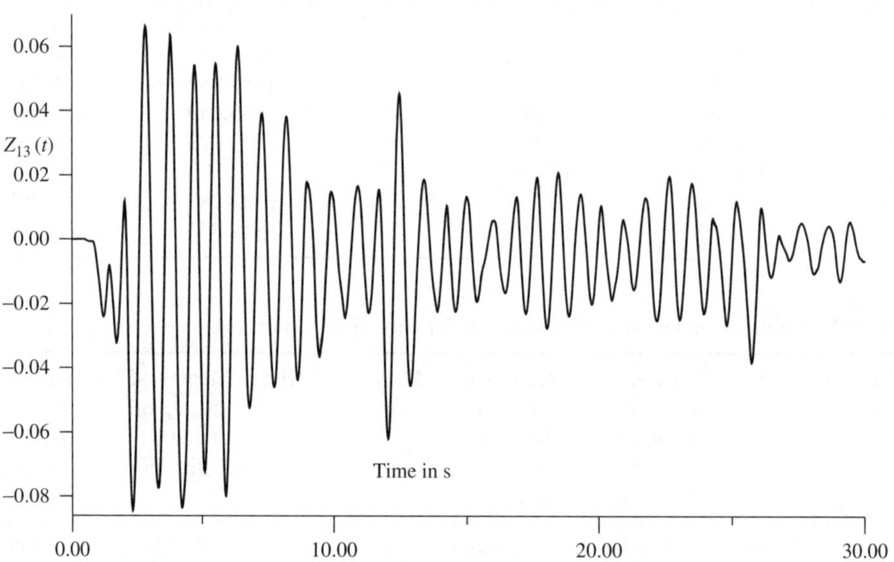

Figure 6.9 Chimney modelled as an MDOF system (by FEM); tip response under El Centro S00E earthquake component; relative displacement by modal summation using i) only the first mode and ii) the first four modes

contribute to the system response. This is indeed confirmed by Figure 6.9 where responses obtained by modal summation (i) considering only the first mode and (ii) using the first four modes, are found to be sufficiently close to each other.

6.4 SDOF Systems under Base Excitation

If one adopts an extreme form of idealisation, either a continuous or a discrete (lumped-parameter) system can be modelled via an single degree of freedom (SDOF) oscillator. For this one degree of freedom system (Figure 6.10), the equation of motion under base excitation takes the form:

$$m\ddot{y} + c(\dot{y} - \dot{g}) + k(y - g) = 0 \tag{6.36}$$

With the absolute displacement $y(t) = z(t) + g(t)$, we rewrite the above equation in terms of the relative displacement $z(t)$ alone as:

$$m\ddot{z} + c\dot{z} + kz = -m\ddot{g} \tag{6.37}$$

The above equation can be solved for $z(t)$ with the help of a convolution integral (Equation 6.19) and later for $y(t)$, the absolute displacement.

Example 6.5: An SDOF model for the chimney under base motion (Figure 6.4)

The SDOF model for the chimney can be obtained, say, by any of the classical methods described in Chapter 3. The Galerkin method has been used (see Example 3.2) to get the fundamental frequency ω_1 of a uniform cantilever beam in terms of the eigenvalue parameter $\mu \left(= \omega_1 \sqrt{\frac{EI}{ml^4}}\right)$. With a one-term approximation (see Equations 3.21–3.23) and the trial function $U_1(x) = \frac{4l^2}{\pi^2}x - \frac{l}{\pi}x^2 - \frac{8l^3}{\pi^3}\sin\frac{\pi x}{2l}$), μ_1 is obtained as 3.5193. For the chimney in Figure 6.4 (with length l = 42 m, diameter = 2.25 m, wall thickness = 6.8 mm, $E = 2.1 \times 10^{11}$ N/m^{2}and m = 500 kg/m), we have $\omega_1 = 7.129$ rad/s. Therefore, the equation of motion is given by:

$$\ddot{z}(t) + \omega_1^2 z(t) = -\ddot{g}(t) \text{ with } \omega_1 = 7.129 \tag{6.38}$$

This value of the fundamental frequency compares well with the result obtained by FEM in Example 6.3. The transient response of the SDOF oscillator under the El Centro 1940 S00E earthquake component is computed from Equation 6.38 and is shown in Figure 6.11.

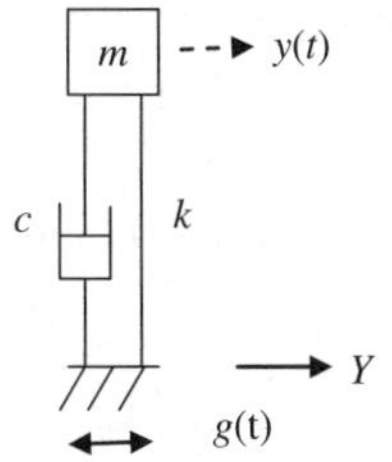

Figure 6.10 SDOF system under base excitation

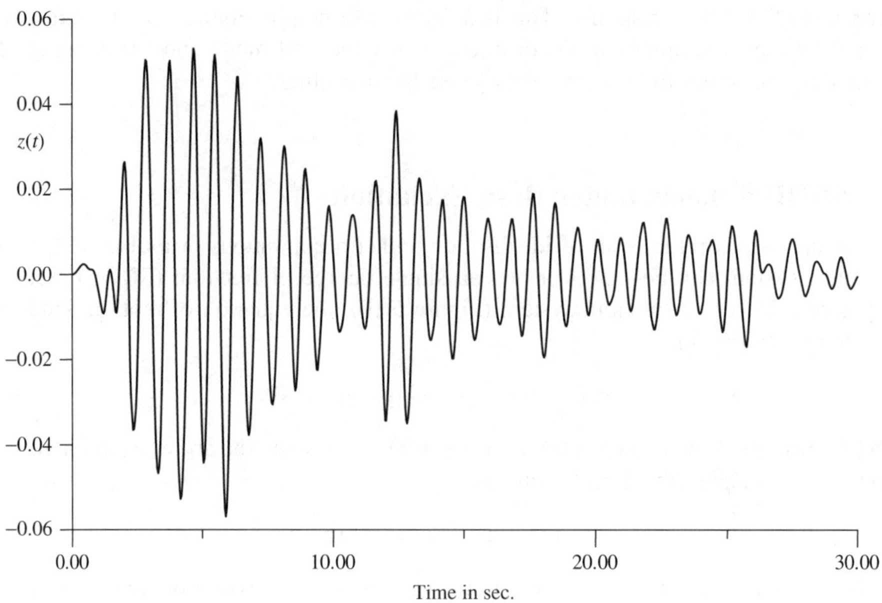

Figure 6.11 Chimney modelled as an SDOF system; relative displacement response at the chimney tip under El Centro S00E earthquake component

6.4.1 Frequency Response of SDOF System under Base Motion

If the base excitation is a harmonic function with $g(t) = G \sin \lambda t$, the Equation 6.37 takes the form:

$$m\ddot{z} + c\dot{z} + kz = m\lambda^2 G \sin \lambda t \Rightarrow \ddot{z} + 2\xi\omega\dot{z} + \omega^2 z = \lambda^2 G \sin \lambda t \tag{6.39}$$

Viscous damping is assumed in the above equation with $\frac{c}{m} = 2\xi\omega$ and $\frac{k}{m} = \omega^2$. Following the same procedure given in Chapter 5 to get the frequency response of an SDOF oscillator, the complex frequency response is obtained in terms of $r = \frac{\lambda}{\omega}$ as:

$$\frac{Z(\lambda)}{G} = M_z(\lambda) = \frac{r^2}{1 - r^2 + i2\xi r} \tag{6.40}$$

The steady-state amplitude is given by:

$$\left|\frac{Z(\lambda)}{G}\right| = \left|M_z(\lambda)\right| = \frac{r^2}{\left[\left(1 - r^2\right)^2 + (2\xi r)^2\right]^{1/2}} \tag{6.41}$$

The steady-state response lags the excitation by an angle θ, given by:

$$\theta_z = \tan^{-1} \frac{2\xi r}{1 - r^2} \tag{6.42}$$

Figure 6.12 shows the frequency response plots – both of amplitude and phase – under the base excitation.

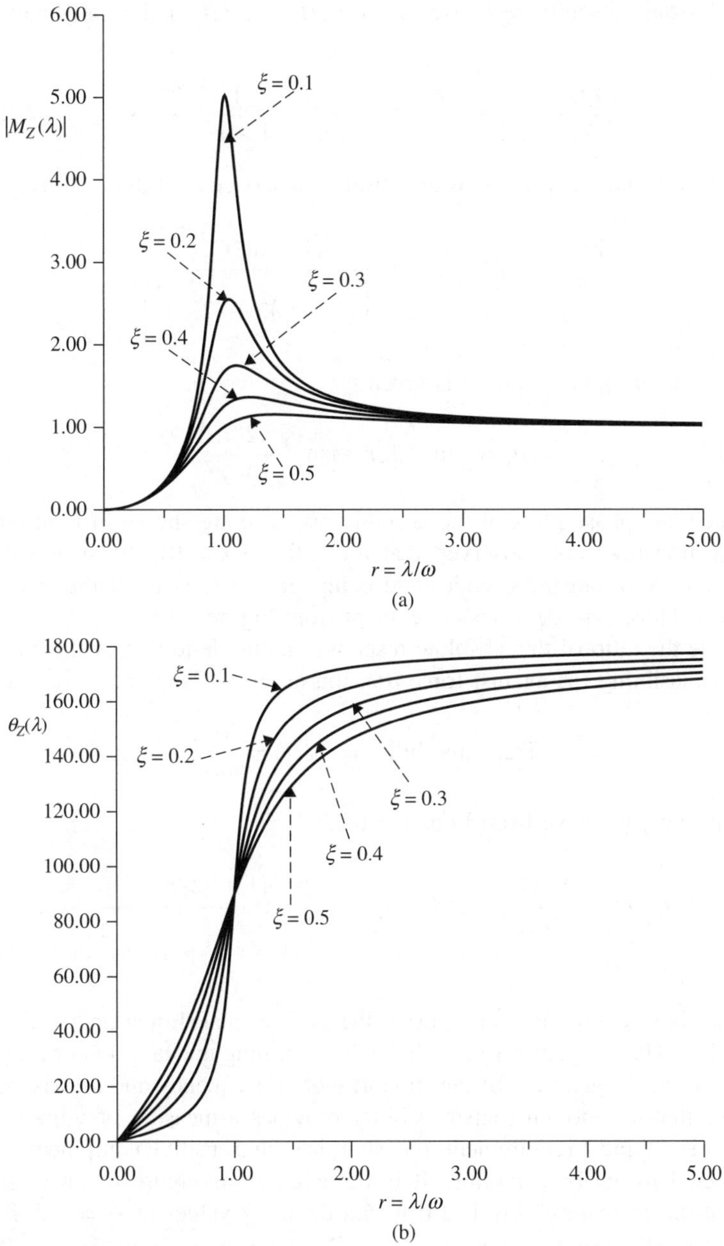

Figure 6.12 An SDOF system under base motion; (a) amplitude plot of the relative displacement response in the steady state and (b) phase plot of the relative displacement response in the steady state

The steady-state absolute response, $y(t)(= z(t) + g(t))$ in the frequency domain is obtained as:

$$\frac{Y(\lambda)}{G} = \frac{r^2}{1 - r^2 + i2\xi r} + 1 = \frac{1 + i2\xi r}{1 - r^2 + i2\xi r} \tag{6.43}$$

From the above equation, we obtain the absolute response amplitude is given by

$$\left|\frac{Y(\lambda)}{G}\right| = \left|\mathcal{M}_y(\lambda)\right| = \frac{\left(1 + (2\xi r)^2\right)^{1/2}}{\left[\left(1 - r^2\right)^2 + (2\xi r)^2\right]^{1/2}} \tag{6.44}$$

The phase of the absolute response is given by

$$\theta_y = \tan^{-1} 2\xi r - \tan^{-1} \frac{2\xi r}{1 - r^2} \tag{6.45}$$

The amplitude and phase plots of the absolute response are shown in Figure 6.13.

From Equation 6.40 it is observed that for $= 0$ ($\Rightarrow \lambda = 0$), the steady-state relative displacement is zero and the pseudostatic component $g(t)$ is itself the system absolute response $y(t)$. Thus, $\mathcal{M}_y(0) = 1$ as is evident from Figure 6.13a.

If we define the ratio of this absolute response amplitude to that of the base motion as motion transmissibility (as against force transmissibility – Chapter 5), we have:

$$\text{Transmissibility} = T_R(t) = \frac{x(t)}{g(t)} \tag{6.46}$$

In the steady state, we have from Equation 6.43:

$$T_R(\lambda) = \frac{Y(\lambda)}{G(\lambda)} \Rightarrow \left|\mathcal{M}_y(\lambda)\right| == \frac{\left(1 + (2\xi r)^2\right)^{1/2}}{\left[\left(1 - r^2\right)^2 + (2\xi r)^2\right]^{1/2}} \tag{6.47}$$

Note that the above expression for $T_R(\lambda)$ is the same as that (Equation 5.125) of the force transmissibility. Thus, Figure 6.13 also helps in describing the variation of transmissibility with respect to the frequency λ of the support motion frequency and the viscous damping ratio ξ. Note that the motion transmissibility provides a measure of vibration isolation. The isolation is a typical requirement to ensure that any sensitive equipments on a system are not effected by its base motion. It is observed from Figure 6.13a of the absolute response that the transmissibility is 1.0 for all damping values at $\frac{\lambda}{\omega} = \sqrt{2}$. Further, it is less than 1.0 for all damping ratios and $\frac{\lambda}{\omega} \geq \sqrt{2}$. These observations are significant if the interest is in reducing the amplitude of motion of the mass point due to the base motion. The effectiveness or efficiency of a vibration isolator is generally denoted by $1 - T_R(\lambda)$. An isolator is thus more effective for a frequency ratio $\geq \sqrt{2}$. Further, the effectiveness in this range ($\frac{\lambda}{\omega} \geq \sqrt{2}$) is unaffected by the presence of viscous damping. For negligible damping, the transmissibility for this range of $\frac{\lambda}{\omega}$ is given by:

$$T_R(\lambda) = \frac{1}{r^2 - 1} \Rightarrow 1 - T_R(\lambda) = \frac{r^2 - 2}{r^2 - 1} \tag{6.48}$$

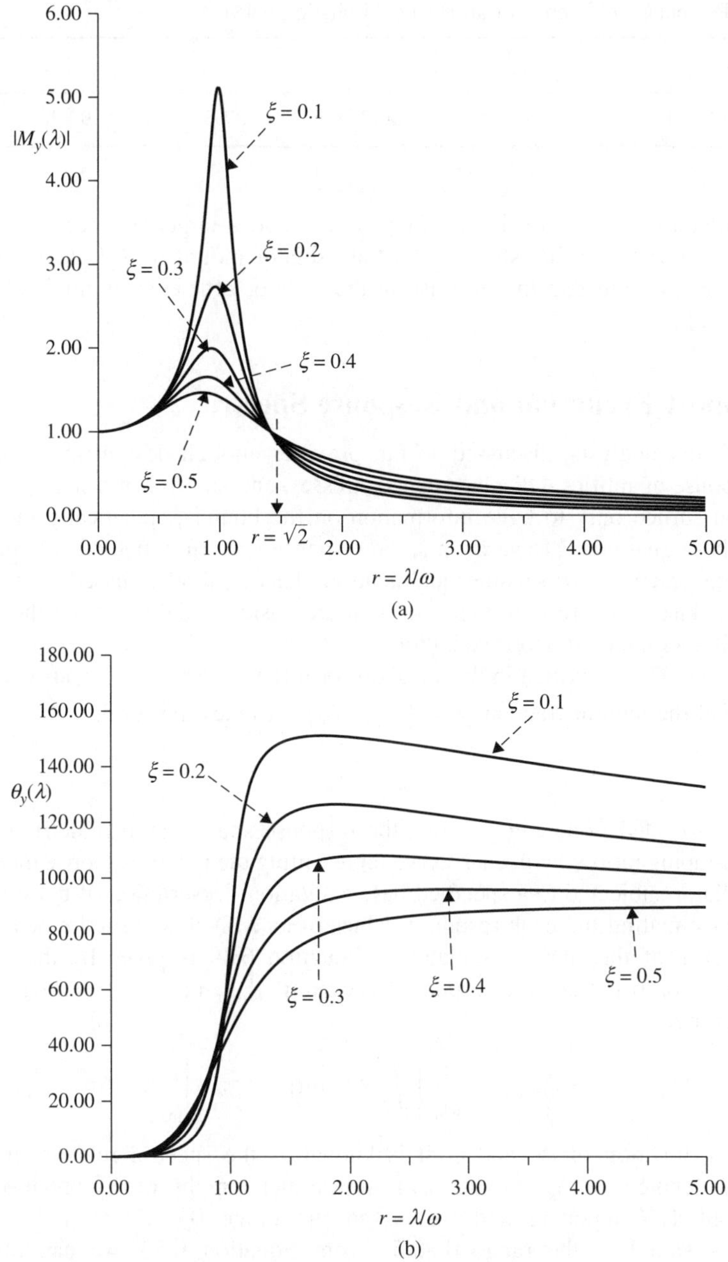

Figure 6.13 An SDOF system under base excitation; (a) amplitude plot of the absolute response in steady state and phase plot of the absolute response in steady state

Table 6.3 Percentage efficiency of an isolator (Equation 6.48)

$r = \frac{\lambda}{\omega}$	$\sqrt{2}$	2	3	4	5
$(1 - T_R(\lambda))x100\%$	0	66.7	87.5	93.3	95.8

Given a structure mass m, excitation frequency λ and a targeted percentage efficiency, one can determine the spring stiffness k of an isolator mount from Equation 6.48 given r. In this context, note that the increase in the isolator efficiency is minimal for $r > 4$ (see Table 6.3).

6.5 Support Excitation and Response Spectra

The time-history analysis, discussed so far, provides enough design information on the system-response quantities – displacements, stresses, and so on. But, for design purposes, it may often suffice only to have information on the bounds, specifically the maximum values, of these quantities. Towards this, the response spectrum method is popularly used in the design practices for seismic qualification. This method is based on the use of a set of graphs known as 'response spectra' that are basically derived from the analysis of SDOF oscillators under support excitation.

If Equation 6.37 is rewritten in the standard form (i.e. in terms of viscous damping ratio $\xi = \frac{c}{2m\omega_n}$ and the natural frequency $\omega_n \left(= \sqrt{k/m}\right)$), it takes the form:

$$\ddot{z} + 2\xi\omega_n\dot{z} + \omega_n^2 z = -\ddot{g} \tag{6.49}$$

From the above ODE, one can generate the response spectra as parametric curves with respect to damping ratio ξ, with each curve representing the loci of response maxima of the SDOF oscillator subjected to a specified base excitation. Description of these spectra and their use in estimating the peak response values of an MDOF system are detailed below.

The response of the SDOF oscillator in Equation 6.49 is given by the convolution integral (Equation 6.19) and the response maximum $|z|_{\max}$ under a given base excitation $g(t)$ is given by:

$$|z(t)|_{\max} = \frac{1}{\omega_n}\left[\int_0^{\tau} \ddot{g}(\tau)h(t-\tau)d\tau\right]_{\max} \tag{6.50}$$

$|z(t)|_{\max}$ is a function of ω_n and ξ. It is known as the spectral relative displacement and usually denoted by S_d. Figure 6.14 shows such graphs of S_d obtained with the El Centro S00E component and for ω_n in the range $0 - 314.16$ rad/s. For these graphs, ξ is varied in the range 0–0.5. From Equation 6.50, we can observe that $\omega_n S_d \left(= \left[\int_0^{\tau} \ddot{g}(\tau)h(t-\tau)d\tau\right]_{\max}\right)$ has the dimension of velocity and it is known as the spectral pseudovelocity response, S_v, of the ground motion $\ddot{g}(t)$. The pseudovelocity needs to be distinguished from the true velocity that can be obtained by differentiating Equation 6.50.

Similarly, the absolute response acceleration is $|\ddot{y}| = |\ddot{z} + \ddot{g}| = 2\xi\omega_n|\dot{z}| + \omega_n^2|z|$. Thus, for low ξ:

$$|\ddot{y}|_{\max} = \omega_n^2|z|_{\max} = \omega_n\left[\int_0^{\tau} \ddot{g}(\tau)h(t-\tau)d\tau\right]_{\max} \tag{6.51}$$

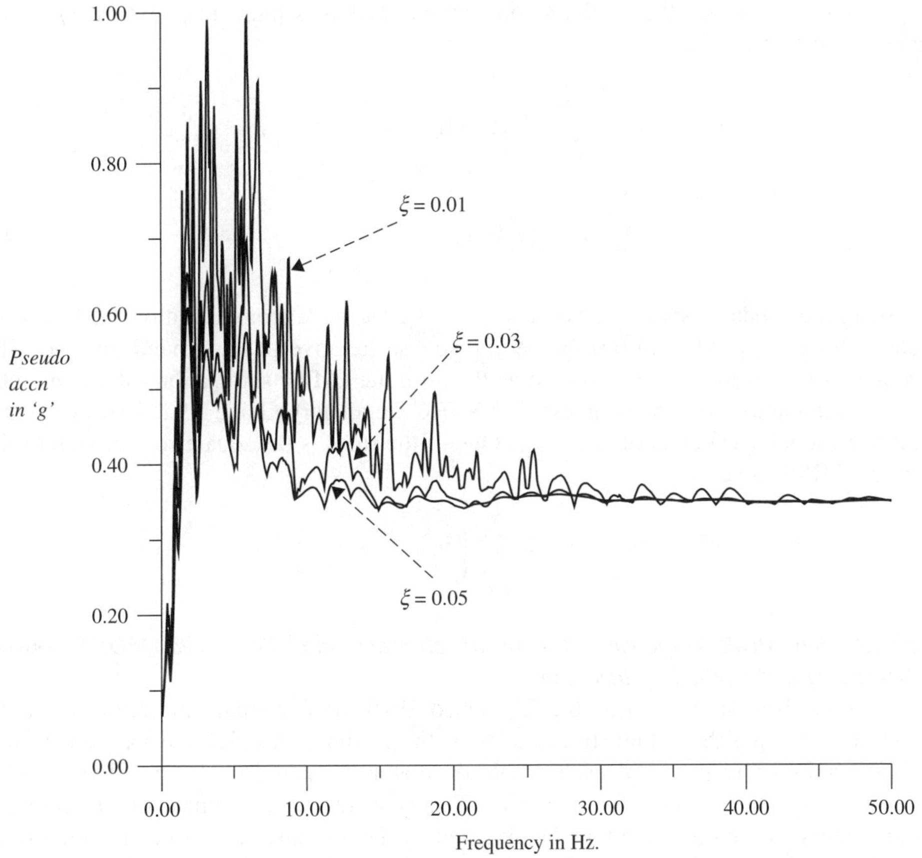

Figure 6.14 Response spectra for El Centro S00E earthquake component for damping ratios, $\xi = 0.01$, 0.03 and 0.05

$|\ddot{y}|_{\max}$ so obtained directly from $|z|_{\max}$ is known as pseudoabsolute acceleration. The response spectra usually recommended by the codes of practice are originally due to Housner (1959). They are generated from previously known seismic recorded time histories and are made smooth (compared to the ones in Figure 6.14) by an averaging or other smoothening process (Clough and Penzien, 1982). The design codes of practice contain such smoothed average earthquake response spectra.

6.5.1 Peak Response Estimates of an MDOF System Using Response Spectra

Each modal Equation 6.34 is an SDOF oscillator. By nodal superposition, the MDOF system response $\boldsymbol{Y}(t)$ is obtained by:

$$\boldsymbol{Y}(t) = \sum_{j=1}^{N} P_{Fj} q_j(t) \tag{6.52}$$

$P_{Fj}, j = 1, 2, .., N$ are the participation factors. The maximum response $|Y(t)|_{\max} := \max_{0<t<T} |Y(t)|$ is given by:

$$|Y(t)|_{\max} = \left| \sum_{j=1}^{N} P_{Fj} q_j(t) \right|_{\max}$$

$$\leq \sum_{j=1}^{N} |P_{Fj}| \left|q_j(t)\right|_{\max} \tag{6.53}$$

Knowing the modal response maxima $\left|q_j(t)\right|_{\max}$, we have from Equation 6.53 an estimate of (or an upper bound on) the peak response (relative to the ground motion). The summation in Equation 6.53 is known as the absolute (ABS) summation rule. A number of other summation rules are suggested (USNRC Regulatory Guide 1.92, 1976) to obtain a more reasonable estimate of $|Y(t)|_{\max}$. One of the rules is the square root of sum of the squares (SRSS) given by:

$$|Y(t)|_{\max} = \sum_{j=1}^{N} \left\{ P_{Fj} \left|q_j(t)\right|_{\max} \right\}^2 \tag{6.54}$$

Example 6.6: Peak response estimate of chimney modelled as a MDOF system (Figure 6.7) and excited by base motion
Consider the base motion to be the El Centro 1940 S00E component acting in the Y-direction. Knowing the natural frequencies of the chimney (see Table 6.1), we can pick up the maximum response of each modal oscillator (Equation 6.34) from Figure 6.14. Table 6.4 lists these values for $\xi = 0.05$. The peak response estimate of the chimney relative to its base motion can now be obtained by the absolute or SRSS summation rules (see also the notes following the Exercise 5 of Chapter 9, on the original derivation of the SRSS and CQC combination rules from the random vibration theory).

Table 6.4 Peak relative response acceleration (pseudo) estimates in 'g' units for the chimney in Figure 6.7 by ABS and SRSS combination rules

Node number	ABS summation rule	SRSS combination rule
1 (bottom)	0.220	0.120
2	0.429	0.235
3	0.496	0.289
4	0.540	0.308
5	0.540	0.316
6	0.540	0.381
7 (top)	1.095	0.610

6.6 Structures under Multi-Support Excitation

For structural systems such as long span bridges and transmission lines whose geometric (length) dimension in the direction of the travelling seismic waves is significant, response could be significantly influenced by the spatial variablity of the ground motion across the structural span [Bogdanoff *et al* 1965, Werner *et al* 1979, Yamamura and Tanaka 1990]. The variation typically appears as time-dependent (non-synchronous or multi-) support motion/boundary condition for such structures. The response of these structures may indeed be considerably different from that caused by uniform ground motion. For instance, one may cite piping systems in power plants that experience multi-support excitation when their primary supporting structures are excited by ground motion during a seismic event. In Chapter 2, treatment of a continuous system under differential (multi-) support motion was briefly presented. Presently, the solution strategy for more general configurations of such systems undergoing multi-support excitations is outlined in the following.

6.6.1 Continuous System Under Multi-Support Excitation

Consider the PDE (6.1) along with the time dependent boundary conditions in Equation 6.2. Proceeding on similar lines as adopted for the case of uniform excitation (see Equations 6.3– 6.5), displacement $\boldsymbol{u}(\boldsymbol{x}, t)$ may be expressed in terms of a dynamic displacement component $\boldsymbol{u}_d(\boldsymbol{x}, t)$ and a pseudo static component (consisting of n_g terms) as:

$$\boldsymbol{u}(\boldsymbol{x}, t) = \boldsymbol{u}_d(\boldsymbol{x}, t) + \sum_{k=1}^{n_g} H_k(\boldsymbol{x})\boldsymbol{h}_k(t) \tag{6.55}$$

$H_k(\boldsymbol{x})$ are unknown functions of $\boldsymbol{x}$ and are so determined as to render the boundary conditions on $\boldsymbol{u}_d(\boldsymbol{x}, t)$ homogeneous. In order to see that the expansion above for the pseudo-static component is feasible, consider a spatially 1D structural system towards a simple elucidation of the underlying concept. Retaining only the time-independent part of the operator, the PDE (actually an ODE here) for determining the pseudo-static component of the displacement $u_s(x, t)$ is given by:

$$-\frac{d^2u_s(x, t)}{dx^2} = \frac{1}{C}\sum_{1}^{n_g} h_k(t)\delta(x - x_k) \tag{6.56}$$

where C is a constant depending on the material and geometric properties, presently assumed to be uniform as an additionally simplifying aid to the exposition. Also, the support locations are given by the set $\{x_k | k \in [1, n_g]\}$. Denoting $w(x)$ to be a typical weight function, the weak form for the above equation is given, following integration by parts, as:

$$\int_\Omega \frac{dw}{dx}\frac{du_s}{dx}\,dx = \sum_{k=1}^{n_g} w_k h_k(t) \tag{6.57}$$

Here $w_k := w(x_k)$. Substituting $u_s(x, t) = \sum_{k=1}^{n_g} H_k(x)h_k(t)$, as assumed in Equation 6.55, in Equation 6.57, one obtains:

$$\sum_{k=1}^{n_g} \int_\Omega \frac{dw}{dx}\frac{dH_k(\boldsymbol{x})}{dx} h_k(t)$$

$$\Rightarrow \sum_{k=1}^{n_g} \boldsymbol{h}_k(t) \int_\Omega \left(\frac{dw}{dx}\frac{dH_k(\boldsymbol{x})}{dx} - w_k\right) = 0 \tag{6.58}$$

Since the support forces $\boldsymbol{h}_k(t)$ could be an arbitrary integrable functions, the above identity implies:

$$\frac{dw}{dx}\frac{dH_k(\boldsymbol{x})}{dx} = w_k, k = 1, 2, , \ldots, n_g \tag{6.59}$$

Moreover, since w is a known function of x, i.e ., letting $\frac{dw}{dx} = \psi(x)$, we have upon integration of the above equations:

$$H_k(x) = \int_0^x \frac{w_k}{\psi(\theta)} d\theta + \alpha \tag{6.60}$$

Here a is a constant of integration. This shows that the expansion $u_s(x, t) = \sum_{k=1}^{n_g} H_k(x)h_k(t)$ is feasible even though $H_k(x)$, which depends on the choice of w (x), could be non-unique. Now returning to the more general case, the functions $H_k(x)$ may be determined by substituting Equation 6.55 in Equation 6.2 resulting in:

$$B_k H_j(\boldsymbol{x}_k) = \delta_{kj}, \quad k, j = 1, 2, ..n_g \text{ and for } B_k\boldsymbol{u}(\boldsymbol{x}_k, t) \neq 0 \tag{6.61}$$

Here $\{B_k\}$ constitutes the linear homogeneous partial differential operator vector $\boldsymbol{B}$ (Equation 6.2) corresponding to the boundary conditions at the n_g support *dof*s. If $B_k\boldsymbol{u}(\boldsymbol{x}_k, t) = \boldsymbol{0}$) for any k, the corresponding $H_k(x)$ may be chosen to be zero for $\boldsymbol{x} \in \Omega$. Since $H_k(\boldsymbol{x}), k = 1, 2, ..n_g$, could be non-unique, one choice for these functions is to assume them to be the static profiles of the structure obtained by imposing in turn unit displacement at each of the n_g support *dof*s and zero displacement at all the other *dof*s. Thus the PDE for $\boldsymbol{u}_d(\boldsymbol{x}, t)$ (as in Equation 6.55) is obtained as:

$$\boldsymbol{L}(\boldsymbol{u}_d(\boldsymbol{x}, t)) = -\sum_{k=1}^{n_g} \boldsymbol{h}_k(t)\boldsymbol{L}(H_k(\boldsymbol{x})) \tag{6.62}$$

The above equation for $\boldsymbol{u}_d(\boldsymbol{x}, t)$ is inhomogeneous with homogeneous boundary conditions, $\boldsymbol{B}\boldsymbol{u}_d(\boldsymbol{x}, t) = \boldsymbol{0}$ and can be solved by the eigen-expansion method. The example to follow describes the application of the above method in obtaining the solution to a continuous system.

The above procedure may however pose difficulties especially for complex systems. The procedure is generally more suitable and computationally adaptable for semi-discretized systems obtained via, say, the FEM and is explained in the section to follow.

Example 6.7: Response of a uniform beam (as a continuous system) under multi-support excitation

For simplicity of illustration, we consider the simply supported uniform beam in Figure 2.20 (Chapter 2), excited by time-dependent boundary conditions at its ends. The governing PDE along with the time-dependent subset of boundary conditions is rewritten hereunder:

$$\frac{\partial^2}{\partial x^2}\left(EI\frac{\partial^2 w}{\partial x^2}\right) + m\frac{\partial^2 w}{\partial t^2} = 0, \quad B_1 w(0, t) = g(t) \text{ and } B_2 w(l, t) = h(t) \tag{6.63}$$

Here $\boldsymbol{L} = \frac{\partial^2}{\partial x^2}\left(EI\frac{\partial^2}{\partial x^2}\right) + m\frac{\partial^2}{\partial t^2}$. Along with the time dependent displacement boundary conditions in Equation 6.63, the end moment boundary conditions for the simply supported beam are $B_3 w(0,t) = EI\frac{d^2 w(0,t)}{dx^2} = 0$ and $B_4 w(l,t) = EI\frac{d^2 w(l,t)}{dx^2} = 0$. With the number of prescribed boundary conditions being 4, we have $B_1 = B_2 = 1.0$ and $B_3 = B_4 = EI\frac{\partial^2}{\partial x^2}$. Further, $x_1 = x_3 = 0$ and $x_2 = x_4 = l$. Also $h_1(t) = g(t), h_2(t) = h(t), h_3(t) = h_4(t) = 0$. By Equation 6.55, we have:

$$w(x,t) = w_d(x,t) + H_1(x)h_1(t) + H_2(x)h_2(t) \tag{6.64}$$

Equations 6.61 lead to the following conditions on $H_1(x)$ and $H_2(x)$:$B_1 H_1(x_1) = 1$, $B_1 H_2(x_1) = 0$, $B_2 H_1(x_2) = 0$, $B_2 H_2(x_2) = 1$, $B_3 H_1(x_3) = 0$, $B_3 H_2(x_3) = 0$, $B_4 H_1(x_4) = 0$, and $B_4 H_2(x_4) = 0$. The choice of $H_1(x) = \left(1 - \frac{x}{l}\right)$ and $H_2(x) = \frac{x}{l}$ (as assumed in section 2.7 satisfies these requirements. We finally obtain the inhomogeneous PDE for $w_d(x,t)$ as:

$$\frac{\partial^2}{\partial x^2}\left(EI\frac{\partial^2 w_d(x,t)}{\partial x^2}\right) + m\frac{\partial^2 w_d(x,t)}{\partial t^2} = P_{eff}(t) \tag{6.65}$$

$P_{eff}(t) = -m[\ddot{g}(t) + \frac{x}{l}\{\ddot{h}(t) - \ddot{g}(t)\}]$ is the effective loading due to the differential support motion and follows from Equation 6.62. Now, following the eigen-expansion method to solve for $w_d(x,t)$, we consider the first m modes in the modal summation (Equation 6.11), and finally obtain the response corresponding to each mode as:

$$Y_j(t) = \frac{1}{\omega_{jD}}\int_0^t P_j(\tau)e^{-\xi_j \omega_j^{(t-\tau)}} \sin \omega_{jD}(t-\tau)\, d\tau \tag{6.66}$$

$\omega_j = \frac{K_j}{M_j}, j = 1, 2, \ldots, m$ is the natural frequency of the jth mode with K_j and M_j respectively being the generalized stiffness and mass associated the mode (Equation 6.13). ξ_j is the modal damping ratio, assumed to be 0.05 for each mode in this example. $P_j(t) = P_{eff}(t)\int_0^l \Phi_j(x)dx/M_j$ is the jth modal generalized force with $\Phi_j(x) = \sin\frac{j\pi}{l}x, j = 1, 2, \ldots, m$ being the eigenfunctions of the simply supported beam. The beam is presently assumed to be of a hollow square cross-section with each side = 0.2 m and thickness = 0.05 m. The other geometric and material properties selected for the beam are: $l = 50$ m, $E = 2/1E11$ N/m^2 and $m = 7850$ kg/m^3. The first few beam eigenfrequencies are obtained as 1.63, 6.50, 14.63, 26.02 rad/sec. etc.

A simple case of multi-support excitation is to assume harmonic functions for $g(t)$ and $\boldsymbol{h}(t)$ with a (constant) time delay introduced between the two as:

$$g(t) = A\sin\lambda t, \quad h(t) = A\sin\lambda(t+\tau) \tag{6.67}$$

The effect of τ on the beam response is shown in Figure 6.15. The figure shows the response evaluated at the beam center for different values of $\tau \in \left(\frac{0, 2\pi}{\lambda}\right)$ with $A = 1$ and $\lambda = 2$rad/sec. It is observed that with $\tau > 0$, the support motion causes a reduction in the response compared to the case with $\tau = 0$. Figure 6.16 shows typical response histories for $\tau = 0$ and $\tau = 0.3 \times \frac{2\pi}{\lambda}$.

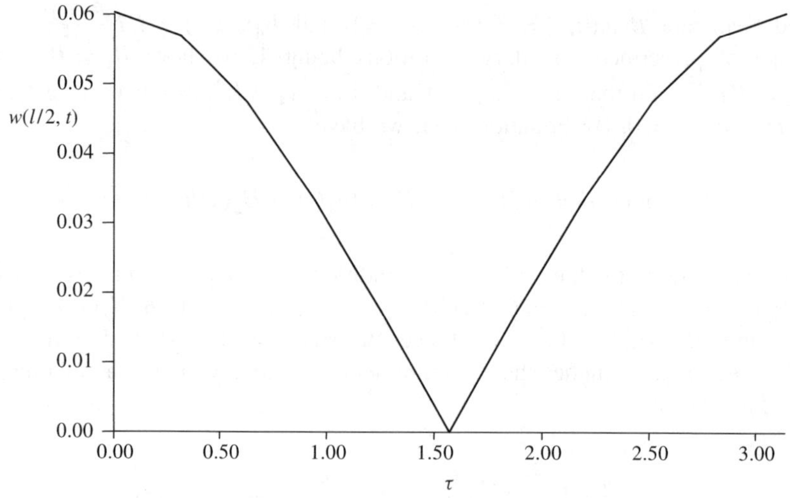

Figure 6.15 Effect of τ on the beam response at $x = \frac{1}{2}$

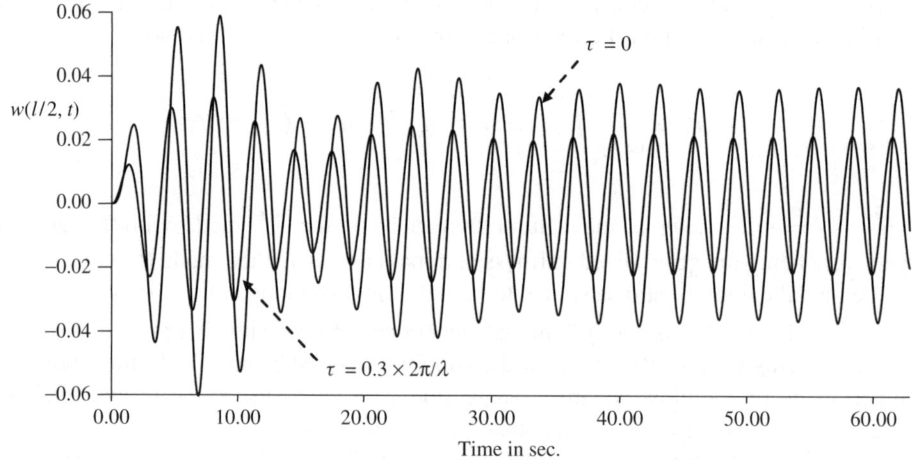

Figure 6.16 Time histories of $w(l/2, t)$ for $\tau = 0$ and $0.3 \times \frac{2\pi}{\lambda}$

6.6.2 *MDOF Systems Under Multi-Support Excitation*

An MDOF system under multi-support excitation can be described [Der Kiureghian 1992, Chopra 2001] by the following ODE-s of motion:

$$\begin{bmatrix} M_s & M_{sg} \\ M_{sg}^T & M_g \end{bmatrix} \begin{Bmatrix} \ddot{x} \\ \ddot{h} \end{Bmatrix} + \begin{bmatrix} C_s & C_{sg} \\ C_{sg}^T & C_g \end{bmatrix} \begin{Bmatrix} \dot{x} \\ \dot{h} \end{Bmatrix} + \begin{bmatrix} K_s & K_{sg} \\ K_{sg}^T & K_g \end{bmatrix} \begin{Bmatrix} x \\ h \end{Bmatrix} = \begin{Bmatrix} 0 \\ F \end{Bmatrix} \tag{6.68}$$

Here $x(t)$ is the absolute displacement vector corresponding to the *dof*s (say, N of them) of the superstructure and $h(t)$, the n_g-dimensional vector of support displacements. M_s, C_s and C_s are the $N \times N$ mass, damping and stiffness matrices of the superstructure. M_g, C_g and K_g are $n_g \times n_g$ are the corresponding matrices associated with the support *dof*s and M_{sg}, C_{sg} and K_{sg} are the coupling matrices between the superstructure and the support *dof*s. $F(t)$ is the vector of reaction forces at the n_g support *dof*s. As in the case with uniform excitation, assume that $x(t) = x_d(t) + x_d(t)$ where $x_d(t)$ and $x_d(t)$ are respectively the dynamic (relative) and pseudo-static displacement components of the superstructure *dof*s. From the first row of the partitioned matrix Equation 6.68, we get the pseudo-static response component as the solution of:

$$[K_s K_{sg}] \begin{Bmatrix} x_s \\ h \end{Bmatrix} = 0 \tag{6.69}$$

The inertia and damping effects are ignored in writing the above equation and we find that:

$$x_s(t) = -K_s^{-1} K_{sg} h(t) \tag{6.70}$$

The equation for the dynamic component $x_d(t)$ can now be written as:

$$M_s \ddot{x}_d + C_s \dot{x}_d + K_s x_d = -[M_s K_s^{-1} K_{sg} + M_{sg}]\ddot{h} - [C_s K_s^{-1} K_{sg} + C_{sg}]\ddot{h} \tag{6.71}$$

The damping terms on the right hand side of the above equation may be smaller than the inertia terms on the same side and are generally ignored. In any case, Equation 6.71 can be solved for the dynamic component of the displacement by, *say*, the eigen-expansion method.

The above procedure is computationally efficient and thus may be readily employed for solving linear structural systems having complex geometry or heterogeneous material properties. (As an exercise, the reader may apply the above method to an FE model of the simply supported beam in Example 6.6 and verify the result *vis-a-vis* the solutions given in Figures 6.15 and 6.16).

6.7 Conclusions

The focus herein has been the response evaluation of structural systems under support motion. Time history and response spectrum approaches are described and illustrated by a few selective examples. The former approach has a universal appeal for dynamic analysis of structures, components or equipments, being subjected to base excitations owing to, say, aseismic effects or during transportation by road, ship or aircraft. Further, general codes of design practice emphasise adopting the time history approach to seismically qualify critical structural components such as those in nuclear installations. However, this approach has the disadvantage that it may not be adequately insightful; that is the numerically evaluated responses pertain only to a specific set of input forcing data and may be of little value for performance assessment against, say, a future earthquake event. Alternatively, one could adopt the response spectrum method that is indeed widely used in practice in view of its simplicity and ready evaluation of peak response estimates for large structural systems. For other interesting aspects specific to earthquake engineering design, such as spectrum compatible accelerograms, response spectrum analysis under multiple support excitations, and so on, the reader may consult the Bibliography provided at the end of this chapter.

Exercises

1. An idealised model of a vehicle is shown in Figure 6.17 along with a rigid body AB (of length l) having mass M and mass moment of inertia I about its centre of mass O. The axles and wheels are simulated by lumped parameters of mass m and stiffness k_1. The tyres are modelled as springs of stiffness k_2. Assuming small-amplitude oscillations of the vehicle, write the equations of motion for vertical vibrations under support excitation at the tyres. Use as coordinates the vertical displacements of two masses (of axles and wheels) and the displacements at A and B. The excitation is identical at both the tyres. The shock absorbers are modelled as viscous dampers (as shown in the figure).
2. Consider the vertical vibrations of the system in Figure 6.18 consisting of a flexible column on a resilient mount. The mount is modelled as a massless oscillator with stiffness k and viscous damping coefficient c. If $x(t)$ is the motion at the base of the column due to the ground movement $g(t)$, find the transmissibility $T(i\lambda) = \frac{X(i\lambda)}{G(i\lambda)}$, where $X(i\lambda)$ and $G(i\lambda)$ are the, respectively, transformed quantities in the frequency domain. Take the properties of the column as: mass per unit length $= m$, height $= l$, area of cross section $= A$ and Young's modulus $= E$. (Hint: The equilibrium equation of the resilient mount is: $c(\dot{x} - \dot{g}) + k(x - g) = f$ and the equation of motion of the column under the support motion $x(t)$ is: $m\frac{\partial^2 y}{\partial t^2} + \bar{c}\frac{\partial y}{\partial t} - EA\frac{\partial^2 y}{\partial x^2} = 0$ with the boundary conditions $y(0, t) = x(t)$ and $\frac{\partial y(l,t)}{\partial z} = 0$. $\bar{c}$ is the viscous damping coefficient per unit length of the column. $f(t)$ is the force transmitted by the column to the mount (Newland, 1989).

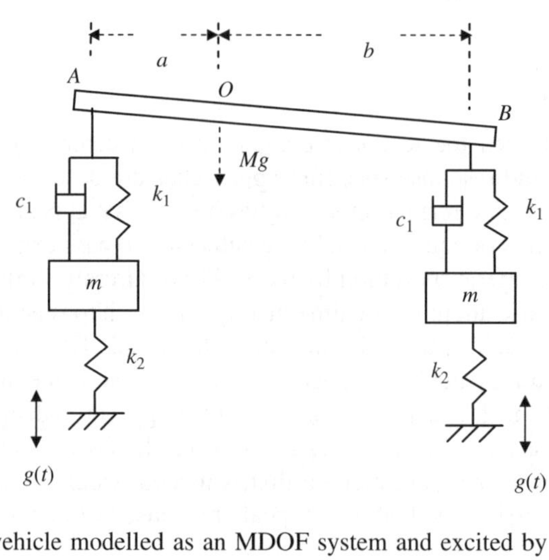

Figure 6.17 A vehicle modelled as an MDOF system and excited by support excitation

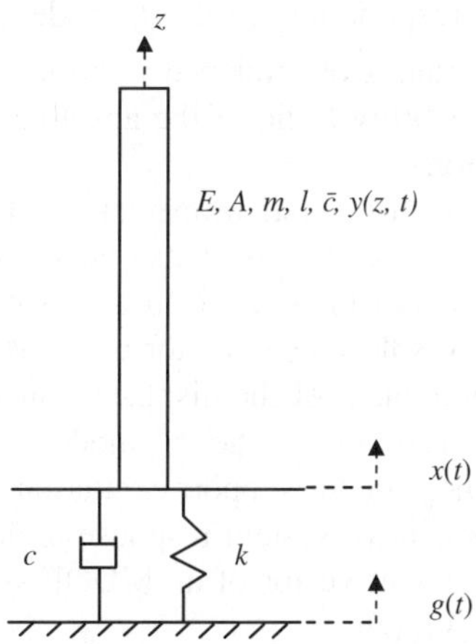

Figure 6.18 Column on a resilient mount; transmissibility from ground to the base of the column

Notations

$A(\mathrm{t}), B(\mathrm{t})$	Integrals in Equation 6.22
$C_{1,}C_2, C_3$	Damping (viscous) values (Example 6.2)
D	Diameter
$f_d(t)$	Effective reaction force on a beam at its base due to support excitation
$\boldsymbol{g}(t)$	Vector of base motions – translational or rotational $\left(= \left(g_1(t), g_2(t), \ldots g_{n_g}(t)\right)^T\right)$
$K_{1,}K_2, K_3$	Stiffnesses (Example 6.2)
K_j	Generalized stiffness
m_1, m_2, m_3	Lumped masses
M_j	Generalized mass
$\mathcal{M}_y(\lambda)$	Complex frequency (absolute) response of an SDOF oscillator under support excitation (Equation 6.44)
$\mathcal{M}_z(\lambda)$	Complex frequency (relative) response of an SDOF oscillator under support excitation (Equation 6.40)
n_g	Number of system degrees of freedom on the boundary at which the base motion is applied
N	Active degrees of freedom (= total dofs – n_g)
P_{fj}	Modal participation factor of mode j (Equation 6.17)
$p_j(t)$	Normalised generalized force associated with the jth mode (Equation 6.16) effective loading vector
$\boldsymbol{p}(t)$	Effective loading vector for MDOF system under support excitation $\left(= \left(p_1(t), p_2(t), \ldots, p_N(t)\right)^T\right)$
$\boldsymbol{P}_{eff}(t)$	Effective loading due to the base motion
S_d, S_v	Spectral relative displacement and spectral pseudovelocity responses respectively under support excitation

T_4	Time period corresponding to the 4th mode of vibration $\left(= \frac{2\pi}{\omega_4}\right)$
$\boldsymbol{T}$	Transformation matrix of influence vectors (Equation 6.25)
$T_R(\lambda)$	Motion transmissibility (ratio of the absolute response amplitude to that of the base motion)
$\boldsymbol{u}_d(\boldsymbol{x}, t)$	Dynamic component of system response under support excitation
$\boldsymbol{u}_s(\boldsymbol{x}, t)$	Static component of system response under support excitation
$\boldsymbol{x}_k$	Spatial coordinates of the excited locations on the boundary
$\boldsymbol{y}(t)$	Total (absolute) displacement vector of an MDOF system
$\boldsymbol{y}_s(t)$	Pseudostatic component of the displacement induced by the base motion
$Y_k(t)$	Time-dependent response of the kth mode
$y(x, t)$	Beam displacement under support excitation
$y_d(x, t)$	Dynamic component of system response under support excitation
$z(t)$	Relative displacement vector of an MDOF system
θ_y	Phase angle of $\mathcal{M}_y(\lambda)$
θ_z	Phase angle of $\mathcal{M}_z(\lambda)$

References

ASCE (2000) Standard 4-98. *Seismic Analysis of Safety-Related Nuclear Structures and Commentary*, American Society of Civil Engineers, Virginia, USA.

Bogdanoff, I.I., Goldberg, J.E. and Schiff, A.J. (1965) The effect of ground transmission time on the response of long structures, *Bulletin of Seismological Society of America*, Vol. 55, pp. 627–640.

Chopra, A.K., (1996) *Dynamics of Structures*, McGraw-Hill, New York.

Clough, R.W. and Penzien, J. (1982) *Dynamics of Structures*, McGraw-Hill, Singapore.

Der Kiureghian, A.D. and Neuenhofer, A. (1992) Response Spectrum Method for Multi-support Seismic Excitations, *Earthquake engineering and structural dynamics*, Vol. 21, pp. 713–740.

EERL (1980) *A Selection of Important Strong Motion Earthquake Records*. Report 80-01, California Institute of Technology, Pasadena.

Housner, G.W. (1959) Behavior of structures during earthquakes. *Proceedings of the ASCE*, **85** (EM-4) pp. 109–129.

Indian Standard 1893 (2002) *Criteria for Earthquake Resistant Design of Structures, Part I – General Provisions and Buildings*, Fifth Revision, Bureau of Indian Standards, New Delhi, India.

International Electrotechnical Commission (1968) Rules for Electric Traction Equipment, Publication 77 (affiliated to International Organization for Standardization) Geneva, Switzerland.

Kreyszig, E. (1999) *Advanced Engineering Mathematics*, John Wiley & Sons, Inc., Singapore.

Newland, D.E. (1989) *Mechanical Vibration Analysis and Computation*, Dover Publications, Inc., New York.

USNRC Regulatory Guide 1.92 (1976) Combining Modal Responses and Spatial Components in Seismic Response Analysis, Washington, D.C., USA.

Werner, S.D., Lee, L.C., Wong, H.L. and Trifunac, M.D. (1979) Structural response to travelling seismic waves, *ASCE, Journal of Structural Engineering*, Vol. 105, pp. 2547–2564.

Yamamura, N. and Tanaka, H. (1990) Response analysis of flexible MDF systems for multiple-support seismic excitations, *Earthquake engineering and structural dynamics*, Vol. 19, pp. 345–357.

Bibliography

On Multiple Support Excitation

Singh, M.P., Burdisso, R.A. and Maldonado, G.O. (1992) Methods used for calculating seismic response of multiply supported Piping Systems. *Transactions of ASME, Journal of Pressure Vessel Technology*, **114**, 46–52.

On Earthquake Response Analysis

Clough, R.W. and Mojtahedi, S. (1976) Earthquake response analysis considering non-proportional Damping. *Earthquake Engineering and Structural Dynamics*, **4**, 489–496.

Igusa, T. and Der Kiureghian, A.D. (1985) Generation of floor response spectra including oscillator structure interaction. *Earthquake Engineering and Structural Dynamics*, **13** (5), 661–676.

Lin, C.W. (1991) Seismic evaluation of systems and components. *Transactions of ASME, Journal of Pressure Vessel Technology*, **113**, 273–283.

Suarez, L.E. and Singh, M.P. (1987) Seismic response of equipment-structure systems. *ASCE, Journal of Engineering Mechanics*, **113** (1), 16–30.

7

Eigensolution Procedures

7.1 Introduction

The eigenvalue problem and its significance in solving linear time-invariant (LTI) partial differential equations (PDEs) and ordinary differential equations (ODEs) are described in detail in the previous chapters. Thus, recall that, for a continuous system (wherein the governing equations are in the form of a system of PDEs), the eigenvalue problem is given by $\boldsymbol{L}\Phi = \lambda\Phi$ with $\boldsymbol{L}$ being a differential operator. The eigensolution consists of an infinite number of eigenvalues, $\lambda_n, n \in N$, and the associated eigenfunctions, $\Phi_n(\boldsymbol{X}) \in \mathcal{H}$, the Hilbert space. For semidiscretized systems, typically derived either by classical methods or the finite element method (FEM), the resulting eigenvalue problem $\boldsymbol{A}\boldsymbol{\Phi}_n = \lambda_n\boldsymbol{\Phi}_n$ is algebraic. Here, the set of eigenvectors $\{\boldsymbol{\Phi}_n\}$ belongs to a finite-dimensional vector space, $\boldsymbol{V}$, so that span$\{\boldsymbol{\Phi}_n\}$ forms an invariant subspace (Appendix G) of $\boldsymbol{V}$. The eigenvalues and eigenvectors (or, alternatively the natural frequencies and mode shapes) thus, in a sense, deconstruct the free-vibration response of a linear, time-invariant system in that the response admits a series expansion, that is the so-called eigenfunction expansion. For self-adjoint systems, the differential operator $\boldsymbol{L}$ (the matrix operator $\boldsymbol{A}$) is symmetric with real eigenvalues and orthogonal eigenfunctions (eigenvectors). The property of orthogonality is desirable as it helps using the eigenfunctions (eigenvectors) as a basis set towards reducing the governing equations into a set of uncoupled second-order ODEs (i.e. single degree of freedom (SDOF) oscillators). This in turn helps to obtain the forced vibration solution by modal superposition, as described in the earlier chapters. Even for nonself-adjoint systems, the right and left eigenfunctions (eigenvectors) possess the biorthogonality property (Equation 2.231) and form a basis set that lends itself to be exploited, whilst obtaining the forced vibration solution.

Of late, concomitant with the phenomenal growth of computing power, the ability to analyse large-scale engineering problems has also increased proportionally. It is here that a discretization strategy as the FEM, which is readily programmable, is useful to arrive at the semidiscretized models corresponding to the governing PDEs. Since cheap computing enables the dimension of the vectors space, wherein the semidiscretized response resides, to be large, it simultaneously creates the need for efficient eigensolution schemes for large system matrices with the generalized eigenvalue problem given by $\boldsymbol{K}\boldsymbol{\Phi}_n = \lambda_n\boldsymbol{M}\boldsymbol{\Phi}_n$ (Equation 5.45). Accordingly, in this chapter, some effort is expended to describe a few

Elements of Structural Dynamics: A New Perspective, First Edition. Debasish Roy and G Visweswara Rao.

of such fundamental eigensolution techniques, whilst not losing sight of their numerical implementation in the context of large systems. These techniques are, in general, iterative. Whilst methods such as inverse iteration, power (forward) iteration and subspace iteration may be looked upon as methods that employ iterations as the generic step, methods such as Jacobi and Lanczos basically utilise similarity transformation to (nearly) diagonalise the system matrices within an iterative framework. From the functionality perspective, on the other hand, inverse, power iteration methods and the Jacobi transformation method are efficient in extracting only the first few eigenvalues of an eigenvalue problem, which can be fairly large sized. These methods, thus, form part of a more general eigensolver such as subspace or Lanczos methods. Specifically, Lanczos method and its variants are known (Ericsson and Ruhe, 1980; Grimes et al. 1991) to be computationally efficient in finding a large number of eigenvalues and the associated eigenvectors. The concepts underlying these methods are described in this chapter.

7.2 Power and Inverse Iteration Methods and Eigensolutions

Methods to handle the standard (canonical) eigenvalue problem $\boldsymbol{A}\boldsymbol{\Phi}_n = \lambda_n \boldsymbol{\Phi}_n$, are relatively straightforward, especially if $\boldsymbol{A}$ is symmetric and positive definite. Note that $[\boldsymbol{\Phi}_n] = \boldsymbol{\Phi}$ is the (column-wise arranged) matrix of eigenvectors. The inverse and power iteration methods are the simplest ones to obtain, respectively, the smallest and largest eigenvalue by iteration. Whilst the power method operates directly on the matrix $\boldsymbol{A}$, the inverse iteration (as the name suggests) operates on its inverse. In the case of the latter, the eigenvalue problem takes the form $\boldsymbol{A}^{-1}\boldsymbol{\Phi}_n = \frac{1}{\lambda_n}\boldsymbol{\Phi}_n$. Both methods iterate with a starting vector, $\boldsymbol{X}_0 \in \boldsymbol{V}$, in order to recursively generate a sequence of vectors (which is expected to converge to the desired eigenvector) according to the following basic strategy at the kth iteration:

- **Power iteration**:

$$\boldsymbol{X}_{k+1} = \boldsymbol{A}\boldsymbol{X}_k = \boldsymbol{A}^k \boldsymbol{X}_0 \tag{7.1}$$

- **Inverse iteration**:

$$\boldsymbol{X}_{k+1} = \boldsymbol{A}^{-1}\boldsymbol{X}_k = \boldsymbol{A}^{-k} \boldsymbol{X}_0 \tag{7.2}$$

Suppose that we express $\boldsymbol{X}_0$ in terms of the system eigenvectors (still unknown) as:

$$\boldsymbol{X}_0 = \sum_{j=1}^{n} \alpha_j \boldsymbol{\Phi}_j \tag{7.3}$$

Presently, restricting attention to systems with no repeated or zero eigenvalues, we order the eigenvalues as $\lambda_1 < \lambda_2 < \ldots\ldots < \lambda_{n-1} < \lambda_n$ and denote the eigenvectors as $\boldsymbol{\Phi}_j$ (with a slight abuse of notation). Then we have from Equation 7.1:

$$\begin{aligned}\boldsymbol{X}_{k+1} = \boldsymbol{A}^k \boldsymbol{X}_0 &= \sum_{j=1}^{n} \alpha_j \lambda_j^k \boldsymbol{\Phi}_j \\ &= \lambda_n^k \sum_{j=1}^{n} \alpha_j \left(\frac{\lambda_j}{\lambda_n}\right)^k \boldsymbol{\Phi}_j \end{aligned} \tag{7.4}$$

It is observed from the above equation that, as the iteration progresses, we have $\left(\frac{\lambda_j}{\lambda_n}\right)^k \to 0$ as $k \to \infty$ and for $j \neq n$. Thus, the generated vector $\boldsymbol{X}_{k+1} \to$ a constant multiple of $\boldsymbol{\Phi}_n$, the eigenvector corresponding to the largest eigenvalue λ_n. Similarly, using the same eigenexpansion for $\boldsymbol{X}_0$, Equation 7.2 of the inverse iteration can be written as:

$$\boldsymbol{X}_{k+1} = \boldsymbol{A}^{-k}X_0 = \sum_{j=1}^{n} \alpha_j \left(\frac{1}{\lambda_j}\right)^k \boldsymbol{\Phi}_j = \frac{1}{\lambda_1^k} \sum_{j=1}^{n} \alpha_j \left(\frac{\lambda_1}{\lambda_j}\right)^k \boldsymbol{\Phi}_j \tag{7.5}$$

The above equation shows that the inverse iteration obtains the vector $\boldsymbol{X}_{k+1}$ such that it approaches the eigenvector $\boldsymbol{\Phi}_1$ corresponding to the smallest eigenvalue λ_1.

Even in the case of a generalized eigenvalue problem $\boldsymbol{K\Phi}_n = \lambda_n \boldsymbol{M\Phi}_n$, if the stiffness and mass matrices are positive definite, reduction to the canonical form is possible by using Cholesky or $\boldsymbol{LDL}^T$ decomposition (Equation 5.46). This would enable the above iterative methods to be applicable to extract the system eigenvalues. For example, by Cholesky decomposition, the symmetric mass matrix $\boldsymbol{M}$ is expressible in the form:

$$\boldsymbol{M} = \boldsymbol{LL}^T \tag{7.6}$$

The generalized eigenvalue problem is then rewritten as:

$$\boldsymbol{K\Phi}_n = \lambda_n \boldsymbol{LL}^T \boldsymbol{\Phi}_n \tag{7.7}$$

With the substitution $\boldsymbol{L}^T \boldsymbol{\Phi}_n = \boldsymbol{\psi}_n$ and premultiplication of both sides of the above equation by $\boldsymbol{L}^{-1}$, one arrives at the canonical form:

$$\bar{\boldsymbol{K}}\boldsymbol{\psi}_n = \lambda \boldsymbol{\psi}_n \tag{7.8}$$

Here, $\bar{\boldsymbol{K}} = \boldsymbol{L}^{-1}\boldsymbol{K}\boldsymbol{L}^{-T}$, which is symmetric. When the mass matrix $\boldsymbol{M}$ is positive semidefinite with zero or nearly zero diagonal elements, the above decomposition can be used on the stiffness matrix $\boldsymbol{K}$. Thus, if $\boldsymbol{K} = \boldsymbol{SS}^T$ and the substitution $\boldsymbol{S}^T\boldsymbol{\Phi}_n = \boldsymbol{\psi}_n$ is used, the resulting canonical form, suitable for inverse iteration, is given by:

$$\bar{\boldsymbol{M}}\boldsymbol{\psi}_n = \frac{1}{\lambda_n}\boldsymbol{\psi}_n \tag{7.9}$$

In this case, $\bar{\boldsymbol{M}} = \boldsymbol{S}^{-1}\boldsymbol{M}\boldsymbol{S}^{-T}$. Reduction to the canonical form (as in Equation 7.8 or 7.9) has the advantage that it enables many available schemes that can only handle this form to be exploited. However, the above vector iteration methods are, in principle, also directly applicable to the generalized eigenvalue problem. The corresponding algorithms with suitable modifications are shown in Table 7.1.

Example 7.1: A cantilever beam: eigensolution by iterative methods

The finite element (FE) model of a cantilever beam is shown in Figure 7.1. The beam is discretized with 10 1-D elements of uniform length, with each node having 4 *dof*s – two translational and two rotational. The size of the system matrices is thus 40 × 40.

Table 7.2 gives the results obtained by the two iteration methods. These results are comparable with the theoretical results (obtained directly from Equation 7.8). The starting vector (40 × 1 size) in both cases is taken as $(1.0, 1.0, \ldots\ldots\ldots, 1.0, 1.0)^T$.

Table 7.1 Generalized eigenvalue problem; algorithms for power and inverse iteration

Inverse iteration		Power iteration	
Initialise X_0 and start iteration: $k = 0, 1, \ldots$		Initialise X_0 and start iteration $k = 0, 1, \ldots$	
$\boldsymbol{K}\bar{\boldsymbol{X}}_{k+1} = \boldsymbol{M}\boldsymbol{X}_k$	(7.10a)	$\boldsymbol{M}\bar{\boldsymbol{X}}_{k+1} = \boldsymbol{K}\boldsymbol{X}_k$	(7.10b)
$\lambda_1^{(k+1)} = \dfrac{\bar{\boldsymbol{X}}_{k+1}^T\boldsymbol{K}\bar{\boldsymbol{X}}_{k+1}}{\bar{\boldsymbol{X}}_{k+1}^T\boldsymbol{M}\bar{\boldsymbol{X}}_{k+1}}$	(7.11a)	$\lambda_n^{k+1} = \dfrac{\bar{\boldsymbol{X}}_{k+1}^T\boldsymbol{K}\bar{\boldsymbol{X}}_{k+1}}{\bar{\boldsymbol{X}}_{k+1}^T\boldsymbol{M}\bar{\boldsymbol{X}}_{k+1}}$	(7.11b)
$\boldsymbol{X}_{k+1} = \dfrac{\bar{\boldsymbol{X}}^{k+1}}{(\bar{\boldsymbol{X}}_{k+1}^T\boldsymbol{M}\bar{\boldsymbol{X}}_{k+1})^{1/2}}$	(7.12a)	$\boldsymbol{X}_{k+1} = \dfrac{\bar{\boldsymbol{X}}^{k+1}}{(\bar{\boldsymbol{X}}_{k+1}^T\boldsymbol{M}\bar{\boldsymbol{X}}_{k+1})^{1/2}}$	(7.12b)
Check convergence:		Check convergence:	
$\dfrac{\lambda_1^{(k+1)} - \lambda_1^{(k)}}{\lambda_1^{(k+1)}} \le \varepsilon_{\text{tol}}$	(7.13a)	$\dfrac{\lambda_n^{(k+1)} - \lambda_n^{(k)}}{\lambda_n^{(k+1)}} \le \varepsilon_{\text{tol}}$	(7.13b)
End		End	

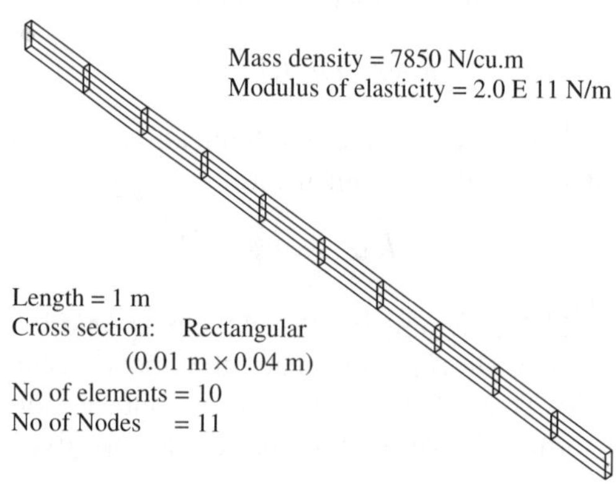

Figure 7.1 Cantilever beam; FE model

Table 7.2 A cantilever beam; eigenvalue result from inverse and power iteration

Method	Eigenvalue	No of iterations
Inverse iteration	$\lambda_1 = 51.23192$	4
Power iteration	$\lambda_{40} = 348990.573$	30

7.2.1 Order and Rate of Convergence – Distinct Eigenvalues

If we use eigenexpansion and express vectors $\boldsymbol{X}_k$ and $\bar{\boldsymbol{X}}_{k+1}$ in Equation 7.10a as $\boldsymbol{\Phi}\boldsymbol{Y}_k$ and $\boldsymbol{\Phi}\boldsymbol{Y}_{k+1}$, respectively, a set of uncoupled equations result (see orthogonality properties in Equation 5.49):

$$\boldsymbol{\Phi}^T\boldsymbol{K}\boldsymbol{\Phi}\boldsymbol{Y}_{k+1} = \boldsymbol{\Phi}^T\boldsymbol{M}\boldsymbol{\Phi}\boldsymbol{Y}_k \Rightarrow \boldsymbol{\Lambda}\boldsymbol{Y}_{k+1} = \boldsymbol{I}\boldsymbol{Y}_k \tag{7.14}$$

Here, $\mathbf{\Lambda} = \text{diag}(\lambda_1, \lambda_2, \ldots\ldots, \lambda_{n-1}, \lambda_n)$ and $\mathbf{I}$ is the identity matrix of size $n \times n$. Equation 7.14 is convenient for studying the convergence characteristics of the iteration process. The eigenvectors of the diagonalised system are the canonical basis in $\mathbb{R}^n$, that is $e_i = (0, 0\ldots, 1, 0\ldots\ldots 0)^T$ with unity at the ith position. Starting the inverse iteration with $\mathbf{Y}_0 = (1, 1, \ldots\ldots 1, 1)^T$, we have from Equation 7.2 the $(k+1)^{th}$ iterate vector as:

$$\mathbf{Y}_{k+1} = \left\{ \left(\frac{1}{\lambda_1}\right)^k, \left(\frac{1}{\lambda_2}\right)^k, \ldots\ldots\ldots, \left(\frac{1}{\lambda_n}\right)^k \right\}^T$$
$$= \left(\frac{1}{\lambda_1}\right)^k \left\{ 1, \left(\frac{\lambda_1}{\lambda_2}\right)^k, \ldots\ldots\ldots, \left(\frac{\lambda_1}{\lambda_n}\right)^k \right\}^T \tag{7.15}$$

It is thus apparent that $\mathbf{Y}_{k+1} \to$ multiple of e_1 as $k \to \infty$. Using Euclidean vector norms, we find the convergence of the iteration process as:

$$\lim_{k\to\infty} \frac{\left\| \lambda_1^k \mathbf{Y}_{k+1} - e_1 \right\|_2}{\left\| \lambda_1^{k-1} \mathbf{Y}_k - e_1 \right\|_2} = \frac{\lambda_1}{\lambda_2} \tag{7.16}$$

The above equation shows that the rate of convergence is $\frac{\lambda_1}{\lambda_2}$ and is linear. A similar argument is possible for power iteration and the rate of convergence will be $\frac{\lambda_{n-1}}{\lambda_n}$. At any iteration k of the two methods, as the vector $\mathbf{Y}_{k+1}$ converges to the true eigenvector $\mathbf{\Phi}_1$ or $\mathbf{\Phi}_n$ as the case may be, the Rayleigh quotient (Equation 7.11a,b) converges to the corresponding eigenvalue at a rate equal to $\left(\frac{\lambda_1}{\lambda_2}\right)^2$ and $\left(\frac{\lambda_{n-1}}{\lambda_n}\right)^2$, respectively, (see Equation 5.82). For the present example, the ratio (theoretical) $\frac{\lambda_{1,\text{exact}}}{\lambda_{2,\text{exact}}} = \frac{51.22619}{204.9050} = 0.25$. Table 7.3 shows the rate of convergence, $\frac{|\lambda_{1,k+1} - \lambda_{1,\text{exact}}|}{|\lambda_{1,k} - \lambda_{1,\text{exact}}|}$ of eigenvalues as against the theoretical minimum value of $\left(\frac{\lambda_1}{\lambda_2}\right)^2 = 0.0625$.

7.2.2 *Shifting and Convergence*

'Shifting' helps in accelerating the eigenvalue extraction. Suppose that the generalized eigenvalue problem $\mathbf{K\Phi}_n = \lambda_n \mathbf{M\Phi}_n$ with distinct eigenvalues is modified as follows:

$$(\mathbf{K} - \sigma \mathbf{M})\,\mathbf{\Phi}_n = \lambda_n \mathbf{M\Phi}_n \tag{7.17}$$

Table 7.3 Inverse iteration; rate of convergence of first eigenvalue

Iteration	Without shift		With shift = 50.0	
	Estimated eigenvalue	Rate of convergence	Estimated eigenvalue	Rate of convergence
1	51.32110	–	51.23193	–
2	51.23226	0.06394	51.23192	1.0
3	51.23191	0.94453	–	–
4	51.23192	0.99977	–	–
5	51.23192	0.99999	–	–

In the above equation, 'σ' stands for the 'shift'. If μ denotes an eigenvalue of the shifted system, we observe that the eigenvalues of the original and shifted problems are related by $\lambda = \mu + \sigma$. The eigenvectors remain the same for both. For the inverse iteration method, the eigenvalues of the original problem are given by $\lambda = \frac{1}{\mu} + \sigma$. For the Example 7.1, if a shift $\sigma = 50$ is used, the first eigenvector converges within two inverse iterations (shown in Table 7.3).

Shifting also helps to obtain higher eigenvalues other than the first one in inverse iteration and similarly lower than the last one in power iteration. For the cantilever beam in Example 7.1, the second theoretical eigenvalue, λ_2 is 204.905. If a shift of 220.0 ($> \lambda_2$) is used, the inverse iteration converges to the second eigenvalue instead of the first one, as shown in Table 7.4, which also shows the convergence process for this eigenvalue.

Generalising these observations, the rate of convergence with the shifting scheme for any higher eigenvalue, $\lambda_j (j > 1)$, can be studied following the steps in Equations 7.14–7.16. The equivalent uncoupled eigenvalue problem is written as

$$(\boldsymbol{\Lambda} - \sigma \boldsymbol{I})\boldsymbol{Y}_{k+1} = \boldsymbol{Y}_k \tag{7.18}$$

If inverse iteration is used with a shift near an eigenvalue, λ_j, the $(k+1)^{th}$ iterate for the eigenvector takes the form:

$$\boldsymbol{Y}_{k+1} = \frac{1}{(\lambda_j - \mu)^k} \left\{ \left(\frac{\lambda_j - \mu}{\lambda_1 - \mu}\right)^k, \ldots \left(\frac{\lambda_j - \mu}{\lambda_{j-1} - \mu}\right)^k, 1, \left(\frac{\lambda_j - \mu}{\lambda_{j+1} - \mu}\right)^k \ldots\ldots\ldots, \left(\frac{\lambda_j - \mu}{\lambda_n - \mu}\right)^k \right\}^T \tag{7.19}$$

Now, the rate of convergence of the eigenvector corresponding to λ_j is either $\left|\frac{\lambda_j - \mu}{\lambda_{j-1} - \mu}\right|$ or $\left|\frac{\lambda_j - \mu}{\lambda_{j+1} - \mu}\right|$, whichever is larger. For the cantilever beam in Example 7.1, the rates of convergence (with shift) corresponding to the second eigenvalue are shown in Table 7.4. With a theoretical result of $\lambda_3 = 321.052$ and a shift of $\mu = 200$, we have $\frac{\lambda_2 - \mu}{\lambda_1 - \mu} = 0.033$ and $\frac{\lambda_2 - \mu}{\lambda_3 - \mu} = 0.041$. Thus, the larger value of 0.041 decides the minimum rate of convergence (Equation 7.19) for the second eigenvalue and it is given by $\left(\frac{\lambda_2 - \mu}{\lambda_3 - \mu}\right)^2 = 0.00164$. With a shift of 220, the minimum rate of convergence is 0.022. Note that if it is possible to predict a shift nearer to an eigenvalue, convergence is accelerated.

Table 7.4 Use of shift in inverse iteration; rate of convergence of second eigenvalue

Iteration	Estimated eigenvalue without shift	Rate of convergence	
		With Shift = 220	With Shift = 200
1	204.58849	0.0516	0.5729
2	204.92134	1.3806	0.9980
3	204.92756	1.0056	1.0
4	204.92768	1.0001	–
5	204.92768	1.0000	–

7.2.3 *Multiple Eigenvalues*

If the cantilever in Example 7.1 above is square in cross section, the eigenvalues are all not distinct. It is possible to have pairs of multiple (repeated) eigenvalues. In this case, these eigenvalues occur in pairs and correspond to identical vibrating mode shapes in the two transverse directions. For each eigenvalue λ of multiplicity m, if the algebraic and geometric multiplicities are the same, recall that the system matrix, $\boldsymbol{A}$ of size $n \times n$, is diagonisable and has n linearly independent eigenvectors (Chapter 5). Also, the associated eigenspace, $E(\lambda)$, is given by the dimension *of* $N(\boldsymbol{A} - \lambda \boldsymbol{I})$ and is spanned by linearly independent eigenvectors. In extracting these eigenvectors of a multiple eigenvalue, it is clear that iterative methods may fail to converge to the second or the subsequent eigenvector, once the first vector is realised. Indeed, irrespective of the choice of the starting vector, the iterative process leads to the same eigenvector. In this case, a corrective strategy is thus needed to deflate the new starting vector, that is to render it orthogonal to the converged eigenvector. Gram–Schmidt orthogonalization is popularly used for this purpose. In the general case, wherein an eigenpair (λ_i, Φ_i) is already obtained, the orthogonalization scheme modifies the new starting vector $\boldsymbol{X}_0$ such that it is $\boldsymbol{M}$-orthonormal with respect to $\boldsymbol{\Phi}_i$. Thus, the scheme obtains the new starting vector as follows:

$$\bar{\boldsymbol{X}}_0 = \boldsymbol{X}_0 - \alpha \boldsymbol{\Phi}_i, \alpha = \boldsymbol{\Phi}_i^T \boldsymbol{M} \boldsymbol{X}_0 \tag{7.20}$$

With the deflation according to Equation 7.20, $\bar{\boldsymbol{X}}_0$ is made to be $\boldsymbol{M}$-orthogonal to the already converged vector, $\boldsymbol{\Phi}_i$, that is $\boldsymbol{\Phi}_i^T \boldsymbol{M} \bar{\boldsymbol{X}}_0 = 0$. The iterative process now starts with this modified vector, $\bar{\boldsymbol{X}}_0$ and converges to the desired next eigenvector.

Table 7.5 Modified algorithm for inverse iteration: generalized eigenvalue problem with multiple eigenvalues

Given: A multiple eigenvalue, λ_i with a converged eigenvector, $\boldsymbol{\Phi}_i$
Aim: To obtain another eigenvector, $\boldsymbol{\Phi}_j$, of the same eigenvalue
Initialise and modify the new starting vector, $\boldsymbol{X}_0$, by Gram–Schmidt orthogonalization:

$$\bar{\boldsymbol{X}}_0 = \boldsymbol{X}_0 - \left(\boldsymbol{\Phi}_i^T \boldsymbol{M} \boldsymbol{X}_0\right) \boldsymbol{\Phi}_i \tag{7.21}$$

Start Iteration: $k = 0, 1, \ldots$

$$\boldsymbol{K} \bar{\boldsymbol{X}}_{k+1} = \boldsymbol{M} \bar{\boldsymbol{X}}_k \tag{7.22}$$

$$\lambda_1^{(k+1)} = \frac{\bar{\boldsymbol{X}}_{k+1}^T \boldsymbol{K} \bar{\boldsymbol{X}}_{k+1}}{\bar{\boldsymbol{X}}_{k+1}^T \boldsymbol{M} \bar{\boldsymbol{X}}_{k+1}} \tag{7.23}$$

$$\bar{\boldsymbol{X}}_{k+1} = \frac{\bar{\boldsymbol{X}}^{k+1}}{(\bar{\boldsymbol{X}}_{k+1}^T \boldsymbol{M} \bar{\boldsymbol{X}}_{k+1})^{1/2}} \tag{7.24}$$

Check convergence:

$$\frac{\lambda_1^{(k+1)} - \lambda_1^{(k)}}{\lambda_1^{(k+1)}} \leq \varepsilon_{\text{tol}} \tag{7.25}$$

Repeat Gram–Schmidt orthogonalization to M-orthogonalize the estimated vector with respect to $\boldsymbol{\Phi}_i$:

$$\bar{X}_{k+1} = \bar{X}_{k+1} - \left(\boldsymbol{\Phi}_i^T \boldsymbol{M} \bar{\boldsymbol{X}}_{k+1}\right) \boldsymbol{\Phi}_i \tag{7.26}$$

End

However, Gram–Schmidt orthogonalization is sensitive to round-off errors and hence, from practical considerations, it is generally required to repeat the orthogonalization process on the estimated eigenvector computed at the end of each iteration as indicated in Table 7.5. To find the rate of convergence, it is only required to extend the procedure earlier outlined for distinct eigenvalues. For an eigenvalue, λ_i of multiplicity 'm', Equation 7.15 takes the form:

$$\boldsymbol{Y}_{k+1} = \left(\frac{1}{\lambda_i}\right)^k \left\{ \left(\frac{\lambda_i}{\lambda_1}\right)^k, \left(\frac{\lambda_i}{\lambda_2}\right)^k, \ldots\ldots, \left(\frac{\lambda_i}{\lambda_{i-1}}\right)^k, 1, 1 \ldots\ldots, 1, \left(\frac{\lambda_i}{\lambda_{i+m}}\right)^k, \ldots\ldots, \left(\frac{\lambda_1}{\lambda_n}\right)^k \right\}^T \lambda_i \text{ with multiplicity } m \tag{7.27}$$

The rate of convergence for the eigenvector is $\frac{\lambda_i}{\lambda_{i+m}}$ and that of the corresponding eigenvalue is $\left(\frac{\lambda_i}{\lambda_{i+m}}\right)^2$. The deflation procedure in Equation 7.20 can be generalized (Figure 7.2) in orthogonalising a given vector $\boldsymbol{X}_0$ with respect to a set of vectors $\{\boldsymbol{\Phi}_i\}$ as

$$\boldsymbol{X}_0 = \boldsymbol{X}_0 - \sum_{i=1}^{m} \left(\boldsymbol{\Phi}_i^T \boldsymbol{M} \boldsymbol{X}_0\right) \boldsymbol{\Phi}_i \tag{7.28}$$

7.2.4 Eigenvalues within an Interval-Shifting Scheme with Gram–Schmidt Orthogonalization and Sturm Sequence Property

Shifting is also used to obtain eigenvalues within a specified interval (σ_1, σ_2), $\sigma_2 > \sigma_1$. Here, for a real symmetric matrix, one can determine by Sylvester's law of inertia (Strang 2006) the number of eigenvalues that are less than a given real number. The law of inertia essentially states that a similarity transformation $\boldsymbol{CDC}^T$ of a real symmetric matrix $\boldsymbol{A}$ (with nonsingular $\boldsymbol{C}$), is a congruent transformation in that the symmetry property is unchanged and the number of positive, negative and zero eigenvalues of $\boldsymbol{A}$ are the same as those of the diagonal matrix $\boldsymbol{D}$. Thus, the signs of the eigenvalues are preserved by

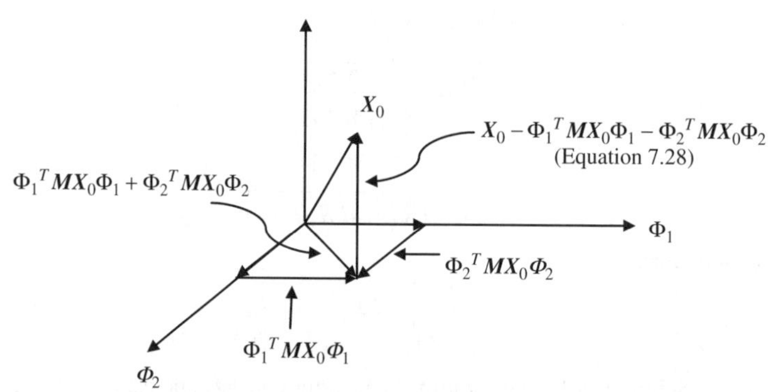

Figure 7.2 Gram–Schmidt orthogonalization scheme. $m = 2$

the transformation. Accordingly, an LDL^T decomposition of $A - \sigma I$ (assuming that the decomposition is possible) gives information on the number of negative eigenvalues of $A - \sigma I$, which is equal to the number of negative diagonal elements in D.

With $\lambda_i, i = 1, 2 \ldots n$ denoting the eigenvalues of A, the eigenvalues of $A - \sigma I$ are $\mu = \{\lambda_i - \sigma\}$. Thus, the LDL^T decomposition gives the number of eigenvalues of A (including the repeated ones) that are less than the shift σ. The so-called Sturm sequence property of characteristic polynomials (Geradin and Rixen, 1994) of matrices also contains this result. A set of polynomials $\{p^{(r)}(\lambda) | r = 0, \ldots, n-1\}$ is said to form a Sturm sequence if every root of $p^{(r+1)}(\lambda)$ 'separates' (i.e. is straddled by) a pair of roots of $p^{(r)}(\lambda)$, as schematically shown in Figure 7.3. If $A^{(r)}$ corresponds to the matrix A with the last 'r' rows and columns removed (representing a constrained system of order '$n - r$'), the eigenvalue separation theorem states that the eigenvalues of $A^{(r+1)}\Phi^{(r+1)} = \lambda^{(r+1)}\Phi^{(r+1)}$ separate (in the same sense as above) those of $A^{(r)}\Phi^{(r)} = \lambda^{(r)}\Phi^{(r)}$ and thus the corresponding characteristic polynomials $p^{(n-r)}(\lambda)$, $r = 0, 1, .., n-1$ form a Sturm sequence. The polynomials $p^{(n-r)}(\lambda)$, in fact correspond to eigenvalue problems associated with the principal minors of the matrix A. Now, if the characteristic polynomials of the matrices A and $A - \sigma I$ are denoted by $p(\lambda)$ and $p_\sigma(\mu)$, they are given by:

$$p(\lambda) = \det(A - \sigma I) = (\lambda - \lambda_1)(\lambda - \lambda_2) \ldots \ldots (\lambda - \lambda_{n-1})(\lambda - \lambda_n) \tag{7.29}$$

$$\begin{aligned} p_\sigma(\mu) &= \det((A - \sigma I) - \mu I) \\ &= (\mu - \mu_1)(\mu - \mu_2) \ldots \ldots (\mu - \mu_{n-1})(\mu - \mu_n) \end{aligned} \tag{7.30}$$

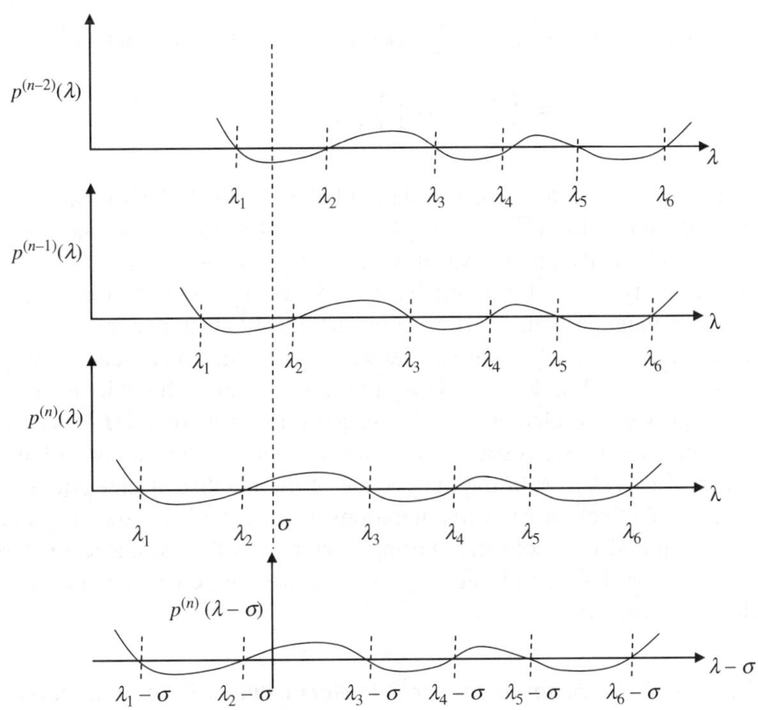

Figure 7.3 Eigenvalue separation property

where $\mu_i = \lambda_i - \sigma$. From Equation 7.30, it is clear that for $\sigma > \lambda_k$, the shifted eigenvalue problem possesses 'k' negative eigenvalues. Figure 7.3 also shows the shifted characteristic polynomial $p_\sigma(\mu)$ having negative eigenvalues (an example case of $k = 2$). It is now needed to establish a correspondence between this observation and the signs of the diagonal elements in $\boldsymbol{D}$ in the $\boldsymbol{LDL}^T$ decomposition of $(\boldsymbol{A} - \sigma\boldsymbol{I})$. From the decomposition of matrix $\boldsymbol{A}$, we have:

$$\det(\boldsymbol{A}) = \det(\boldsymbol{L}) \cdot \det(\boldsymbol{D}) \cdot \det(\boldsymbol{L}^T) = \det(\boldsymbol{D}) \text{ (since } \det(L) = \det(\boldsymbol{L}^T) = 1)$$
$$\Rightarrow \prod_{i=1}^{n} d_{ii} = \prod_{i=1}^{n} \lambda_i \tag{7.31}$$

with d_{ii} being the diagonal elements of $\boldsymbol{D}$. For a positive definite matrix, all $d_{ii} > 0$ (Appendix G). The equality $\prod_{i=1}^{n} d_{ii} = \prod_{i=1}^{n} \lambda_i$ follows from the fact that the product of the eigenvalues (i.e. roots of the characteristic polynomial $p(\lambda)$) equals the determinant of $\boldsymbol{A}$. Now, via the $\boldsymbol{LDL}^T$ decomposition of $\boldsymbol{A} - \sigma\boldsymbol{I}$, we have:

$$\det(\boldsymbol{A} - \sigma\boldsymbol{I}) = \det(\bar{\boldsymbol{L}}) \cdot \det(\bar{\boldsymbol{D}}) \cdot \det(\bar{\boldsymbol{L}}^T) = \det(\bar{\boldsymbol{D}})$$
$$\Rightarrow \prod_{i=1}^{n} \bar{d}_{ii} = \prod_{i=1}^{n} (\lambda_i - \sigma) \tag{7.32}$$

From the above equation, it follows that:

$$\det(\boldsymbol{A}^{(r)} - \sigma\boldsymbol{I}) = \det(\bar{\boldsymbol{L}}^{(r)}) \cdot \det(\bar{\boldsymbol{D}}^{(r)}) \cdot \det(\bar{\boldsymbol{L}}^{(r)T}) = \det(\bar{\boldsymbol{D}}^{(r)})$$
$$\Rightarrow \prod_{i=1}^{n-r} \bar{d}_{ii}^{(r)} = \prod_{i=1}^{n-r} \left(\bar{\lambda}_i^{(r)} - \sigma\right) \tag{7.33}$$

As 'r' reduces from $n - 1$ to zero, the signs of the diagonal elements $\bar{d}_{ii}$ get adjusted so as to satisfy the condition $\prod_{i=1}^{n-r} \bar{d}_{ii} = \prod_{i=1}^{n-r} \left(\bar{\lambda}_i - \sigma\right)$ at each stage and the number of negative diagonal elements equals that of negative eigenvalues of $\boldsymbol{A}$.

The above property of a Sturm sequence or Sylvester's law of inertia also applies to the generalized eigenproblem. In this case, the $\boldsymbol{LDL}^T$ factorisation is performed on $\boldsymbol{K} - \sigma\boldsymbol{M}$ and the number of negative elements in the diagonal matrix $\boldsymbol{D}$ yields the number of eigenvalues of $\boldsymbol{K\Phi}_n = \lambda_n \boldsymbol{M\Phi}_n$ that are less than the shift σ. If n_1 and n_2 are the number of negative elements in the diagonal matrices of $\boldsymbol{LDL}^T$ decompositions of $\boldsymbol{K} - \sigma_1\boldsymbol{M}$ and $\boldsymbol{K} - \sigma_2\boldsymbol{M}$, respectively, then $n_2 - n_1$ is the number of eigenvalues within the interval (σ_1, σ_2). With the knowledge of the number of eigenvalues within an interval, one can effectively use inverse iterations to extract all the associated eigenpairs. It is, however, required that both the starting vectors and the estimated vectors at each iteration 'k', are to be deflated (Table 7.5) of the converged eigenvectors for computing the remaining eigenvectors.

Example 7.2: Cantilever beam in Figure 7.1; eigenvalues within an interval

For the cantilever beam in Example 7.1, we intend to compute, say, the frequencies within the interval (200.0, 900.0 rad/s). A step-by-step procedure to obtain these frequencies by inverse iteration is given below.

- **Step 1:** The $\boldsymbol{LDL}^T$ decomposition of $\boldsymbol{K} - \sigma \boldsymbol{M}$ with $\sigma = 200$ rad/s gives the number of negative diagonal elements in D as 1. With $\sigma = 900.0$, on the other hand, the number of negative diagonals turns out to be 4. Thus, we infer that there are three frequencies within the specified interval. These are to be computed along with the corresponding eigenvectors.
- **Step 2:** Divide the given interval into a few subintervals (=15 in this case) and let the jth subinterval be denoted as $\left(\sigma_1^{(j)}, \sigma_2^{(j)}\right)$.
- **Step 3:** Loop over the number of subintervals. For the jth subinterval, first find the frequencies in that interval by $\boldsymbol{LDL}^T$ decomposition (Sturm sequence check) with shifts $= \sigma_1^{(j)}$ and $\sigma_2^{(j)}$. In the present case, the sum of frequencies lying within all the subintervals must be equal to 3.
- **Step 4:** In each subinterval, there may exist (i) no frequencies, (ii) a single frequency and/or (iii) multiple frequencies.

For case (i), skip to the next subinterval and for the other two cases, follow the iterations in Table 7.5 to compute the frequencies existing in that subinterval. Take the starting vector as $\boldsymbol{X}_0 = (1.0, 1.0\ldots\ldots, 1.0)^T$ (40-dimensional) and mass-orthonormalise with respect to all the previously converged eigenvectors according to Equation 7.28. In the present example, the case of multiple frequencies does not exist.

Figure 7.4 shows the convergence of each of the three required frequencies obtained by the inverse iteration. These three frequencies almost agree with the corresponding (second, third and fourth) obtained from the frequency Equation 2.129, which gives the first few natural frequencies (in rad/s) of the cantilever as 51.24282, 204.92768, 321.07392, 899.21937 and 1284.3012.

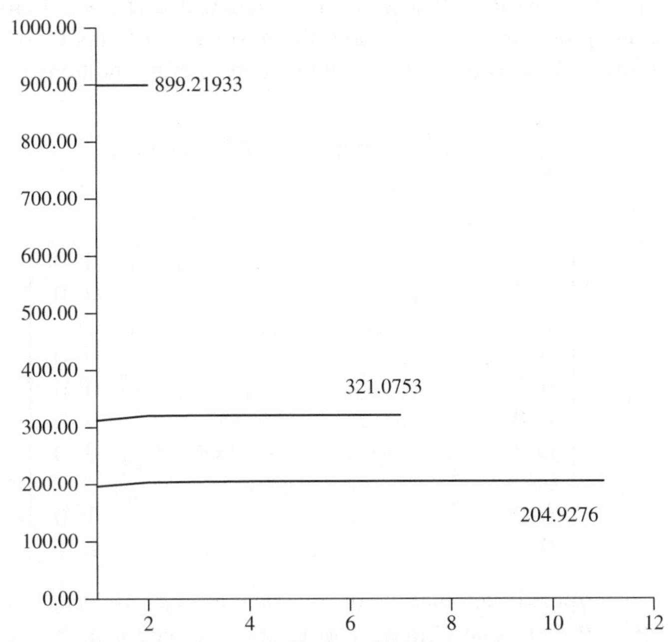

Figure 7.4 Natural frequencies (rad/s) within a specified interval by inverse iteration

7.3 Jacobi, Householder, QR Transformation Methods and Eigensolutions

Before other iterative methods such as the subspace method is discussed, it is pertinent here to describe some of the transformation methods. For the standard eigenvalue problem, these transformation methods use a series of similarity transformations $\boldsymbol{P}^T\boldsymbol{A}\boldsymbol{P}$ to reduce the matrix $\boldsymbol{A}$ to a diagonal form $\boldsymbol{\Lambda}$ (containing the eigenvalues λ on the diagonal). The associated recursive relation is of the form:

$$\boldsymbol{A}_{k+1} = \boldsymbol{P}_k^T \boldsymbol{A}_k \boldsymbol{P}_k, \text{ with } \boldsymbol{A}_1 = \boldsymbol{A} \tag{7.34}$$

The basic concept lies in constructing suitable unimodular transformation matrices $\boldsymbol{P}_j$, $j = 1, 2 \ldots$ such that $\boldsymbol{A}_{k+1} \to \boldsymbol{\Lambda}$ and $\hat{\boldsymbol{\Phi}}_k \left(= \boldsymbol{P}_1\boldsymbol{P}_2 \ldots\ldots \boldsymbol{P}_j \ldots\ldots P_{k-1}\boldsymbol{P}_k\right) \to \boldsymbol{\Phi}$ as $k \to \infty$. In the case of a generalized eigenvalue problem, the transformations are simultaneously carried out on $\boldsymbol{K}$ and $\boldsymbol{M}$ and the recursive relations are of the form:

$$\boldsymbol{K}_{k+1} = \boldsymbol{P}_k^T \boldsymbol{K}_k \boldsymbol{P}_k, \text{ with } \boldsymbol{K}_1 = \boldsymbol{K} \text{ and}$$

$$\boldsymbol{M}_{k+1} = \boldsymbol{P}_k^T \boldsymbol{M}_k \boldsymbol{P}_k, \text{ with } \boldsymbol{M}_1 = \boldsymbol{M} \tag{7.35a,b}$$

The sequence of transformations converge to $\boldsymbol{K}_{k+1} \to \boldsymbol{\Lambda}$, $\boldsymbol{M}_{k+1} \to \boldsymbol{I}$ and $\hat{\boldsymbol{\Phi}}_k$ $\left(= \boldsymbol{P}_1\boldsymbol{P}_2 \ldots\ldots \boldsymbol{P}_j \ldots\ldots \boldsymbol{P}_{k-1}\boldsymbol{P}_k\right) \to \boldsymbol{\Phi}$ as $k \to \infty$.

7.3.1 *Jacobi Method*

The Jacobi transformation method is the oldest and widely used technique for eigensolutions of real symmetric matrices. It was originally developed (Jacobi, 1846) to solve the standard eigenvalue problem. In this method, the off-diagonal terms of $\boldsymbol{A}$ are systematically zeroed row-by-row or column-by-column by constructing the transformation matrix $\boldsymbol{P}$ in the form:

$$\begin{array}{c} \quad\quad\quad i^{th}\text{ column} \quad\quad j^{th}\text{ column} \\ \quad\quad\quad \downarrow \quad\quad\quad\quad\quad \downarrow \end{array}$$

$$\boldsymbol{P} = \begin{bmatrix} 1\;0 & \ldots\ldots\ldots\ldots\ldots\ldots\ldots\ldots\ldots\ldots & 0\;0 \\ 0\;1 & \ldots\ldots\ldots\ldots\ldots\ldots\ldots\ldots\ldots\ldots & 0\;0 \\ \ldots & \ldots\ldots\ldots\ldots\ldots\ldots\ldots\ldots\ldots\ldots & \ldots \\ 0\;0 & \ldots\ldots \cos\theta \ldots\ldots -\sin\theta \ldots\ldots & 0\;0 \\ 0\;0 & \ldots\ldots\ldots\ldots\ldots\ldots\ldots\ldots\ldots\ldots & 0\;0 \\ 0\;0 & \ldots\ldots\ldots\ldots\ldots\ldots\ldots\ldots\ldots\ldots & 0\;0 \\ 0\;0 & \ldots\ldots \sin\theta \ldots\ldots -\cos\theta \ldots\ldots & 0\;0 \\ \ldots & \ldots\ldots\ldots\ldots\ldots\ldots\ldots\ldots\ldots\ldots & \ldots \\ 0\;0 & \ldots\ldots\ldots\ldots\ldots\ldots\ldots\ldots\ldots\ldots & 1\;0 \\ 0\;0 & \ldots\ldots\ldots\ldots\ldots\ldots\ldots\ldots\ldots\ldots & 0\;1 \end{bmatrix} \tag{7.36}$$

The angle 'θ' is determined such that the off-diagonal term a_{ij} reduces to zero under the transformation $\boldsymbol{P}^T\boldsymbol{A}\boldsymbol{P}$. Note that $\boldsymbol{P}$ in the above form is orthogonal, that is $\boldsymbol{P}^T\boldsymbol{P} = \boldsymbol{I}$. In geometrical terms, the transformation rotates the ith and j^{th} row and column vectors by an angle θ in the hyperplane spanned by them. If the orthogonal transformation $\boldsymbol{P}^T\boldsymbol{A}\boldsymbol{P}$ results in a matrix $\boldsymbol{A}'$, the particular choice of $\boldsymbol{P}$, as above, affects only the elements in i^{th} and j^{th} rows and i^{th} and j^{th} columns of $\boldsymbol{A}$. In particular, the off-diagonal terms a'_{ij} are of the form:

$$a'_{ij} = a'_{ji} = \frac{1}{2}(a_{ii} - a_{jj})\sin 2\theta - a_{ij}\cos 2\theta \tag{7.37}$$

By imposing the condition $a'_{ij} = 0$, we obtain

$$\theta = \frac{1}{2}\tan^{-1}\frac{2a_{ij}}{a_{ii} - a_{jj}} \tag{7.38}$$

For $a_{ii} = a_{jj}$, one has $\theta = \pm\frac{\pi}{4}$ depending on whether a_{ij} is positive or not. If transformations are carried out one-by-one on each of the $n(n-1)/2$ off-diagonal elements row-by-row (keeping in mind that $\boldsymbol{A}$ is symmetric), they are counted as one sweep (or a Jacobi iteration). Whilst a specific transformation applied once on $\boldsymbol{A}$ may render the associated off-diagonal term zero, the same terms may become nonzero over subsequent transformations. Accordingly, the number of sweeps that are, in principle, required to drive all the off-diagonal entries to zero is infinite. If these sweeps are carried out systematically row-by-row or column-by-column regardless of whether an off-diagonal entry is already zero (modulo a specified tolerance), it is referred to as the cyclic Jacobi method. In the threshold Jacobi method, the transformation is applied only when an off-diagonal term (absolute value) is larger than the tolerance.

7.3.1.1 Convergence of the Jacobi Method

Consider an orthogonal transformation under which the off-diagonal element a_{ij} becomes $a'_{ij} = 0$. The Frobenius norms of the matrices $\boldsymbol{A}$ and $\boldsymbol{A}'$ are, respectively, given by $\|\boldsymbol{A}\|_F\left(=\sqrt{tr\left(\boldsymbol{A}^T\boldsymbol{A}\right)}\right)$ and $\|\boldsymbol{A}'\|_F\left(=\sqrt{tr\left(\boldsymbol{A}'^T\boldsymbol{A}'\right)}\right)$. Thus, we have

$$\|\boldsymbol{A}\|_F^2 = \sum_{l=1}^{n}\sum_{m=1}^{n} a_{lm}^2 \text{ and } \|\boldsymbol{A}'\|_F^2 = \sum_{l=1}^{n}\sum_{m=1}^{n} {a'_{lm}}^2 \tag{7.39}$$

Since the Frobenius norm is invariant under a similarity transformation, $\|\boldsymbol{A}\|_F^2 = \|\boldsymbol{A}'\|_F^2$. Further, let the sums of the squares of the off-diagonal elements in the two cases be denoted by S_A and $S_{A'}$, respectively. Then, we have:

$$S_A = \|\boldsymbol{A}\|_F^2 - \sum_{l=1}^{n} a_{ll}^2 \text{ and } S_{A'} = \|\boldsymbol{A}'\|_F^2 - \sum_{l=1}^{n} {a'_{ll}}^2 \tag{7.40}$$

In expanded forms, S_A and $S_{A'}$ are given by:

$$S_A = \sum_{\substack{l=1 \\ l\neq i,j}}^{n}\ \sum_{\substack{m=1 \\ l\neq i,j \\ l\neq m}}^{n} a_{lm}^2 + \sum_{\substack{l=1 \\ l\neq i,j}}^{n}\left(a_{li}^2 + a_{lj}^2\right) + \sum_{\substack{m=1 \\ m\neq i,j}}^{n}\left(a_{im}^2 + a_{jm}^2\right) + 2a_{ij}^2$$

$$S_{A'} = \sum_{\substack{l=1 \\ l \neq i,j}}^{n} \sum_{\substack{m=1 \\ m \neq i,j \\ m \neq l}}^{n} {a'_{lm}}^2 + \sum_{\substack{l=1 \\ l \neq i,j}}^{n} \left({a'_{li}}^2 + {a'_{lj}}^2\right)$$

$$+ \sum_{\substack{m=1 \\ m \neq i,j}}^{n} \left({a'_{im}}^2 + {a'_{jm}}^2\right) + 2{a'_{ij}}^2 \qquad (7.41\text{a,b})$$

Owing to the transformations being rotational over appropriate 2D hyperplanes (Equation 7.38), we also have the following relations:

$$\begin{aligned} a'_{lm} &= a_{lm}, lm \neq ij \\ a'_{li} &= a'_{il} = a_{li} \cos\theta - a_{lj} \sin\theta,\ l \neq i, j \\ a'_{lj} &= a'_{jl} = a_{li} \sin\theta + a_{lj} \cos\theta,\ l \neq i, j \end{aligned} \qquad (7.42\text{a–c})$$

From the above equations, we have

$${a'_{li}}^2 + {a'_{lj}}^2 = {a'_{li}}^2 + {a'_{lj}}^2 \text{ and } {a'_{im}}^2 + {a'_{jm}}^2 = a_{im}^2 + a_{jm}^2 \qquad (7.43)$$

In view of Equations 7.41 and 7.42 and the fact that $a'_{ij} = 0$, we have:

$$S_{A'} = S_A - 2a_{ij}^2 \qquad (7.44)$$

For a certain i and j, if $|a_{ij}|$ is the largest out of $n(n-1)$ off-diagonal elements, then at the start of a sweep, $S_A^{(0)} \leq n(n-1)a_{ij}^2$. Hence, at the end of a transformation, we have:

$$S_{A'} \leq \left(1 - \frac{2}{n(n-1)}\right) S_A \qquad (7.45)$$

If $S_A^{1/2}$ stands for the norm of the matrix with off-diagonal terms of $\boldsymbol{A}$, the above equation shows that the sequence of Jacobi transformations converges the norm to zero linearly by a factor $\left(1 - \frac{2}{n(n-1)}\right)^{1/2}$. Thus, with progressive transformations, $S_{A'}$ shows a strictly decreasing trend implying that these elements tend to zero as the number of transformations approaches infinity, which establishes the convergence of the method. Further, if it is required to have the convergence within a tolerance of 10^{-s} within k transformations with s denoting the number of significant digits, then we have:

$$\left(1 - \frac{2}{n(n-1)}\right)^{k/2} < 10^{-s} \Rightarrow k \approx \frac{n^2}{4} s \qquad (7.46)$$

Thus, the number of transformations required is proportional to the square of the size of system matrix. From a perspective of practicality, this method is applicable for reasonably smaller size systems. Accordingly, the method typically finds application as a part of other eigensolvers, for example subspace iteration and Lanczos eigensolvers, which use reduced-order eigensystems constructed from the original one.

7.3.1.2 Jacobi Method for the Generalized Eigenvalue Problem

In this case, the transformation matrices $\boldsymbol{P}_k, k = 1, 2 \ldots$ are taken to be of the form:

$$\boldsymbol{P}_k = \begin{bmatrix} 1 & 0 & \cdots & & & 0 & 0 \\ 0 & 1 & \cdots & & & 0 & 0 \\ \cdots & & & & & & \cdots \\ 0 & 0 & \cdots & \overset{\overset{i^{th}\text{ column}}{\downarrow}}{1} \cdots & \overset{\overset{j^{th}\text{ column}}{\downarrow}}{\alpha} \cdots & 0 & 0 \\ 0 & 0 & \cdots & & & 0 & 0 \\ 0 & 0 & \cdots & & & 0 & 0 \\ 0 & 0 & \cdots & \gamma \cdots & 1 \cdots & 0 & 0 \\ \cdots & & & & & & \cdots \\ 0 & 0 & \cdots & & & 1 & 0 \\ 0 & 0 & \cdots & & & 0 & 1 \end{bmatrix} \tag{7.47}$$

The parameters α and γ are to be so chosen that each transformation reduces the off-diagonal elements $\boldsymbol{K}_{ij}^{(k)}$ and $\boldsymbol{M}_{ij}^{(k)}$ to zero. With such transformed matrices, $\boldsymbol{K}^{(k+1)} = \boldsymbol{P}_k^T \boldsymbol{K}^{(k)} \boldsymbol{P}_k$ and $\boldsymbol{M}^{(k+1)} = \boldsymbol{P}_k^T \boldsymbol{M}^{(k)} \boldsymbol{P}_k$, we obtain the following nonlinear coupled equations in α and γ:

$$\alpha \boldsymbol{K}_{ii}^{(k)} + (1 + \alpha\gamma) \boldsymbol{K}_{ij}^{(k)} + \boldsymbol{K}_{jj}^{(k)} = 0$$

$$\alpha \boldsymbol{M}_{ii}^{(k)} + (1 + \alpha\gamma) \boldsymbol{M}_{ij}^{(k)} + \boldsymbol{M}_{jj}^{(k)} = 0 \tag{7.48a,b}$$

Following elimination of one of the parameters, we have a quadratic equation in the other (i.e. either α or γ). In terms of the off-diagonal stiffness and mass elements involved in the above equations, these parameters are given by Nguyen (2006) and Bathe (1996).

$$\alpha = \frac{R_1}{R}, \gamma = -\frac{R_2}{R} \tag{7.49}$$

Here, $R_1 = K_{jj}^{(k)} M_{ij}^{(k)} - M_{jj}^{(k)} K_{ij}^{(k)}$, $R_2 = K_{ii}^{(k)} M_{ij}^{(k)} - M_{ii}^{(k)} K_{ij}^{(k)}$ and $R = \frac{b}{2} + \frac{sign(b)}{2} \left(b^2 + 4R_1R_2\right)^{1/2}$ with $b = K_{ii}^{(k)} M_{jj}^{(k)} - M_{ii}^{(k)} K_{jj}^{(k)}$.

Table 7.6 details the computational steps involved in the Jacobi method for the generalized eigenvalue problem.

The thresholds $\left(\frac{K_{ij}^2}{K_{ii}K_{jj}}\right)^{1/2}$ and $\left(\frac{M_{ij}^2}{M_{ii}M_{jj}}\right)^{1/2}$ in Equation 7.50 indicate a measure of coupling between the ith and j^{th} degrees of freedom and are used to proceed with or skip the transformation connected with the off-diagonal elements K_{ij} and M_{ij}.

Example 7.3: The cantilever in Figure 7.1: eigensolution by the Jacobi method

The system matrices $[\boldsymbol{K}]_{n\times n}$ and $[\boldsymbol{M}]_{n\times n}$ are symmetric ($n = 40$) with bandwidth = 8. In each sweep, the number of maximum transformations required (= the number of nonzero off-diagonal elements) = (bandwidth – 1) $\times \left(n - \frac{\text{bandwidth}}{2}\right) = 252$. On each

Table 7.6 Jacobi method (with threshold) and its algorithm

Generalized eigenvalue problem: $\boldsymbol{K\Phi}_n = \lambda_n \boldsymbol{M\Phi}_n$, $\boldsymbol{K}$ and $\boldsymbol{M}$ symmetric matrices of size $n \times n$
sweep = 0
start sweep = sweep + 1
Start loop over $k = 1, n(n-1)/2$ (number of off-diagonal terms)
Identify row number, i and column number, j
Fix thresholds $\left(\dfrac{K_{ij}^2}{K_{ii}K_{jj}}\right)^{1/2}$ and $\left(\dfrac{M_{ij}^2}{M_{ii}M_{jj}}\right)^{1/2}$
Proceed with transformation if thresholds $> \varepsilon_{th}$, otherwise skip
Find α and γ (Equation 7.49) and construct transformation matrix $\boldsymbol{P}$ (Equation 7.47)
Find transformed matrices:

$$\boldsymbol{K}^{(k+1)}\left(= \boldsymbol{P}_k^T \boldsymbol{K}^{(k)} \boldsymbol{P}_k\right) \text{ and } \boldsymbol{M}^{(k+1)}\left(= \boldsymbol{P}_k^T \boldsymbol{M}^{(k)} \boldsymbol{P}_k\right) \tag{7.50}$$

Find eigenvalues $\lambda_l^k = \dfrac{K_{ll}^k}{M_{ll}^k}$ and $\lambda_l^{k+1} = \dfrac{K_{ll}^{k+1}}{M_{ll}^{k+1}}$
Compute eigenvectors:

$$\boldsymbol{\Phi}^{k+1} = \boldsymbol{\Phi}^k \boldsymbol{P}^k \tag{7.51}$$

End loop over k
Check convergence on eigenvalues $\dfrac{\left|\lambda_l^{k+1} - \lambda_l^k\right|}{\lambda_l^{k+1}} \leq \varepsilon_{\text{tol}}, l = 1, 2 \ldots, n$
End sweeps

of these off-diagonal elements of $\boldsymbol{K}$ and $\boldsymbol{M}$, the threshold checks are performed with $\varepsilon_{th} = 1.0e-3$. At the end of each sweep, the convergence of eigenvalues is checked with $\varepsilon_{tol} = 1.0\text{E}{-3}$. Table 7.7 gives details of the convergence of the first eight eigenvalues against the number of iterations (sweeps). Figure 7.5 shows the convergence with regard to $\left|S_{A'} - S_A\right|$ (Equation 7.44).

7.3.2 Householder and QR Transformation Methods

The Householder method is suitable to reduce a general matrix to an upper or lower Hessenberg form (as defined below) by a finite number of unitary similarity transformations without iterations. A matrix in the upper Hessenberg form contains all zero entries below the first subdiagonal and, if the zeros occur above the first subdiagonal, it is referred to as a lower Hessenberg matrix. QR transformation method iteratively reduces the Hessenberg matrix to a triangular matrix or a nearly triangular matrix. In the triangular case, the diagonal elements are the eigenvalues of $\boldsymbol{A}$. The nearly triangular matrix, on the other hand, arises if the eigenvalues of $\boldsymbol{A}$ are complex. Each pair of complex conjugate eigenvalues occupy a 2×2 submatrix on the diagonal line of the nearly triangular matrix (corresponding to a Jordan canonical form). Complex eigensolutions, arising in the case of real unsymmetric matrices, are discussed in a subsequent section. Note that whilst the QR method is capable of directly obtaining the eigensolution for a matrix, it is more effective if the matrix is reduced to the Hessenberg or the tridiagonal form prior to iterations.

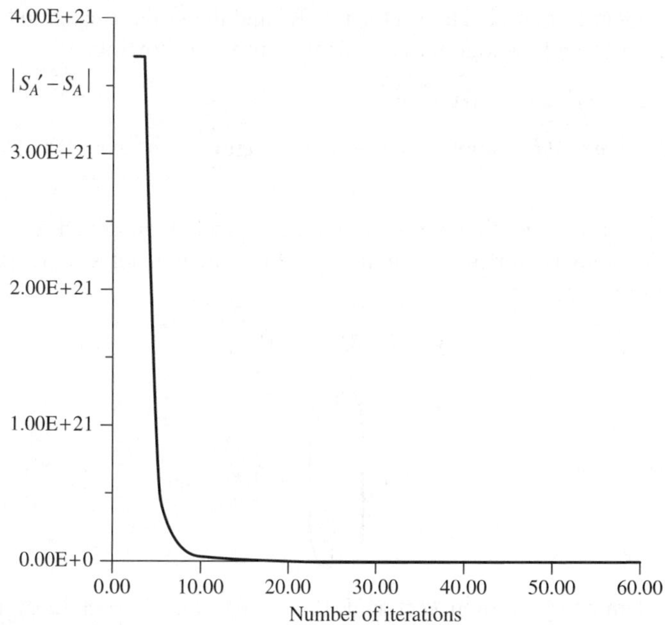

Figure 7.5 Cantilever in Figure 7.1; convergence of $\left|S_{A'} - S_A\right|$ (Equation 7.44) in Jacobi method

Table 7.7 Cantilever in Figure 7.1; convergence of eigenvalues by the Jacobi method, $\varepsilon_{\text{tol}} = 1.0\text{E}{-3}$

Mode number	At the end of 40 iterations	At the end of 50 iterations	At the end of 60 iterations
1	463.197	51.231	51.232
2	720.160	204.928	204.928
3	921.850	321.074	321.075
4	1074.609	899.219	899.219
5	1219.895	1284.301	1284.301
6	1850.835	1763.341	1763.341
7	2752.032	2919.495	2919.495
8	3489.229	3596.877	3596.877

7.3.2.1 Householder Transformation Method

For a real symmetric matrix, the Householder transformation results in a tridiagonal form and finally to a diagonal form by QR transformation. In this case, the Householder method uses a similarity transformation $\boldsymbol{P} \in \mathbb{R}^{n \times n}$ in the form:

$$\boldsymbol{P} = \boldsymbol{I} - \alpha \boldsymbol{v}\boldsymbol{v}^T, \alpha = \frac{2}{v^T v} \tag{7.52}$$

$\boldsymbol{I} \in \mathbb{R}^{n\times n}$ is the identity matrix. The vector $\boldsymbol{v} \in \mathbb{R}^n$ and the scalar α need to be determined. That $\boldsymbol{P}$ is symmetric and orthogonal is evident from the following:

$$\boldsymbol{P}^T = \boldsymbol{I} - \alpha(\boldsymbol{v}\boldsymbol{v}^T)^T = \boldsymbol{I} - \alpha\boldsymbol{v}\boldsymbol{v}^T = P$$
$$\boldsymbol{P}^T\boldsymbol{P} = (\boldsymbol{I} - \alpha\boldsymbol{v}\boldsymbol{v}^T)(\boldsymbol{I} - \alpha\boldsymbol{v}\boldsymbol{v}^T) = \boldsymbol{I} - \alpha\boldsymbol{v}\boldsymbol{v}^T - \alpha\boldsymbol{v}\boldsymbol{v}^T + \alpha^2\boldsymbol{v}(\boldsymbol{v}^T\boldsymbol{v})\boldsymbol{v}^T = \boldsymbol{I} = \boldsymbol{P}\boldsymbol{P}^T \tag{7.53}$$

The Householder transformation aims at finding $\boldsymbol{v}$ such that, given $\boldsymbol{x} \in \mathbb{R}^n$, we have $\boldsymbol{P}\boldsymbol{x} = \boldsymbol{y} \in \mathbb{R}^n$, another vector of a desired form. In particular, let $\boldsymbol{x} \neq \boldsymbol{y}$, $\|\boldsymbol{x}\| = \|\boldsymbol{y}\|$ and $\boldsymbol{y}$ to be of the form:

$$\boldsymbol{y} = (\Upsilon, 0, \ldots., 0)^T$$
$$= \Upsilon \begin{pmatrix} 1 \\ 0 \\ \cdot \\ \cdot \\ \cdot \\ 0 \end{pmatrix} = \Upsilon \boldsymbol{e}_1 \tag{7.54}$$

Here, $\boldsymbol{e}_1$ is an elementary column vector $(1, 0, \ldots., 0)^T$ and $\Upsilon = \pm \|\boldsymbol{x}\|$. Thus:

$$\left(\boldsymbol{I} - \alpha\boldsymbol{v}\boldsymbol{v}^T\right)\boldsymbol{x} = \Upsilon\boldsymbol{e}_1 \Rightarrow \boldsymbol{x} - \alpha\boldsymbol{v}\boldsymbol{v}^T\boldsymbol{x} = \Upsilon\boldsymbol{e}_1$$
$$\Rightarrow \boldsymbol{x} - \left(\alpha\boldsymbol{v}^T\boldsymbol{x}\right)\boldsymbol{v} = \Upsilon\boldsymbol{e}_1 \tag{7.55}$$

From the above equation, $\boldsymbol{v}$ is obtained as:

$$\boldsymbol{v} = \frac{1}{\alpha\boldsymbol{v}^T\boldsymbol{x}}\left(\boldsymbol{x} - \Upsilon\boldsymbol{e}_1\right) \tag{7.56}$$

The scalar factor $\frac{1}{\alpha\boldsymbol{v}^T\boldsymbol{x}}$ is cancelled out when the transformation matrix $\boldsymbol{P}$ is formed and thus we have:

$$\boldsymbol{v} = \left(\boldsymbol{x} - \Upsilon\boldsymbol{e}_1\right) \tag{7.57}$$

The uniqueness of $\boldsymbol{v}$ (except for the sign) follows from:

$$\left(\boldsymbol{I} - \alpha\boldsymbol{v}\boldsymbol{v}^T\right)\boldsymbol{x} = \left(\boldsymbol{I} - \alpha\boldsymbol{w}\boldsymbol{w}^T\right)\boldsymbol{x} \Rightarrow \boldsymbol{v}\left(\boldsymbol{v}^T\boldsymbol{x}\right) = \boldsymbol{w}\left(\boldsymbol{w}^T\boldsymbol{x}\right) \Rightarrow \boldsymbol{v} = \pm\boldsymbol{w} \tag{7.58}$$

To observe how the transformation works on a symmetric matrix $\boldsymbol{A}$, let us consider $\boldsymbol{x}$ to be the first column of $\boldsymbol{A}$, that is $\boldsymbol{x} = \left(a_{11}, a_{21}, \ldots\ldots., a_{n1}\right)^T$. This column is to be reduced to the vector y of the form $\left(a_{11}, y_1, 0, 0, \ldots\ldots.0\right)^T$. Since $\|\boldsymbol{x}\| = \|\boldsymbol{y}\|$, we have:

$$y_1^2 = \sum_{i=2}^{n} a_{i1}^2 \tag{7.59}$$

From Equation 7.57, the vector v is given by:

$$\boldsymbol{v} = \boldsymbol{x} - y = \left(0, a_{21} \pm \sqrt{\sum_{i=2}^{n} a_{i1}^2}, a_{31}, \ldots\ldots, a_{n1}\right)^T \tag{7.60}$$

The $\pm$ sign in the above equation is generally replaced by the sign of a_{21}. With $\boldsymbol{v}$ thus determined, the first transformation $\boldsymbol{P}^T\boldsymbol{A}\boldsymbol{P}$ results in:

$$A^{(1)} = P^T A P = \begin{bmatrix} a_{11} & \text{sign}(a_{21})y_1 & 0 & \ldots\ldots & 0 \\ \text{sign}(a_{21})y_1 & * & * & \ldots\ldots & * \\ 0 & * & * & \ldots\ldots & * \\ \ldots & \ldots & \ldots & \ldots & \ldots \\ 0 & * & * & \ldots\ldots & * \end{bmatrix} \tag{7.61}$$

By the above transformation, the elements below and above the first subdiagonal are zeroed and the elements in the submatrix, principal minor of dimension $n-1 \times n-1$, get altered, as expected of the transformation. The stars within the matrix above stand for nonzero elements in $A^{(1)}$. The transformations are continued $n-2$ times on the matrix A to zero the elements of the $n-2$ columns and rows below and above the first subdiagonal and finally reduce A to a tridiagonal form. Note that, as transformations progress, the first $r-1$ rows and columns of $A^{(r-1)}$ remain unaltered at the rth stage ($r \le n-2$). It is relevant to mention that P is known as a reflection matrix and its action on any vector is that the component of the vector normal to the plane of reflection gets its direction reversed and the component in the plane of reflection remains unaltered.

Example 7.4: The cantilever in Figure 7.1; Tridiagonal from by Householder transformations

The generalized eigenvalue problem $K\Phi_n = \lambda_n M\Phi_n$ of the cantilever is first transformed to a standard eigenvalue problem using Cholesky decomposition. Thus, according to Equation 7.8, the symmetric matrix $\hat{K}$ is generated and Householder transformations are carried out $n-2$ (= 38 in the present case) times to reduce the matrix to a tridiagonal form. The tridiagonal matrix is shown in Table 7.8 (with only some of its elements explicitly shown).

7.3.2.2 QR Transformation Method

In this method (Francis, 1961; Wilkinson, 1965), at any iteration, a matrix A (with real entries) is decomposed into a product of a unitary matrix Q and an upper triangular matrix R (all with real entries). This is followed by a similarity unitary transformation as given below:

$$A_k = Q_k R_k, \qquad A_{k+1} = Q_k^T A_k Q_k \tag{7.62}$$

Table 7.8 Tridiagonal matrix for cantilever in Figure 7.1; Householder transformations

	Col 1	Col 2	Col 3	Col 4	Col 38	Col 39	Col 40
Row 1	0.686E8	0.173E9	−0.20E−7	0.14E−7	0.0	0.0	–
Row 2	0.173E9	0.270E10	0.149E10	0.16E−6	0.0	0.0	–
Row 3	−0.20E−7	0.149E10	0.214E10	0.144E10	0.0	0.0	–
–	–	–	–	–	–	–	–
–	–	–	–	–	–	–	–
–	–	–	–	–	–	–	–
Row 39	0.0	0.0	0.0	0.0	−0.167E8	0.921E7	−0.525E6
Row 40	0.0	0.0	0.0	0.0	−0.19E−8	−0.525E6	0.154E6

Since $\boldsymbol{Q}_k^T \boldsymbol{Q}_k = \boldsymbol{I}$, the transformed matrix $\boldsymbol{A}_{k+1}$ can also be expressed as:

$$\boldsymbol{A}_{k+1} = \boldsymbol{Q}_k^T \left(\boldsymbol{Q}_k \boldsymbol{R}_k\right) \boldsymbol{Q}_k = \boldsymbol{R}_k \boldsymbol{Q}_k \tag{7.63}$$

If $\boldsymbol{A}_1 = \boldsymbol{A}$, then $\boldsymbol{A}_{k+1} = \boldsymbol{Q}_k^T \boldsymbol{Q}_{k-1}^T \ldots\ldots \boldsymbol{Q}_2^T \boldsymbol{Q}_1^T \boldsymbol{A} \boldsymbol{Q}_1 \boldsymbol{Q}_2 \ldots\ldots \boldsymbol{Q}_{k-1} \boldsymbol{Q}_k$. As the iterations progress, in the limit as $k \to \infty$, $\boldsymbol{A}_{k+1}$ tends to a (nearly) triangular matrix. For a real symmetric matrix, $\boldsymbol{A}_{k+1} \to \boldsymbol{\Lambda}$, the diagonal matrix containing the eigenvalues on the diagonal. Further, $\boldsymbol{Q}_1 \boldsymbol{Q}_2 \ldots\ldots \boldsymbol{Q}_{k-1} \boldsymbol{Q}_k \to \boldsymbol{\Phi}$, the associated eigenvector matrix.

Convergence of the QR method

Consider the nonunitary transformation of a symmetric matrix $\boldsymbol{B}$ using Cholesky decomposition at the kth iteration:

$$\boldsymbol{B}_k = \boldsymbol{L}_k \boldsymbol{U}_k = \boldsymbol{U}_k^T \boldsymbol{U}_k,\ \boldsymbol{B}'_k = \boldsymbol{B}_{k+1} = \boldsymbol{L}_k^{-1} \boldsymbol{B}_k \boldsymbol{L}_k = \boldsymbol{L}_k^{-1} \boldsymbol{L}_k \boldsymbol{U}_k \boldsymbol{L}_k = \boldsymbol{U}_k \boldsymbol{L}_k = \boldsymbol{U}_k \boldsymbol{U}_k^T \tag{7.64a,b}$$

The iterative process involving the above transformation is referred to as Cholesky iteration/*LR* method (Rutishauser, 1958). The convergence of the *QR* method is closely related to that of the *LR* method, which is clarified below. From Equations 7.62 and 7.63 we have:

$$\boldsymbol{A}_k^2 = \boldsymbol{A}_k^T \boldsymbol{A}_k = \boldsymbol{R}_k^T \boldsymbol{Q}_k^T \boldsymbol{Q}_k \boldsymbol{R}_k = \boldsymbol{R}_k^T \boldsymbol{R}_k,\ \boldsymbol{A}_{k+1}^2 = \boldsymbol{A}_{k+1} \boldsymbol{A}_{k+1}^T = \boldsymbol{R}_k \boldsymbol{Q}_k \boldsymbol{Q}_k^T \boldsymbol{R}_k^T = \boldsymbol{R}_k \boldsymbol{R}_k^T \tag{7.65a,b}$$

The above equation shows that whilst $\boldsymbol{A}_k$ and $\boldsymbol{A}_{k+1}$ are related by *QR* decomposition, $\boldsymbol{A}_k^2$ and $\boldsymbol{A}_{k+1}^2$ by Cholesky decomposition (similar to $\boldsymbol{B}_k$ and $\boldsymbol{B}_{k+1}$ in Equation 7.64. With $\mathcal{M}$ representing a set of positive definite real symmetric matrices, consider the square-root mapping, $f : \mathcal{M} \to \mathcal{M}$, $f(\boldsymbol{B}) = \sqrt{\boldsymbol{B}}$. If $\boldsymbol{B}$ is diagonal, so is $\sqrt{\boldsymbol{B}}$. If the convergence of the Cholesky iteration method is achieved by reducing a matrix to the diagonal form, it ensures the same for *QR* iterations via the square root mapping analogy. With this in view, it is proved in the following steps that Cholesky iterations reduce a symmetric matrix $\boldsymbol{B}$ (equivalently $\boldsymbol{A}^2$) to a diagonal matrix as $k \to \infty$.

For clarity, the matrix multiplications leading to $\boldsymbol{B}_k$ and $\boldsymbol{B}'_k$ (Equation 7.64) are shown below in the expanded form:

$$\boldsymbol{B}_k = \begin{bmatrix} U_{11} & 0 & \ldots & & \ldots \\ U_{12} & U_{22} & 0 & \ldots & \\ U_{13} & U_{23} & U_{33} & 0 & \ldots \\ \ldots & \ldots & \ldots & \ldots & \ldots \\ U_{1n} & U_{2n} & U_{3n} & \ldots & U_{nn} \end{bmatrix} \begin{bmatrix} U_{11} & U_{12} & \ldots & & & U_{1n} \\ 0 & U_{22} & U_{23} & \ldots & & U_{2n} \\ 0 & 0 & U_{33} & U_{34} & \ldots & U_{3n} \\ \ldots & \ldots & \ldots & \ldots & \ldots & \ldots \\ 0 & 0 & 0 & \ldots & 0 & U_{nn} \end{bmatrix}$$

$$\boldsymbol{B}'_k = \begin{bmatrix} U_{11} & U_{12} & \ldots & & & U_{1n} \\ 0 & U_{22} & U_{23} & \ldots & & U_{2n} \\ 0 & 0 & U_{33} & U_{34} & \ldots & U_{3n} \\ \ldots & \ldots & \ldots & \ldots & \ldots & \ldots \\ 0 & 0 & 0 & \ldots & 0 & U_{nn} \end{bmatrix} \begin{bmatrix} U_{11} & 0 & \ldots & & \ldots \\ U_{12} & U_{22} & 0 & \ldots & \\ U_{13} & U_{23} & U_{33} & 0 & \ldots \\ \ldots & \ldots & \ldots & \ldots & \ldots \\ U_{1n} & U_{2n} & U_{3n} & \ldots & U_{nn} \end{bmatrix} \tag{7.66}$$

From the above, typical elements in $\boldsymbol{B}_k$ and $\boldsymbol{B'}_k$ are written as:

$$B_{k,ij} = \sum_{l=1}^{i} U_{k,li} U_{k,lj}, \quad B'_{k,ij} = B_{k+1,ij} = \sum_{l=j}^{n} U_{k,il} U_{k,jl} \tag{7.67}$$

Let us form the sequences, S_{ii} and S'_{ii}, each member of which is a cumulative sum of the diagonal elements in $\boldsymbol{B}_k$ and $\boldsymbol{B}'_k$, respectively, (up to a certain i; no sum intended). Denoting the members of the two sequences as $S_{ii}^{(k)}$ and $S_{ii}^{\prime(k)}$ (respectively) for the kth iteration, we have:

$$S_{ii}^{(k)} = \sum_{m=1}^{i} B_{k,mm} \text{ and } S'^{(k)}_{ii} = \sum_{m=1}^{i} B'_{k,mm} \tag{7.68}$$

In terms of the elements in the upper triangular matrix, $\boldsymbol{U}_k$, these are expressible as:

$$S_{ii}^{(k)} = \sum_{m=1}^{i} \sum_{l=1}^{i} U_{k,lm}^2 \text{ and } S'^{(k)}_{ii} = \sum_{m=1}^{i} \sum_{l=m}^{n} U_{k,ml}^2 \tag{7.69}$$

Specifically, for $i = 1, 2$ and n , the two sequences are given by:

$$\begin{aligned}
S_{11} &= \left\{B_{1,11}, B_{2,11}, \ldots\ldots, B_{k,11}\right\} = \left\{U_{1,11}^2, U_{2,11}^2, \ldots\ldots\ldots, U_{k,11}^2\right\} \\
S_{22} &= \left\{(B_{1,11} + B_{1,22}), (B_{2,11} + B_{2,22}), \ldots\ldots (B_{k,11} + B_{k,22})\right\} \\
&= \left\{\sum_{i=1}^{2}\sum_{l=1}^{i} U_{1,li}^2, \sum_{i=1}^{2}\sum_{l=1}^{i} U_{2,li}^2, \ldots\ldots, \sum_{i=1}^{2}\sum_{l=1}^{i} U_{k,li}^2\right\} \\
S_{nn} &= \left\{\operatorname{tr}\left(\boldsymbol{B}^{(1)}\right), \operatorname{tr}\left(\boldsymbol{B}^{(2)}\right), \ldots\ldots\ldots, \operatorname{tr}\left(\boldsymbol{B}^{(k)}\right)\right\} \\
&= \left\{\sum_{i=1}^{n}\sum_{l=1}^{i} U_{1,li}^2, \sum_{i=1}^{n}\sum_{l=1}^{i} U_{2,li}^2, \ldots\ldots\ldots\ldots\ldots, \sum_{i=1}^{n}\sum_{l=1}^{i} U_{k,li}^2\right\}
\end{aligned} \tag{7.70}$$

$$\begin{aligned}
S'_{11} &= \left\{B'_{1,11}, B'_{2,11}, \ldots\ldots, B'_{k,11}\right\} = \left\{\sum_{l=1}^{n} U_{1,1l}^2, \sum_{l=1}^{n} U_{2,1l}^2, \ldots\ldots\ldots, \sum_{l=1}^{n} U_{k,1l}^2\right\} \\
S'_{22} &= \left\{\left(B'_{1,11} + B'_{1,22}\right), \left(B'_{2,11} + B'_{2,22}\right), \ldots\ldots \left(B'_{k,11} + B'_{k,22}\right)\right\} \\
&= \left\{\left(\sum_{i=i}^{2}\sum_{l=i}^{n} U_{1,il}^2\right), \left(\sum_{i=i}^{2}\sum_{l=i}^{n} U_{2,il}^2\right), \ldots\ldots, \left(\sum_{i=i}^{2}\sum_{l=i}^{n} U_{k,il}^2\right)\right\} \\
S'_{nn} &= \left\{\operatorname{tr}\left(\boldsymbol{B}'^{(1)}\right), \operatorname{tr}\left(\boldsymbol{B}'^{(2)}\right), \ldots\ldots\ldots, \operatorname{tr}\left(\boldsymbol{B}'^{(k)}\right)\right\} \\
&= \left\{\sum_{i=i}^{n}\sum_{l=i}^{n} U_{1,il}^2, \left(\sum_{i=i}^{n}\sum_{l=i}^{n} U_{2,il}^2\right) \ldots\ldots\ldots, \left(\sum_{i=i}^{n}\sum_{l=i}^{n} U_{k,il}^2\right)\right\}
\end{aligned} \tag{7.71}$$

In the above equations, $\operatorname{tr}(\cdot)$ stands for the trace of a square matrix. Equations 7.70 and 7.71 indicate that

$$S'^{(k)}_{ii} - S^{(k)}_{ii} \geq 0, \ \forall k, i \tag{7.72}$$

and that

$$S^{(k+1)}_{ii} = S'^{(k)}_{ii}, \ \forall k, i \tag{7.73}$$

The equality sign in Equation 7.72 applies for $i = n$ at any iteration $k \geq 1$. In this case, each member of the sequences S_{nn} and S'_{nn} is identical and equal to the trace of $\boldsymbol{B}$ or $\boldsymbol{B}'$ (A_k^2 or $\boldsymbol{A}_{k+1}^2$), which is invariant under the similarity transformation. Owing to the invariance property and orthogonal nature of the transformations involved, the sequences S_{ii} and S'_{ii} remain bounded. Thus, as iterations progress, the diagonal terms of $\boldsymbol{B}$ or $\boldsymbol{B}'$ increase in their modulus at the expense of the off-diagonal terms and tend to a limit as $k \to \infty$. $\boldsymbol{B}_k$ (equivalently $\boldsymbol{A}_k^2$) finally tends to a diagonal matrix with the diagonal terms equalling the squares of its eigenvalues, $\lambda_i, i = 1, 2, \ldots, n$. With the convergence of $\boldsymbol{A}_k^2$ thus established, the convergence of $\boldsymbol{A}_k$ via the QR transformation follows from the square-root mapping, $f(\boldsymbol{B})$ which maps diagonal matrices into diagonal matrices.

Example 7.5: The cantilever in Figure 7.1 and eigensolution by* QR *transformation method

The standard eigenvalue problem corresponding to the cantilever example is derived in Example 7.4. The eigensolution is obtained by the QR method by transforming the equivalent stiffness matrix $\hat{\boldsymbol{K}}$ (in Equation 7.8) to a diagonal matrix. Table 7.9 shows the first 12 eigenvalues obtained from the method. A tolerance of 1.0E–6 is used to check the convergence.

Implementation Issues with the QR Method

The QR method is more effective if used along with Householder transformations that first reduce the original system matrix into a tridiagonal form. Further, it is possible to accelerate the convergence of the QR method by using the shifting strategy. Specifically, with a shift σ, the steps involved in Equations 7.62 and 7.63 take the form:

$$\begin{aligned} \boldsymbol{A}_k - \sigma \boldsymbol{I} &= \boldsymbol{Q}_k \boldsymbol{R}_k \\ \boldsymbol{A}_{k+1} &= \boldsymbol{Q}_k^T \boldsymbol{A}_k \boldsymbol{Q}_k \\ &= \boldsymbol{Q}_k^T \left(\boldsymbol{Q}_k \boldsymbol{R}_k + \sigma I \right) \boldsymbol{Q}_k \\ &= \boldsymbol{R}_k \boldsymbol{Q}_k + \sigma \boldsymbol{I} \end{aligned} \tag{7.74a,b}$$

Table 7.9 Eigensolution for cantilever in Figure 7.1; QR transformations

Mode number	At the end of first iteration	At the end of 5 iterations	At the end of 10 iterations	At the end of 22 iterations
1	0.6192E+02	0.5123E+02	0.5123E+02	51.232
2	0.2607E+03	0.2049E+03	0.2049E+03	204.928
3	0.1559E+04	0.3211E+03	0.3211E+03	321.075
4	0.1568E+04	0.8992E+03	0.8992E+03	899.219
5	0.1881E+04	0.1284E+04	0.1284E+04	1284.301
6	0.5450E+04	0.1764E+04	0.1763E+04	1763.341
7	0.5989E+04	0.2931E+04	0.2919E+04	2919.495
8	0.6717E+04	0.3597E+04	0.3597E+04	3596.877
9	0.7312E+04	0.4413E+04	0.4374E+04	4373.736
10	0.8100E+04	0.6191E+04	0.6137E+04	6136.563
11	0.8698E+04	0.7056E+04	0.7053E+04	7053.365
12	0.1253E+05	0.8182E+04	0.8221E+04	8221.303

7.4 Subspace Iteration

Subspace iteration is a simultaneous vector iteration method, a generalisation of iteration methods discussed above. The method is computationally efficient in extracting the lower-order eigenvalues of large systems (even those with millions of degrees of freedom). It enables one to obtain several eigenpairs simultaneously by iteration. To extract the first $m(\ll n)$ eigenvalues and eigenvectors, the iteration starts with (instead of a single vector as in inverse and power iterations) a subspace $E_0 \subset \boldsymbol{V}$ containing a set of linearly independent vectors (basis vectors), $\boldsymbol{X}_0$, spanning E_0. Further iterations recursively build a sequence of basis sets, X_k, that, respectively, span subspaces, $E_k, k = 1, 2\ldots..$ As, for instance, with the inversion iteration scheme, these subspaces (which are presently not one-dimensional) approach the desired invariant subspace E_∞ spanned by the first m eigenvectors $\boldsymbol{\Phi}_i, i = 1, 2 \ldots m$. Note that a subspace $E \subset \boldsymbol{V}$ is invariant under a linear transformation $\boldsymbol{T}$, if $\boldsymbol{Tx} \in E, \forall \boldsymbol{x} \in E$. The transformation $\boldsymbol{T}$ in the present context corresponds to the standard eigenproblem matrix operator $\boldsymbol{A}$ or the similar operator in Equation 7.8 describing the generalized eigenvalue problem.

Table 7.10 details the computational steps involved in the subspace iteration algorithm. Equation 7.75 is an inverse iteration step, starting with the basis set $\boldsymbol{X}_k$ at each iteration k. The new (improved) basis set, $\bar{\boldsymbol{X}}_{k+1}$, is orthonormalised in Equation 7.78 before starting the next iteration. The orthonormalisation is needed to guide each element vector in $\bar{\boldsymbol{X}}_{k+1}$ to converge to a different eigenvector, $\boldsymbol{\Phi}_i, i = 1, 2 \ldots m$, thereby avoiding convergence towards the first one only. If each basis set $\boldsymbol{X}_k$ is treated as a set of trial mode shapes in the discrete form, the reduced eigenvalue problem in Equation 7.77 is reminiscent of

Table 7.10 Subspace Iteration method and its algorithm

Assume the starting subspace: $E_0 = \boldsymbol{X}_0, n \times r$, with $\boldsymbol{X}_0^T \boldsymbol{M} \boldsymbol{X}_0 = \boldsymbol{I}$
Start iterations $k = 1, 2 \ldots.$
Solve

$$\boldsymbol{K}\bar{\boldsymbol{X}}_{k+1} = \boldsymbol{M}\boldsymbol{X}_k \tag{7.75}$$

Project $\boldsymbol{K}$ and $\boldsymbol{M}$ matrices on to the subspace of dimension $= r$:

$$\boldsymbol{K}_{k+1} = \bar{\boldsymbol{X}}_{k+1}^T \boldsymbol{K} \bar{\boldsymbol{X}}_{k+1}\ \boldsymbol{M}_{k+1} = \bar{\boldsymbol{X}}_{k+1}^T \boldsymbol{M} \bar{\boldsymbol{X}}_{k+1} \tag{7.76}$$

Solve reduced-order eigenvalue problem:

$$\boldsymbol{K}_{k+1}\boldsymbol{\psi}_{k+1} = \boldsymbol{M}_{k+1}\boldsymbol{\psi}_{k+1}\boldsymbol{\Lambda}_{k+1}, \boldsymbol{\Lambda}_{k+1} = \mathrm{diag}\,[\lambda_j^{(k+1)}] \tag{7.77}$$

Generate improved eigenvector matrix and mass-orthonormalisation:

$$\boldsymbol{X}_{k+1} = \bar{\boldsymbol{X}}_{k+1}\boldsymbol{\psi}_{k+1},\ \boldsymbol{X}_{k+1}^T \boldsymbol{M} \boldsymbol{X}_{k+1} = \boldsymbol{I} \tag{7.78}$$

Convergence test on eigenvalues:

$$\max \left| \frac{\lambda_j^{(k+1)} - \lambda_j^{(k)}}{\lambda_j^{(k+1)}} \right| \le \varepsilon_{\mathrm{tol}}, j = 1, 2 \ldots m \tag{7.79}$$

Perform Sturm sequence check:
End

the Rayleigh–Ritz method (Chapters 2 and 5). The generalized Jacobi method can be conveniently used to solve the reduced eigenvalue problem.

It is necessary to have more vectors ('r' in Table 7.10) than 'm' in each $\boldsymbol{X}_k$ to maintain reasonable accuracy in extracting all the desired eigenpairs, especially those towards the end of the spectrum. However, this dimensional augmentation of the basis set may in turn affect the efficacy of the iterative process, which now involves more arithmetic operations. In this respect, a suitable choice could be either $r = m + 8$ or $r = 2m$ (as typically used with some of the commercially available eigenvalue solvers).

7.4.1 Convergence in Subspace Iteration

The convergence check (Equation 7.79) is based on the maximum relative difference of the approximated eigenvalues obtained over two successive iterations, k and $k + 1$. Note that the iterative process in subspace iteration is meant to converge in the limit to the entire eigenspace, $E_\infty (= span\{\boldsymbol{\Phi}_1, \boldsymbol{\Phi}_2, \ldots\ldots, \boldsymbol{\Phi}_m\})$. This depends largely on two factors. The first is the choice of the starting vectors (see Geradin and Rixen, 1994, Bathe, 1996) in $\boldsymbol{X}_0$ such that they are not orthogonal to any element in E_∞. The other is the orthonormalisation of the approximated eigenvectors at every iteration. Since, during the iteration process each of the vectors in $\boldsymbol{X}_{k+1}$ (Equation 7.78) attempts to converge towards the corresponding true eigenvector, the rate of convergence is the same as that of the inverse iteration method. However, the global convergence rate of a vector $\boldsymbol{X}_{j,k+1}$ in E_{k+1} is equal to $\frac{\lambda_j}{\lambda_{m+1}}$. The last is the ratio of the jth eigenvalue to the one immediately outside the subspace. If the number of initial vectors is taken to be $r > m$, the rate of convergence is $\frac{\lambda_j}{\lambda_{r+1}}$ in which case there is an improvement in the convergence at the cost of an additional overhead in arithmetic operations.

Example 7.6: The cantilever in Figure 7.1; eigensolution by subspace iteration
The eigensolution is sought for $m = 12$. The starting vector $\boldsymbol{X}_0$ is chosen to contain $r = m + 8$ vectors and thus is of size 40×20. Keeping in view the standard practice (Bathe, 1996), $\boldsymbol{X}_0$ is selected such that the degrees of freedom associated with small stiffness and large mass are predominantly represented. Table 7.11 gives the convergence

Table 7.11 Eigensolution by subspace iteration method and convergence

Iteration number	Natural frequencies in rad/s					
	ω_1	ω_2	ω_3	ω_4	ω_5	ω_6
1	51.2319	204.928	321.075	899.220	1284.750	1763.360
2	52.2319	204.928	321.075	899.219	1284.300	1763.340
3	–	–	–	899.219	1284.300	1763.340
Iteration number	ω_7	ω_8	ω_9	ω_{10}	ω_{11}	ω_{12}
1	2919.620	3603.910	4374.910	6145.460	7152.400	8253.130
2	2919.500	3596.880	4373.740	6136.580	7053.590	8221.830
3	2919.500	3596.880	4373.740	6136.560	7053.370	8221.310
4	–	–	–	6136.560	7053.360	8221.300
5	–	–	–	–	7053.360	8221.300

of the first 12 eigenvalues using the subspace iteration method. The convergence tolerance ε_{tol} is taken as 10^{-6}.

As seen from the table, the global convergence of the initial subspace to E_∞ (spanned by the first 12 eigenvectors) is reached within five iterations.

7.5 Lanczos Transformation Method

The Lanczos method transforms the system matrix $\boldsymbol{A}$ to a tridiagonal matrix that is in turn solved by methods such as Jacobi or QR. Considering the generalized eigenvalue problem $\boldsymbol{K\Phi} = \lambda\boldsymbol{M\Phi}$ (suffix dropped in the vector $\boldsymbol{\Phi}$ for convenience) the principle underlying the transformation is as follows.

The generalized eigenvalue problem is rewritten as:

$$\boldsymbol{MK}^{-1}\boldsymbol{M\Phi} = {}^{1}\!/\!_{\lambda}\boldsymbol{M\Phi} \tag{7.80}$$

Using the Lanczos transformation, $\boldsymbol{\Phi} = \boldsymbol{X\varphi}$ with $\boldsymbol{X}^T\boldsymbol{MX} = \boldsymbol{I}$ and φ a vector unknown at this stage, Equation 7.80 takes the form:

$$\boldsymbol{X}^T\boldsymbol{MK}^{-1}\boldsymbol{MX\varphi} = ({}^{1}\!/\!_{\lambda})\,\boldsymbol{\varphi}$$

$$\Rightarrow \boldsymbol{T\varphi} = \mu\boldsymbol{\varphi} \text{ with } \boldsymbol{T} = \boldsymbol{X}^T\boldsymbol{MK}^{-1}\boldsymbol{MX} \tag{7.81}$$

$\boldsymbol{X} = \{\boldsymbol{x}_1, \boldsymbol{x}_2, \ldots\ldots\boldsymbol{x}_n\}$ is known as the matrix of Lanczos vectors, with each $\boldsymbol{x}_j \in \boldsymbol{R}^n$. Here, $\boldsymbol{\varphi}$ is recognised as the eigenvector of $\boldsymbol{T}$ associated with the eigenvalue μ. Equation 7.81 reveals an orthogonal transformation that reduces the generalized eigenvalue problem to a standard one $\boldsymbol{T\varphi} = \mu\boldsymbol{\varphi}$. In particular, if the resulting $\boldsymbol{T}$ matrix is to attain a tridiagonal form, it is interesting to determine the implications of the equality in Equation 7.81. To this end, let us define $\boldsymbol{T}$ as:

$$\boldsymbol{T} = \begin{bmatrix} \alpha_1 & \beta_2 & 0 & . & . & . & . & . & . & . \\ \beta_2 & \alpha_2 & \beta_3 & . & . & . & . & . & . & . \\ 0 & \beta_3 & \alpha_3 & . & . & . & . & . & . & . \\ & & & & \cdots\cdot & & & & & \\ 0 & 0 & 0 & & \beta_{n-1} & & \alpha_{n-1} & & \beta_n & \\ 0 & 0 & 0 & & \cdots\cdots & & \beta_n & & \alpha_n & \end{bmatrix} \tag{7.82}$$

α_j and β_j are scalars to be determined during the transformation. Starting with an arbitrary vector $\boldsymbol{x}_0 \in \boldsymbol{R}^n$, we generate the Lanczos vectors sequentially as follows:

$$\boldsymbol{x}_1 = {}^{\boldsymbol{x}_0}\!/\!_{\boldsymbol{x}_0^T\boldsymbol{Mx}_0} \tag{7.83}$$

The above equation involves mass orthogonalization of $\boldsymbol{x}_0$. With $\beta_1 = 0$, for any $\boldsymbol{x}_j$, $j = 2, \ldots n+1$, we have:

$$K\bar{\boldsymbol{x}}_j = \boldsymbol{Mx}_{j-1} \tag{7.84}$$

$$\hat{\boldsymbol{x}}_j = \bar{\boldsymbol{x}}_j - \alpha_{j-1}\boldsymbol{x}_{j-1} - \beta_{j-1}\boldsymbol{x}_{j-2} \tag{7.85}$$

Here, $\alpha_{j-1} = \bar{\boldsymbol{x}}_j^T\boldsymbol{Mx}_{j-1}$. The vector $\hat{\boldsymbol{x}}_j$ is normalised with respect to $\boldsymbol{M}$ to obtain:

$$\boldsymbol{x}_j = \beta_j^{-1}\hat{\boldsymbol{x}}_j, \text{ with } \beta_j = \left(\hat{\boldsymbol{x}}_j^T\boldsymbol{M}\hat{\boldsymbol{x}}_j\right)^{1/2} \tag{7.86}$$

Combining Equations 7.84 and 7.85, we obtain the three-parameter recursive relationship:

$$K^{-1}Mx_{j-1} = \alpha_{j-1}x_{j-1} + \beta_{j-1}x_{j-2} + \beta_j x_j \tag{7.87}$$

The above equation is rewritten in the matrix form as:

$$K^{-1}M\left\{x_1, x_2, \ldots, x_{j-1}, \ldots .x_n\right\} = \left\{x_1, x_2, \ldots \ldots .x_n\right\}$$

$$\times \begin{bmatrix} \alpha_1 & \beta_2 & 0 & \ldots\ldots\ldots & & \\ \beta_2 & \alpha_2 & . & \ldots\ldots\ldots & & \\ 0 & \beta_3 & \alpha_3 & \ldots\ldots\ldots & & \\ \ldots.. & & & \beta_{n-1} & 0 & \\ 0 & 0 & 0 & \beta_{n-1} & \alpha_{n-1} & \beta_n \\ 0 & 0 & 0 & \ldots\ldots & \beta_n & \alpha_n \end{bmatrix} + [\mathbf{0}\ \mathbf{0}\ \ldots\ldots\ldots, \beta_{n+1}x_{n+1}] \tag{7.88}$$

where $\mathbf{0} \in \boldsymbol{R}^n$ is the zero vector. Equation 7.88 can be expressed in matrix form as:

$$K^{-1}MX = XT + \beta_{n+1}x_{n+1}e_n^T \tag{7.89}$$

where $e_n^T = \{0, 0, \ldots\ldots\ldots, 1\}$ is a vector with only the nth element being nonzero and equal to unity. Premultiplying the above equation by $X^T M$ gives:

$$X^T MK^{-1}MX = X^T MXT + \beta_{n+1}X^T Mx_{n+1}e_n^T \tag{7.90}$$

From the orthogonality of the Lanczos vectors contained in the recursive relationship (Equation 7.87), we have $X^T MX = I$ and $X^T Mx_{j+1} = \mathbf{0}$. The equation above shows the orthogonality transformation involved in the Lanczos method (originally stated in Equation 7.81), as clarified below:

$$X^T MK^{-1}MX = T \Rightarrow Q^T K^{-1}Q = T \Rightarrow K^{-1}Q = QT \text{ with } Q = MX \tag{7.91}$$

Note that for $j = n + 1$, $\hat{x}_j$ from Equation 7.85 becomes zero, consistent with the fact that the Euclidean vector space $\mathbb{R}^n$ is spanned by the column vectors of $X \in \mathbb{R}^{n\times n}$ and no more vectors exist that are mass-orthogonal to the column vectors of X. The eigenvalues of $K\Phi = \lambda M\Phi$ are given by the reciprocals of the eigenvalues of the matrix T and the eigenvectors by $X\varphi$.

The Lanczos method suffers from the shortcoming that the orthogonality of the vectors x_j, as implied in the recursive relationship (Equation 7.87), is lost due to round-off errors during computations. With this loss of orthogonality, some eigenvalues may get spuriously repeated, thereby adversely affecting the otherwise elegant scheme of Lanczos transformations. It is specifically ineffective for the tridiagonalisation of the original system, which is probably a large-dimensional one. However, the method is more effective than those discussed earlier in solving for a fairly large number of eigenpairs (λ, Φ), say m of them so that one still has $m \ll n$. In this context, it is customary to generate more than m (say $2m$) Lanczos vectors and thus construct the matrix $\in \mathbb{R}^{2m\times 2m}$. This eventually helps in obtaining a better set of the first m eigenpairs (with the eigenvalues arranged in the ascending order). In addition, the Gram–Schmidt orthogonalization scheme is incorporated into the recursive strategy such that the Lanczos vector x_j is orthogonal to all the previously generated vectors $\{x_{j-1}\}$ $j \leq n + 1$. The Lanczos transformations involved in Equations 7.83–7.87 comprises of one Lanczos stage. Due to the loss of orthogonality and the eventual approximations in the computed eigenvalues and eigenvectors at the end of a Lanczos stage, it may be necessary to take recourse to multiple stages/iterations for convergence. For a check on the convergence of the eigensolution, the error involved must be suitably quantified.

7.5.1 Lanczos Method and Error Analysis

Assume that the first m eigenpairs have already been extracted by the Lanczos transformation. With Cholesky decomposition of the mass matrix, $\boldsymbol{M} = \boldsymbol{L}\boldsymbol{L}^T$, Equation 7.80 can be recast as a standard eigenvalue problem as:

$$\boldsymbol{L}^T\boldsymbol{K}^{-1}\boldsymbol{L}\boldsymbol{\psi} = \frac{1}{\lambda}\boldsymbol{\psi} \text{ with } \boldsymbol{\psi} = \boldsymbol{L}^T\boldsymbol{\Phi} \tag{7.92}$$

If $\bar{\boldsymbol{\Phi}}_j$ denotes the jth computed eigenvector, letting $\bar{\boldsymbol{\psi}}_j = \boldsymbol{L}^T\bar{\boldsymbol{\Phi}}_j$ (with overbars indicating computed quantities) the error is given by:

$$\begin{aligned}
\|\boldsymbol{r}_j\| &= \left\|\boldsymbol{L}^T\boldsymbol{K}^{-1}\boldsymbol{L}\bar{\boldsymbol{\psi}}_j - \mu_j\bar{\boldsymbol{\psi}}_j\right\| \left(\text{since } \frac{1}{\lambda_j} = \mu_j\right) \\
&= \left\|\boldsymbol{L}^T\boldsymbol{K}^{-1}\boldsymbol{L}\boldsymbol{L}^T\bar{\boldsymbol{\Phi}}_j - \mu_j\boldsymbol{L}^T\bar{\boldsymbol{\Phi}}_j\right\| \\
&= \left\|\boldsymbol{L}^T\boldsymbol{K}^{-1}\boldsymbol{L}\boldsymbol{L}^T\boldsymbol{X}\bar{\boldsymbol{\varphi}}_j - \mu_j\boldsymbol{L}^T\boldsymbol{X}\bar{\boldsymbol{\varphi}}_j\right\| \\
&= \left\|\boldsymbol{L}^T\left(\boldsymbol{K}^{-1}\boldsymbol{M}\boldsymbol{X} - \mu_j\boldsymbol{X}\right)\bar{\boldsymbol{\varphi}}_j\right\|
\end{aligned} \tag{7.93}$$

From the eigenvalue problem $\boldsymbol{T}\boldsymbol{\varphi}_j = \mu_j\boldsymbol{\varphi}_j$ for the tridiagonal matrix $\boldsymbol{T}$, we have $\boldsymbol{X}\boldsymbol{T}\boldsymbol{\varphi}_j = \mu_j\boldsymbol{X}\boldsymbol{\varphi}_j$ and thus Equation 7.93 leads to

$$\|\boldsymbol{r}_j\| = \left\|\boldsymbol{L}^T\left(\boldsymbol{K}^{-1}\boldsymbol{M}\boldsymbol{X} - \boldsymbol{X}\boldsymbol{T}\right)\bar{\boldsymbol{\varphi}}_j\right\| \tag{7.94}$$

From Equation 7.89, we have $\boldsymbol{K}^{-1}\boldsymbol{M}\boldsymbol{X} - \boldsymbol{X}\boldsymbol{T} = \beta_{m+1}\boldsymbol{x}_{m+1}\boldsymbol{e}_m^T$ and thus the error norm is given by

$$\|\boldsymbol{r}_j\| = \left\|\boldsymbol{L}^T\beta_{m+1}\boldsymbol{x}_{m+1}\boldsymbol{e}_m^T\bar{\boldsymbol{\varphi}}_j\right\| \tag{7.95}$$

Now $\boldsymbol{X}^T\boldsymbol{M}\boldsymbol{X} = \boldsymbol{I} \Rightarrow \boldsymbol{X}^T\boldsymbol{L}\boldsymbol{L}^T\boldsymbol{X} = \boldsymbol{I} \Rightarrow \left(\boldsymbol{L}^T\boldsymbol{X}\right)^T\boldsymbol{L}^T\boldsymbol{X} = \boldsymbol{I} \Rightarrow \left\|\boldsymbol{L}^T\boldsymbol{x}_{m+1}\right\| = 1$. Also $\boldsymbol{e}_m^T\bar{\boldsymbol{\varphi}}_j = \bar{\varphi}_{mj}$, the mth element of the jth computed eigenvector. Hence

$$\|\boldsymbol{r}_j\| \le \left\|\beta_{m+1}\bar{\varphi}_{m,j}\right\| \tag{7.96}$$

On the other hand, the error is also given by

$$\boldsymbol{r}_j = \boldsymbol{T}\bar{\boldsymbol{\varphi}}_j - \bar{\mu}\bar{\boldsymbol{\varphi}}_j \tag{7.97}$$

If $\boldsymbol{\Sigma}$ is the diagonal matrix containing the eigenvalues of $\boldsymbol{T}$ on its diagonal, we have $\boldsymbol{T} = \boldsymbol{\varphi}\boldsymbol{\Sigma}\boldsymbol{\varphi}^T$ (here $\boldsymbol{\varphi}$ denotes the matrix of eigenvectors of $\boldsymbol{T}$). Then we have, from Equation 7.97:

$$\begin{aligned}
\boldsymbol{r}_j &= \boldsymbol{\varphi}\boldsymbol{\Sigma}\boldsymbol{\varphi}^T\bar{\boldsymbol{\varphi}}_j - \bar{\mu}\boldsymbol{\varphi}\boldsymbol{\varphi}^T\bar{\boldsymbol{\varphi}}_j \text{ (since } \boldsymbol{\varphi}\boldsymbol{\varphi}^T = \boldsymbol{I}) \\
&\Rightarrow \boldsymbol{r}_j = \boldsymbol{\varphi}\left(\boldsymbol{\Sigma} - \bar{\mu}\boldsymbol{I}\right)\boldsymbol{\varphi}^T\bar{\boldsymbol{\varphi}}_j \\
&\Rightarrow \boldsymbol{\varphi}^T\boldsymbol{r}_j = \left(\boldsymbol{\Sigma} - \bar{\mu}\boldsymbol{I}\right)\boldsymbol{\varphi}^T\bar{\boldsymbol{\varphi}}_j \text{ (since } \boldsymbol{\varphi}\boldsymbol{\varphi}^T = \boldsymbol{I})
\end{aligned} \tag{7.98}$$

$$\Rightarrow (\boldsymbol{\Sigma} - \bar{\mu}\boldsymbol{I})^{-1}\boldsymbol{\varphi}^T\boldsymbol{r}_j = \boldsymbol{\varphi}^T\bar{\boldsymbol{\varphi}}$$
$$\Rightarrow \boldsymbol{\varphi}(\boldsymbol{\Sigma} - \bar{\mu}\boldsymbol{I})^{-1}\boldsymbol{\varphi}^T\boldsymbol{r}_j = \bar{\boldsymbol{\varphi}}_j$$
$$\Rightarrow \left\|\boldsymbol{\varphi}(\boldsymbol{\Sigma} - \bar{\mu}\boldsymbol{I})^{-1}\boldsymbol{\varphi}^T\boldsymbol{r}_j\right\| = \left\|\bar{\boldsymbol{\varphi}}_j\right\|$$
$$\Rightarrow \left\|\bar{\boldsymbol{\varphi}}_j\right\| \le \|\boldsymbol{\varphi}\| \left\|(\boldsymbol{\Sigma} - \bar{\mu}\boldsymbol{I})^{-1}\right\| \left\|\boldsymbol{\varphi}^T\right\| \left\|\boldsymbol{r}_j\right\| \tag{7.99}$$

Given that the eigenvector can be normalised so that $\|\bar{\boldsymbol{\varphi}}_j\| = 1$ and that $\|\boldsymbol{\varphi}\boldsymbol{\varphi}^T\| = 1$, the above equation becomes

$$1 \le \|(\boldsymbol{\Sigma} - \bar{\mu}\boldsymbol{I})^{-1}\|\|\boldsymbol{r}_j\| \tag{7.100}$$

Since $\left\|(\boldsymbol{\Sigma} - \bar{\mu}I)^{-1}\right\| = \max \frac{1}{|\mu_j - \bar{\mu}|}$, we have, for some j:

$$\min\left|\mu_j - \bar{\mu}\right| \le \left\|\boldsymbol{r}_j\right\| \tag{7.101}$$

The above equation gives the accuracy with which the computed eigenvalue $\bar{\mu}$ approaches an eigenvalue μ_j. Combining Equations 7.96 and 7.101, we finally have

$$\min\left|\frac{1}{\lambda_j} - \bar{\mu}\right| \le \left\|\beta_{m+1}\bar{\varphi}_{mj}\right\| \tag{7.102}$$

The fact that a true system eigenvalue λ_j is unknown *a priori* shows that the above equation may not be practicable to check the convergence of eigenvalues. However, since the error norm $\left\|\beta_{m+1}\bar{\varphi}_{mj}\right\|$ is available during the Lanczos stages, the check is made based on the following criterion with a specified tolerance ε_{tol}:

$$\left\|\beta_{m+1}\bar{\varphi}_{m,j}\right\| \le \varepsilon_{\text{tol}} \tag{7.103}$$

Equation 7.102 helps in obtaining the bounds on λ_j, which is 'close' to the computed value $1/\bar{\mu}$ in the form:

$$\bar{\mu} - \left\|\beta_{m+1}\bar{\varphi}_{mj}\right\| \le \frac{1}{\lambda_j} \le \bar{\mu} + \left\|\beta_{m+1}\bar{\varphi}_{mj}\right\| \tag{7.104}$$

A Sturm sequence check can be performed at the end of each Lanczos iteration (one stage) using the above bounds on λ_j as shifts to see if all the desired eigenvalues are realised or not. Table 7.12 details the computational steps involved in the Lanczos algorithm.

Example 7.7: The cantilever in Figure 7.1 – eigensolution by Lanczos method
The eigensolution is obtained with $m = 12$ and with $r = 2 \times m = 24$. The starting vector $\boldsymbol{x}_0$ is randomly generated. Table 7.13 gives the convergence of the first 12 eigenvalues in terms of $\beta_{m+1}\bar{\varphi}_{m,j}$, $j = 1, 2, .., 12$ (Equation 7.103). The convergence tolerance ε_{tol} is taken as 10^{-8}. The convergence for the 12 eigenvalues of the present example is achieved only in one iteration.

With increasing demand for eigensolutions of large-sized engineering systems, several new variants of the Lanczos method, for example the block Lanczos and the shifted block Lanczos have been developed (see Bibliography). The Lanczos method described in Table 7.12 uses a single vector $\boldsymbol{x}_k$ at each iteration, in generating the tridiagonal matrix, $\boldsymbol{T}$. A computationally faster variant is provided by the block Lanczos method in which

Table 7.12 The algorithm for the Lanczos transformation method

Aim: To obtain the lowest m eigenpairs of the Generalized eigenvalue problem: $\boldsymbol{KX} = \lambda \boldsymbol{MX}$

Step 1. Start of a Lanczos stage with a randomly generated vector $\boldsymbol{x}_1$, with $\boldsymbol{x}_1^T \boldsymbol{M} \boldsymbol{x}_1 = 1$

Step 2. Start of Lanczos transformation
For $j = 2, 3 \ldots\ldots, r\ (= 2 \times m)$, Set $\beta_1 = 0$

Step 3. Solve

$$\boldsymbol{K}\bar{\boldsymbol{x}}_j = \boldsymbol{M}\boldsymbol{x}_{j-1} \tag{7.105}$$

Step 4. Find the parameter,

$$\alpha_j = \bar{\boldsymbol{x}}_j^T \boldsymbol{M} \boldsymbol{x}_{j-1} \tag{7.106}$$

Step 5. Compute

$$\hat{\boldsymbol{x}}_j = \boldsymbol{K}^{-1}\boldsymbol{M}\boldsymbol{x}_{j-1} - \alpha_{j-1}\boldsymbol{x}_{j-1} - \beta_{j-1}\boldsymbol{x}_{j-2} \tag{7.107}$$

Step 6. Gram–Schmidt orthogonalization of $\hat{\boldsymbol{x}}_j$ with respect to the previous Lanczos vectors (of step 7):

$$\hat{\boldsymbol{x}}_j = \hat{\boldsymbol{x}}_j - \sum_{k=1}^{j-1} \left(\hat{\boldsymbol{x}}_j^T \boldsymbol{M} \boldsymbol{x}_k \right) \boldsymbol{x}_k \tag{7.108}$$

Step 7. Find the parameter, $\beta_j = \left(\hat{\boldsymbol{x}}_j^T \boldsymbol{M} \hat{\boldsymbol{x}}_j \right)^{1/2}$ and generate the new Lanczos vector:

$$\boldsymbol{x}_j = \beta_j^{-1} \hat{\boldsymbol{x}}_j \tag{7.109}$$

Step 8. Update the tridiagonal matrix, $\boldsymbol{T}$ with, α_j and β_j
End of a Lanczos stage

Step 9. Solve the standard eigenvalue problem :

$$\boldsymbol{T}\boldsymbol{\varphi} = \mu \boldsymbol{\varphi} \tag{7.110}$$

Step 10. Find eigenvalues and eigenvector matrix:

$$\lambda_j = \frac{1}{\mu_j},\ j = 1, 2, \ldots m \ \text{ and } \ \boldsymbol{\Phi}_{n\times m} = \boldsymbol{X}_{n\times r}\boldsymbol{\varphi}_{r\times m} \tag{7.111}$$

Here, $\boldsymbol{X}_{n\times r} = \{x_1, x_2, \ldots, x_r\}$

Step 11. Check for convergence (Equation 7.103) and perform Sturm sequence to see if all the desired m eigenpairs are realised (Equation 7.104)

the computations are performed with blocks of Lanczos vectors, $\{\boldsymbol{x}_k\}$ at each iteration. Details on these alternative versions of Lanczos method are available in Grimes et al. (1991) and references given in the Bibliography.

7.6 Systems with Unsymmetric Matrices

Nonself-adjoint systems and the associated properties of loss of symmetry and positive definiteness are described in Chapter 2 (for continuous systems). These characteristics are also manifest in the corresponding discretized multidegree-of-freedom (MDOF) models.

Table 7.13 Eigensolution by the Lanczos method and convergence; $m = 12$

Mode number, j	Natural frequency in rad/s	Convergence in a Lanczos stage ($\lvert\beta_{m+1}\bar{\varphi}_{m,j}\rvert$)
1	51.232	2.073E−19
2	204.928	1.323E−17
3	321.075	5.682E−17
4	899.219	2.509E−16
5	1284.301	2.087E−14
6	1763.341	7.305E−13
7	2919.495	1.104E−10
8	3596.877	1.824E−09
9	4373.736	4.107E−09
10	6136.563	1.925E−09
11	7053.365	3.336E−09
12	8221.303	7.943E−10

Accordingly, the resulting matrices (the stiffness and damping matrices, in particular) are unsymmetric. The mathematical models describing the dynamics of structures with translating and/or rotating parts belong to this category. For example, vibrating power transmission belts and aerial cables, often modelled as one-dimensional continuous systems, are typically associated with the axial transport of mass. The translation speed significantly changes (Mote, 1966; Perkins, 1990) the dynamic behaviour of these systems. For rotating turbine shafts or helicopter blades or spinning satellites, centrifugal and coriolis forces (Meirovitch, 1970) are known to affect bending vibrations (Likins, 1972; Meirovitch, 1974). Before the techniques for the eigensolution of such unsymmetric systems are discussed, it is prudent to have an understanding on how such unsymmetric matrices arise in the discretized equations of motion. To this end, consider a system undergoing rigid-body motion under a simultaneous translation $\boldsymbol{c}(t)$ and rotation $\boldsymbol{Q}(t)$. $\boldsymbol{Q}(t)$ is orthogonal with $\boldsymbol{Q}^T\boldsymbol{Q} = \boldsymbol{I}$. We introduce two (body-fixed) frames O and O^* corresponding to the positions of the body at two time instants, namely t and t^* respectively (Figure 7.6).

It is assumed (without a loss of generality) that the two frames coincide at $t_0 = 0$. At these two instants, reference coordinates of a material point within the moving body, are related by:

$$\boldsymbol{x}^*(\boldsymbol{X}, t^*) = \boldsymbol{Q}(t)\boldsymbol{x}(\boldsymbol{X}, t) + \boldsymbol{c}(t) \tag{7.112}$$

$\boldsymbol{X}$ is the reference coordinate of the material body at $t = 0$. Note that the (scalar) mass density ρ is unchanged by the above transformation, that is $\rho^*(\boldsymbol{x}^*, t^*) = \rho(\boldsymbol{x}, t)$. A scalar field Φ or a vector field $\mathbf{a}$, or a higher-order tensor field $\boldsymbol{A}$ (see Bibliography in Chapter 1 for definition of tensors and their notation) is referred to as frame indifferent if the following identities are satisfied:

$$\Phi^*(\boldsymbol{x}^*, t^*) = \Phi(\boldsymbol{x}, t), \boldsymbol{a}^*(\boldsymbol{x}^*, t^*) = \boldsymbol{Q}(t)\boldsymbol{a}(\boldsymbol{x}, t), \boldsymbol{A}^*(\boldsymbol{x}^*, t^*) = \boldsymbol{Q}\boldsymbol{A}(\boldsymbol{x}, t)\boldsymbol{Q}^T \tag{7.113a–c}$$

Whilst some quantities like the mass density are frame indifferent, some are not. For example, from Equation 7.112, the velocity $\frac{d\boldsymbol{x}^*}{dt}$ is related to $\frac{d\boldsymbol{x}}{dt}$ as:

$$\dot{\boldsymbol{x}}^* = \boldsymbol{Q}\dot{\boldsymbol{x}} + \dot{\boldsymbol{Q}}\boldsymbol{x} + \dot{\boldsymbol{c}} \tag{7.114}$$

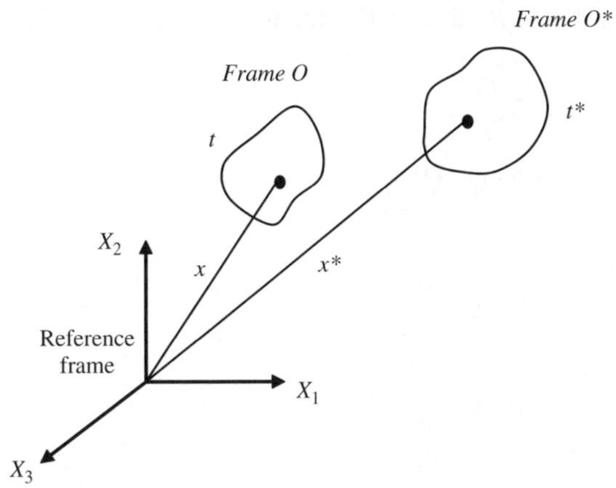

Figure 7.6 Moving frames relative to reference coordinate system

Thus, (see Equation 7.113b) the velocity vector is not frame independent. The case with accelerations is similar. From Equation 7.114, we get the relation:

$$\ddot{\boldsymbol{x}}^* = \boldsymbol{Q}\ddot{\boldsymbol{x}} + 2\dot{\boldsymbol{Q}}\dot{\boldsymbol{x}} + \ddot{\boldsymbol{Q}}\boldsymbol{x} + \ddot{\boldsymbol{c}} \tag{7.115}$$

If the principle of frame indifference holds, then the governing laws of motion must be of the same form in all frames. Consider first, the law of mass balance (Chapter 1), that is $\frac{\partial \rho}{\partial t} + \rho \nabla.\boldsymbol{v} = 0$. Here, $\boldsymbol{v} = \frac{dx}{dt}$. From Equation 1.55, we have:

$$\rho_0\left(\boldsymbol{X}\right) = J\rho = J^*\rho^* \tag{7.116}$$

With $\boldsymbol{F}$ as the deformation gradient, $J = \det \boldsymbol{F}$. Since $\boldsymbol{F}^* = \boldsymbol{Q}F$ and $\det \boldsymbol{Q} = 1$, $J^* = \det \boldsymbol{F}^* = \det \boldsymbol{F} = J$. Hence, from Equation 7.116, $\rho = \rho^*$. Further, one may show that (Appendix G):

$$\nabla_{\boldsymbol{x}^*}.\boldsymbol{v}^* = \text{tr } \boldsymbol{L}^* = \text{tr } \boldsymbol{L} = \nabla_{\boldsymbol{x}} \cdot \boldsymbol{v} \tag{7.117}$$

Here $\boldsymbol{L}$ and $\boldsymbol{L}^*$ are the velocity gradients.

Hence, we obtain the frame indifference of the mass-balance equation:

$$\frac{\partial \rho^*}{\partial t^*} + \rho^* \nabla_{\boldsymbol{x}^*} \cdot \boldsymbol{v}^* = 0 = \frac{\partial \rho}{\partial t} + \rho \nabla_{\boldsymbol{x}} \cdot \boldsymbol{v}$$

Regarding the linear momentum balance (Equation 1.65), the frame indifference requires:

$$\rho \frac{d\boldsymbol{v}_{\boldsymbol{x}}}{dt} - \nabla_x \cdot \boldsymbol{\tau} - \rho \boldsymbol{f}_{bx} = 0 \text{ and}$$

$$\rho^* \frac{d\boldsymbol{v}_{\boldsymbol{x}^*}}{dt^*} - \nabla_{x^*} \cdot \boldsymbol{\tau}^* - \rho^* \boldsymbol{f}_{bx^*} = 0 \tag{7.118a,b}$$

Here, we postulate the axiomatic requirement that the stress tensor is frame indifferent. From Equation 7.113c, this requirement implies:

$$\boldsymbol{\tau}^* = \boldsymbol{Q}\boldsymbol{\tau}\boldsymbol{Q}^T \tag{7.119}$$

In view of the above, the second term in Equation 7.118b is simplified as:

$$
\begin{aligned}
\nabla_{x^*} \cdot \tau^* &= \frac{\partial \tau^*_{ij}}{\partial x^*_j} \\
&= \frac{\partial \left(Q \tau Q^T\right)_{ij}}{\partial x^*_j} \\
&= Q_{ir} Q_{js} \frac{\partial \tau_{rs}}{\partial x^*_j} \\
&= Q_{ir} Q_{js} \frac{\partial \tau_{rs}}{\partial x_m} Q_{jm} \\
&= Q_{ir} \delta_{ms} \frac{\partial \tau_{rs}}{\partial x_m} \ (\text{since } Q^T Q = I,\ Q_{js} Q_{jm} = \delta_{ms}) \\
&= Q_{ir} \frac{\partial \tau_{rs}}{\partial x_s} \\
&= Q \nabla_x \cdot \tau
\end{aligned}
\tag{7.120}
$$

Substituting the above result in Equation 7.118b, we have:

$$\rho^* \left(\frac{d v_{x^*}}{dt^*} - f_{bx^*} \right) = Q \nabla_x \cdot \tau \tag{7.121}$$

If the equation of motion in frame O^* is to retain the same form as Equation 7.118a corresponding to frame O, it is required that

$$\rho^* \left(\frac{d v_{x^*}}{dt^*} - f_{bx^*} \right) = Q \rho \left(\frac{d v_x}{dt} - f_{bx} \right) \tag{7.122}$$

With $\rho^* = \rho$ and the acceleration $\frac{d v_{x^*}}{dt^*}$ given by Equation 7.115, the above condition yields the following relation between the body-force vectors f_{bx^*} of frame O^* and f_{bx} of frame O:

$$
\begin{aligned}
f_{bx^*} &= Q f_{bx} + 2\dot{Q}\dot{x} + \ddot{Q}x + \ddot{c} \\
&\Rightarrow f_{bx} = Q^T \left(f_{bx^*} - \ddot{c} - 2\dot{Q}\dot{x} - \ddot{Q}x \right)
\end{aligned}
\tag{7.123}
$$

The above equation shows that the body forces vary from frame to frame so as to keep the equations of motion frame invariant. Note that, unlike f_{bx^*}, f_{bx} contains additional 'fictitious' terms that arise due to the acceleration $\frac{d v_x}{dt^*}$ of the relatively moving frame. These fictitious force terms are variously categorised as coriolis, centrifugal or Euler forces, as explained below.

Let $W = \dot{Q} Q^T$, that is $\dot{Q} = W Q$ and $\ddot{Q} = \left(\dot{W} + W^2\right) Q$. Note that since Q is orthogonal, W is skew-symmetric (Appendix G). Using the above identities in Equation 7.123, we get

$$f_{bx} = Q^T (f_{bx^*} - \ddot{c}) - 2 Q^T W Q \dot{x} - Q^T \left(\dot{W} + W^2\right) Q x \tag{7.124}$$

If Ω is the angular velocity vector of the body-fixed xyz-axes in frame O and ω, the angular velocity vector of the axes with respect to the frame O^*, we have $\omega(t) = Q(t)\Omega(t)$.

Further, $\boldsymbol{\omega}$ is the unique axial vector (Appendix G) of the second-order skew-symmetric tensor $\boldsymbol{W}$, that is $\boldsymbol{W\omega} = \boldsymbol{0}$, and given any vector $\boldsymbol{u} \in \boldsymbol{V}$, $\boldsymbol{Wu} = \boldsymbol{\omega} \times \boldsymbol{u}$. It follows that $\boldsymbol{\Omega}$ is the axial (or dual) vector of $\boldsymbol{Q}^T\boldsymbol{W}\boldsymbol{Q}$ (note that $\boldsymbol{Q}^T\boldsymbol{W}\boldsymbol{Q}$ is skew-symmetric). In view of these identities, Equation 7.124 can be modified as follows. The term $\boldsymbol{Q}^T\boldsymbol{W}\boldsymbol{Q}\dot{\boldsymbol{x}} = \boldsymbol{\Omega}\dot{\boldsymbol{x}} = \boldsymbol{\Omega}v = \boldsymbol{\Omega} \times v$. Similarly $\boldsymbol{Q}^T\dot{\boldsymbol{W}}\boldsymbol{Q}\boldsymbol{x} = \dot{\boldsymbol{\Omega}}\boldsymbol{x} = \dot{\boldsymbol{\Omega}} \times \boldsymbol{x}$ and $\boldsymbol{Q}^T\boldsymbol{W}^2\boldsymbol{Q}\boldsymbol{x} = \boldsymbol{Q}^T\boldsymbol{W}\boldsymbol{W}\boldsymbol{Q}\boldsymbol{x} = \boldsymbol{Q}^T\boldsymbol{W}\left(\boldsymbol{Q}\boldsymbol{Q}^T\right)\boldsymbol{W}\boldsymbol{Q}\boldsymbol{x} = \boldsymbol{Q}^T\boldsymbol{W}\boldsymbol{Q}(\boldsymbol{Q}^T\boldsymbol{W}\boldsymbol{Q}\boldsymbol{x}) = \boldsymbol{Q}^T\boldsymbol{W}\boldsymbol{Q}(\boldsymbol{\Omega} \times \boldsymbol{x}) = \boldsymbol{\Omega} \times \boldsymbol{\Omega} \times \boldsymbol{x}$. Equation 7.124 can be rewritten as:

$$\boldsymbol{f}_{bx} = \boldsymbol{Q}^T(\boldsymbol{f}_{bx^*} - \ddot{\boldsymbol{c}}) - \dot{\boldsymbol{\Omega}} \times \boldsymbol{x} - \boldsymbol{\Omega} \times \boldsymbol{\Omega} \times \boldsymbol{x} - 2\boldsymbol{\Omega} \times \boldsymbol{v} \tag{7.125}$$

With the body force defined by the above equation in terms of $\boldsymbol{\Omega}$, the momentum balance Equation 7.118a represents the equations of motion with respect to the moving frame O.

Example 7.8: A rotating shaft and equations of motion

The rotating shaft is shown in Figure 7.7. A simplified 2-dof model is taken with its mass, m lumped at its centre of gravity, C. The stiffness of the shaft (in the two transverse directions) is simulated by equivalent spring stiffnesses $\frac{1}{2}K_y$ and $\frac{1}{2}K_z$ associated with each of the two halves. These stiffness parameters are assumed to be constant $\forall t \geq 0$. The rotation speed Ω_x is with respect to the body-fixed longitudinal axis of the beam and is time invariant. The body-fixed longitudinal axis, x of the beam is so oriented as to be coincident with the inertially fixed X-axis. The shaft motion is described by two transverse displacement coordinates $v(t)$ and $w(t)$ at C.

The angular velocity vector $\boldsymbol{\Omega}$ of the rotating (body-fixed) frame O is given by $(\Omega_x, 0, 0)^T$. With respect to the coordinate axes (xyz) in this frame, the body-force vector is given by Equation 7.125:

$$f_{bx} = K_y vj + K_z wk - \boldsymbol{\Omega} \times \boldsymbol{\Omega} \times \begin{pmatrix} v \\ w \end{pmatrix} - 2\boldsymbol{\Omega} \times \begin{pmatrix} \dot{v} \\ \dot{w} \end{pmatrix} \tag{7.126}$$

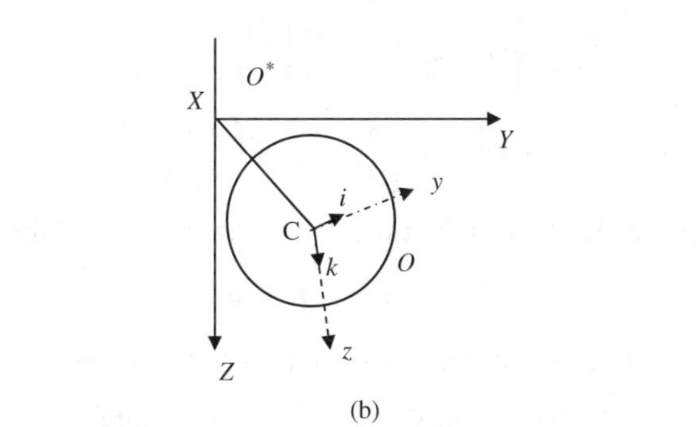

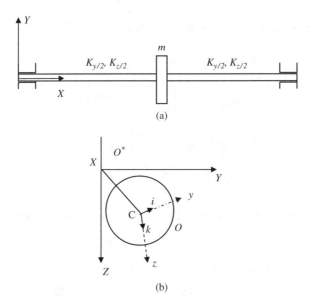

Figure 7.7 (a) Rotating shaft with lumped parameters and (b) 2-dof discrete model

i, j and k are the unit vectors along the x, y and z-axes, respectively. Here, the spring forces are taken to be frame indifferent. Simplifying the above equation and adding the inertia terms ($\rho \frac{dv_x}{dt}$), we obtain the equations of motion as:

$$\begin{bmatrix} m & 0 \\ 0 & m \end{bmatrix} \begin{pmatrix} \ddot{v} \\ \ddot{w} \end{pmatrix} + 2m\Omega_x \begin{bmatrix} 0 & -1 \\ 1 & 0 \end{bmatrix} \begin{pmatrix} \dot{v} \\ \dot{w} \end{pmatrix} + \left\{ \begin{bmatrix} K_y & 0 \\ 0 & K_x \end{bmatrix} - m\Omega_x^2 \begin{bmatrix} 1 & 0 \\ 0 & 1 \end{bmatrix} \right\} \begin{pmatrix} v \\ w \end{pmatrix} = \begin{pmatrix} 0 \\ 0 \end{pmatrix} \tag{7.127}$$

The second term in the above equation is referred to as the coriolis force and can be viewed as the fictitious damping force associated with a skew-symmetric matrix. The centrifugal force is represented by the fourth term and affects the stiffness matrix. However, a more straightforward alternative to the derivation of this equation is through Hamilton's principle, as shown below. The position and velocity vectors of C are:

$$r = vj + wk$$
$$\dot{r} = \dot{v}j + \dot{w}k + \Omega_x i \times (vj + wk) \Rightarrow \left(\dot{v} - \Omega_x w\right) j + \left(\dot{w} + \Omega_x v\right) k \tag{7.128a,b}$$

Now, the expression for the kinetic energy is given by:

$$\begin{aligned} T &= \frac{1}{2} m \left\{ \left(\dot{v} - \Omega_x w\right)^2 + \left(\dot{w} + \Omega_x v\right)^2 \right\} \\ &= T_0 + T_1 + T_2 \end{aligned} \tag{7.129a}$$

where $T_0 = \frac{1}{2} m\Omega_x^2 \left(w^2 + v^2\right)$ is independent of the velocities $\dot{v}$, $\dot{w}$ and represents the centrifugal force. $T_1 = \frac{1}{2}\left\{2\Omega_x m \left(\dot{w}v - \dot{v}w\right)\right\}$ represents the coriolis term, which is linear in $\dot{v}$, $\dot{w}$. $T_2 = \frac{1}{2} m(\dot{v}^2 + \dot{w}^2)$ is the homogeneous quadratic function in terms of the velocity coordinates. The potential energy is given by:

$$V = \frac{1}{2} K_y v^2 + \frac{1}{2} K_z w^2 \tag{7.129b}$$

These energy terms can be rewritten in the following form:

$$\begin{aligned} T_0 &= \frac{1}{2} \Omega_x^2 (v \quad w) \begin{bmatrix} m & 0 \\ 0 & m \end{bmatrix} \begin{pmatrix} v \\ w \end{pmatrix} \\ T_1 &= \frac{1}{2} (\dot{v} \quad \dot{w}) \begin{bmatrix} 0 & -m\Omega_x \\ m\Omega_x & 0 \end{bmatrix} \begin{pmatrix} v \\ w \end{pmatrix} \\ T_2 &= \frac{1}{2} (\dot{v} \quad \dot{w}) \begin{bmatrix} m & 0 \\ 0 & m \end{bmatrix} \begin{pmatrix} \dot{v} \\ \dot{w} \end{pmatrix} \\ V &= \frac{1}{2} (v \quad w) \begin{bmatrix} K_y & 0 \\ 0 & K_z \end{bmatrix} \begin{pmatrix} v \\ w \end{pmatrix} \end{aligned} \tag{7.130a–d}$$

Using Hamilton's principle, we arrive at the following coupled ODEs:

$$\boldsymbol{M}\ddot{\boldsymbol{X}} + \boldsymbol{G}\dot{\boldsymbol{X}} + \boldsymbol{K}\boldsymbol{X} = 0 \tag{7.131}$$

Here, $\boldsymbol{X} = (v \quad w)^T$. $\boldsymbol{M}$, $\boldsymbol{G}$ and $\boldsymbol{K}$ are the system matrices given by:

$$\boldsymbol{M} = \begin{bmatrix} m & 0 \\ 0 & m \end{bmatrix}, \boldsymbol{G} = \begin{bmatrix} 0 & -2m\Omega \\ 2m\Omega & 0 \end{bmatrix} \text{ and } \boldsymbol{K} = \begin{bmatrix} k_y - m\Omega^2 & 0 \\ 0 & k_z - m\Omega^2 \end{bmatrix} \tag{7.132}$$

The equations of motion (Equation 7.131) are identical to Equations 7.127. Note that the coriolis term couples the two equations in v and w. Further, it induces (fictitious) damping into the system that could lead to a potential loss of stability (see Example 7.10). The skew-symmetric matrix $\boldsymbol{G}$ is known as the gyroscopic matrix.

Premultiplying Equation 7.131 by $\dot{\boldsymbol{X}}^T$, we have:

$$\dot{\boldsymbol{X}}^T \boldsymbol{M}\ddot{\boldsymbol{X}} + \dot{\boldsymbol{X}}^T \boldsymbol{G}\dot{\boldsymbol{X}} + \dot{\boldsymbol{X}}^T \boldsymbol{K}\boldsymbol{X} = 0 \tag{7.133}$$

Owing to the skew-symmetry of $\dot{\boldsymbol{X}}^T \boldsymbol{G}\dot{\boldsymbol{X}} = 0$. Further, $\dot{\boldsymbol{X}}^T \boldsymbol{M}\ddot{\boldsymbol{X}} = \frac{d}{dt}\left(\dot{\boldsymbol{X}}^T \boldsymbol{M}\dot{\boldsymbol{X}}\right) = 2\frac{d}{dt}(T_2)$ and $\dot{\boldsymbol{X}}^T \boldsymbol{K}\boldsymbol{X} = \frac{d}{dt}\left(\boldsymbol{X}^T \boldsymbol{K}\boldsymbol{X}\right) = 2\frac{d}{dt}(V - T_0)$. Equation 7.133 can be rewritten as:

$$\frac{d}{dt}\left(T_2 + V - T_0\right) = 0 \tag{7.134}$$

This equation corresponds to the energy conservation of the linear gyroscopic system where $T_2 + V - T_0$ is the Hamiltonian.

Example 7.9: An axially moving string on elastic supports and equations of motion

In Example 2.6, the axially moving string was analysed as a nonself-adjoint continuous system. In the present case, the interest is to demonstrate the nonself-adjoint characteristics of the translating string via spatial discretization using a FEM. As shown in Figure 7.8

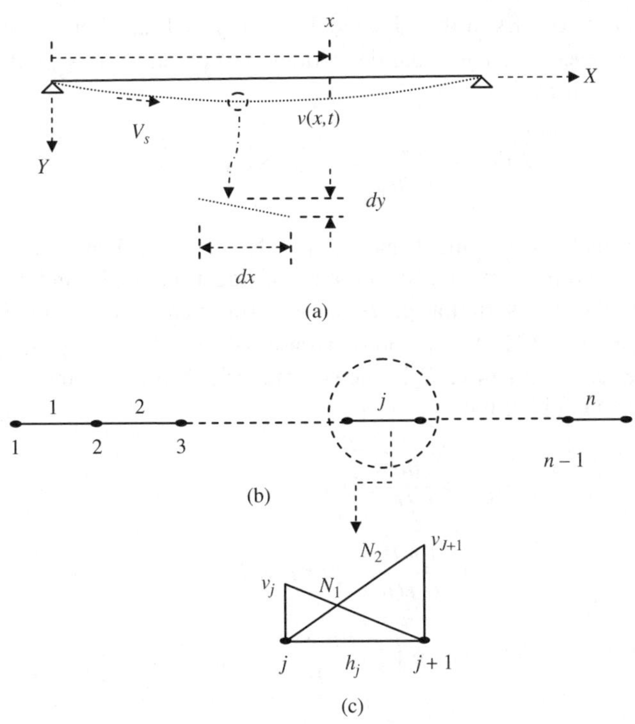

Figure 7.8 Axially moving string; (a) geometry, (b) finite element model and (c) jth element (zoomed view) and the shape functions

below, the string of length l, subjected to an axial tension H, is simply supported and moves with a constant (translational) speed V_S. It is of uniform mass, m per unit length. The speed parameter, V_S, again renders the present system gyroscopic as described below.

The velocity $\dot{v}$ of any material point on the string with respect to the fixed frame of reference (XYZ) is given by $\dot{v} = \frac{\partial v}{\partial t} + V_S \frac{\partial v}{\partial x}$. In this frame of reference, the expressions for the kinetic and potential energies are given by:

$$T = \frac{1}{2}\int_0^l m\left(\frac{\partial v}{\partial t} + V_S\frac{\partial v}{\partial t}\right)^2 dx \text{ and } V = \frac{1}{2}\int_0^l H\left(\frac{\partial v}{\partial x}\right)^2 dx \qquad (7.135)$$

Analogous to Example 7.8, we presently identify $T_0 = \frac{1}{2}\int_0^l mV_S^2\left(\frac{\partial v}{\partial x}\right)^2 dx$, $T_1 = \frac{1}{2}\int_0^l 2mV_S\frac{\partial v}{\partial t}\frac{\partial v}{\partial t}dx$ and $T_2 = \frac{1}{2}\int_0^l m\left(\frac{\partial v}{\partial t}\right)^2 dx$. With the Lagrangian $L = T - V$, the equations of motion can be obtained via Hamilton's principle as:

$$\delta\int_{t_1}^{t_2} \boldsymbol{L}\left(v, \frac{\partial v}{\partial t}, \frac{\partial v}{\partial t}, t\right) dt = 0 \qquad (7.136)$$

Adopting the FE discretization (Figure 7.8) with the vertical deflection as the only unknown degree of freedom per node, the approximation to $v(x, t)$ within an element j is given as (allowing for a slight abuse of notations):

$$v(x, t) = v_j(t)N_1(x) + v_{j+1}(t)N_2(x), \; j = 1, 2, ...n \qquad (7.137)$$

n is the total number of FEs in the FE model. $v_j(t)$, $j = 1, 2, ...n + 1$ are the unknown functions of time describing the nodal deflections. $N_1(x)$ and $N_2(x)$ are the (globally) C^0 shape functions given by

$$N_1(x) = \frac{x_{j+1} - x}{h_j} \text{ and } N_2(x) = \frac{x - x_j}{h_j} \qquad (7.138)$$

h_j is the length of the jth element and x_j, $j = 1, 2, ..., n + 1$ are the x-coordinates of the nodes in the discretized model. Substitution of Equations 7.137 and 7.138 in Equation 7.136 and taking the first variation gives the governing equations of motion in the same form as in Equation 7.131. In the present case, $X(t) = \{v_1(t), v_2(t), \ldots \ldots, v_n(t)\}^T$. Whilst $\boldsymbol{M}$ and $\boldsymbol{K}$ are symmetric, G is skew-symmetric. Moreover, the matrices are tridiagonal with elements given by:

$$\begin{aligned} M_{ij} &= \frac{2}{3}\frac{ml^2}{Hn}, \quad for\ i = j \\ &= \frac{1}{6}\frac{ml^2}{Hn}, \quad for\ i \neq j \end{aligned} \qquad (7.139)$$

$$\begin{aligned} K_{ij} &= 2n\left(1 - \frac{mS^2}{H}\right), \quad for\ i = j \\ &= -n\left(1 - \frac{mS^2}{H}\right), \quad for\ i \neq j \end{aligned} \qquad (7.140)$$

$$
\begin{aligned}
G_{ij} &= 0, \quad \mathit{for}\ i = j \\
&= \left(\frac{mS^2}{H}\right)^{\frac{1}{2}}, \quad \mathit{for}\ i < j \\
&= -\left(\frac{mS^2}{H}\right)^{1/2}, \quad \mathit{for}\ i > j
\end{aligned} \tag{7.141}
$$

7.6.1 Skew-Symmetric Matrices and Eigensolution

An eigensolution of the skew-symmetric system (of the type in Equation 7.131) can be obtained by transforming the problem into two independent symmetric eigenvalue problems. To this end, we define a $2n$-dimensional state vector, $\boldsymbol{Y}(t) = (\boldsymbol{X}^T, \dot{\boldsymbol{X}}^T,)^T$ and Equation 7.131 is recast in the first-order state vector form as:

$$\hat{\boldsymbol{M}}\dot{\boldsymbol{Y}}(t) + \hat{\boldsymbol{K}}\boldsymbol{Y}(t) = 0 \tag{7.142}$$

with $\hat{\boldsymbol{K}} = \begin{bmatrix} \boldsymbol{0} & -\boldsymbol{M} \\ \boldsymbol{K} & \boldsymbol{G} \end{bmatrix}$ and $\hat{\boldsymbol{M}} = \begin{bmatrix} \boldsymbol{M} & 0 \\ 0 & \boldsymbol{M} \end{bmatrix}$.

$\hat{\boldsymbol{M}}$ and $\hat{\hat{\boldsymbol{K}}} \in \mathbb{R}^{2n\times 2n}$. Whilst $\hat{\boldsymbol{M}}$ is symmetric, $\hat{\boldsymbol{K}}$ is skew-symmetric. The eigenvalue problem associated with Equation 7.142 is given by:

$$\lambda\hat{\boldsymbol{M}}\dot{\boldsymbol{Y}} + \hat{\boldsymbol{K}}\boldsymbol{Y} = 0 \tag{7.143}$$

Using the symmetry and skew-symmetry of $\hat{\boldsymbol{M}}$ and $\hat{\boldsymbol{K}}$, respectively, the above eigenvalue problem is reducible to two symmetric eigenvalue problems (Meirovitch, 1974, 1975) details of which are given below. Further, the eigensolution consists of n pairs of pure imaginary complex conjugates $\lambda_r = \pm i\omega_r$ and n pairs of associated complex conjugate eigenvectors $\boldsymbol{Y}_r = \boldsymbol{y}_r + i z_r$ and $\boldsymbol{Y}_r^* = \boldsymbol{y}_r - i z_r$, $r = 1, 2, \ldots, n$. Inserting λ_r and $\boldsymbol{Y}_r$ into Equation 7.143, we obtain:

$$-\omega_r\hat{\boldsymbol{M}}z_r + \hat{\boldsymbol{K}}\boldsymbol{y}_r = 0 \text{ and } \omega_r\hat{\boldsymbol{M}}\boldsymbol{y}_r + \hat{\boldsymbol{K}}z_r = 0 \tag{7.144}$$

The above two equations further reduce to two independent eigenvalue problems:

$$\omega_r^2\hat{\boldsymbol{M}}\boldsymbol{y}_r = \hat{\boldsymbol{K}}^T\hat{\boldsymbol{M}}^{-1}\hat{\boldsymbol{K}}\boldsymbol{y}_r \text{ and } \omega_r^2\hat{\boldsymbol{M}}z_r = \hat{\boldsymbol{K}}^T\hat{\boldsymbol{M}}^{-1}\hat{\boldsymbol{K}}\boldsymbol{y}_r = 0 \tag{7.145}$$

Both $\hat{\boldsymbol{M}}$ and $\hat{\boldsymbol{K}}^T\hat{\boldsymbol{M}}^{-1}\hat{\boldsymbol{K}}$ being symmetric, Equation 7.145 represents two symmetric eigenvalue problems. These can be solved by any of the methods described earlier, for the n pairs of repeated eigenvalues ω_r and n pairs of associated eigenvectors $\boldsymbol{y}_r$ and z_r. The orthogonality properties exhibited by the n pairs of $\boldsymbol{y}_r$ and z_r are given by:

$$\boldsymbol{y}_r^T\hat{\boldsymbol{M}}\,\boldsymbol{y}_s = z_r^T\hat{\boldsymbol{M}}\,z_s = \delta_{rs},\ \boldsymbol{y}_r^T\hat{\boldsymbol{M}}\,z_s = z_r^T\hat{\boldsymbol{M}}\boldsymbol{y}_s = 0$$

and

$$z_s^T\hat{\boldsymbol{K}}\boldsymbol{y}_r = -\boldsymbol{y}_s^T\hat{\boldsymbol{K}}z_r = \omega_r\delta_{rs},\ \boldsymbol{y}_s^T\hat{\boldsymbol{K}}\boldsymbol{y}_r = z_s^T\hat{\boldsymbol{K}}z_r = 0, r, s = 1, 2 \ldots .n \tag{7.146a,b}$$

The modal superposition procedure detailed in Chapter 5 is applicable to obtain the forced vibration solution under an external forcing function $\boldsymbol{F}(x,t)$ as:

$$\boldsymbol{Y}(t) = \sum_{r=1}^{n} q_{1r}(t)\boldsymbol{y}_r + \sum_{r=1}^{n} q_{2r}(t)\boldsymbol{z}_r \tag{7.147}$$

In the above equation, $q_{1r}(t)$ and $q_{2r}(t)$ are the n pairs of modal coordinates. Using the orthogonality properties of the eigenvectors $\boldsymbol{y}_r$ and $\boldsymbol{z}_r$, we uncouple the system equations into n pairs of modal equations:

$$\dot{q}_{1r}(t) - \omega_r q_{2r}(t) = f_{1r}(t) \quad \text{and} \quad \dot{q}_{2r}(t) + \omega_r q_{1r}(t) = f_{2r}(t) \tag{7.148}$$

with $r = 1, 2 \ldots n$ where $f_{1r}(t) = \boldsymbol{y}_r^T \boldsymbol{F}(t)$ and $f_{2r}(t) = \boldsymbol{z}_r^T \boldsymbol{F}(t)$.

Solving the above modal equations for $q_{1r}(t)$ and $q_{2r}(t)$, one obtains the forced vibration response via modal summation as in Equation 7.147.

Example 7.10: Eigensolution of the axially moving string in Figure 7.8

The FE model of the string consists of 1D bar elements with the vertical displacement coordinate functions $\{v_j(t)\}$ being the only nodal degrees of freedom. Twenty elements, associated with 19 dofs, are presently employed. For presentation of numerical results, the parameter $\frac{ml^2}{H}$ in Equation 7.139 is chosen as unity. The parameter $\sqrt{\frac{mV_S^2}{H}}$ in Equations 7.141 and 7.142 is varied in the range 0.0–1.0 and its effect on the first five natural frequencies is shown in Figure 7.9. The eigensolution is obtained by the subspace iteration method. The results match with those in Chapter 2 (Figure 2.22) by a continuum approach. As the parameter $\sqrt{\frac{mV_S^2}{H}}$ approaches unity, all the natural frequencies of the string approach zero indicating buckling instability corresponding to the critical speed, $V_c = \sqrt{\frac{H}{m}}$.

The effect of the speed parameter $\sqrt{\frac{mV_S^2}{H}}$ on the mode shape corresponding to the first natural frequency of the string is shown in Figure 7.10. For different V_S, deflection profiles are captured at a few intermediate time instants within one time period of the natural frequency. Whilst the mode shapes for $V_S = 0$ correspond to the usual normal modes of symmetric matrices (in that each material point follows a harmonic motion about the equilibrium position), the snapshots become antisymmetric for $V_S > 0$.

7.6.2 Unsymmetric Matrices – A Rotor Bearing System

Figure 7.11 shows a flexible rotor bearing system (Nelson and McVaugh, 1976). The shaft is of circular cross section with uniformly distributed mass and stiffness. It carries a rigid disk of mass m_D, diametrical moment of inertia I_D and polar moment of inertia I_P. The FE model (Figure 7.12) of the shaft consists of $1D$ beam elements with 4 dof per node – two translational and two rotational. The translations (transverse displacements) are denoted by v and w and the rotations by $\theta_Y \left(= \frac{\partial w}{\partial x}\right)$ and $\theta_Z \left(= -\frac{\partial v}{\partial x}\right)$. These displacements and translations are with reference to the fixed frame of reference O^* with axes XYZ. A rotating frame of reference O is attached to the deformed cross section with axes xyz. Both the frames are initially coincident with X represented along the undeformed shaft longitudinal axis.

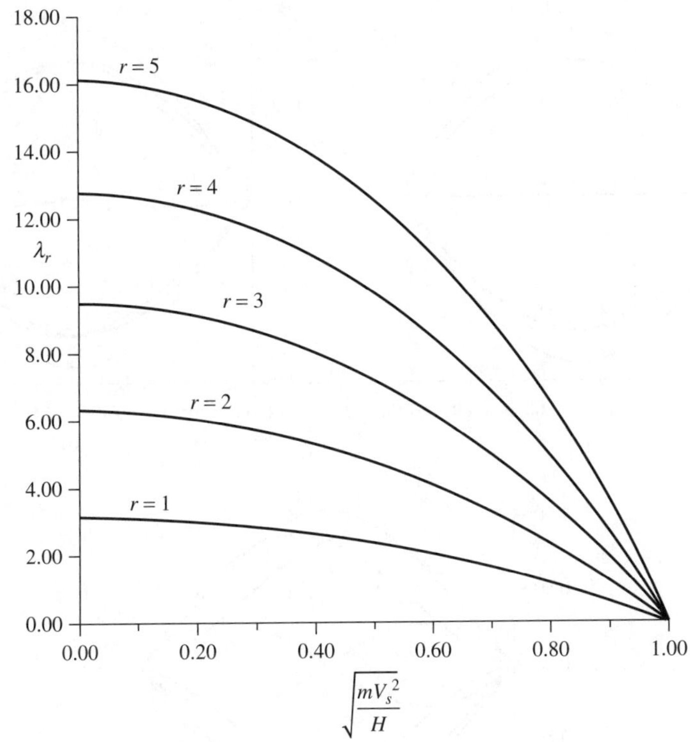

Figure 7.9 Effect of speed parameter, V_s (in terms of $\sqrt{mV_s^2/H}$) on the first five natural frequencies of the axially moving string

Assuming that the reference frame O^* rotates with a constant speed Ω about the X axis, the angular velocity vector $(\omega_x, \omega_y, \omega_z)^T$ of frame O is given by (Appendix G):

$$\begin{pmatrix} \omega_x \\ \omega_y \\ \omega_z \end{pmatrix} = \begin{bmatrix} -\sin\theta_z & 1 & 0 \\ \cos\theta_z \sin\Omega t & 0 & \cos\Omega t \\ \cos\theta_z \cos\Omega t & 0 & -\sin\Omega t \end{bmatrix} \begin{pmatrix} \dot{\theta}_Y \\ \Omega \\ \dot{\theta}_z \end{pmatrix} \tag{7.149}$$

For small rotations θ_Y and θ_Z, the above transformation gives:

$$\omega_x = -\theta_Z \dot{\theta}_Y + \Omega, \begin{pmatrix} \omega_y \\ \omega_z \end{pmatrix} = \begin{bmatrix} \sin\Omega t & \cos\Omega t \\ \cos\Omega t & -\sin\Omega t \end{bmatrix} \begin{pmatrix} \dot{\theta}_Y \\ \dot{\theta}_Z \end{pmatrix} \tag{7.150}$$

7.6.2.1 Rotor Shaft Element Matrices

With the beam shape functions N_i, $i = 1, 2, 3$ and 4 given by Chapter 3 $N_1 = 1 - 3\xi^2 + 2\xi^3$, $N_2 = L_e(1 - 4\xi^2 + 3\xi^3)$, $N_3 = 3\xi^2 - 2\xi^3$, $N_4 = L_e(\xi^2 - \xi^3)$ where $\xi = X/L_e$ and L_e, the element length, the shaft deformations at each node relative to the frame O^* are taken as:

$$(v, w, \theta_Z, \theta_Y)^T = [\boldsymbol{\psi}]\{\boldsymbol{u}^e\} \tag{7.151}$$

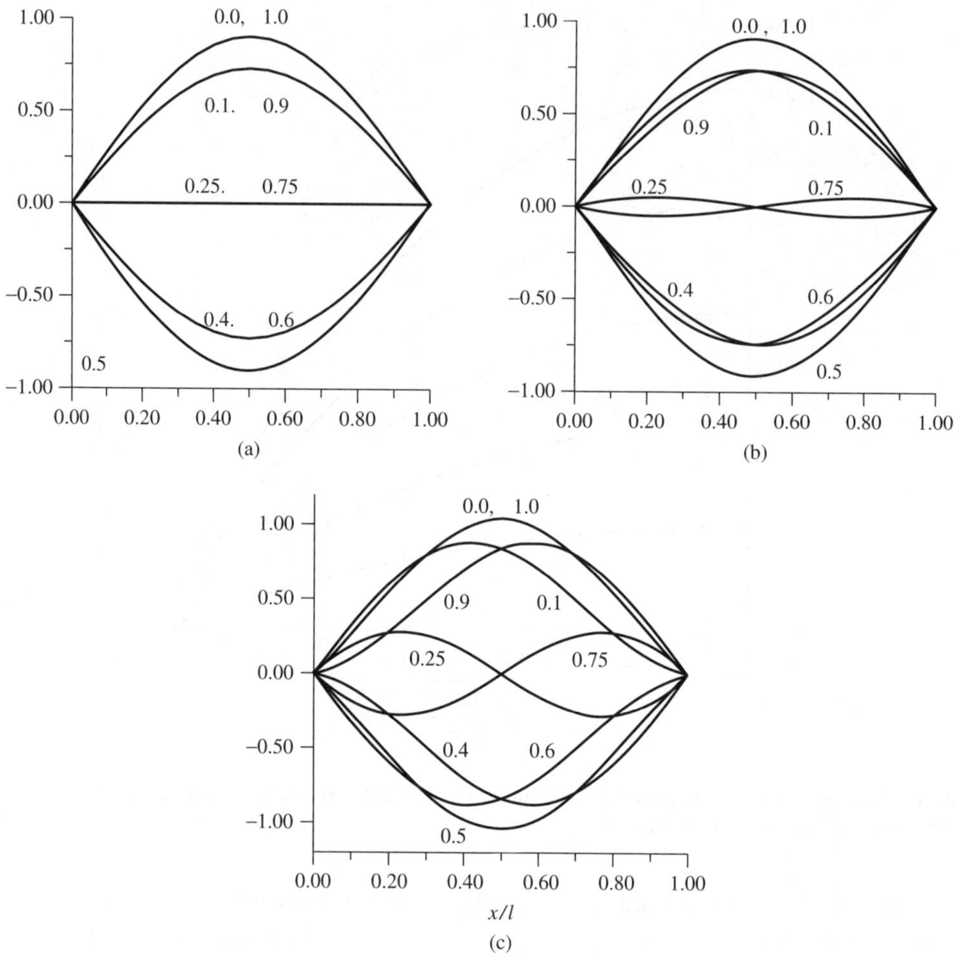

Figure 7.10 Mode shapes of a translating string; first mode; (a) $\sqrt{mS^2/H} = 0.0$, (b) $\sqrt{mS^2/H} = 0.1$ and (c) $\sqrt{mV_s^2/H} = 0.5$. Vibrating profiles are shown at nine time instants within one period of oscillation

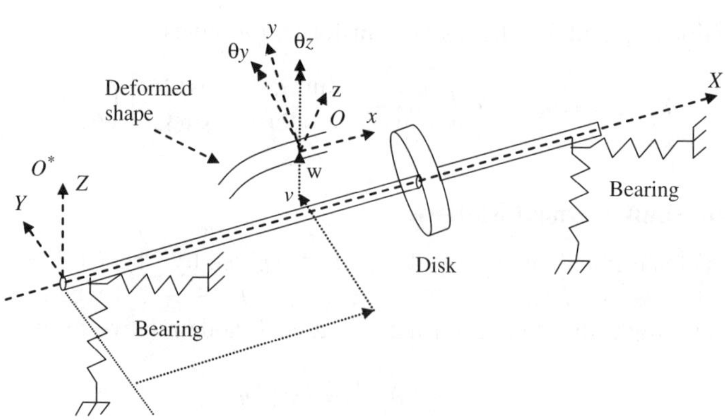

Figure 7.11 A rotor bearing system

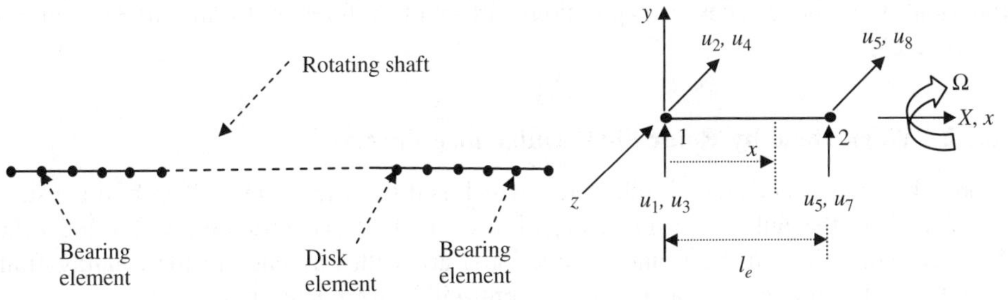

Figure 7.12 Finite element model of the rotor bearing system and a rotor finite element with 4 *dof* per node

Here, $\boldsymbol{\psi} = \begin{pmatrix} \boldsymbol{\psi}_t \\ \boldsymbol{\psi}_r \end{pmatrix}$, with $\boldsymbol{\psi}_t = \begin{bmatrix} N_1 & 0 & 0 & N_2 & N_3 & 0 & 0 & N_4 \\ 0 & N_1 & -N_2 & 0 & 0 & N_3 & -N_4 & 0 \end{bmatrix}$ and $\boldsymbol{\psi}_r = \begin{bmatrix} 0 & -N_1' & N_2' & 0 & 0 & -N_3' & N_4' & 0 \\ N_1' & 0 & 0 & N_2' & N_3' & 0 & 0 & N_4' \end{bmatrix}$

$\boldsymbol{u}^e = (u_1, u_2, \ldots\ldots, u_8)^T$ is the vector of nodal displacements per element, which are unknown functions of time. Here, the pairs (u_1, u_2) and (u_5, u_6) denote, respectively, the translational degrees of freedom at the left and right ends of an element and the other pairs (u_3, u_4) and (u_7, u_8) the corresponding rotational degrees of freedom $\boldsymbol{\psi}_t$ and $\boldsymbol{\psi}_r$ are shape-function matrices of size 2×8, corresponding to the translational and rotational degrees of freedom, respectively. The energy expressions for the rotor shaft element in terms of the nodal displacement and velocities are given by:

Kinetic energy:

$$T_e = \int_0^{l_e} \left\{ \frac{1}{2} \begin{pmatrix} \dot{v} \\ \dot{w} \end{pmatrix}^T \begin{bmatrix} m & 0 \\ 0 & m \end{bmatrix} \begin{pmatrix} \dot{v} \\ \dot{w} \end{pmatrix} + \frac{1}{2} \begin{pmatrix} \omega_y \\ \omega_z \end{pmatrix}^T \begin{bmatrix} I_{De} & 0 \\ 0 & I_{De} \end{bmatrix} \begin{pmatrix} \omega_y \\ \omega_z \end{pmatrix} - \Omega I_{Pe} \theta_Z \dot{\theta}_Y \right\} dX \tag{7.152}$$

Following Equation 7.150, the expression for T_e takes the form (with respect to O^*):

$$\begin{aligned} T_e &= \int_0^{l_e} \left\{ \frac{1}{2} \begin{pmatrix} \dot{v} \\ \dot{w} \end{pmatrix}^T \begin{bmatrix} m & 0 \\ 0 & m \end{bmatrix} \begin{pmatrix} \dot{v} \\ \dot{w} \end{pmatrix} + \frac{1}{2} \begin{pmatrix} \dot{\theta}_Y \\ \dot{\theta}_Z \end{pmatrix}^T \begin{bmatrix} I_{De} & 0 \\ 0 & I_{De} \end{bmatrix} \begin{pmatrix} \dot{\theta}_Y \\ \dot{\theta}_Z \end{pmatrix} - \Omega I_{Pe} \theta_Z \dot{\theta}_Y \right\} dX \\ &= \int_0^1 \left(\frac{1}{2} m \dot{\boldsymbol{u}}^T \boldsymbol{\psi}_t^T \boldsymbol{\psi}_t \dot{\boldsymbol{u}} + \frac{1}{2} I_{De} \dot{\boldsymbol{u}}^T \boldsymbol{\psi}_r^T \boldsymbol{\psi}_r \dot{\boldsymbol{u}} - \Omega I_{Pe} \dot{\boldsymbol{u}}^T \boldsymbol{\psi}_{ry}^T \boldsymbol{\psi}_{rz} \boldsymbol{u} \right) d\boldsymbol{\xi} \end{aligned} \tag{7.153}$$

I_{De} and I_{Pe} are the diametrical and polar moments of inertia of the shaft element. $\boldsymbol{\psi}_{ry}$ and $\boldsymbol{\psi}_{rz}$ are the first and second rows in $\boldsymbol{\psi}_r$. The first two terms in the above equation denote the contribution to kinetic energy from the translational and rotational inertia and the third term corresponds to the kinetic energy due to the gyroscopic moment. The potential energy due to bending is given by:

$$\begin{aligned} V_e &= \frac{1}{2} \int_0^{l_e} \begin{pmatrix} v'' \\ w'' \end{pmatrix}^T E \begin{bmatrix} I_{De} & 0 \\ 0 & I_{De} \end{bmatrix} \begin{pmatrix} v'' \\ w'' \end{pmatrix} dX \\ &= \frac{1}{2} \int_0^1 E I_{De} \boldsymbol{u}^T \boldsymbol{\psi}_t''^T \boldsymbol{\psi}_t'' \boldsymbol{u} d\boldsymbol{\xi} \end{aligned} \tag{7.154}$$

Note that the above energy expressions are with reference to the fixed frame of reference O^*.

7.6.2.2 Work Done by Rotor Shaft Unbalance Forces

A rotor shaft always has radial imbalance, which is distributed over its length. It is usually represented by the unbalance mass, m_u, times the shaft eccentricities, $\varepsilon_y(x)$ and $\varepsilon_z(x)$. These eccentricities (in the y and z directions) are with reference to the rotating frame O. Between the frames O and O^*, the eccentricities are related by:

$$\begin{Bmatrix} \varepsilon_Y(X) \\ \varepsilon_Z(X) \end{Bmatrix} = \begin{bmatrix} \cos \Omega t & -\sin \Omega t \\ \sin \Omega t & \cos \Omega t \end{bmatrix} \begin{Bmatrix} \varepsilon_y(x) \\ \varepsilon_z(x) \end{Bmatrix} \tag{7.155}$$

The unbalance force due to the above eccentricities in frame O is given by

$$\begin{pmatrix} Q_y \\ Q_z \end{pmatrix} = m_u \Omega^2 \begin{Bmatrix} \varepsilon_y(x) \\ \varepsilon_z(x) \end{Bmatrix} \tag{7.156}$$

In the context of a FE model (based on linear shape functions), the eccentricities at the two end nodes of each element suffice to describe the eccentricity fields within the element interior. These functions are given by:

$$\begin{aligned} \varepsilon_Y(\xi) &= \varepsilon_Y(0)\,(1-\xi) + \varepsilon_Y(1)\xi \\ \varepsilon_Z(\xi) &= \varepsilon_Z(0)(1-\xi) + \varepsilon_Z(1)\xi \end{aligned} \tag{7.157}$$

The associated work done by the unbalance forces relative to the fixed frame of reference is given by

$$W_e = \int_0^1 m_u \Omega^2 \boldsymbol{u}^T \boldsymbol{\psi}_t^T \left[\begin{Bmatrix} \varepsilon_Y(\xi) \\ \varepsilon_Z(\xi) \end{Bmatrix} \cos \Omega t + \begin{Bmatrix} \varepsilon_Z(\xi) \\ -\varepsilon_Y(\xi) \end{Bmatrix} \sin \Omega t \right] d\xi \tag{7.158}$$

By Hamilton's principle, the first variation of the Lagrangian ($= T_e - V_e + W_e$) gives the following equations of motion for the rotating-shaft element:

$$\left\{[\boldsymbol{M}_t^e] + [\boldsymbol{M}_r^e]\right\}\{\ddot{\boldsymbol{u}}^e\} - \Omega[\boldsymbol{G}^e]\{\dot{\boldsymbol{u}}\} + \left[\boldsymbol{K}^e\right]\{\boldsymbol{u}\} = \left\{\boldsymbol{Q}^e\right\} \tag{7.159}$$

$\boldsymbol{Q}^e$ represents the discretized nodal force vector (8×1) due to the mass imbalance in the shaft. The matrices in the above equation are given by:

$$\boldsymbol{M}_t^e = \int_0^1 m_e \boldsymbol{\psi}^T \boldsymbol{\psi} d\xi, \boldsymbol{M}_r^e = \int_0^1 I_{De} \boldsymbol{\phi}^T \boldsymbol{\phi} d\xi$$

$$\boldsymbol{G}^e = \boldsymbol{N}_e - \boldsymbol{N}_e^T \quad \text{where} \quad \boldsymbol{N}_e = I_{Pe} \int_0^1 \boldsymbol{\phi}^T \boldsymbol{N} \boldsymbol{\phi} d\xi$$

$$\text{and } \boldsymbol{K}^e = \int_0^1 E I_{de} \boldsymbol{\psi}''^T \boldsymbol{\psi}'' d\xi \tag{7.160a–c}$$

Here, $\boldsymbol{G}^e$ is the gyroscopic damping matrix and $\boldsymbol{K}^e$ the bending stiffness matrix. In the long hand, these element matrices are obtainable as:

$$
\boldsymbol{M}_t^e = m\frac{L_e}{420}\begin{bmatrix}
156 & 0 & 0 & 22L_e & 54 & 0 & 0 & -13L_e \\
 & 156 & -22L_e & 0 & 0 & 54 & 13L_e & 0 \\
 & & 4L_e^2 & 0 & 0 & -13L_e & -3L_e^2 & 0 \\
 & & & 4L_e^2 & 13L_e & 0 & 0 & -3L_e^2 \\
 & & & & 156 & 0 & 0 & -22L_e \\
 & & & & & 156 & 22L_e & 0 \\
 & & & & & & 4L_e^2 & 0 \\
 & & & & & & & 4L_e^2
\end{bmatrix}
$$

$$
\boldsymbol{M}_r^e = \frac{mr^2}{120L_e}\begin{bmatrix}
36 & 0 & 0 & 22L_e & 54 & 0 & 0 & -13L_e \\
 & 36 & -3L_e & 0 & 0 & -36 & -3L_e & 0 \\
 & & 4L_e^2 & 0 & 0 & 3L_e & -L_e^2 & 0 \\
 & & & 4L_e^2 & -3L_e & 0 & 0 & -L_e^2 \\
 & & & & 36 & 0 & 0 & -3L_e \\
 & & & & & 36 & 3L_e & 0 \\
 & & & & & & 4L_e^2 & 0 \\
 & & & & & & & 4L_e^2
\end{bmatrix}
$$

$$
\boldsymbol{K}^e = \frac{EI_{de}}{L_e}\begin{bmatrix}
12 & 0 & 0 & 6L_e & -12 & 0 & 0 & 6L_e \\
 & 12 & -6L_e & 0 & 0 & -12 & 6L_e & 0 \\
 & & 4L_e^2 & 0 & 0 & 6L_e & 2L_e^2 & 0 \\
 & & & 4L_e^2 & -6L_e & 0 & 0 & 2L_e^2 \\
 & & & & 12 & 0 & 0 & -6L_e \\
 & & & & & 12 & 6L_e & 0 \\
 & & & & & & 4L_e^2 & 0 \\
 & & & & & & & 4L_e^2
\end{bmatrix}
$$

$$
\boldsymbol{G}^e = \frac{2mr^2}{120L_e}\begin{bmatrix}
0 & 36 & -3L_e & 0 & 0 & -36 & -3L_e & 0 \\
 & 0 & 0 & -3L_e & 36 & 0 & 0 & -3L_e \\
 & & 0 & 4L_e^2 & 0 & 0 & 3L_e & -L_e^2 \\
 & & & 0 & 0 & -3L_e & L_e^2 & 0 \\
 & & & & 0 & 36 & 3L_e & 0 \\
 & & & & & 0 & 0 & 3L_e \\
 & & & & & & 0 & 4L_e^2 \\
 & & & & & & & 0
\end{bmatrix} \tag{7.161}
$$

7.6.2.3 Disk Element Matrices

For a disk element with mass M_D and with diametrical and polar moments of inertia given by I_D and I_P, the expression for kinetic energy is of the form:

$$
\begin{aligned}
T_D &= \frac{1}{2}\begin{pmatrix}\dot{v}\\ \dot{w}\end{pmatrix}^T \begin{bmatrix}M_D & 0\\ 0 & M_D\end{bmatrix}\begin{pmatrix}\dot{v}\\ \dot{w}\end{pmatrix} + \frac{1}{2}\begin{pmatrix}\omega_y\\ \omega_z\end{pmatrix}^T \begin{bmatrix}I_D & 0\\ 0 & I_D\end{bmatrix}\begin{pmatrix}\omega_y\\ \omega_z\end{pmatrix} - \Omega I_P \theta_z \dot{\theta}_Y \\
&= \frac{1}{2}\begin{pmatrix}\dot{v}\\ \dot{w}\end{pmatrix}^T \begin{bmatrix}M_D & 0\\ 0 & M_D\end{bmatrix}\begin{pmatrix}\dot{v}\\ \dot{w}\end{pmatrix} + \frac{1}{2}\begin{pmatrix}\dot{\theta}_Y\\ \dot{\theta}_z\end{pmatrix}^T \begin{bmatrix}I_D & 0\\ 0 & I_D\end{bmatrix}\begin{pmatrix}\dot{\theta}_Y\\ \dot{\theta}_z\end{pmatrix} - \Omega I_P \theta_z \dot{\theta}_Y
\end{aligned} \tag{7.162}
$$

Similar to the shaft element, the disk may also have mass imbalance with eccentricities ε_{Dy} and ε_{Dz}. The work done, W_{De}, by the disk unbalance forces in frame O^* are similarly obtained as:

$$W_{De} = M_D\Omega^2 \boldsymbol{u}^T \boldsymbol{\psi}_t^T \left[\begin{Bmatrix} \varepsilon_{Dy} \\ \varepsilon_{Dz} \end{Bmatrix} \cos\Omega t + \begin{Bmatrix} \varepsilon_{Dz} \\ -\varepsilon_{Dy} \end{Bmatrix} \sin\Omega t\right] \tag{7.163}$$

By Hamilton's principle, the equations of motion of the rigid disk are given by:

$$\left\{[M_t^D] + [M_r^D]\right\}\{\ddot{u}^D\} - \Omega[G^D]\{\dot{u}^D\} = \{Q^D\} \tag{7.164}$$

The element matrices for the rigid disk are obtained as:

$$[M_t^D] + [M_r^D] = \begin{bmatrix} M_D & 0 & 0 & 0 \\ 0 & M_d & 0 & 0 \\ 0 & 0 & I_D & 0 \\ 0 & 0 & 0 & I_D \end{bmatrix} \text{ and } G^D = \begin{bmatrix} 0 & 0 & 0 & 0 \\ 0 & 0 & 0 & 0 \\ 0 & 0 & 0 & -I_p \\ 0 & 0 & I_p & I_d \end{bmatrix} \tag{7.165}$$

The forcing term Q^D is due to the disk mass imbalance forces and is given by:

$$Q^D = \left\{\varepsilon_{Dy},\ \varepsilon_{Dz},\ 0,\ 0\right\}^T \cos\Omega t + \left\{-\varepsilon_{Dz},\ \varepsilon_{Dy},\ 0,\ 0\right\}^T \sin\Omega t \tag{7.166}$$

7.6.2.4 Bearing Element

Bearings are commonly modelled by linear springs and dampers. The simplest model of, say, a fluid-film bearing consists of a combination of spring and dash-pot elements (Figure 7.13). This model, in particular, involves four stiffness and four damping coefficients that are consistent with the steady-state operating characteristics of the bearing (Lund, 1974; Gunter, 1967). They are, however, dependent on the shaft rotation speed. They form a 2-dof model for each bearing. It is described by 2×2 stiffness and damping matrices which are, in general, unsymmetric. The governing equation of motion for the bearing is:

$$C_b\dot{q}_e + K_b q_e = Q_b \tag{7.167}$$

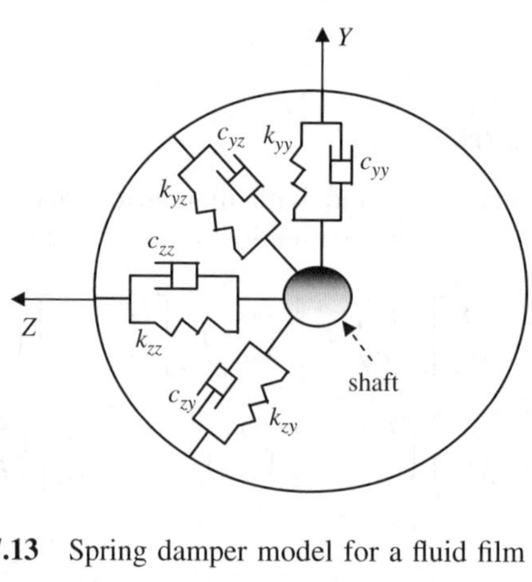

Figure 7.13 Spring damper model for a fluid film bearing

7.6.2.5 Rotor Bearing System Matrices

Assembly of the above element matrices for the shaft and disk/bearing components yields the system equation of motion in the form:

$$\boldsymbol{M}\ddot{\boldsymbol{u}} + (\boldsymbol{C} + \boldsymbol{G})\,\dot{\boldsymbol{u}} + \boldsymbol{K}u = \boldsymbol{F}(t) \tag{7.168}$$

C is the unsymmetric part of the damping matrix owing to the presence of bearings. It may also contain a symmetric counterpart due to a viscous damping mechanism (if present) in the shaft. $\boldsymbol{G}$ is the skew-symmetric matrix accounting for the gyroscopic coupling. The unsymmetric bearing stiffness coefficients render the stiffness matrix $\boldsymbol{K}$ also unsymmetric.

7.6.3 Unsymmetric Systems and Eigensolutions

For unsymmetric systems governed by equations such as Equation 7.168, the eigensolution involves transforming the equations into a $2N$-dimensional state-space form as:

$$[\bar{\boldsymbol{M}}]\begin{Bmatrix}\ddot{\boldsymbol{u}}\\ \dot{\boldsymbol{u}}\end{Bmatrix} + [\bar{\boldsymbol{K}}]\begin{Bmatrix}\dot{\boldsymbol{u}}\\ \boldsymbol{u}\end{Bmatrix} = \begin{Bmatrix}\boldsymbol{0}\\ \boldsymbol{F}(t)\end{Bmatrix} \tag{7.169}$$

where $\bar{\boldsymbol{M}} = \begin{bmatrix}\boldsymbol{0} & -\boldsymbol{M}\\ \boldsymbol{M} & \boldsymbol{C}\end{bmatrix}$ and $\bar{\boldsymbol{K}} = \begin{bmatrix}\boldsymbol{M} & \boldsymbol{0}\\ \boldsymbol{0} & \boldsymbol{K}\end{bmatrix}$.

With $\bar{\boldsymbol{K}}$ and $\bar{\boldsymbol{M}}$ unsymmetric and (λ, ϕ) an eigenpair, a generalized unsymmetric eigenvalue problem is thus given by:

$$\bar{\boldsymbol{K}}\boldsymbol{\phi} = \lambda\bar{\boldsymbol{M}}\boldsymbol{\phi} \tag{7.170}$$

A corresponding canonical form is $\boldsymbol{A}\boldsymbol{x} = \lambda\boldsymbol{x}$ with $\boldsymbol{A} = \bar{\boldsymbol{M}}^{-1}\bar{\boldsymbol{K}}$ (say), which is unsymmetric. The loss of symmetry no longer guarantees positive definiteness and this is a characteristic of a nonself-adjoint system. Analogous to the discussion on nonself-adjoint differential operators (Chapter 2), the adjoint $\boldsymbol{A}^*$ of the matrix $\boldsymbol{A}$ is defined by the relation $\langle \boldsymbol{y}, \boldsymbol{A}\boldsymbol{x}\rangle = \langle \boldsymbol{A}^*\boldsymbol{y}, \boldsymbol{x}\rangle$. Thus, for real vector spaces, $\boldsymbol{A}^* = \boldsymbol{A}^T$. If $\boldsymbol{A} \neq \boldsymbol{A}^*$, the matrix is said to be nonself-adjoint. In a real vector space, the matrix (i.e. the linear transformation operator) is real (i.e. all components are real numbers), but the eigenvalues and eigenvectors may not necessarily be real. For example, if $(\lambda, \boldsymbol{x})$ is an eigenpair of real unsymmetric matrix $\boldsymbol{A}$, we have:

$$\boldsymbol{A}\boldsymbol{x} = \lambda\boldsymbol{x} \tag{7.171}$$

Taking complex conjugates on both sides gives:

$$\boldsymbol{A}\boldsymbol{x}^* = \lambda^*\boldsymbol{x}^* \tag{7.172}$$

From the above equations, we also have

$$\boldsymbol{x}^{*T}\boldsymbol{A}\boldsymbol{x} = \lambda\boldsymbol{x}^{*T}\boldsymbol{x},\ \boldsymbol{x}^T\boldsymbol{A}\boldsymbol{x}^* = \lambda^*\boldsymbol{x}^T\boldsymbol{x}^* \tag{7.173a,b}$$

Subtracting transpose of Equation 7.173b from Equation 7.173a yields:

$$\boldsymbol{x}^{*T}(\boldsymbol{A} - \boldsymbol{A}^T)\boldsymbol{x} = (\lambda - \lambda^*)\boldsymbol{x}^{*T}\boldsymbol{x} \tag{7.174}$$

Unless $\boldsymbol{A}$ is symmetric, the above equation shows that $\lambda \neq \lambda^*$ implying that the eigenvalues of an unsymmetric matrix may be complex and the corresponding eigenvectors are also complex. It also follows that, if $\lambda = \sigma + \omega i$ is an eigenvalue, its complex conjugate $\lambda^* = \sigma - \omega i$ is also an eigenvalue. The eigenvector corresponding to $\sigma - \omega i$ is the complex conjugate, $\boldsymbol{x}^*$. The eigenvalue problem corresponding to $\boldsymbol{A}^T$ is known as the adjoint eigenvalue problem and is given by:

$$A^T y = \lambda y \tag{7.175}$$

Since $\det(A) = \det(A^T)$, the eigenvalues of A^T are identical to those of A. However, the eigenvectors differ from each other. Equation 7.175 may also be written as $y^T A = \lambda y^T$ and hence, whilst x is referred to as a right eigenvector, y is called a left eigenvector of A. If all the eigenvalues λ are distinct, these two vectors x and y are biorthogonal. Specifically, if Y and X contain the left and right eigenvectors, respectively, they can be so normalised that the following biorthogonality property is satisfied:

$$Y^T X = I, \text{ and } Y^T A X = \Lambda \tag{7.176}$$

I in the above equation is the identity matrix and $\Lambda = \mathrm{diag}[\lambda_1, \lambda_2, \ldots\ldots, \lambda_n]$. Referring to the generalized eigenvalue problem in Equation 7.170, the adjoint eigenvalue problems is given by:

$$\psi^T \bar{K} = \lambda \psi^T \bar{M} \tag{7.177}$$

The left and right eigenvectors ψ and Φ possess the biorthogonality property:

$$\psi_i^T \bar{M} \Phi_j = \delta_{ij} \quad \text{and} \quad \psi_i^T \bar{K} \Phi_j = \lambda_i \delta_{ij} \tag{7.178}$$

7.6.3.1 Unsymmetric Eigenvalue Problems and Solution Methods

Subspace iteration and Lanczos transformation methods, described earlier in the context of symmetric eigenvalue problems are also suited to handle unsymmetric eigenvalue problems. The methods, however, need modifications, the major one being the use of two sets of starting vectors to finally converge to the left and right eigenvectors. As with the symmetric problems, these two methods are also known for their efficiency in solving generalized unsymmetric eigenvalue problems involving large matrices. The relevant algorithms are discussed in the following sections.

7.6.3.2 Two-Sided Lanczos Transformation Method

Consider two independent sets of Lanczos vectors, $X = \{x_1, x_2, \ldots\ldots x_m\}$ and $Y = \{y_1, y_2, \ldots\ldots y_m\}$ with $Y^T \bar{M} X = I$. The Lanczos transformations, $\Phi = X\varphi$ and $\psi = Y\theta$ substituted in Equations 7.170 and 7.177 result in a standard eigenvalue problem, $T\varphi = \mu\varphi$ and its adjoint $\theta^T T = \theta^T \mu$ with

$$T = Y^T \bar{M} \bar{K}^{-1} \bar{M} X \tag{7.179}$$

Here, $\mu = \frac{1}{\lambda}$. φ and θ are the right and left eigenvectors of the matrix T. In the present case, the unsymmetric matrix T, which is unknown, is assumed to be of the tridiagonal form:

$$T = \begin{bmatrix} \alpha_1 & \beta_2 & 0 & \cdots & & \\ \delta_2 & \alpha_2 & \beta_3 & \cdots & & \\ 0 & \delta_3 & \alpha_3 & \cdots & & \\ & & & \cdots & & \\ 0 & 0 & 0 & \delta_{n-1} & \alpha_{n-1} & \beta_n \\ 0 & 0 & 0 & \cdots & \delta_n & \alpha_n \end{bmatrix} \tag{7.180}$$

The elements α_j, β_{j+1} and δ_{j+1i}, $j = 1, 2 \ldots$. are to be computed with $\beta_1 = \delta_1 = 0$. From Equation 7.179, we obtain the two identities:

$$\bar{K}^{-1}\bar{M}X = XT$$
$$\bar{K}^{-T}\bar{M}^{T}Y = YT \tag{7.181a,b}$$

With T defined as in Equation 7.180, the above equations result in the following two recurrence relationships (analogous to Equation 7.87 of the symmetric case):

$$\delta_{j+1}x_{j+1} = \bar{K}^{-1}\bar{M}x_j - \alpha_j x_j - \beta_j x_{j-1}$$
$$\beta_{j+1}y_{j+1} = \bar{K}^{-T}\bar{M}^{T}y_j - \alpha_j y_j - \delta_j y_{j-1},\ j = 1, 2, \ldots\ldots m \tag{7.182a,b}$$

From the orthogonality property $y_r^T\bar{M}x_s = \delta_{rs}$, Equation 7.182a gives $\alpha_j = y_j^T\bar{M}\bar{K}^{-1}\bar{M}x_j$ and (Equation 7.185b) gives $\alpha_j = x_j^T M^T K^{-T} M^T y_j$. Since $y_j^T\bar{M}\bar{K}^{-1}\bar{M}x_j = (x_j^T\bar{M}^T\bar{K}^{-T}\bar{M}^T y_j)^T$ is a scalar quantity, α_j can be obtained from any one of these expressions. To obtain δ_{j+1} and β_{j+1} for $j > 1$, let the expressions on the right-hand sides of Equation 7.182a,b be denoted by r_{j+1} and $\bar{r}_{j+1}$, respectively. The product $\bar{r}_{j+1}^T\bar{M}r_{j+1}$ gives:

$$\bar{r}_{j+1}^T\bar{M}r_{j+1} = \beta_{j+1}y_{j+1}^T\bar{M}\delta_{j+1}x_{j+1} = \beta_{j+1}\delta_{j+1}y_{j+1}^T\bar{M}x_{j+1} = \beta_{j+1}\delta_{j+1} \tag{7.183}$$

Equation 7.183 provides a choice in choosing the values for β_{j+1} and δ_{j+1}. If $\bar{r}_{j+1}^T\bar{M}r_{j+1} > 0$, $\beta_{j+1} = (\bar{r}_{j+1}^T\bar{M}r_{j+1})^{1/2} = \delta_{j+1}$ and if $\bar{r}_{j+1}^T\bar{M}r_{j+1} < 0$, $\beta_{j+1} = (|\bar{r}_{j+1}^T\bar{M}r_{j+1}|)^{\frac{1}{2}} = -\delta_{j+1}$. With $m \ll n$, $T\varphi = \mu\varphi$ is a reduced-order eigenvalue problem and can be solved, among others, by the QR transformation method (described in the section to follow). The two-sided Lanczos algorithm is detailed in Table 7.14.

Example 7.11: Eigensolution of a Nonuniform Rotor Bearing System

The shaft is of nonuniform circular cross section and is supported on hydrodynamic bearings. It carries a concentrated disk at a distance of 0.0889 m from left end. The geometric configuration is shown in Figure 7.14 and the relevant geometric details are given in Table 7.15. The same figure also shows the FE model of the rotor bearing system using beam elements. The disk element (added mass/inertia) is attached at node 5 and the two bearings at nodes 11 and 15, respectively. It is required to find the natural frequencies of the rotor bearing system.

Two sets of bearing coefficients are considered. In both of them, the damping coefficients are taken to be zero. The bearing stiffness coefficients matrices for the two sets are given below:

$$\text{Set 1: } K_b = \begin{bmatrix} 4.378 & E07 & 0.0 \\ 0.0 & 4.378 & E07 \end{bmatrix} \text{ N/m,}$$

$$\text{Set 2: } K = \begin{bmatrix} 3.503 & E07 & -8.756E06 \\ -8.756E06 & 3.503 & E07 \end{bmatrix} \text{ N/m}$$

The type of bearing, as in Set 1, is known as isotropic and that in Set 2 as orthotropic. Since the stiffness coefficient matrices in both the cases are symmetric, the only source of asymmetry in the system equations of motion is due to the skew-symmetric damping

Table 7.14 Two-sided Lanczos algorithm for systems with unsymmetric matrices

Aim: To obtain the lowest m eigenpairs of the generalized eigenvalue Problem (unsymmetric system): $\boldsymbol{K}\phi = \lambda \boldsymbol{M}\phi$ of size n

Step 1. Start with two randomly generated vectors $\boldsymbol{x}_0$ and $\boldsymbol{y}_0$; mass-orthogonalize the starting vectors

$$\boldsymbol{x}_1 = \frac{\boldsymbol{x}_0}{\left(\boldsymbol{y}_0^T \bar{\boldsymbol{M}} \boldsymbol{x}_0\right)^{1/2}} \quad \text{and} \quad \boldsymbol{y}_1 = \frac{y_0}{\left(\boldsymbol{y}_0^T \bar{\boldsymbol{M}} \boldsymbol{x}_0\right)^{1/2}} \tag{7.184}$$

Step 2. Set $\beta_1, \delta_1 = 0$.
Start Lanczos stages: For $j = 2, 3 \ldots\ldots, r(= 2 \times m < n)$

Step 3. Solve

$$\bar{\boldsymbol{K}}\bar{\boldsymbol{x}}_j = \bar{\boldsymbol{M}}\boldsymbol{x}_j, \bar{\boldsymbol{K}}^T \bar{\boldsymbol{y}}_j = \bar{\boldsymbol{M}}^T \boldsymbol{y}_j \tag{7.185}$$

Find $\alpha_j = \bar{\boldsymbol{y}}_j^T \bar{\boldsymbol{M}} \bar{\boldsymbol{x}}_j$

Step 4. Compute

$$\hat{\boldsymbol{x}}_j = \bar{\boldsymbol{x}}_{j-1} - \alpha_{j-1}\boldsymbol{x}_{j-1} - \beta_{j-1}\boldsymbol{x}_{j-2}$$

$$\hat{\boldsymbol{y}}_j = \bar{\boldsymbol{y}}_{j-1} - \alpha_{j-1}\boldsymbol{y}_{j-1} - \delta_{j-1}\boldsymbol{y}_{j-2} \tag{7.186a,b}$$

Step 5. Gram–Schmidt orthogonalization of $\hat{\boldsymbol{x}}_j$ and $\hat{\boldsymbol{y}}_j$ with respect to the previous Lanczos vectors:

$$\hat{\boldsymbol{x}}_j = \hat{\boldsymbol{x}}_j - \sum_{k=1}^{j-1} \left(\hat{\boldsymbol{y}}_j^T \boldsymbol{M} \boldsymbol{x}_k\right) \boldsymbol{x}_k$$

$$\hat{\boldsymbol{y}}_j = \hat{\boldsymbol{y}}_j - \sum_{k=1}^{j-1} \left(\hat{\boldsymbol{x}}_j^T \boldsymbol{M} \boldsymbol{y}_k\right) \boldsymbol{y}_k \tag{7.187a,b}$$

Step 6.

$$\text{If } \hat{\boldsymbol{y}}_j^T \boldsymbol{M}\boldsymbol{x}_k > 0, \beta_j = (\hat{\boldsymbol{y}}_j^T \boldsymbol{M}\boldsymbol{x}_k)^{\frac{1}{2}} = \delta_j$$

$$\text{If } \hat{\boldsymbol{y}}_j^T \boldsymbol{M}\boldsymbol{x}_k < 0, \beta_j = (|\hat{\boldsymbol{y}}_j^T \boldsymbol{M}\boldsymbol{x}_k|)^{\frac{1}{2}} = -\delta_j \tag{7.188a,b}$$

Step 7. Update Lanczos vectors:

$$\boldsymbol{x}_j = \frac{\hat{x}_j}{\beta_j} \text{ and } \boldsymbol{y}_j = \frac{\hat{y}_j}{\delta_j} \tag{7.189}$$

End of a Lanczos stage

Step 9: Form the tridiagonal matrix,

$$\boldsymbol{T} = \begin{bmatrix} \alpha_1 & \beta_2 & 0 & \ldots\ldots\ldots & & \\ \delta_2 & \alpha_2 & \beta_3 & \ldots\ldots\ldots & & \\ 0 & \delta_3 & \alpha_3 & \ldots\ldots\ldots & & \\ & & & \ldots\ldots & & \\ 0 & 0 & 0 & \delta_{r-1} & \alpha_{r-1} & \beta_r \\ 0 & 0 & 0 & \ldots\ldots & \delta_r & \alpha_r \end{bmatrix} \tag{7.190}$$

and solve $\boldsymbol{T}$ for its complex eigenvalues μ and (left and right) eigenvectors $\boldsymbol{\varphi}$ and $\boldsymbol{\theta}$

Step 10: Obtain eigensolution for the original system:

$$\lambda = \frac{1}{\mu}, \boldsymbol{\Phi} = \boldsymbol{X}\boldsymbol{\varphi} \text{ and } \boldsymbol{\psi} = \boldsymbol{Y}\boldsymbol{\theta} \tag{7.191}$$

Here $\boldsymbol{X}_{nxr} = \{\boldsymbol{X}_1, \boldsymbol{X}_2, \ldots \boldsymbol{X}_r\}$ and $\boldsymbol{Y}_{nxr} = \{\boldsymbol{Y}_1, \boldsymbol{Y}_2, \ldots \boldsymbol{Y}_r\}$

Step 11. Test convergence, If not convergent, repeat iterations on Lanczos stages

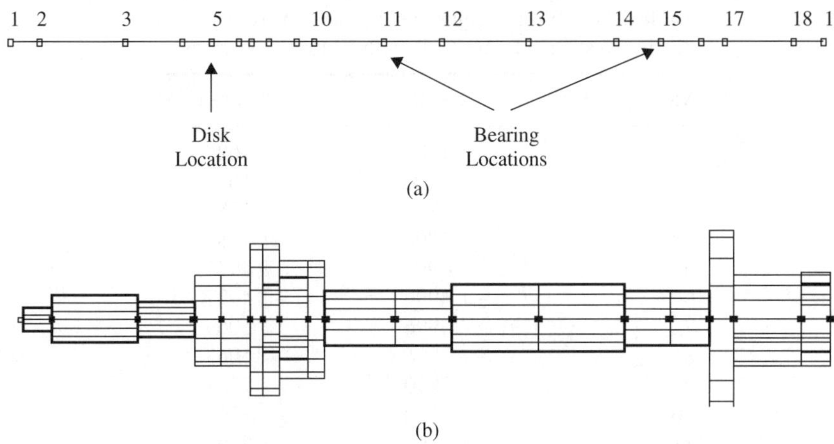

Figure 7.14 Rotor bearing system (Nelson and McVaugh, 1976); (a) FE model and (b) geometric configuration. Modified from Nelson and McVaugh, 1976

Table 7.15 Shaft geometry

Node number	Axial distance from left end (m)	Inner diameter (m)	Outer diameter (m)	Node number	Axial distance from left end (m)	Inner diameter (m)	Outer diameter (m)
1	0.0	–	0.0051	11	0.1651	–	0.0127
2	0.0127	–	0.0102	12	0.1905	–	0.0152
3	0.0508	–	0.0076	13	0.2286	–	0.0152
4	0.0762	–	0.0203	14	0.2667	–	0.0127
5	0.0899	–	0.0203	15	0.2870	–	0.0127
6	0.1016	–	0.0330	16	0.3048	–	0.0381
7	0.1067	0.0152	0.0330	17	0.3150	–	0.0203
8	0.1141	0.0178	0.0254	18	0.3454	0.0152	0.0203
9	0.1270	–	0.0254	19	0.3581	0.0152	0.0203
10	0.1346	–	0.0127	–	–	–	–

matrix (G). The procedure of reducing the equations into two symmetric eigenvalue problems, as described earlier, could also have been followed. However, for purposes of numerical illustration, the solution is presently obtained by the two-sided Lanczos method. Tables 7.16 and 7.17, respectively, show the first few eigenvalues (also known as whirl speeds) of the rotor bearing system at rotation speeds 0 and 20 000 RPM for both the bearing types. As the system is undamped and skew-symmetric, the eigenvalues are purely imaginary.

Once more, note that the eigenvalues occur in conjugate pairs. Figures 7.15 and 7.16 show the variation of the whirl speeds with respect to the rotation speed, Ω. These figures are referred to as Campbell diagrams. Note that these figures are drawn with frequencies corresponding to displacement degree of freedom only (i.e. with alternate frequencies listed in Tables 7.16 and 7.17). It can be observed that as the rotation speed increases, each pair of whirl frequencies separate out and, in rotor dynamics terminology, one is known as the backward whirl and the other, the forward whirl. The nature of the

Table 7.16 Whirl frequencies in rad/s for rotor bearing system (Figure 7.14) with isotropic bearings

Mode number	$\Omega = 0$	$\Omega = 20\,000$ RPM
1	i 1704	i 1627
2	i 1704	$-i$ 1627
3	i 1704	i 1780
4	$-i$ 1704	$-i$ 1780
5	i 4996	i 4927
6	$-i$ 4996	$-i$ 4927
7	i 4996	i 5063
8	$-i$ 4996	$-i$ 5063
9	i 7990	i 7764
10	$-i$ 7990	$-i$ 7764
11	i 7990	i 8207
12	$-i$ 7990	$-i$ 8207

Table 7.17 Whirl frequencies in rad/s for rotor bearing system (Figure 7.14) with orthotropic bearings

Mode number	$\Omega = 0$	$\Omega = 20\,000$ RPM
1	i 1480	i 1461
2	$-i$ 1480	$-i$ 1461
3	i 1702	i 1721
4	$-i$ 1702	$-i$ 1721
5	i 4156	i 4152
6	$-i$ 4156	$-i$ 4152
7	i 4990	i 4991
8	$-i$ 4990	$-i$ 4991
9	i 6988	i 6937
10	$-i$ 6988	$-i$ 6937
11	i 7981	i 8025
12	$-i$ 7981	$-i$ 8025

whirl – forward or backward (Appendix G) – depends on the direction of the whirl orbit. In this context, the rotation speeds coinciding with the whirl speeds are known as the critical speeds of rotation. These critical speeds are obtained from the intersections of the synchronous speed line (a 45° line) with the whirl-speed curves. Table 7.18 shows these critical speeds for the rotor bearing system.

7.6.3.3 QR Transformation Method and Unsymmetric Matrices

The Lanczos method involves, at each iteration, reduction of the original size, n, of the eigenvalue problem to one of size $m \ll n$. This reduced eigenvalue problem needs to be solved to obtain the eigensolution at the end of each iteration. The QR transformation method, described earlier in the context of symmetric eigenvalue problems, is also applicable to extract the complex eigenvalues and eigenvectors of an unsymmetric

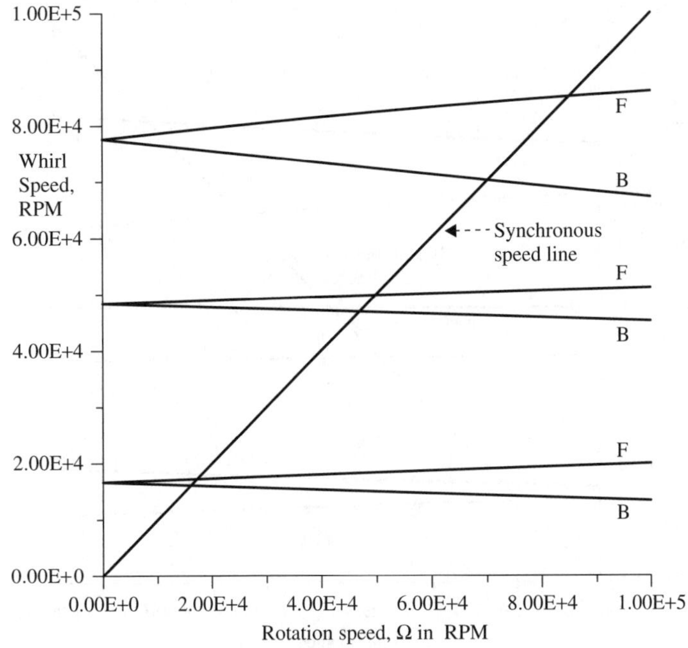

Figure 7.15 Rotor bearing system (Figure 7.14) with isotropic bearings: Campbell diagram at the disk node 5

Table 7.18 Critical speeds in RPM of rotor bearing system in Figure 7.14

Serial no.	Isotropic bearings		Orthotropic bearings	
	Backward	Forward	Backward	Forward
1	15 550	17 100	14 240	16 740
2	46 500	49 750	39 820	48 440
3	68 250	82 100	64 700	81 600

matrix. The QR method is especially effective if the matrix is already in a Hessenberg or tridiagonal form. Consider the eigenvalue problem $\boldsymbol{AX} = \lambda\boldsymbol{X}$, where $\boldsymbol{A}$ is unsymmetric. The QR method employs an orthogonal similarity transformation and reduces $\boldsymbol{A}$ to a nearly triangular form (Jordan's canonical form). The method is iterative and, at the iteration k, the matrix $\boldsymbol{A}_k$ is first decomposed as $\boldsymbol{A}_k = \boldsymbol{Q}_k\boldsymbol{R}_k$, where $\boldsymbol{Q}_k$ is orthogonal and $\boldsymbol{R}_k$ upper triangular. Here, $\boldsymbol{A}_0 = \boldsymbol{A}$. By the orthogonal similarity transformation, the transformed matrix $\boldsymbol{A}_{k+1}$ is obtained as $\boldsymbol{A}_{k+1}= \boldsymbol{Q}_k^T \boldsymbol{A}_k \boldsymbol{Q}_k = \boldsymbol{Q}_k^T \left(\boldsymbol{Q}_k \boldsymbol{R}_k\right) \boldsymbol{Q}_k = \boldsymbol{R}_k \boldsymbol{Q}_k$. In the limit as $k \to \infty$, $\boldsymbol{A}_{k+1}$ tends to a nearly triangular matrix. The complex eigenvalues can be extracted from each of the 2×2 submatrices across the diagonal line of the nearly triangular matrix (i.e. with some nonzero elements on the subdiagonal line). Further, whilst $\boldsymbol{Q}_1 \boldsymbol{Q}_2 \ldots \ldots \boldsymbol{Q}_{k-1} \boldsymbol{Q}_k \to \boldsymbol{X}$, the associated (right) eigenvector matrix, the transpose $\boldsymbol{X}^T$ yields the left eigenvector matrix $\boldsymbol{Y}$, associated with the adjoint eigenvalue problem $\boldsymbol{Y}^T \boldsymbol{A} = \boldsymbol{Y}^T \lambda$.

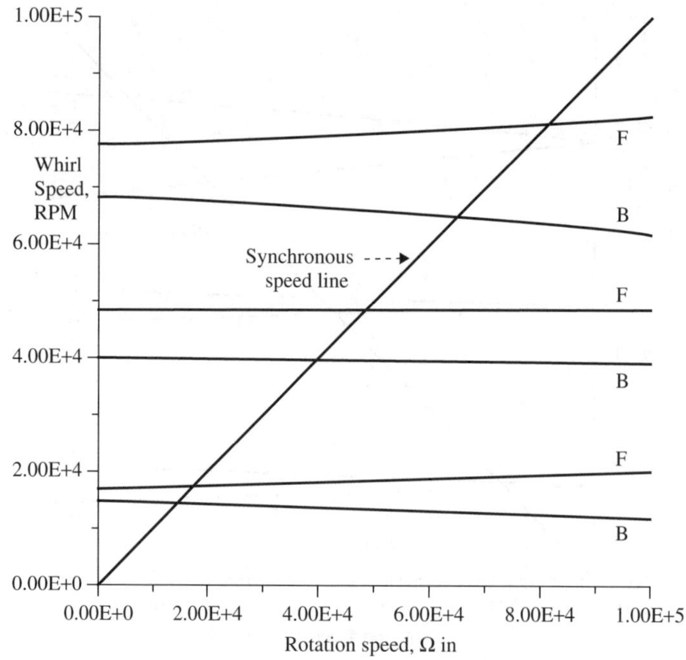

Figure 7.16 Rotor bearing system (Figure 7.14) with orthotropic bearings: Campbell diagram at the disk node 5. F – Forward whirl, B – Backward whirl.

Example 7.12: Eigensolution of a tridiagonal matrix by the QR method

The QR method is now applied to the tridiagonal matrix, $\boldsymbol{T}$ corresponding to the rotor bearing system in Figure 7.14. Specifically, the case with isotropic bearings and rotation speed, $\Omega = 20\,000$ RPM is considered. Whilst the size $2n$ of the state-space matrices $\bar{\boldsymbol{K}}$ and $\bar{\boldsymbol{M}}$ is 152, the number of eigenvalues required (m) is 12. Thus, the tridiagonal matrix $\boldsymbol{T}$ is generated corresponding to size $2m = 24$. Reduction of the matrix $\boldsymbol{T}$ into a nearly triangular form by QR transformation is shown in Table 7.19. The 2×2 submatrices around the diagonal of the nearly triangular matrix are highlighted in the table. This result is obtained at the end of 300 iterations via the orthogonal similarity transformation $\boldsymbol{T}_{k+1} = \boldsymbol{Q}_k^T \boldsymbol{T}_k \boldsymbol{Q}_k$. Pairs of complex eigenvalues are obtained from each of these submatrices and are shown in Table 7.20. Since the original system matrices are only skew-symmetric, the eigenvlaues are purely imaginary. The first six pairs of eigenvalues are obtained with high accuracy at the end of 300 iterations (see Table 7.16 for comparison).

7.7 Dynamic Condensation and Eigensolution

The eigensolution methods (symmetric or otherwise) described in the preceding sections involve floating-point operations on full system matrices. In the process, these matrices may need assembling at every stage of computation. This invariably demands enormous computational and storage overheads. For such large systems, it is often useful to work with reduced-order models. Dynamic condensation methods constitute a family of such

Table 7.19 Nearly triangular matrix by QR transformation (after 300 iterations)

Column 1	Column 2	Column 3	Column 4	Column 5	Column 6
0.13E−4	**−0.85E−3**	0.39E−3	−0.11E−2	0.18E−4	0.34E−4
0.45E−3	**−0.13E−4**	−0.77E−3	−0.43E−4	−0.25E−3	−0.26E−4
0.000	0.0	**0.27E−3**	**−0.48E−3**	−0.22E−3	0.27E−4
0.000	0.0	**0.81E−3**	**−0.27E−3**	0.11E−3	0.95E−5
0.000	0.0	0.	0.	**0.56E−4**	**−0.24E−3**
0.000	0.0	0.	0.	**0.19E−3**	**−0.56E−4**
0.0	0.0	0.	0.	0.	0.
0.0	0.0	0.	0.	0.	0.
0.0	0.0	0.	0.	0.	0.
0.0	0.0	0.	0.	0.	0.
0.0	0.0	0.	0.	0.	0.
0.0	0.0	0.	0.	0.	0.

Column 7	Column 8	Column 9	Column 10	Column 11	Column 12
−0.52E−4	0.40E−4	0.91E−5	0.43E−4	−0.26E−4	0.10E−4
−0.20E−3	0.22E−4	−0.15E−4	−0.24E−4	0.97E−6	−0.93E−5
−0.21E−3	0.59E−4	−0.12E−3	0.73E−6	−0.94E−4	0.25E−5
0.39E−4	0.58E−5	0.10E−3	0.43E−4	0.43E−4	0.71E−5
0.31E−3	−0.20E−3	0.19E−3	0.37E−4	0.13E−3	0.18E−4
0.84E−4	−0.22E−3	−0.74E−4	−0.25E−3	0.85E−4	−0.12E−3
0.13E−3	**−0.20E−3**	−0.52E−4	−0.37E−5	−0.66E−4	−0.55E−5
0.29E−3	**−0.13E−3**	0.44E−4	0.11E−3	−0.54E−4	0.57E−4
0.	0.	**0.29E−4**	**−0.14E−3**	0.13E−3	0.26E−4
0.	0.	**0.12E−3**	**−0.29E−4**	0.62E−4	−0.87E−4
0.	0.	0.	0.	**−0.14E−4**	**−0.11E−3**
0.	0.	0.	0.	**0.14E−3**	**0.14E−4**

Table 7.20 Eigenvalues of tridiagonal matrix by *QR* transformation; (see the result for the rotor bearing system with isotropic bearings at $\Omega = 20\,000$ RPM in Table 7.16)

Mode no.	At the end of 100 iterations		At the end of 200 iterations		At the end of 300 iterations	
	Real part	Imaginary part	Real part	Imaginary part	Real part	Imaginary part
1	−8.65E−3	1627.12	−1.06E−6	1627.13	−1.21E−10	1627.13
2	−8.65E−3	−1627.12	−1.06E−6	−1627.13	−1.21E−10	−1627.13
3	1.04E−2	1780.59	1.26E−6	1780.55	1.61E−10	1780.55
4	1.04E−2	−1780.59	1.26E−6	−1780.55	1.61E−10	−1780.55
5	0.04	4914.71	1.28E−2	4923.43	7.64E−4	4923.92
6	0.04	−4914.71	1.28E−2	−4923.43	7.64E−4	−4923.92
7	−0.26	5073.80	−1.35E−2	5064.84	−8.08E−4	5064.34
8	−0.26	−5073.80	−1.35E−2	−5064.84	−8.08E−4	−5064.34
9	0.38	7826.21	1.52E−3	7827.52	6.07E−6	7827.52
10	0.38	−7826.21	1.52E−3	7827.52	6.07E−6	−7827.52
11	−0.43	8273.20	−1.70E−3	8272.04	−6.77E−6	8272.04
12	−0.43	−8273.20	−1.70E−3	−8272.04	−6.77E−6	−8272.04

schemes that provide such models. In the following, we briefly look at a few of the dynamic condensation methods (Surez and Singh, 1992; Qu and Fu, 1998; Kane and Torby, 1991) suitable for both symmetric and unsymmetric systems.

7.7.1 Symmetric Systems and Dynamic Condensation

Consider the generalized symmetric eigenvalue problem $\boldsymbol{K\Phi} = \boldsymbol{M\Phi\Lambda}$ where $\boldsymbol{\Lambda} = \mathrm{diag}\,[\lambda_1, \lambda_2, \ldots\ldots, \lambda_n]$ and $\boldsymbol{\Phi}$ is the matrix of eigenvectors. Let m ($\ll n$) be the number of the so-called 'master' degrees of freedom to be retained in the reduced-order model, which consequently eliminates the so-called 'slave' degrees of freedom ($n - m$ of them). The choice of these master degrees of freedom is mostly problem dependent and often driven by the semidiscretized model size, response locations of interest and the boundary constraints present in the system. Let a generic partition of the symmetric eigenvalue problem be as follows:

$$\boldsymbol{K\Phi} = \boldsymbol{M\Phi\Lambda} \Rightarrow \begin{bmatrix} \boldsymbol{K}_{mm} & \boldsymbol{K}_{ms} \\ \boldsymbol{K}_{sm} & \boldsymbol{K}_{ss} \end{bmatrix} \begin{bmatrix} \boldsymbol{\Phi}_m \\ \boldsymbol{\Phi}_s \end{bmatrix} = \begin{bmatrix} \boldsymbol{M}_{mm} & \boldsymbol{M}_{ms} \\ \boldsymbol{M}_{sm} & \boldsymbol{M}_{ss} \end{bmatrix} \begin{bmatrix} \boldsymbol{\Phi}_m \\ \boldsymbol{\Phi}_s \end{bmatrix} [\boldsymbol{\Lambda}] \tag{7.192}$$

The partitioned forms of the associated orthogonality properties $\boldsymbol{\Phi}^T\boldsymbol{M\Phi} = \boldsymbol{I}$ and $\boldsymbol{\Phi}^T\boldsymbol{K\Phi} = \boldsymbol{\Lambda}$ are:

$$\begin{bmatrix} \boldsymbol{\Phi}_m^T & \boldsymbol{\Phi}_s^T \end{bmatrix} \begin{bmatrix} \boldsymbol{M}_{mm} & \boldsymbol{M}_{ms} \\ \boldsymbol{M}_{sm} & \boldsymbol{M}_{ss} \end{bmatrix} \begin{bmatrix} \boldsymbol{\Phi}_m \\ \boldsymbol{\Phi}_s \end{bmatrix} = [\boldsymbol{I}] \quad \text{and}$$

$$\begin{bmatrix} \boldsymbol{\Phi}_m^T & \boldsymbol{\Phi}_s^T \end{bmatrix} \begin{bmatrix} \boldsymbol{K}_{mm} & \boldsymbol{K}_{ms} \\ \boldsymbol{K}_{sm} & \boldsymbol{K}_{ss} \end{bmatrix} \begin{bmatrix} \boldsymbol{\Phi}_m \\ \boldsymbol{\Phi}_s \end{bmatrix} = [\boldsymbol{\Lambda}] \tag{7.193a,b}$$

Expanding the lower part of Equation 7.192, we have:

$$\boldsymbol{K}_{ss}\boldsymbol{\Phi}_s = -\boldsymbol{K}_{sm}\boldsymbol{\Phi}_m + \left[\boldsymbol{M}_{sm}\boldsymbol{\Phi}_m + \boldsymbol{M}_{ss}\boldsymbol{\Phi}_s\right][\boldsymbol{\Lambda}] \tag{7.194}$$

Now, let

$$\boldsymbol{\Phi}_s = \boldsymbol{R\Phi}_m \tag{7.195}$$

$\boldsymbol{R}$ is an unknown transformation matrix linking the so-called 'slave' eigenvectors with the 'master' eigenvectors. Substituting $\boldsymbol{R}\Phi_m$ for Φ_s in Equation 7.194, one obtains an implicit expression for the unknown $\boldsymbol{R}$ matrix as

$$\boldsymbol{R} = \boldsymbol{K}_{ss}^{-1}\left\{-\boldsymbol{K}_{sm} + \left[\boldsymbol{M}_{sm} + \boldsymbol{M}_{ss}\boldsymbol{R}\right]\left[\boldsymbol{\Phi}_m\right][\boldsymbol{\Lambda}]\left[\boldsymbol{\Phi}_m^{-1}\right]\right\} \tag{7.196}$$

Equation 7.193 also transform to:

$$\boldsymbol{\Phi}_m^T\left[\boldsymbol{M}_{mm} + \boldsymbol{M}_{ms}\boldsymbol{R} + \boldsymbol{R}^T\boldsymbol{M}_{sm} + \boldsymbol{R}^T\boldsymbol{M}_{ss}\boldsymbol{R}\right]\boldsymbol{\Phi}_m = \boldsymbol{I}$$

$$\boldsymbol{\Phi}_m^T\left[\boldsymbol{K}_{mm} + \boldsymbol{K}_{ms}\boldsymbol{R} + \boldsymbol{R}^T\boldsymbol{K}_{sm} + \boldsymbol{R}^T\boldsymbol{K}_{ss}\boldsymbol{R}\right]\boldsymbol{\Phi}_m = \boldsymbol{\Lambda} \tag{7.197a,b}$$

Expressions inside the square brackets in the above equations show that the required reduced-order model is represented by the equivalent mass and stiffness matrices:

$$[\boldsymbol{M}_R]_{m\times m} = \boldsymbol{M}_{mm} + \boldsymbol{M}_{ms}\boldsymbol{R} + \boldsymbol{R}^T\boldsymbol{M}_{sm} + \boldsymbol{R}^T\boldsymbol{M}_{ss}\boldsymbol{R} \quad \text{and}$$

$$[\boldsymbol{K}_R]_{m\times m} = \boldsymbol{K}_{mm} + \boldsymbol{K}_{ms}\boldsymbol{R} + \boldsymbol{R}^T\boldsymbol{K}_{sm} + \boldsymbol{R}^T\boldsymbol{K}_{ss}\boldsymbol{R} \tag{7.198a,b}$$

Equation 7.197 describes the orthogonality properties for the reduced-order model. The objective is to obtain the reduced-order model represented by $\boldsymbol{M}_R$ and $\boldsymbol{K}_R$. This requires that Equations 7.196 and 7.197 are solved simultaneously. To this end, an iterative procedure is considered starting with an initial approximation for $\boldsymbol{R}$ as $\boldsymbol{K}_{ss}^{-1}\boldsymbol{K}_{sm}$ (first term on the right-hand side of Equation 7.196). The rest of the iterative scheme is detailed in Table 7.21. It is assumed that only m eigenvalues (equal to the number of selected 'master' degree of freedom) are required to be extracted and the prior selection of the 'master' degrees of freedom must be consistent with this requirement.

Following Equation 7.196, it seems necessary that an eigensolution is needed at each iteration. However, from the two orthogonality relationships in Equation 7.197, the eigensolution can be avoided if the following substitution for $\boldsymbol{\Phi}_m\boldsymbol{\Lambda}\boldsymbol{\Phi}_m^{-1}$ is effected in Equation 7.196:

$$\boldsymbol{\Phi}_m\boldsymbol{\Lambda}\boldsymbol{\Phi}_m^{-1} = \boldsymbol{\Phi}_m\left[\boldsymbol{\Phi}_m^T\boldsymbol{K}_R\boldsymbol{\Phi}_m\right]\boldsymbol{\Phi}_m^{-1} = \boldsymbol{\Phi}_m\boldsymbol{\Phi}_m^T\boldsymbol{K}_R = \left[\boldsymbol{\Phi}_m^T\boldsymbol{M}_R\right]^{-1}\boldsymbol{\Phi}_m^T\boldsymbol{K}_R = \boldsymbol{M}_R^{-1}\boldsymbol{K}_R \tag{7.199}$$

For checking the convergence of the eigensolution, the reduced eigenvalue problem Equation 7.201 is solved at intermediate stages of iterations by any of the extraction methods described earlier.

Example 7.13: Eigensolution by dynamic condensation of a symmetric system

The rotor bearing system in Figure 7.14 is considered for this example with zero rotation speed so that the resulting matrices are symmetric. The bearings are considered to be isotropic and no other sources of damping (except the bearing damping) are assumed to be present. Presently, the dynamic condensation is effected directly on the system matrices

Table 7.21 Iterative technique for dynamic condensation of a symmetric system

Partition the stiffness and mass matrices with m 'master' degrees of freedom and $n - m$ 'slave' degrees of freedom

Initially assume $\boldsymbol{R}_0 = -\boldsymbol{K}_{ss}^{-1}\boldsymbol{K}_{sm}$

Start Iterations For $k = 1, 2 \ldots$.

Step 1. Compute:

$$\boldsymbol{M}_R^{(k)} = \boldsymbol{M}_{mm} + \boldsymbol{M}_{ms}\boldsymbol{R}_{k-1} + \boldsymbol{R}_{k-1}^T\boldsymbol{M}_{sm} + \boldsymbol{R}_{k-1}^T\boldsymbol{M}_{ss}\boldsymbol{R}_{k-1}$$
$$\boldsymbol{K}_R^{(k)} = \boldsymbol{K}_{mm} + \boldsymbol{K}_{ms}\boldsymbol{R}_{k-1} + \boldsymbol{R}_{k-1}^T\boldsymbol{K}_{sm} + \boldsymbol{R}_{k-1}^T\boldsymbol{K}_{ss}\boldsymbol{R}_{k-1} \tag{7.200a,b}$$

Step 2.
solve the reduced-order eigenvalue problem:

$$\boldsymbol{K}_R^{(k)}\boldsymbol{\Phi}_m^{(k)} = \boldsymbol{M}_R^{(k)}\boldsymbol{\Phi}_m^{(k)}\boldsymbol{\Lambda}_m^{(k)} \tag{7.201}$$

to check for convergence of eigenvalues:

Step 4. If convergence is not achieved, revise the transformation matrix R :

$$\boldsymbol{R}_k = \boldsymbol{K}_{ss}^{-1}\left\{-\boldsymbol{K}_{sm} + \left[\boldsymbol{M}_{sm} + \boldsymbol{M}_{ss}\boldsymbol{R}_{k-1}\right]\boldsymbol{M}_R^{-1}\boldsymbol{K}_R\right\} \tag{7.202}$$

and repeat steps 1–3.

Table 7.22 Dynamic condensation of a symmetric system; eigensolution for the rotor bearing system (Figure 7.14) with zero rotation and isotropic bearings (see Table 7.16)

Mode no.	Eigenvalue solution of full system	Dynamic condensation scheme (Master dof = 12)
1	1704.08	1704.08
2	1704.08	1704.08
3	4996.25	4996.25
4	4996.25	4996.25
5	7990.47	7990.48
6	7990.47	7990.48
7	12736.28	12736.27
8	12736.28	12736.27

K and M without rewriting the system equations in state-space form. Considering the FE model in Figure 7.14, we have the total number of degrees of freedom, $n = 76$. The transverse degrees of freedom of the disk and bearing nodes are included in the 'master' degrees of freedom. Thus, $m = 12$.

The results presented in Table 7.22 match well with those in Table 7.16 obtained for the full system.

7.7.2 Unsymmetric Systems and Dynamic Condensation

Consider the unsymmetric eigenvalue problem $\bar{\boldsymbol{K}}\boldsymbol{\Phi} = \bar{\boldsymbol{M}}\boldsymbol{\Phi}\boldsymbol{\Lambda}$ and its adjoint $\bar{\boldsymbol{K}}^T\boldsymbol{\psi} = \bar{\boldsymbol{M}}^T\boldsymbol{\psi}^T\boldsymbol{\Lambda}^T$ with the system matrices $\bar{\boldsymbol{M}} = \begin{bmatrix} \boldsymbol{0} & -\boldsymbol{M} \\ \boldsymbol{M} & \boldsymbol{C} \end{bmatrix}$ and $\bar{\boldsymbol{K}} = \begin{bmatrix} \boldsymbol{M} & \boldsymbol{0} \\ 0 & \boldsymbol{K} \end{bmatrix}$. Here, $\boldsymbol{\Lambda} = \text{diag}\,[\lambda_1, \lambda_2, \ldots \ldots, \lambda_n]$ with λ_i, $i = 1, 2 \ldots n$ being complex. In partitioned form with respect to the master and slave degrees of freedom, the two eigenvalue problems are written as:

$$\begin{bmatrix} \bar{\boldsymbol{K}}_{mm} & \bar{\boldsymbol{K}}_{ms} \\ \bar{\boldsymbol{K}}_{sm} & \bar{\boldsymbol{K}}_{ss} \end{bmatrix} \begin{bmatrix} \boldsymbol{\Phi}_m \\ \boldsymbol{\Phi}_s \end{bmatrix} = \begin{bmatrix} \bar{\boldsymbol{M}}_{mm} & \bar{\boldsymbol{M}}_{ms} \\ \bar{\boldsymbol{M}}_{sm} & \bar{\boldsymbol{M}}_{ss} \end{bmatrix} \begin{bmatrix} \boldsymbol{\Phi}_m \\ \boldsymbol{\Phi}_s \end{bmatrix} [\boldsymbol{\Lambda}]$$

$$\begin{bmatrix} \bar{\boldsymbol{K}}_{mm} & \bar{\boldsymbol{K}}_{ms} \\ \bar{\boldsymbol{K}}_{sm} & \bar{\boldsymbol{K}}_{ss} \end{bmatrix}^T \begin{bmatrix} \boldsymbol{\psi}_m \\ \boldsymbol{\psi}_s \end{bmatrix} = \begin{bmatrix} \bar{\boldsymbol{M}}_{mm} & \bar{\boldsymbol{M}}_{ms} \\ \bar{\boldsymbol{M}}_{sm} & \bar{\boldsymbol{M}}_{ss} \end{bmatrix}^T \begin{bmatrix} \boldsymbol{\psi}_m \\ \boldsymbol{\psi}_s \end{bmatrix} [\boldsymbol{\Lambda}]^T \tag{7.203a,b}$$

Here, $\bar{\boldsymbol{K}}_{mm} = \begin{bmatrix} \boldsymbol{M}_{mm} & \boldsymbol{0} \\ 0 & \boldsymbol{K}_{mm} \end{bmatrix}$, $\bar{\boldsymbol{M}}_{mm} = \begin{bmatrix} \boldsymbol{0} & -\boldsymbol{M}_{mm} \\ \boldsymbol{M}_{mm} & \boldsymbol{C}_{mm} \end{bmatrix}$,

$$\bar{\boldsymbol{K}}_{ms} = \begin{bmatrix} \boldsymbol{M}_{ms} & \boldsymbol{0} \\ 0 & \boldsymbol{K}_{ms} \end{bmatrix}, \bar{\boldsymbol{M}}_{ms} = \begin{bmatrix} \boldsymbol{0} & -\boldsymbol{M}_{ms} \\ \boldsymbol{M}_{ms} & \boldsymbol{C}_{ms} \end{bmatrix},$$

$$\bar{\boldsymbol{K}}_{sm} = \begin{bmatrix} \boldsymbol{M}_{sm} & \boldsymbol{0} \\ 0 & \boldsymbol{K}_{sm} \end{bmatrix}, \bar{\boldsymbol{M}}_{sm} = \begin{bmatrix} \boldsymbol{0} & -\boldsymbol{M}_{sm} \\ \boldsymbol{M}_{sm} & \boldsymbol{C}_{sm} \end{bmatrix} \text{ and}$$

$$\bar{\boldsymbol{K}}_{ss} = \begin{bmatrix} \boldsymbol{M}_{ss} & \boldsymbol{0} \\ 0 & \boldsymbol{K}_{ss} \end{bmatrix}, \bar{\boldsymbol{M}}_{ms} = \begin{bmatrix} \boldsymbol{0} & -\boldsymbol{M}_{ss} \\ \boldsymbol{M}_{ss} & \boldsymbol{C}_{ss} \end{bmatrix}$$

The associated biorthogonality properties $\boldsymbol{\psi}^T \boldsymbol{M}\boldsymbol{\Phi} = \boldsymbol{I}$ and $\boldsymbol{\psi}^T \boldsymbol{K}\boldsymbol{\Phi} = \boldsymbol{\Lambda}$ are also partitioned as:

$$[\boldsymbol{\psi}_m^T \quad \boldsymbol{\psi}_s^T]\begin{bmatrix}\bar{\boldsymbol{M}}_{mm} & \bar{\boldsymbol{M}}_{ms}\\ \bar{\boldsymbol{M}}_{sm} & \bar{\boldsymbol{M}}_{ss}\end{bmatrix}\begin{bmatrix}\boldsymbol{\Phi}_m\\ \boldsymbol{\Phi}_s\end{bmatrix} = [\boldsymbol{I}] \text{ and}$$

$$[\boldsymbol{\psi}_m^T \quad \boldsymbol{\psi}_s^T]\begin{bmatrix}\bar{\boldsymbol{K}}_{mm} & \bar{\boldsymbol{K}}_{ms}\\ \bar{\boldsymbol{K}}_{sm} & \bar{\boldsymbol{K}}_{ss}\end{bmatrix}\begin{bmatrix}\boldsymbol{\Phi}_m\\ \boldsymbol{\Phi}_s\end{bmatrix} = [\boldsymbol{\Lambda}] \qquad (7.204\text{a,b})$$

Expanding the lower part of the Equation 7.203, we have:

$$\bar{\boldsymbol{K}}_{ss}\boldsymbol{\Phi}_s = -\bar{\boldsymbol{K}}_{sm}\boldsymbol{\Phi}_m + \left[\bar{\boldsymbol{M}}_{sm}\boldsymbol{\Phi}_m + \bar{\boldsymbol{M}}_{ss}\boldsymbol{\Phi}_s\right][\boldsymbol{\Lambda}]$$

$$\bar{\boldsymbol{K}}_{ss}^T\boldsymbol{\psi}_s = -\bar{\boldsymbol{K}}_{sm}^T\boldsymbol{\psi}_m + \left[\bar{\boldsymbol{M}}_{sm}^T\boldsymbol{\psi}_m + \bar{\boldsymbol{M}}_{ss}^T\boldsymbol{\Phi}_s\right][\boldsymbol{\Lambda}]^T \qquad (7.205\text{a,b})$$

The dynamic condensation in the present case also follows a procedure (Rao, 2002) similar to the one adopted for the symmetric eigenvalue problem. However, the transformation between the master and slave degrees of freedom is two-sided in that two transformation matrices are defined with respect to right and left eigenvectors in the form:

$$\boldsymbol{\Phi}_s = \boldsymbol{R}\boldsymbol{\Phi}_m \text{ and } \boldsymbol{\psi}_s = \boldsymbol{S}\boldsymbol{\psi}_m \qquad (7.206\text{a,b})$$

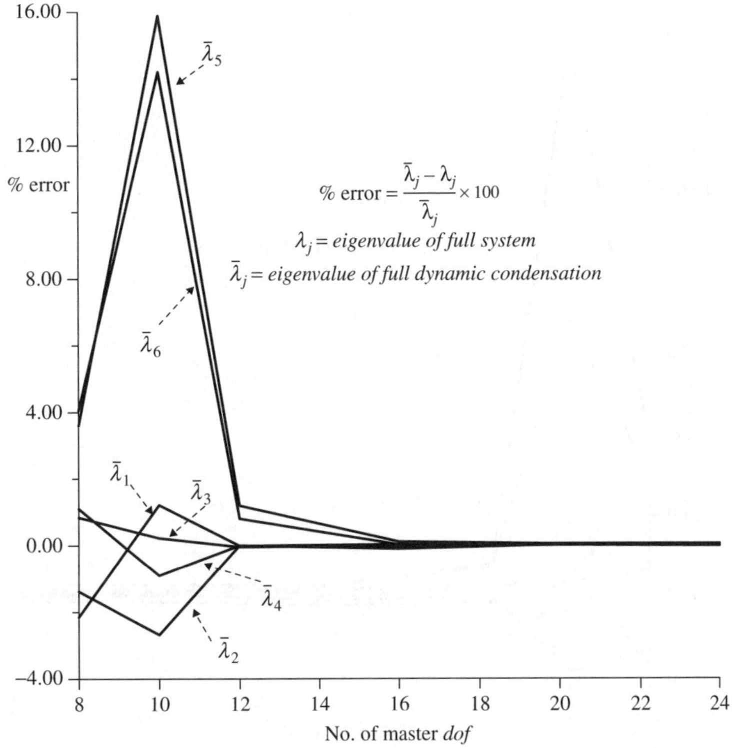

Figure 7.17 Percentage error between eigenvalue results of full model and reduced-order model for rotor bearing system (Figure 7.14) with isotropic bearings

Substituting the above transformations in Equation 7.205 gives:

$$\boldsymbol{R} = \bar{\boldsymbol{K}}_{ss}^{-1}\left\{-\bar{\boldsymbol{K}}_{sm} + \left[\bar{\boldsymbol{M}}_{sm} + \bar{\boldsymbol{M}}_{ss}\boldsymbol{R}\right][\boldsymbol{\Phi}_m][\boldsymbol{\Lambda}][\boldsymbol{\Phi}_m^{-1}]\right\}$$
$$\boldsymbol{S} = \bar{\boldsymbol{K}}_{ss}^{-T}\left\{-\bar{\boldsymbol{K}}_{sm}^{T} + [\bar{\boldsymbol{M}}_{sm}^{T} + \bar{\boldsymbol{M}}_{ss}^{T}\boldsymbol{S}][\boldsymbol{\psi}_m][\boldsymbol{\Lambda}]^T[\boldsymbol{\psi}_m^{-1}]\right\} \tag{7.207a,b}$$

In terms of $\boldsymbol{R}$ and $\boldsymbol{S}$, the biorthogonality relationships take the form:

$$\boldsymbol{\psi}_m^T\left[\bar{\boldsymbol{M}}_{mm} + \bar{\boldsymbol{M}}_{ms}\boldsymbol{R} + \boldsymbol{S}^T\bar{\boldsymbol{M}}_{sm} + \boldsymbol{S}^T\bar{\boldsymbol{M}}_{ss}\boldsymbol{R}\right]\boldsymbol{\Phi}_m = \boldsymbol{I}$$
$$\boldsymbol{\psi}_m^T\left[\bar{\boldsymbol{K}}_{mm} + \bar{\boldsymbol{K}}_{ms}\boldsymbol{R} + \boldsymbol{S}^T\bar{\boldsymbol{K}}_{sm} + \boldsymbol{S}^T\bar{\boldsymbol{K}}_{ss}\boldsymbol{R}\right]\boldsymbol{\Phi}_m = \boldsymbol{\Lambda} \tag{7.208a,b}$$

The above equations yield the reduced-order mass and stiffness matrices:

$$\bar{\boldsymbol{M}}_R = \bar{\boldsymbol{M}}_{mm} + \bar{\boldsymbol{M}}_{ms}\boldsymbol{R} + \boldsymbol{S}^T\bar{\boldsymbol{M}}_{sm} + \boldsymbol{S}^T\bar{\boldsymbol{M}}_{ss}\boldsymbol{R}$$
$$\bar{\boldsymbol{K}}_R = \bar{\boldsymbol{K}}_{mm} + \bar{\boldsymbol{K}}_{ms}\boldsymbol{R} + \boldsymbol{S}^T\bar{\boldsymbol{K}}_{sm} + \boldsymbol{S}^T\bar{\boldsymbol{K}}_{ss}\boldsymbol{R} \tag{7.209a,b}$$

An iterative procedure can be adopted to obtain the matrices $\boldsymbol{R}$ and $\boldsymbol{S}$ (Equation 7.207) and in turn the reduced-order matrices $\bar{\boldsymbol{M}}_R$ and $\bar{\boldsymbol{K}}_R$ with the starting approximations $\boldsymbol{R} = -\bar{\boldsymbol{K}}_{ss}^{-1}\bar{\boldsymbol{K}}_{sm}$ and $\boldsymbol{S} = -\bar{\boldsymbol{K}}_{ss}^{-T}\bar{\boldsymbol{K}}_{sm}^{T}$. The orthogonality relationships (Equation 7.208) of the reduced-order model helps in avoiding the eigensolution at each step by substituting $\bar{\boldsymbol{M}}_R^{-1}\bar{\boldsymbol{K}}_R$ for $\boldsymbol{\Phi}_m\boldsymbol{\Lambda}\boldsymbol{\Phi}_m^{-1}$ and $\bar{\boldsymbol{M}}_R^{-T}\bar{\boldsymbol{K}}_R^{T}$ for $\boldsymbol{\psi}_m\boldsymbol{\Lambda}^T\boldsymbol{\psi}_m^{-1}$ in Equation 7.207. It is necessary to obtain eigensolution of the reduced-order eigenvalue problem $\bar{\boldsymbol{K}}_R\boldsymbol{\Phi}_m = \bar{\boldsymbol{M}}_R\boldsymbol{\Phi}_m\boldsymbol{\Lambda}$

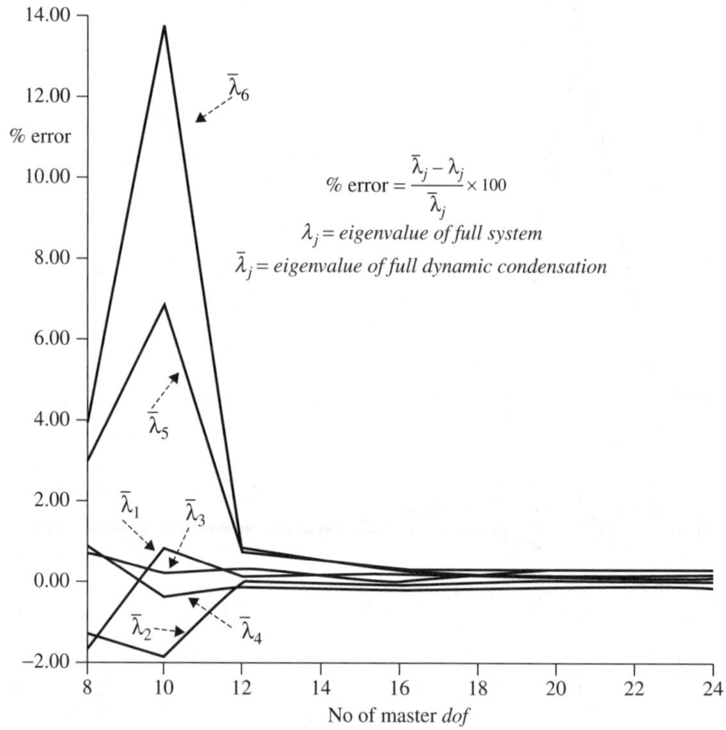

Figure 7.18 Percentage error (between eigenvalue results of full model and reduced-order model) for rotor bearing system (Figure 7.14) with isotropic bearings

and its adjoint $\boldsymbol{\psi}_m^T \bar{\boldsymbol{K}}_R = \boldsymbol{\psi}_m^T \bar{\boldsymbol{M}}_R \boldsymbol{\Lambda}$ only at few intermediate steps of iteration (to check convergence).

Example 7.14: Dynamic condensation of a unsymmetric system

The reduced-order model and its eigensolution are obtained for the rotor bearing system in Example 7.11. Referring to the FE model in Figure 7.14, the order of the system matrices $\bar{\boldsymbol{K}}$ and $\bar{\boldsymbol{M}}$ is 152. The effectiveness of the dynamic condensation with respect to the number of master degree of freedom is shown in Figures 7.17 and 7.18. m is varied from 6 to 24 and the percentage error in eigenvalues of the reduced-order model with reference to those of the full system is plotted in these figures. Specifically, the transverse degrees of freedom of the disk and bearing nodes are included in each case of the selected master degrees of freedom. The rotation speed, Ω is fixed at 10 000 RPM (1047.2 rad/s) in obtaining the results. $m = 12$ seems to provide enough accuracy in obtaining the

Table 7.23 Dynamic condensation of unsymmetric system; eigen solution for the rotor bearing system (Figure 7.14) with isotropic bearings and $\Omega = 20\,000$ RPM (see Table 7.16)

Mode number	Iteration 1	Iteration 5	Iteration 10
1	2.09 + i1621.21	−0.10 + i1622.06	−3.3E−11+ i1627.39
2	2.09 − i1621.21	−0.10 − i1622.06	−3.3E−11 − i1627.39
3	−2.08 + i1774.67	0.11 + i1775.40	−3.4E−11 + i1780.60
4	−2.08 – i1774.67	0.11 − i1775.40	−3.4E−11 − i1780.60
5	−0.22 + i4926.75	−0.01 + i4926.70	9.1E−13 + i4927.11
6	−0.22 − i4926.75	−0.01 − i4926.70	9.1E−13 − i4927.11
7	0.21 + i5063.11	0.01 + i5063.05	9.9E−13 + i5063.45
8	0.21 − i5063.11	0.01 − i5063.05	9.9E−13 – i5063.45
9	−0.36 + i7759.74	−0.007 + i7759.73	1.9E−12 + i7764.04
10	−0.36 – i7759.74	−0.007 − i7759.73	1.9E−12 − i7764.04
11	0.36 + i8202.77	0.007 + i8202.89	3.1E−12 + i8207.15
12	0.36 − i8202.77	0.007 − i8202.89	3.1E−12 − i8207.15

Table 7.24 Dynamic condensation of unsymmetric system; eigensolution for the rotor bearing system (Figure 7.14) with orthotropic bearings and $\Omega = 20\,000$ RPM (see Table 7.17)

Mode number	Iteration 1	Iteration 5	Iteration 10
1	0.3 + i1462.04	−0.07 + i1460.87	2.4E−09 + i1460.73
2	−0.3 − i1462.04	−0.07 − i1460.87	2.4E−09 − i1460.73
3	0.33 + i1721.08	0.06 + i1720.90	3.6E−09 + i1720.91
4	0.33 − i1721.08	0.06 − i1720.90	3.6E−09 − i1720.91
5	0.18 + i4151.51	−0.004 + i4151.63	7.4E−09 + i4151.63
6	0.18 − i4151.51	−0.004 – i4151.63	7.4E−09 − i4151.63
7	−0.21 + i4993.16	0.007 + i4991.19	−1.9E−09 + i4991.34
8	−0.21 − i4993.16	0.007 − i4991.19	−1.9E−09 − i4991.34
9	−0.03 + i6937.31	−0.005 + i6937.30	−1.1E−10 + i6937.33
10	−0.03 − i6937.31	−0.005 − i6937.30	−1.1E−10 − i6937.33
11	−0.04 + i8024.63	0.007 + i8024.66	2.7E−10 + i8024.67
12	−0.04 − i8024.63	0.007 − i8024.66	2.7E−1 − i8024.67

reduced-order model for the system considered. Tables 7.23 and 7.24 give the eigenvalues (whirl frequencies)) for the particular case of $m = 12$ and $\Omega = 20\,000$ RPM.

7.8 Conclusions

Since the eigensolution is central to the modal superposition method for MDOF systems, eigenextraction procedures have been described in this chapter in some detail. Note that for large system dimensions, the eigenvalue extraction methods are iterative and hence inherently approximate. As such, the exposition includes discussion on the accuracy and relative merits of various methods that are of typical contemporary interest. Methods for both symmetric and unsymmetric eigenproblems are presented and followed up by illustrative examples. In the backdrop of the cheap availability of high-end computing resources, the material in this chapter is deemed to be a useful link to the related area of computational structural dynamics. Specifically, for large structural systems involving FEM-based semidiscretization with, say, millions of degrees of freedom, a majority of practical situations requires only the first few eigenvectors in the modal summand for an adequately accurate response evaluation. This could, for instance, be the case when the excitation frequency spectrum and the band of system frequencies in the modal summand overlap each other. Accordingly, the computational eigensolution methods, including those covered herein, that attend to this requirement remain efficient irrespective of the original eigenproblem dimension.

An alternative route to the computation of the system time histories (applicable even for non-LTI and nonlinear dynamic systems) is through direct integration methods. Since the solutions we seek using this route are approximate, an insightful understanding of their formal accuracy and stability (in a sense made more precise later) assume great importance. We are now ready to dwell upon some of these issues in the chapter to follow.

Exercises

1. Analogous to the *QR* method, one may use a *QL* method (Wilkinson, 1965) where the matrix $\boldsymbol{A}$ is factorised (see Equation 7.62 of the QR method) at the kth iteration as $\boldsymbol{A}_k = \boldsymbol{Q}_k \boldsymbol{L}_k$ with $\boldsymbol{L}_k$ being a lower triangular matrix. This is followed by the construction of the transformed matrix $\boldsymbol{A}_{k+1} = \boldsymbol{Q}_k^T \boldsymbol{A}_k \ \boldsymbol{Q}_k = \boldsymbol{Q}_k^T \left(\boldsymbol{Q}_k \boldsymbol{L}_k\right) \boldsymbol{Q}_k = \boldsymbol{L}_k \boldsymbol{Q}_k$. Use this *QL* method and solve the cantilever in Figure 7.1 for its first few eigenvalues.
2. Use the subspace iteration (Table 7.10) along with a shifting strategy and thus obtain the eigensolution for the cantilever in Figure 7.1.
3. Obtain the convergence criteria in terms of error bounds for the two-sided Lanczos algorithm in Table 7.14 (similar to Equation 7.104 of 'standard' Lanczos method).
4. Whilst a $2N$-dimensional state-space form in Equation 7.142 is specifically suitable for handling systems with skew-symmetric matrices, Equation 7.169 is meant for generally unsymmetric systems. Note that, in writing these first-order equations, we utilise N dummy equations of the form $\boldsymbol{M}\dot{\boldsymbol{u}} = \boldsymbol{M}\dot{\boldsymbol{u}}$. Instead, if the dummy equations are $\boldsymbol{I}\dot{\boldsymbol{u}} = \boldsymbol{I}\dot{\boldsymbol{u}}$, one gets:

$$[\bar{\boldsymbol{M}}]\begin{Bmatrix}\ddot{u}\\ \dot{u}\end{Bmatrix} + [\bar{\boldsymbol{K}}]\begin{Bmatrix}\dot{u}\\ u\end{Bmatrix} = \begin{Bmatrix}0\\ \boldsymbol{F}(t)\end{Bmatrix} \tag{7.210}$$

where $\bar{\boldsymbol{M}} = \begin{bmatrix}\boldsymbol{M} & 0\\ 0 & \boldsymbol{I}\end{bmatrix}$ and $\bar{\boldsymbol{K}} = \begin{bmatrix}\boldsymbol{C} & \boldsymbol{K}\\ -\boldsymbol{I} & \boldsymbol{0}\end{bmatrix}$.

Using the system matrices for the axially moving string in Example 7.9, crosscheck if the two formulations in Equations 7.169 and 7.210 yield identical results (eigensolution).

Notations

a, a_1, a_2	Real constants
a_{ij}	Elements in the matrix $\boldsymbol{A}$
a'_{ij}	Elements in the matrix $\boldsymbol{A}'$
$\boldsymbol{A}, \boldsymbol{A}'$	Matrices
$\boldsymbol{A}^*$	Adjoint of $\boldsymbol{A}$
$\boldsymbol{A}^{(r)}$	Matrix formed by removing the last 'r' rows and columns from $\boldsymbol{A}$
b	Real constant
$\boldsymbol{B}_k, \boldsymbol{B}'_k$	Matrices at k^{th} iteration in QR method
$\boldsymbol{c}(t)$	Translation of the position vector $\boldsymbol{x}^*$ relative to $\boldsymbol{x}$
$\boldsymbol{C}$	Nonsingular matrix
C_b	2×2 bearing damping matrix
$\boldsymbol{D}$	Diagonal matrix (in $\boldsymbol{LDL}^T$ decomposition)
$\boldsymbol{e}_1$	Elementary column vector $(1, 0, \ldots., 0)^T$
E, E_0, E_k	Set of linearly independent vectors (basis vectors) in subspace iteration method
E_∞	Invariant subspace, $\subset V$
$\boldsymbol{f}(t)$	External forcing function
$f_{1r}(t), f_{2r}(t)$	Modal forces
$\boldsymbol{f}_{bx}$ and $\boldsymbol{f}_{bx^*}$	Body force vectors in the two frames O and O^*
$\boldsymbol{F}, \boldsymbol{F}^*$	Deformation gradients in body-fixed frames O and O^*
$\boldsymbol{G}$	Gyroscopic matrix
$\boldsymbol{G}^e$	Elemental gyroscopic damping matrix
h_j	Element length (Example 7.9)
I_D, I_P	Diametrical and polar moments of inertia of a disk element
I_{De}, I_{pe}	Diametrical and polar moments of inertia of shaft element
$\boldsymbol{I}$	Unit matrix
K_b	2×2 bearing stiffness matrix
$\bar{\boldsymbol{K}}$	Unsymmetric matrix $\in \mathbb{R}^{2n \times 2n}$
$\hat{\boldsymbol{K}}$	Skew-symmetric matrix $\in \mathbb{R}^{2n \times 2n}$
$\boldsymbol{K}^e$	Elemental bending stiffness
$\boldsymbol{K}_R$	Condensed stiffness matrix
$\boldsymbol{K}_{mm}, \boldsymbol{K}_{ms}, \boldsymbol{K}_{sm}, \boldsymbol{K}_{ss}$	Partitions of the stiffness matrix $\boldsymbol{K}$ (Equation 7.192)
L_e	Element length
$\boldsymbol{L}, \boldsymbol{L}^*$	Velocity gradients (see Equation 7.117)
$\boldsymbol{L}_k$	Lower triangular matrix at k^{th} iteration
m	Integer
m_e	Multiplicity of an eigenvalue, an integer
$\bar{\boldsymbol{M}}$	Unsymmetric matrix $\in \mathbb{R}^{2n \times 2n}$
$\hat{\boldsymbol{M}}$	Symmetric matrix $\in \mathbb{R}^{2n \times 2n}$
$\boldsymbol{M}^e_t, \boldsymbol{M}^e_r$	Element mass matrices corresponding to translational and rotational degrees of freedom

M_D	Mass of a disk element
$\boldsymbol{M}_R$	Condensed mass matrix
$\boldsymbol{M}_{mm}, \boldsymbol{M}_{ms}, \boldsymbol{M}_{sm}, \boldsymbol{M}_{ss}$	Partitions of the stiffness matrix $\boldsymbol{M}$ (Equation 7.192)
n	Number of degrees of freedom in a system
N_1, N_2, N_3, N_3	Shape functions
O, O^*	Body-fixed frames at two time instants t and t^*, respectively
$\boldsymbol{P}$	Transformation matrix (see Equation 7.36)
$p(\lambda)$	Characteristic polynomial of $\boldsymbol{A}$
$p_\sigma(\mu)$	Characteristic polynomial of $\boldsymbol{A} - \sigma \boldsymbol{I}$
$\boldsymbol{P}_j, j = 1, 2 \ldots.$	Transformation matrices in Jacobi iteration
$p^{(n)}(\lambda)$	Characteristic polynomial of an $n \times n$ matrix
$q_{1r}(t), q_{2r}(t)$	n pairs of modal coordinates (Equation 7.147)
$\boldsymbol{Q}$	Unitary matrix in QR method
$\boldsymbol{Q}(t)$	Rotation of the position vector $\boldsymbol{x}^*$ relative to $\boldsymbol{x}$ Equation 7.112
$\boldsymbol{Q}_k$	Unitary matrix at k^{th} iteration in QR method
Q_z, Q_y	Unbalance forces in the Y and Z directions
$\boldsymbol{Q}^e$	Nodal force vector
Q^D	Unbalance force at a disk element
r	Integer (used in subspace method)
$\boldsymbol{r}_j$	Error vector (see Lanczos method)
$\boldsymbol{R}$	Upper triangular matrix in QR method , transformation matrix in Equation 7.195
$\boldsymbol{R}, \boldsymbol{S}$	Transformation matrices (Equation 7.206)
$\boldsymbol{R}_k$	Upper triangular matrix at k^{th} iteration in QR method
s	Integer (Equation 7.46)
$\boldsymbol{S}$	Lower triangular matrix (see Section 7.2)
$S_{\boldsymbol{A}}, S'_{\boldsymbol{A}}$	Sums of the squares of the off-diagonal elements in A and $\boldsymbol{A}'$ (Jacobi method)
S_{ii} and S'_{ii},	Matrices in QR method (each member of these matrices is a cumulative sum of the diagonal elements in $\boldsymbol{B}_k$ and $\boldsymbol{B}'_k$)
T_0, T_1, T_2	Components of kinetic energy
T_e	Kinetic energy of an element (Equation 7.153)
$\boldsymbol{T}_{\boldsymbol{D}}$	Kinetic energy of a disk element (Equation 7.162)
T	Linear transformation matrix
$\boldsymbol{T}$	Tridiagonal matrix in Lanczos method
$\boldsymbol{T}_k$	Tridiagonal matrix at k^{th} iteration in QR method
$\boldsymbol{u}^{\boldsymbol{e}}$	Vector of nodal displacements per element (Equation 7.151) $(= (u_1, u_2, \ldots\ldots, u_8)^T)$
$\boldsymbol{U}_k$	Upper triangular matrix at k^{th} iteration
v	Translational degree of freedom (in the Y direction)
$\boldsymbol{v}$	Vector Velocity $\left(= \frac{dx}{dt}\right)$
$v(x, t)$	Transverse displacement (Example 7.9)
$v_j(t),\ j = 1, 2, \ldots.. n$	Nodal displacements (Example 7.9)
V_c	Critical speed of translation

V_e	Potential energy of an element
V_s	Translation speed of string
w	Translational degree of freedom (in the Z direction)
$\boldsymbol{W}$	Skew-symmetric matrix ($= \dot{\boldsymbol{Q}}\boldsymbol{Q}^T$)
$\boldsymbol{x}, \boldsymbol{x}^*$	Time dependent position vectors of a material point in two body-fixed frames
$\hat{\boldsymbol{x}}_j, \bar{\boldsymbol{x}}_j$	Vectors
$\boldsymbol{X}$	Matrix of Lanczos vectors ($= \{\boldsymbol{x}_1, \boldsymbol{x}_2, \ldots \ldots \boldsymbol{x}_n\}$), reference coordinate of the material body at $t = 0$ (see Equation 7.112), matrix of right eigenvectors (Equation 7.176)
$\boldsymbol{X}_1$	Matrix of Lanczos vectors
$\boldsymbol{X}_0, \bar{\boldsymbol{X}}_0$	Starting vectors for iteration
$\boldsymbol{X}_k, \bar{\boldsymbol{X}}_k$	Vectors at k^{th} vector iteration
$\boldsymbol{y}$	Vector
$\boldsymbol{y}_r$	Vector – real part of $\boldsymbol{Y}_r$
$\boldsymbol{Y}$	Matrix of left eigenvectors (Equation 7.176)
$\boldsymbol{Y}_1$	Matrix of Lanczos vectors
$\boldsymbol{Y}(t)$	Vector function ($= (\boldsymbol{Z}^T, \dot{\boldsymbol{Z}}^T,)^T$ in Equation 7.142)
$Y_j(t)$	Response of jth degree of freedom (Equation 7.147)
$\boldsymbol{Y}_r, \boldsymbol{Y}_r^*$	Complex eigenvectors
$\boldsymbol{Y}_k$	Vector
z	Vector
z_r	Vector – imaginary part of $\boldsymbol{Y}_r$ in r^{th} mode
$\boldsymbol{Z}(t)$	Vector function ($= (v(t)\ w(t))^T$ in Equation 7.132)
$\alpha, \alpha_j, j = 1, 2, .., n$	Real constants
β_j	Real constants
γ	Real constant
μ	Eigenvalue of tridiagonal matrix $\boldsymbol{T}$ (and reciprocal of λ)
ε_{th}	Error threshold in Jacobi method
ε_{tol}	Error tolerance for eigenvalues
$\varepsilon_y(x), \varepsilon_z(x)$	Shaft eccentricities in the Y and Z directions
$\varepsilon_{Dy}, \varepsilon_{Dz}$	Disk eccentricities in the Y and Z directions
θ	Rotation angle in Jacobi method
θ_Y, θ_Z	Rotational degree of freedom in fixed frame of reference
λ_r	Eigenvalues (pure imaginary)
ξ	Local coordinate ($=x/L_e$)
$\rho\,(\boldsymbol{x}, t)$	Mass density
σ	Shift parameter
$\boldsymbol{\tau}$ and $\boldsymbol{\tau}^*$	Stress tensors in the two frames of reference O and
$\boldsymbol{\phi}$	Eigenvector matrix of the tridiagonal matrix $\boldsymbol{T}$
$\boldsymbol{\psi}$	Matrix (eigenvector matrix or shape function matrix)
$\boldsymbol{\psi}_t, \boldsymbol{\psi}_r$	Shape function matrices corresponding to the translational and rotational degrees of freedom
$\boldsymbol{\psi}_{ry}, \boldsymbol{\psi}_{rz}$	First and second rows in $\boldsymbol{\psi}_r$
ω_r	Eigenvalues
$\omega_x, \omega_y, \omega_z$	Angular velocities in rotating frame of reference

$\boldsymbol{\Phi}$	Eigenvector matrix
$\boldsymbol{\Phi}_m$, $\boldsymbol{\Phi}_s$	Partitions of $\boldsymbol{\Phi}$ according to the master and slave degrees of freedom (Equation 7.192)
$\boldsymbol{\Sigma}$	Diagonal matrix containing the eigenvalues of the tridiagonal matrix, $\boldsymbol{T}$ on its diagonal
$\Omega(\Omega_x)$	Rotation speed (angular velocity) about X axis
$\boldsymbol{\Omega}$	Angular velocity vector

References

Bathe, E.L. (1996) *Finite Element Procedures*, Prentice-Hall, Englewood Cliffs, NJ.

Ericsson, T. and Ruhe, A. (1980) The spectral transformation Lanczos method for the numerical solution of large sparse generalized symmetric eigenvalue problems. *Mathematics Computation*, **35** (152), 1251–1268.

Francis, J.G.F. (1961) The QR transformation, parts 1 and 2. *The Computer Journal*, **4**, 265–271, 332–345.

Geradin, M. and Rixen, D. (1994) *Mechanical Vibrations – Theory and application to Structural Dynamics*, John Wiley & Sons, Inc., Chichester, UK.

Grimes, R.G., Lewis, J.G. and Simon, H.D. (1991) A Shifted Block Lanczos Algorithm for Solving Sparse Symmetric Generalized Eigenproblems. Report RNR-91-012, NASA Ames Research Center, Moffett Field, CA.

Gunter, E.J. (1967) The influence of internal friction on the stability of high speed rotors. *Transactions of ASME, Journal of Engineering for Industry, Series B*, **89**, 683–688.

Jacobi, C.G.J. (1846) Über ein leichtes Verfahren die in der Theorie der Säcularstörungen vorkommenden Gleichungen numerisch aufzulösen *Crelle's Journal*, **30**, 51–94.

Kane, K. and Torby, B.J. (1991) The extended modal reduction method applied to rotor dynamic problems. *Transactions of ASME, Journal of Vibration and Acoustics*, **113**, 79–84.

Likins, P.W. (1972) Finite element appendage equations for hybrid coordinate dynamic analysis. *International Journal of Solids and Structures*, **8**, 709–731.

Lund, J.W. (1974) Modal response of a flexible rotor in fluid film bearings. *Transactions of ASME, Journal of Engineering for Industry*, **96**, 525–533.

Meirovitch, L. (1970) *Methods of Analytical Dynamics*, McGraw-Hill, New York.

Meirovitch, L. (1974) A new method of solution of the eigenvalue problem for gyroscopic systems. *AIAA Journal*, **12** (10), 1337–1342.

Meirovitch, L. (1975) A modal analysis for the response of linear gyroscopic systems. *Transactions of ASME Journal of Applied Mechanics*, **42**, 446–450.

Mote Jr., C.D. (1966) On the nonlinear oscillations of an axially moving string. *Transactions of ASME, Journal of Applied Mechanics*, **33**, 463–464.

Nelson, H.D. and McVaugh, J.M. (1976) The dynamics of rotor bearing systems using finite elements. *Transactions of ASME Journal of Engineering for Industry*, **98**, 593–600.

Nguyen, D.T. (2006) *Finite Element Methods: Parallel-Sparse Statics and Eigen-solutions*, Springer Science+Business Media, Inc., New York, USA.

Perkins, N.C. (1990) Linear dynamics of a translating string on an elastic foundation. *Transactions of ASME, Journal of Vibration and Acoustics*, **112**, 2–7.

Qu, Z.Q. and Fu, Z.F. (1998) New structural dynamic condensation method for finite element models. *AIAA Journal*, **36**, 1320–1324.

Rao, G.V. (2002) Dynamic condensation and synthesis of unsymmetric structural systems. *Transactions of ASME, Journal of Applied Mechanics*, **69**, 610–616.

Rutishauser, H. (1958) Solution of Eigenvalue Problems with the LR Transformation. *Nat. Bur. Stan. Appl. Math. Ser.*, **49**, 47–81.

Strang, G. (2006) *Linear Algebra and its Applications*, 4th edn, Thomson Brooks/Cole.

Surez, L.E. and Singh, M.P. (1992) Dynamic condensation method for eigenvalue analysis. *AIAA Journal*, **30**, 1045–1054.

Wilkinson, J.H. (1965) *The Algebraic Eigenvalue Problem*, Oxford University Press, Oxford, UK.

Bibliography

On Nonclassically Damped Systems

Cronin, D.L. (1990) Eigenvalue and eigenvector determination of non-classically damped dynamic systems. *Computers and Structures*, **36** (1), 133–138.

On Modal Testing

Ewins, D.J. (1984) *Modal Testing: Theory and Practice*, Research Studies Press Ltd., Taunton, UK.

On Eigenextraction Methods

Meirovitch, L. (1980) *Computational Methods in Structural Dynamics*, Sijthoff & Noordohoff, Netherlands.

On Dynamic Condensation Methods

Rouch, K.E. and Kao, J.S. (1980) Dynamic reduction in rotor dynamics by the finite element method. *Transactions of ASME Journal of Mechanical Design*, **102**, 360–367.

On Subspace Iteration Method

Cheu, T., Johnson, C.P. and Craig, R.R. Jr (1987) Computer algorithms for calculating efficient initial vectors for subspace iteration method. *International Journal of Numerical Methods in Engineering*, **24**, 1841–1848.

On Other Variants of Lanczos Method

Gupta, K.K. and Lawson, C.I. (1988) Development of a Block lanczos algorithm for free vibration analysis of spinning structures. *International Journal of Numerical Methods in Engineering*, **26**, 1029–1037.

Gupta, K.K., Lawson, C.I. and Ahmadi, A.R. (1992) On development of a finite element and solution of associated eigenproblem by a Block Lanczos procedure. *International Journal of Numerical Methods in Engineering*, **33**, 1611–1623.

Simon, H.D. (1984) The Lanczos algorithm with partial reorthogonalization. *Mathematics of Computation*, **42** (165), 115–142.

8

Direct Integration Methods

8.1 Introduction

The modal superposition technique that finds mention in the preceding chapters is applicable only for linear time-invariant (LTI) systems. The method uses an expansion of eigenfunctions (in the case of continuous systems) or eigenvectors (in the case of discrete or discretized systems). The acceptability of the eigensolution via any of the eigenvalue extraction methods is generally restricted to the lower end of the eigenspectrum. Considering discretized systems, most often encountered in practical applications, the use of a large number of shape functions (say, corresponding to a finite element scheme) may generate spurious high-frequency modes. Thus, the solution based on modal summation would be accurate when the system response can be well approximated by only a few lower order modes. Under impact/impulse loads (or, more generally, for broad-banded inputs) containing high-frequency components, the modal sum may be too oscillatory or may fail to converge. In such cases, an alternative family of time-marching schemes, referred to as the direct integration methods (Bathe, 1996), provides a more robust numerical alternative. Moreover, for LTI or nonlinear structural systems wherein the modal summation is inapplicable, direct integration methods remain the only viable and universal route for response evaluation. Here, the dynamic response (the time history) is obtained by recursively integrating the coupled equation over small time steps, starting with the initial time, t_0 and marching forward with a chosen step size Δt. Following the finite element method (FEM)-based semidiscretization discussed in Chapter 4, the equations of motion take the following form:

$$\boldsymbol{M}\ddot{\boldsymbol{u}}(t) + \boldsymbol{C}\dot{\boldsymbol{u}}(t) + \boldsymbol{K}\boldsymbol{u}(t) + \boldsymbol{R}(\boldsymbol{u}(t), \dot{\boldsymbol{u}}(t), \ddot{u}(t), t) = \boldsymbol{F}(t) \tag{8.1}$$

provided that the structural material does not exhibit history dependence. Here, $\boldsymbol{R}(\boldsymbol{u}, \dot{\boldsymbol{u}}, \ddot{\boldsymbol{u}}, t)$ represents the nonlinear restoring force vector that may arise due to geometric and/or material nonlinearities. Thus, a direct integration method can be viewed as a numerical tool for the temporal discretization of Equation 8.1 so that the fully discretized form of the equilibrium equations at a time instant $t_n > t_0$ is given by:

$$\boldsymbol{M}\ddot{\boldsymbol{u}}_n + \boldsymbol{C}\dot{\boldsymbol{u}}_n + \boldsymbol{K}\boldsymbol{u}_n + \boldsymbol{R}_n = \boldsymbol{F}_n \tag{8.2}$$

The vectors $\boldsymbol{u}_n$, $\dot{\boldsymbol{u}}_n$ and $\ddot{\boldsymbol{u}}_n$, respectively, describe the approximated displacement, velocity and acceleration at t_n. We assume that the initial displacement and velocity vectors, $\boldsymbol{u}_0(:=\boldsymbol{u}(t_0))$ and $\dot{\boldsymbol{u}}_0(:=\dot{\boldsymbol{u}}(t_0))$, respectively, are known. Direct integration methods differ among themselves based on the kind of time-marching formulation employed. More specifically, the methods may be broadly classified as explicit and implicit. In the former case, the response at time $t_{n+1}(= t_n + \Delta t)$ is obtained as a function of the response already available at the preceding time instants $t_j < t_{n+1}$. Thus, the latest state $\boldsymbol{u}_{n+1}$ is obtained via the general form:

$$\boldsymbol{u}_{n+1} = \boldsymbol{f}\{\boldsymbol{u}_n, \dot{\boldsymbol{u}}_n, \ddot{\boldsymbol{u}}_n, \boldsymbol{u}_{n-1}, \dot{\boldsymbol{u}}_{n-1}, \ddot{\boldsymbol{u}}_{n-1}, \ldots\ldots\ldots, \boldsymbol{u}_0, \dot{\boldsymbol{u}}_0, \ddot{\boldsymbol{u}}_0, t_n\} \tag{8.3}$$

On the other hand, an implicit method requires the function on the RHS of the equation above to include $\boldsymbol{u}_{n+1}$ also as an argument. The resulting circularity in the formulation needs special techniques to be adopted for response evaluation. Further, if the response only at the immediately preceding instant t_n is utilised to obtain a solution at t_{n+1}, the corresponding explicit method is referred to as single step. A multistep method, in contrast, utilises available response information at more than just one preceding instant. A similar classification also applied to the implicit methods. Before we discuss some of the well-known methods for direct integration of these (semidiscretized) initial-value problems, it is convenient to write Equation 8.1 in the form of first-order ordinary differential equations (ODEs) (state-space representation):

$$\dot{\boldsymbol{y}} = \boldsymbol{f}(\boldsymbol{y}, t) \text{ with } \boldsymbol{y}(t) = (\boldsymbol{u}^T(t), \dot{\boldsymbol{u}}^T(t))^T \tag{8.4}$$

Here, $\boldsymbol{f}(\boldsymbol{y}, t) = \{\dot{\boldsymbol{u}}^T(t), (\boldsymbol{M}^{-1}(\boldsymbol{F} - \boldsymbol{C}\dot{\boldsymbol{u}} + \boldsymbol{K}\boldsymbol{u} + \boldsymbol{R}))^T\}^T$. The solution to the above first-order equation is unique if $\boldsymbol{f}$ is continuous in some interval $[a \le t \le b]$, and satisfies the Lipschitz condition: $|\boldsymbol{f}(\boldsymbol{y}_1, t) - \boldsymbol{f}(\boldsymbol{y}_2, t)| \le K|\boldsymbol{y}_1 - \boldsymbol{y}_2|$, K being a constant (independent of y and t). Further, if $\boldsymbol{y}(t)$ and its derivatives $\boldsymbol{y}^{n+1}(t)$, $n = 1, 2 \ldots$ are continuous over the interval $[a, b]$, $\boldsymbol{y}(t_{n+1})$ is given by Taylor's theorem as:

$$\boldsymbol{y}(t_{n+1}) = \boldsymbol{y}(t_n + \Delta t) = \boldsymbol{y}(t_n) + \Delta t\boldsymbol{f} + \frac{\Delta t^2}{2!}\boldsymbol{f}' + \frac{\Delta t^3}{3!}\boldsymbol{f}'' + \ldots \frac{\Delta t^n}{n!}\boldsymbol{f}^n + \frac{\Delta t^{n+1}}{n+1!}\boldsymbol{f}^{n+1}(\xi) \tag{8.5}$$

for some $\xi \in [t_n, t_{n+1}] \subset [a, b]$. Here, $\boldsymbol{f}^k$ stands for k^{th} total derivative of $\boldsymbol{f}(\boldsymbol{y}, t)$ with respect to t. These derivatives $\frac{d^k\boldsymbol{f}}{dt^k}$, k = 1, 2 ... are given by:

$$\frac{d\boldsymbol{f}}{dt} = \boldsymbol{f}_t + \boldsymbol{f}\boldsymbol{f}_y, \frac{d^2\boldsymbol{f}}{dt^2} = \boldsymbol{f}_{tt} + 2\boldsymbol{f}\boldsymbol{f}_{ty} + \boldsymbol{f}^2\boldsymbol{f}_{yy} + \boldsymbol{f}_t\boldsymbol{f}_y + \boldsymbol{f}\boldsymbol{f}_y^2, \ldots\ldots \tag{8.6}$$

In the above equation, the subscripts t and $\boldsymbol{y}$, respectively, indicate partial derivatives with respect to t and $\boldsymbol{y}$. The Taylor series in Equation 8.5 may be directly used to evolve a family of direct integration methods, depending on the order of truncation. With the approximate numerical estimate y_n replacing the true y (t_n), the solution at $t_n + \Delta t$ is thus given by

$$\boldsymbol{y}_{n+1} = \boldsymbol{y}_n + \Delta t\boldsymbol{f} + \frac{\Delta t^2}{2!}\frac{d\boldsymbol{f}}{dt} + \frac{\Delta t^3}{3!}\frac{d^2\boldsymbol{f}}{dt^2} + \cdots \frac{\Delta t^k}{k!}\frac{d^k\boldsymbol{f}}{dt^k} + \frac{\Delta t^{k+1}}{k+1!}\frac{d^{k+1}\boldsymbol{f}}{dt^{k+1}} + \cdots \tag{8.7}$$

With the initial value $\boldsymbol{y}_0$ known, the expression for $\boldsymbol{f}(\boldsymbol{y}, t)$ given and the total derivatives $\frac{d^k\boldsymbol{f}}{dt^k}$, k = 1, 2 ... determined from this expression, Equation 8.7 yields the numerical solution at t_{n+1} and thus defines a recursive strategy to obtain solutions at discrete time instants, $t_0 + \Delta t$, $t_0 + 2\Delta t, \ldots t_0 + n\Delta t \ldots$. Such schemes may also be readily modified for nonuniform step sizes. Consistency of a direct integration method requires that the discretized Equation 9.7 tends to the original differential equation as $\Delta t \to 0$. If we retain only k terms, for example in expansion (8.7), the discretization leads to a truncation error = $\boldsymbol{y}(t_{n+1}) - \boldsymbol{y}_{n+1} = \frac{\Delta t^{k+1}}{k+1!}\frac{d^{k+1}f}{dt^{k+1}} + \cdots$. This truncation error is said to be of order $k+1$ (Appendix F) and indicated by $\boldsymbol{O}(\Delta t^{k+1})$. Since this error (i.e. the remainder) reduces to zero as $\Delta t \to 0$, the consistency of the method is ensured. Such a method is said to be of order k (which provides a formal, *a priori* measure of numerical accuracy). In view of Equation 8.4, Equation 8.7 may be used to yield the following maps:

$$\boldsymbol{u}_{n+1} = \boldsymbol{u}_n + \Delta t\dot{\boldsymbol{u}}_n + \frac{\Delta t^2}{2!}\ddot{\boldsymbol{u}}_n + \frac{\Delta t^3}{3!}\dddot{\boldsymbol{u}}_n + \cdots\cdots + \boldsymbol{O}(\Delta t^{k+1})$$

$$\dot{\boldsymbol{u}}_{n+1} = \dot{\boldsymbol{u}}_n + \Delta t\ddot{\boldsymbol{u}}_n + \frac{\Delta t^2}{2!}\dddot{\boldsymbol{u}}_n + \cdots\cdots + \boldsymbol{O}(\Delta t^{k+1}) \tag{8.8a,b}$$

The computation of the total derivatives in Equation 8.6 is tedious and more so with increase in the order of the method. In view of this limitation, many variants of the Taylor-series method have emerged with the objective of minimising the derivative computations whilst maintaining the desired order of accuracy. For example, there are finite difference methods that express the first-order derivative in Equation 8.4 by forward, backward or central difference approximations within the interval $[t_n, t_{n+1}]$:

$$\dot{\boldsymbol{y}}(t) = \frac{\boldsymbol{y}(t_{n+1}) - \boldsymbol{y}(t_n)}{\Delta t} = \boldsymbol{f}(\boldsymbol{y}, t) \tag{8.9a}$$

$$\dot{\boldsymbol{y}}(t) = \frac{\boldsymbol{y}(t_n) - \boldsymbol{y}(t_{n-1})}{\Delta t} = \boldsymbol{f}(\boldsymbol{y}, t) \tag{8.9b}$$

$$\dot{\boldsymbol{y}}(t) = \frac{\boldsymbol{y}(t_{n+1}) - \boldsymbol{y}(t_{n-1})}{2\Delta t} = \boldsymbol{f}(\boldsymbol{y}, t) \tag{8.9c}$$

The forward difference approximation in Equation 8.9a results from Taylor-series expansion of $\boldsymbol{y}(t_{n+1})$ in Equation 8.7 after neglecting the terms containing Δt to the powers ≥ 2. The case with the backward difference is similar and these two methods are known as forward and backward Euler methods with first-order accuracy (i.e. with truncation error $\equiv O(\Delta t^2)$). Expanding $\boldsymbol{y}(t_{n+1})$ and $\boldsymbol{y}(t_{n-1})$ about $\boldsymbol{y}(t_n)$ and subtracting one from the other result in the central difference method represented by Equation 8.9c. The truncation error in this method is of $O(\Delta t^3)$. In other words, the last method is of second order or of formal order 2 (higher than those of the former two methods).

Whilst direct integration methods should in general be consistent, the other desirable (or even necessary) computational features are those of stability and convergence. Broadly speaking, stability of the numerical solution is deemed to have been achieved if it is not sensitively dependent upon the temporal discretization (i.e. for instance, discretization errors do not exponentially vary with linear variations in the time step size within a 'reasonable' range). It is expedient to have Δt as large as possible, evidently to save computational time. It is in this context that Shannon's sampling theorem (Shannon, 1949) helps us develop an understanding on an upper bound for the sampling interval (e.g. the step size Δt) that forms a crucial element whilst replacing a continuous signal (e.g. the solution of an ODE) by that of a discrete map (e.g. the ones in Equation 8.8). The

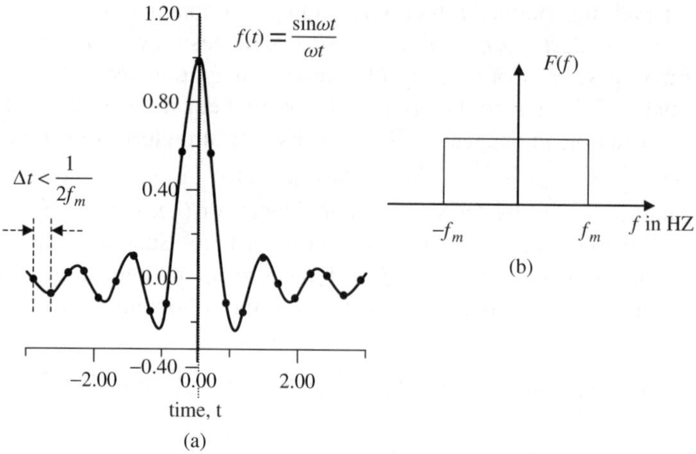

Figure 8.1 (a) Sync function, $\frac{\sin\omega t}{\omega t}$ with $\omega = 10$ and on illustration of its sampling and (b) Fourier transform of the sync function – band limited signal

theorem states that if a continuous function $y(t)$ is band-limited with its Fourier transform $F(f) = 0$ for $|f| > f_m$, it is completely determined by an infinite sequence of sampled data $(n\Delta t), n = 0, 1, 2, \ldots,$ provided $\Delta t \leq \frac{1}{2f_m}$. Δt is called the sampling period and $2f_m$, the sampling frequency (also known as the Nyquist frequency). Figure 8.1 shows a function commonly known as the 'sinc' function (sinc(x)) and its Fourier transform. The reconstructed function via the sampling using Nyquist frequency is also shown (in dots) in Figure 8.1a.

We find a direct application of Shannon's sampling theorem in solving LTI systems by modal superposition. Whilst solving the modal equation (5.101), the integration step size is decided by the frequency, ω_m (in rad-s^{-1}) of the highest mode considered in the modal summation. The step size is typically taken to be smaller than the sampling period $\frac{1}{2}(\frac{2\pi}{\omega_m}) = \frac{\pi}{\omega_m}$.

Note that the mere choice of a step size according to the above sampling theorem by no means ensures stability of a discretization scheme. In this regard, the temporal discretization schemes such as those in Equation 8.9 may in fact build up the solution. Consider, for example the second-order ODE of motion for an single degree of freedom (SDOF) oscillator $\ddot{y} + 2\xi\omega\dot{y} + \omega^2 y = 0$ with the initial condition $y(0) = y_0$. The equation admits a bounded solution $y(t) = y(0)\exp(-\xi\omega t)\sin(\omega\sqrt{1-\xi^2 t})$, $\forall t > 0$ (Figure 8.2a) provided $\xi > 0$. Suppose that we take $\Delta t = 0.0628 \ll$ sampling period $= \frac{\pi}{\omega} = 0.31415$. If, now, the forward Euler method – Equation 8.9a – is used, it results in an incorrectly divergent (unstable) solution (Figure 8.2b). The selected parameter values are $\omega = 10.0$ and $\xi = 0.02$. The divergent solution, so obtained, highlights the importance to understand the so-called stability features of such discretization schemes against variations in Δt.

The recursive mapping corresponding to an integration method, for instance to an SDOF system, is expressible as:

$$\boldsymbol{x}_{n+1} = \boldsymbol{A}\boldsymbol{x}_n + \boldsymbol{R}_n \tag{8.10}$$

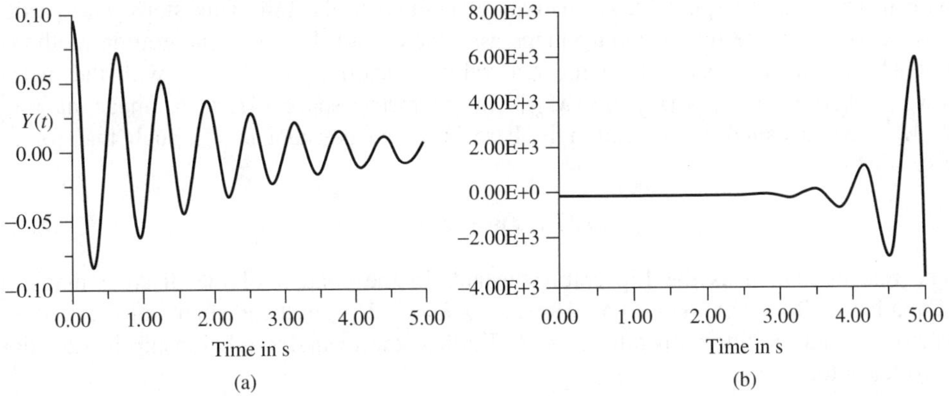

Figure 8.2 An unforced SDOF oscillator $\omega = 10$, $\xi = 0.05$; initial condition $y_0 = 0.1$. (a) Exact solution and (b) solution by forward Euler method with $\Delta t = 0.0628 \ll \pi/\omega$ (sampling period)

Here, $\boldsymbol{x}_n$ represents the system state vector comprising of $u_n, \dot{u}_n, \ddot{u}_n$ at $t_n, n = 0, 1, 2\ldots$ and $\boldsymbol{A}$ is known as the amplification matrix. $\boldsymbol{R}_n$ is the load vector. For ease of further exposition, consider the homogeneous part of Equation 8.10. If $(\lambda_j, \boldsymbol{\Phi}_j)$, $j = 1, 2, 3$ are the eigenpairs of $\boldsymbol{A}$, the eigenexpansion of the initial system state $\boldsymbol{x_0} = \sum_{j=1}^{3} a_j \boldsymbol{\Phi}_j$ transforms Equation 8.10 as:

$$\boldsymbol{x}_{n+1} = \boldsymbol{A}^{n+1} \sum_{j=1}^{3} a_j \boldsymbol{\Phi}_j = \sum_{j=1}^{3} a_j \boldsymbol{A}^{n+1} \boldsymbol{\Phi}_j = \sum_{j=1}^{3} a_j \lambda_j^{n+1} \boldsymbol{\Phi}_j \tag{8.11}$$

The above equation is indicative of the magnification or attenuation (as the case may be) of an initial state for $n > 0$ being governed by the spectral radius of $\boldsymbol{A}$ defined by $\rho(A) = \max_{i=1,2,3} |\lambda_i|$. If λ_j $(= \sigma_j + j\omega_j)$ is complex, then the discretization scheme is asymptotically stable if $\rho(A) < 1$, which means that all the eigenvalues lie within the unit circle in the $\sigma - \omega$ plane (Figure 8.3). An integration scheme is stable if the magnitude of some of the eigenvalues, $|\lambda_j| = 1$ and all others lie within the unit circle. It is unstable if at least one eigenvalue is such that $|\lambda_j| > 1$.

For a more rigorous understanding, the above concept of stability may be linked to the contraction mapping theorem (Kreyszig, 2001; Brooks and Schmitt, 2009) associated

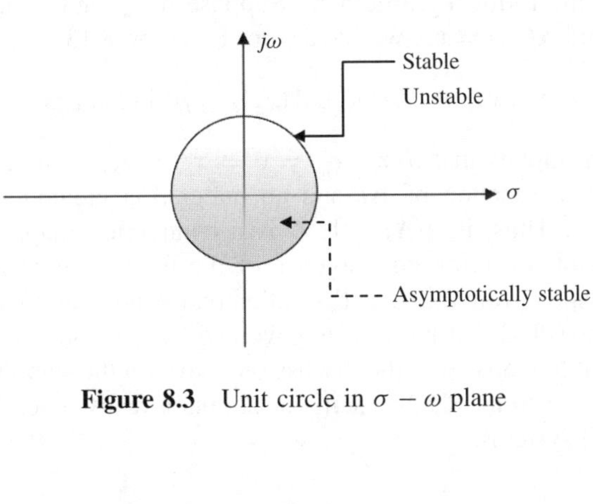

Figure 8.3 Unit circle in $\sigma - \omega$ plane

with metric spaces (Appendix H). This is elaborated in the following steps with particular reference to the time-stepping maps associated with the direct integration methods. Let M be a metric space of all the real-valued vectors $\boldsymbol{x}_n \in M \subset R^3$ with the metric $d(\boldsymbol{x}_{m+1}, \boldsymbol{x}_m) = \max_{1 \le i \le 3} \|x_{i,m+1} - x_{i,m}\|$. In the metric space (M, d), a linear mapping $\boldsymbol{T}: M \to M$ is called a contraction if there is a real constant $\alpha < 1$ such that for all $\boldsymbol{x}, \boldsymbol{y} \in M$,

$$d(\boldsymbol{Tx}, \boldsymbol{Ty}) \le \alpha\, d(\boldsymbol{x}, \boldsymbol{y}) \tag{8.12}$$

The real number α is the Lipschitz constant. In the context of the discrete mapping (Equation 8.10), denote $\alpha \equiv \rho(A)$. Here, $x_{i,m}, i = 1, 2, 3$ are the elements of the solution vector $\boldsymbol{x}_m$ and the linear operator $A \equiv T$. Further, the recursive relationship in Equation 8.10 leads to:

$$d(\boldsymbol{x}_{m+1}, \boldsymbol{x}_m) = d(\boldsymbol{A}\boldsymbol{x}_m, \boldsymbol{A}\boldsymbol{x}_{m-1}) \tag{8.13}$$

If $\rho(A) \le 1$, we have from the Equation 8.12:

$$d(\boldsymbol{A}\boldsymbol{x}_m, \boldsymbol{A}\boldsymbol{x}_{m-1}) \le \rho(\boldsymbol{A}) d(\boldsymbol{x}_m, \boldsymbol{x}_{m-1}) \tag{8.14}$$

Equations 8.13 and 8.14 finally result in $d(\boldsymbol{x}_{m+1}, \boldsymbol{x}_m) \le \rho^m(A) d(\boldsymbol{x}_1, \boldsymbol{x}_0)$. Using this result and the triangular inequality, we have, for $n > m$,:

$$\begin{aligned} d(\boldsymbol{x}_m, \boldsymbol{x}_n) &\le d(\boldsymbol{x}_m, \boldsymbol{x}_{m+1}) + d(\boldsymbol{x}_{m+1}, \boldsymbol{x}_{m+2}) + \cdots\cdots + d(\boldsymbol{x}_{n-1}, \boldsymbol{x}_n) \\ &\le (\rho^m(A) + \rho^{m+1}(A) + \cdots\cdots\cdot + \rho^{n-1}(A)) d(\boldsymbol{x}_0, \boldsymbol{x}_1) \\ &= \rho^m(A) \frac{1 - \rho^{n-m}(A)}{1 - \rho(A)} d(\boldsymbol{x}_0, \boldsymbol{x}_1) \end{aligned} \tag{8.15}$$

If $\rho(\boldsymbol{A}) < 1$, we have $1 - \rho^{n-m}(\boldsymbol{A}) < 1$ and $d(\boldsymbol{x}_m, \boldsymbol{x}_n) \le \frac{\rho^m(\boldsymbol{A})}{1-\rho(\boldsymbol{A})} d(\boldsymbol{x}_0, \boldsymbol{x}_1)$. For sufficiently large m, $\frac{\rho^m(\boldsymbol{A})}{1-\rho(\boldsymbol{A})}$ can be made as small as possible and hence $\{\boldsymbol{x}_m\}$ is a Cauchy (or fundamental) sequence in $\boldsymbol{M}$ (see Appendix B for a definition of a Cauchy sequence). Moreover, if $\boldsymbol{M}$ is complete (see Appendix B), every Cauchy sequence in $\boldsymbol{M}$ converges to a limit (a fixed point) in $\boldsymbol{M}$ with $d(\boldsymbol{x}_m, \boldsymbol{A}(\boldsymbol{x}_m)) = 0 \Rightarrow \boldsymbol{A}(\boldsymbol{x}_m) = \boldsymbol{x}_m$. With a contraction mapping in $\boldsymbol{M}$ that is complete, the contraction mapping theorem (Banach fixed point theorem) states that the mapping has a 'unique' fixed point. Specifically, the uniqueness of the point follows from the following argument. Suppose if $\boldsymbol{x}_m$ and $\boldsymbol{x}_n$ are two fixed points with $\boldsymbol{A}(\boldsymbol{x}_m) = \boldsymbol{x}_m$ and $\boldsymbol{A}(\boldsymbol{x}_n) = \boldsymbol{x}_n$, we have from Equation 8.13:

$$d(\boldsymbol{x}_m, \boldsymbol{x}_n) = d(\boldsymbol{A}(\boldsymbol{x}_m), \boldsymbol{A}(\boldsymbol{x}_n)) \le \rho(\boldsymbol{A})\, d(\boldsymbol{x}_m, \boldsymbol{x}_n) \tag{8.16}$$

The above equation implies that $d(\boldsymbol{x}_m, \boldsymbol{x}_n) = 0 \Rightarrow \boldsymbol{x}_m = \boldsymbol{x}_n$, a unique fixed point. Note that the matrix $\boldsymbol{A}$ is a function of Δt and an integration method is stable if the step size yields $\rho(\boldsymbol{A}) \le 1$. Thus, if $\rho(\boldsymbol{A}) < 1$, $\boldsymbol{A}$ is a contraction mapping and the method is asymptotically stable ensuring convergence of the time-stepping map to a numerical solution whose 'distance' from the 'exact' solution (corresponding to the governing partial differential equations (PDEs) of motion) is governed by the discretization and truncation errors. It is important to realise that the discussion above on the stability of the integration methods with reference to an SDOF oscillator admits a ready extension to multidegree-of-freedom (MDOF) systems.

Another factor that may affect the quality of the numerical solution is the contribution from spurious high frequency modes resulting from the semidiscretization. This often necessitates that an integration scheme be equipped with a so-called 'numerical dissipation' strategy to suppress the undesirable contributions from these higher modes. In other words, algorithmic damping in the high-frequency regime is often sought, whilst least affecting the contributions of the important low-frequency modes. Some of these issues pertaining to the numerical integration methods are thus discussed in this chapter with specific applications to linear structural dynamic systems.

8.2 Forward and Backward Euler Methods

Just as the finite difference approximations (as in Equation 8.9) are derivable from truncated Taylor's expansions of the response quantities of interest (e.g. displacements and velocities), the most popularly used integrations schemes for structural dynamic systems also follow from such expansions, implicit or explicit. By now it is clear that these approximations render the system ODEs discrete and lead to coupled difference equations that need to be solved recursively.

8.2.1 Forward Euler Method

With $\boldsymbol{y}(t) = (\boldsymbol{u}(t), \dot{\boldsymbol{u}}(t))^T$, the forward Euler integration method (Equation 8.9a) may be readily connected up with finite difference equations for the displacement and velocity vectors as follows:

$$\frac{\boldsymbol{u}(t+\Delta t) - \boldsymbol{u}(t)}{\Delta t} = \dot{\boldsymbol{u}}(t) \Rightarrow \boldsymbol{u}_{n+1} = \boldsymbol{u}_n + \Delta t \dot{\boldsymbol{u}}_n$$

$$\text{and} \quad \frac{\dot{\boldsymbol{u}}(t+\Delta t) - \dot{\boldsymbol{u}}(t)}{\Delta t} = \ddot{\boldsymbol{u}}(t) \Rightarrow \dot{\boldsymbol{u}}_{n+1} = \dot{\boldsymbol{u}}_n + \Delta t \ddot{\boldsymbol{u}}_n \tag{8.17a,b}$$

The above equations along with the system equilibrium ODE-s (Equation 8.2) (in the semidiscrete form) yield the acceleration response at t_{n+1} as:

$$\ddot{\boldsymbol{u}}_{n+1} = \boldsymbol{M}^{-1}\{\boldsymbol{F}_{n+1} - \boldsymbol{C}(\dot{\boldsymbol{u}}_n + \Delta t \ddot{\boldsymbol{u}}_n) - \boldsymbol{K}(\boldsymbol{u}_n + \Delta t \dot{\boldsymbol{u}}_n) - \boldsymbol{R}_{n+1}\} \tag{8.18}$$

For a linear system ($\boldsymbol{R}_{n+1} = \boldsymbol{0}$), Equations 8.17 and 8.18, if solved recursively, yield the system state at any instant t_{n+1}. The equations also show that the Euler forward integration method is explicit. Further, if the lumped-mass approach is adopted in arriving at the mass matrix, it is diagonal, which means that its inversion is straightforward. The stability properties of the discretization scheme are examined in the following steps. Once more, for expositional convenience, the SDOF oscillator equation $\ddot{u} + 2\xi\omega\dot{u} + \omega^2 u = 0$ is taken for assessing the stability of the numerical scheme. As noted before, a system of such SDOF oscillators typically represents the different eigenmodes for an LTI structural system. Using Equations 8.17 and 8.18, the amplification matrix $\boldsymbol{A}$ (Equation 8.10) for the explicit Euler method can be described by a 2×2 matrix in terms of the displacement and velocity as:

$$\begin{pmatrix} u_{n+1} \\ \dot{u}_{n+1} \end{pmatrix} = \boldsymbol{A} \begin{pmatrix} u_n \\ \dot{u}_n \end{pmatrix} + \boldsymbol{L}_n \tag{8.19}$$

where L_n is the load term. The explicit Euler map (Equation 8.17) for the test equation $\ddot{u} + 2\xi\omega\dot{u} + \omega^2 u = 0$ is given by:

$$u_{n+1} = u_n + \Delta t \dot{u}_n$$
$$\dot{u}_{n+1} = \dot{u}_n + \Delta t(-2\xi\omega\dot{u}_n - \omega^2 u_n) \tag{8.20a,b}$$

Thus, the amplification matrix A takes the form

$$A = \begin{bmatrix} 1 & \Delta t \\ -\omega^2 \Delta t & 1 - 2\xi\omega\Delta t \end{bmatrix} \tag{8.21}$$

The eigenvalues of A are given by $\lambda_{1,2} = 1 + \omega\Delta t(-\xi \pm \sqrt{\xi^2 - 1})$. Since $\xi < 1$, the eigenvalues are complex and are given by

$$\lambda_{1,2} = 1 + \omega\Delta t(-\xi \pm i\sqrt{1 - \xi^2}), \quad i = \sqrt{-1} \tag{8.22}$$

The spectral radius, $\rho(A)$ is given by

$$\rho(A) = \max_{j=1,2} |\lambda_j| = 1 + \omega^2\Delta t^2 - 2\xi\omega\Delta t \tag{8.23}$$

For an undamped system ($\xi = 0$), $\rho(A) > 1$ and the explicit Euler method is unstable for any step size. For a damped system, however, the numerical scheme is stable, if the choice of Δt satisfies the condition:

$$\Delta t \le \frac{2\xi}{\omega} \tag{8.24}$$

Strict inequality in the above equation ensures asymptotic stability of the integration scheme.

Example 8.1: The SDOF oscillator and solution by the explicit Euler method

The oscillator parameters are $\omega = 10$ rad/s, $\xi = 0.05$. The initial displacement is $u_0 = 0.1$. The stable solution is obtained with $\Delta t = \frac{2\xi}{\omega} (= 0.01)$ and an asymptotically stable solution is reported with $\Delta t = 0.002 < \frac{2\xi}{\omega}$ in Figure 8.4. The numerical method is unstable for all $\Delta t > \frac{2\xi}{\omega}$ and one such unstable solution obtained with $\Delta t = 0.0628$ is shown in Figure 8.2.

Clearly, the truncation error in the explicit Euler method is $O(\Delta t^2)$ and the method is convergent with $O(\Delta t)$ (i.e. order 1). Considering a scalar ODE $\dot{y} = f(t, y)$, the truncation error e_{n+1} $(= y(t_{n+1}) - y_{n+1})$ in the explicit Euler method is directly obtained from Equations 8.7 and 8.9a as $\frac{\Delta t^2}{2!}\frac{df(\varsigma)}{dt}, t_n \le \varsigma \le t_{n+1}$. This error is referred to as the local truncation error. The error is 'local', since it is based on the assumption that $e_n = 0$, that is that y_n and $f(t_n, y_n)$ at the previous step are known exactly. In general, the assumption that $e_n = 0$ seldom holds and the discretization error, which is propagated from t_0 to the current step t_{n+1}, is known as the global (discretization) error. In terms of e_n, the global error in the explicit Euler method is given by:

$$e_{n+1} = \left(y(t_n) + \Delta t f(t_n, y(t_n)) + \frac{\Delta t^2}{2!}\frac{df(\varsigma)}{dt} \right) - (y_n + \Delta t f(t_n, y_n), \quad t_n \le \varsigma \le t_{n+1}$$
$$= e_n + \Delta t(f(t_n, y(t_n)) - f(t_n, y_n)) + \frac{\Delta t^2}{2!}\frac{df(\varsigma)}{dt}, \quad t_n \le \varsigma \le t_{n+1} \tag{8.25}$$

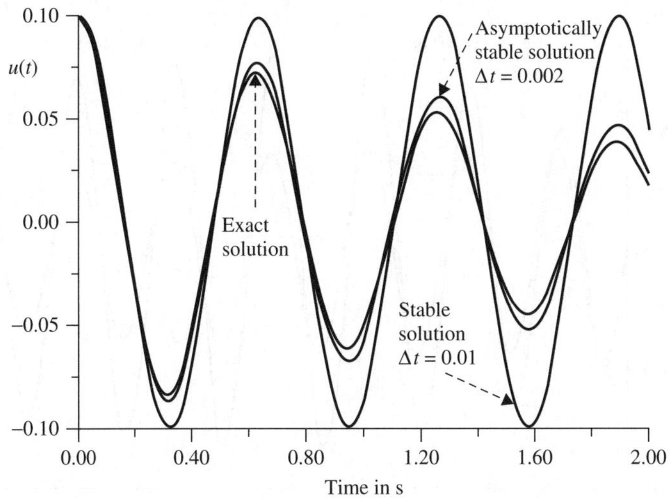

Figure 8.4 SDOF oscillator and solution by forward Euler method. Stable and asymptotically stable solutions vis-a-vis the exact solution; $\omega = 10\,\text{rad/s}, \xi = 0.05, u_0 = 0.1$

Assuming that f is continuous and second-order differentiable and denoting $\frac{\partial f}{\partial y}$ by f_y, we simplify the above equation to:

$$e_{n+1} = e_n + \Delta t f_y(t_n, \zeta) e_n + \frac{\Delta t^2}{2!}\frac{df(\varsigma)}{dt}, \quad t_n \le \varsigma \le t_{n+1}, y(t_n) \le \zeta \le y_n$$

$$= e_n(1 + \Delta t f_y(t_n, \zeta)) + \frac{\Delta t^2}{2!}\frac{df(\varsigma)}{dt} \tag{8.26}$$

If $\sup_{-\infty \le y \le \infty,\, t_0 \le t \le t_{n+1}} |f_y(t, y)| \le \wp < \infty$ and $\sup_{t_0 \le t \le t_{n+1}} \left|\frac{df}{dt}\right| \le Q$, we have

$$|e_{n+1}| \le |e_n|(1 + \Delta t \wp) + \frac{\Delta t^2}{2!} Q \tag{8.27}$$

Considering the map $z_{n+1} = z_n(1 + \Delta t \wp) + \frac{\Delta t^2}{2!} Q$, we have the solution (Conte and Boor, 1981; Atkinson and Han, 2004) as

$$z_n = \mathrm{c}(1 + \Delta t \wp)^n + \frac{Q\Delta t}{2\wp} \tag{8.28}$$

Since $e_0 = 0$, let us have $z_0 = 0$ to determine the constant c. This leads to the solution:

$$z_n = \frac{Q\Delta t}{2\wp}\{(1 + \Delta t \wp)^n - 1\} \le \frac{Q\Delta t}{2\wp}(e^{n\Delta t \wp} - 1) = \frac{Q\Delta t}{2\wp}(e^{(t_n - t_0)\wp} - 1) \tag{8.29}$$

From Equations 8.27 and 8.29, we have the following upper bound to the global error:

$$|e_n| \le \frac{Q\Delta t}{2\wp}(e^{t_n - t_0)\wp} - 1) \tag{8.30}$$

Thus, whilst the local truncation error is $\boldsymbol{O}(\Delta t^2)$, the global error is $\boldsymbol{O}(\Delta t)$ and is bounded by a constant times the step size. Since the global error is an aggregate of the local errors

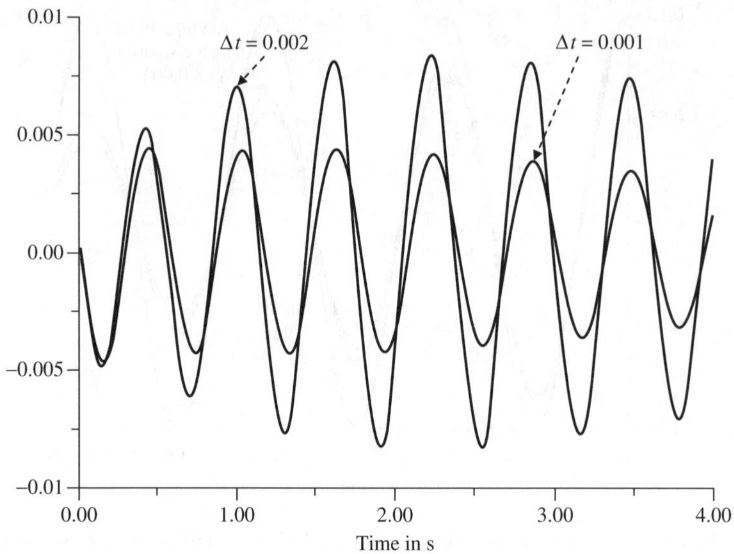

Figure 8.5 SDOF oscillator. Error histories by forward Euler method for $\Delta t = 0.001, 0.002$

that accumulate over the total number of time steps, it is typically one order Δt less than that of the local truncation error. With the approximation $(1 + \Delta t\wp)^n \approx \mathbf{1} + n\Delta t\wp$, Equation 8.29 directly gives an estimate for the upper bound as $|e_n| \leq \frac{Q\Delta t(t_n - t_0)}{2} = \boldsymbol{O}(\Delta t)$. In any case, it needs emphasis that the error estimate provides only a crude (possibly unrealistic) upper bound. Figure 8.5 shows the history of the error $y(t_n) - y_n$ numerically computed for Example 8.1 for which the exact solution $y(t_n)$ is known. The estimated error bound from $\frac{Q\Delta t(t_n - t_0)}{2}$ is generally found to be higher than the actual error, e_n.

8.2.2 Backward (Implicit) Euler Method

In this scheme, the finite difference equations for the displacement and velocity vectors are obtained as:

$$\frac{u(t+\Delta t) - u(t)}{\Delta t} = \dot{\boldsymbol{u}}(t+\Delta t) \Rightarrow \boldsymbol{u}_{n+1} = \boldsymbol{u}_n + \Delta t \dot{\boldsymbol{u}}_{n+1} \quad \text{and}$$

$$\frac{\dot{u}(t+\Delta t) - \dot{u}(t)}{\Delta t} = \ddot{\boldsymbol{u}}(t+\Delta t) \Rightarrow \dot{\boldsymbol{u}}_{n+1} = \dot{\boldsymbol{u}}_n + \Delta t \ddot{\boldsymbol{u}}_{n+1} \tag{8.31a,b}$$

The above equations are implicit as the unknowns $\dot{\boldsymbol{u}}_{n+1}$ and $\ddot{\boldsymbol{u}}_{n+1}$ are present on both sides of the equations. Substituting them in the system equilibrium equations (8.2) yields the matrix algebraic equations in the unknown acceleration vector at t_{n+1} as:

$$[\boldsymbol{M} + \Delta t\, \boldsymbol{C} + \Delta t^2 \boldsymbol{K}]\ddot{\boldsymbol{u}}_{n+1} = \{\boldsymbol{F}_{n+1} - \boldsymbol{C}\dot{\boldsymbol{u}}_n - \boldsymbol{K}(\boldsymbol{u}_n + \Delta t \dot{\boldsymbol{u}}_n)\} \tag{8.32}$$

Note that the above equation carries no special advantages through the lumped-mass approach. It admits solution, at every time step, by matrix decomposition of $\boldsymbol{M} + \Delta t\, \boldsymbol{C} + \Delta t^2 \boldsymbol{K}$, the coefficient matrix to the unknown vector, $\ddot{\boldsymbol{u}}_{n+1}$. With accelerations so evaluated, the displacement and velocity components are obtained from Equation 8.31. To study the stability of the implicit method, the following two-dimensional linear map corresponding to the test equation $\ddot{u} + 2\xi\omega\dot{u} + \omega^2 u = 0$ is utilised:

$$u_{n+1} = u_n + \Delta t\dot{u}_{n+1}$$
$$\dot{u}_{n+1} = \dot{u}_n + \Delta t(-2\xi\omega\dot{u}_{n+1} - \omega^2 u_{n+1}) \tag{8.33a,b}$$

The amplification matrix $\boldsymbol{A}$ is presently given by:

$$\boldsymbol{A} = \begin{bmatrix} 1 & -\Delta t \\ \omega^2\Delta t & 1 + 2\xi\omega\Delta t \end{bmatrix}^{-1} \tag{8.34a}$$

The eigenvalues of $\boldsymbol{A}$ are the reciprocals of those of the matrix $\begin{bmatrix} 1 & -\Delta t \\ \omega^2\Delta t & 1 + 2\xi\omega\Delta t \end{bmatrix}$ and are given by the complex conjugates (since $\xi < 1$):

$$\lambda_{1,2} = \frac{1 + \xi\omega\Delta t \pm j\omega\Delta t\sqrt{1-\xi^2}}{(1+\xi\omega\Delta t)^2 + \omega^2\Delta t^2(1-\xi^2)} \tag{8.34b}$$

The spectral radius of $\boldsymbol{A}$ is given by $\rho(\boldsymbol{A}) = \frac{1}{\{(1+\xi\omega\Delta t)^2+\omega^2\Delta t^2(1-\xi^2)\}^{1./2}}$, which remains less than unity for any $\Delta t > 0$. Thus, the scheme, as applied to the SDOF oscillator, is asymptotically stable.

Example 8.2: SDOF oscillator and solution by the implicit Euler method

The oscillator parameters and the initial condition are the same as those in Example 8.1. The oscillator response is obtained with $\Delta t = 0.0314, 0.01$ and 0.002. The computed solutions, which are asymptotically stable, are shown in Figure 8.6 along with the exact solution.

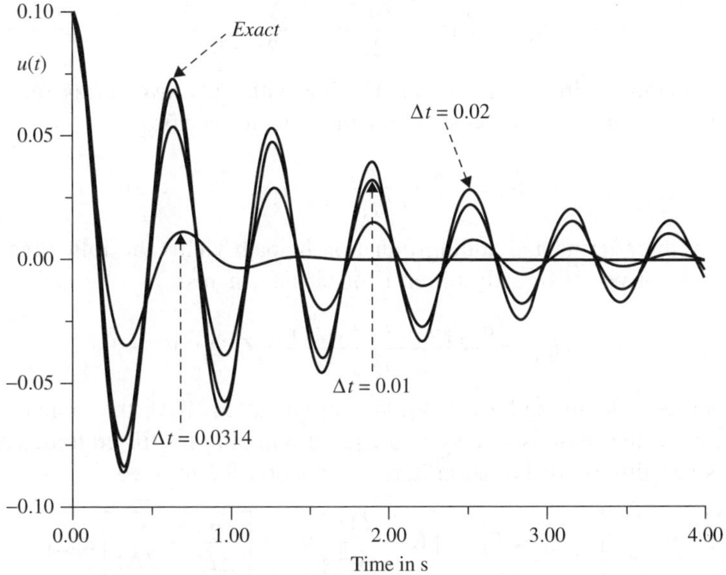

Figure 8.6 SDOF oscillator: asymptotically stable solutions by backward (implicit) Euler method; $\omega = 10$ rad/s, $\xi = 0.05$ and $u_0 = 0.1$

In the backward Euler method, the local truncation error e_{n+1} is obtained as (Equations 9.7 and 9.9b):

$$e_{n+1} = y(t_{n+1}) - y_{n+1}$$
$$= \left(y(t_n) + \Delta t f(t_n, y(t_n)) + \frac{\Delta t^2}{2!}\frac{df(\zeta)}{dt} \right) - (y_n + \Delta t f(t_{n+1}, y_{n+1}), \quad t_n \le \zeta \le t_{n+1} \tag{8.35}$$

By Taylor's expansion, we have $f(t_{n+1}, y_{n+1}) = f(t_n, y_n) + \Delta t \frac{df}{dt} + (y_{n+1} - y_n)\frac{df}{dy} + \cdots$. Substitution of this expansion in Equation 8.35 and simplification yield the local truncation error for the implicit Euler method as $-\frac{\Delta t^2}{2!}\frac{df(\zeta)}{dt}, t_n \le \zeta \le t_{n+1}$, which is of the same order ($O(\Delta t^2)$) as that of the explicit Euler method. However, the former has the crucial advantage of being comparatively more stable, as we have demonstrated for an SDOF oscillator. One observation from Figure 8.6 is that the numerical accuracy (as against stability) of the implicit solution depends on the step size, which controls the truncation error. It is often the practice to choose 'sufficiently' small Δt to compute solutions and check the 'distance' between two such solutions, with one being obtained, say, with Δt halved, so that it is within a specified tolerance.

8.3 Central Difference Method

This scheme utilises Taylor's expansions of $\boldsymbol{u}_{n+1}$ and $\boldsymbol{u}_{n-1}$ about $\boldsymbol{u}_n$, to arrive at finite difference approximations for velocity and acceleration. We have:

$$\boldsymbol{u}_{n+1} = \boldsymbol{u}_n + \Delta t \dot{\boldsymbol{u}}_n + \frac{\Delta t^2}{2!}\ddot{\boldsymbol{u}}_n + \frac{\Delta t^3}{3!}\dddot{\boldsymbol{u}}_n + \cdots \text{ and}$$
$$\boldsymbol{u}_{n-1} = \boldsymbol{u}_n - \Delta t \dot{\boldsymbol{u}}_n + \frac{\Delta t^2}{2!}\ddot{\boldsymbol{u}}_n - \frac{\Delta t^3}{3!}\dddot{\boldsymbol{u}}_n + \cdots \tag{8.36a,b}$$

Subtracting Equation 8.36b from Equation 8.36a with only two terms retained on the right-hand sides, we have the expression for the velocity vector as:

$$\dot{\boldsymbol{u}}_n = \frac{u_{n+1} - u_{n-1}}{2\Delta t} + \boldsymbol{O}(\Delta t^2) \tag{8.37a}$$

Retaining only three terms and adding Equation 8.36a,b yields the following expression for the acceleration vector directly in terms of displacements:

$$\ddot{\boldsymbol{u}}_n = \frac{u_{n+1} - 2u_n + u_{n-1}}{\Delta t^2} + \boldsymbol{O}(\Delta t^2) \tag{8.37b}$$

Note that both the velocity and acceleration approximations have the same order of truncation error. Now, the semidiscrete system equilibrium at t_n is utilised to determine $\boldsymbol{u}_{n+1}$. This requires substitution of Equation 8.37 in Equation 8.2 to have:

$$\left[\frac{\boldsymbol{M}}{\Delta t^2} + \frac{\boldsymbol{C}}{2\Delta t}\right] u_{n+1} = \boldsymbol{F}_n - \left[\boldsymbol{K} - \frac{2\boldsymbol{M}}{\Delta t^2}\right] u_n - \left[\frac{\boldsymbol{M}}{\Delta t^2} - \frac{\boldsymbol{C}}{2\Delta t}\right] u_{n-1} \tag{8.38}$$

Equation 8.38 along with the Equation 8.37 constitutes the central difference scheme. This is an explicit numerical scheme and requires extra information on $\boldsymbol{u}_{-1}$ in addition to the commonly specified initial conditions $\boldsymbol{u}_0$ and $\dot{\boldsymbol{u}}_0$. To determine $\boldsymbol{u}_{-1}$, Equation 8.2 is utilised to first have $\ddot{\boldsymbol{u}}_0 = \boldsymbol{F}_0 - \boldsymbol{C}\dot{\boldsymbol{u}}_0 - \boldsymbol{K}\boldsymbol{u}_0$. Equation 8.37 is then used with $n = 0$ to get $\boldsymbol{u}_{-1}$. With $\boldsymbol{u}_{-1}$ and $\boldsymbol{u}_0$ available, further computations are straightforward in recursively obtaining the displacement response at any t_n. The velocity and acceleration components are derivable from Equation 8.37.

The stability of the numerical scheme is again examined with the help of the test equation $\ddot{u} + 2\xi\omega\dot{u} + \omega^2 u = 0$. The velocity and acceleration expressions in Equation 8.37, when substituted in the test equation, give the following finite difference equation:

$$(1 + \xi\omega\Delta t)u_{n+1} + (\omega^2\Delta t^2 - 2)u_n + (1 - \xi\omega\Delta t)u_{n-1} = 0 \tag{8.39}$$

To identify the amplification matrix A, we use the transformations $\boldsymbol{A}u_n = u_{n+1}$ and $\boldsymbol{A}u_{n-1} = u_n$ in Equation (8.39) (assuming Δt to be uniform over the two steps). This yields $\{(1 + \xi\omega\Delta t)\boldsymbol{A}^2 + (\omega^2\Delta t^2 - 2)\boldsymbol{A} + (1 - \xi\omega\Delta t)\}u_{n-1} = 0$. Since $\boldsymbol{A}$ satisfies its own characteristic equation, its eigenvalues are the roots of the equation:

$$(1 + \xi\omega\Delta t)\lambda^2 + (\omega^2\Delta t^2 - 2)\lambda + (1 - \xi\omega\Delta t) = 0 \tag{8.40}$$

It is instructive here to highlight the use of Z-transform (Appendix H) in the study of the stability of the time-stepping maps. Denoting the Z-transform of the sequence $\{u_n\}$ as $Z(u_n) = U(z)$ and noting that $Z(u_{n+1}) = zU(z)$, $Z(u_{n-1}) = \frac{1}{z}U(z)$, we obtain the characteristic equation in terms of the z (complex) variable as $\mathcal{F}(z) = (1 + \xi\omega\Delta t)z^2 + (\omega^2\Delta t^2 - 2)z + (1 - \xi\omega\Delta t) = 0$, which is the same as Equation 8.40. Jury's stability criterion (Jury, 1974) provides a means to determine whether or not the roots of the characteristic equation lie within the unit circle (in the z-plane) without actually evaluating the roots. This criterion is equivalent to knowing whether or not $\rho(\boldsymbol{A}) < 1$. To meet the last requirement, it stipulates the necessary and sufficient conditions in terms of the coefficients of the characteristic polynomial $\mathcal{F}(z)$. In the present case, the degree of the polynomial is given by $n = 2$ and $\mathcal{F}(z)$ is expressed as $a_2 z^2 + a_1 z + a_0$ with $a_2 (= 1 + \xi\omega\Delta t) > 0$. Then, Jury's criteria provide the necessary conditions as:

$$\mathcal{F}(1) > 0; \quad (-1)^n \mathcal{F}(-1) > 0 \tag{8.41}$$

These conditions lead to:

$$(1 + \xi\omega\Delta t) + (\omega^2\Delta t^2 - 2) + (1 - \xi\omega\Delta t) > 0 \Rightarrow \omega^2\Delta t^2 > 0 \quad \text{and}$$

$$(1 + \xi\omega\Delta t) - (\omega^2\Delta t^2 - 2) + (1 - \xi\omega\Delta t) > 0 \Rightarrow 4 - \omega^2\Delta t^2 > 0 \tag{8.42a,b}$$

Whilst the condition $\omega^2\Delta t^2 > 0$ is obviously satisfied, Equation 8.42b requires $\omega\Delta t < 2$ or $\Delta t < \frac{T}{\pi}$, where $T\left(= \frac{2\pi}{\omega}\right)$ is the time period of the oscillator. The number of sufficient conditions as per Jury's stability criterion is $n - 1$ and, in the present case, the only sufficient condition is $|a_0| < |a_n|$ which is equivalent to $|1 - \xi\omega\Delta t| < |1 + \xi\omega\Delta t|$. As this condition is also satisfied, we have the only condition for asymptotic stability given by $\omega\Delta t < 2 \Rightarrow \Delta t < \frac{T}{\pi}$. Thus, the scheme is stable for $\Delta t \le \frac{T}{\pi}$.

Example 8.3: SDOF oscillator and solution by the central difference scheme

With the same parameters as in Example 8.1 adopted for the oscillator, the stable solutions obtained by the central difference scheme are shown in Figure 8.7 along with the exact solution. An unstable solution obtained with $\Delta t > \frac{T}{\pi}$ is shown in Figure 8.8.

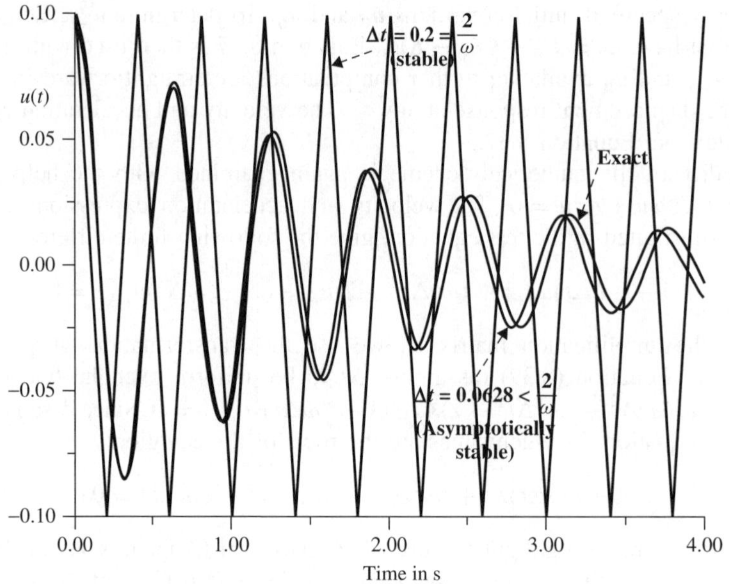

Figure 8.7 SDOF oscillator: stable solutions by central difference method; $\omega = 10\,\text{rad/s}$, $\xi = 0.05$ and $u_0 = 0.1$

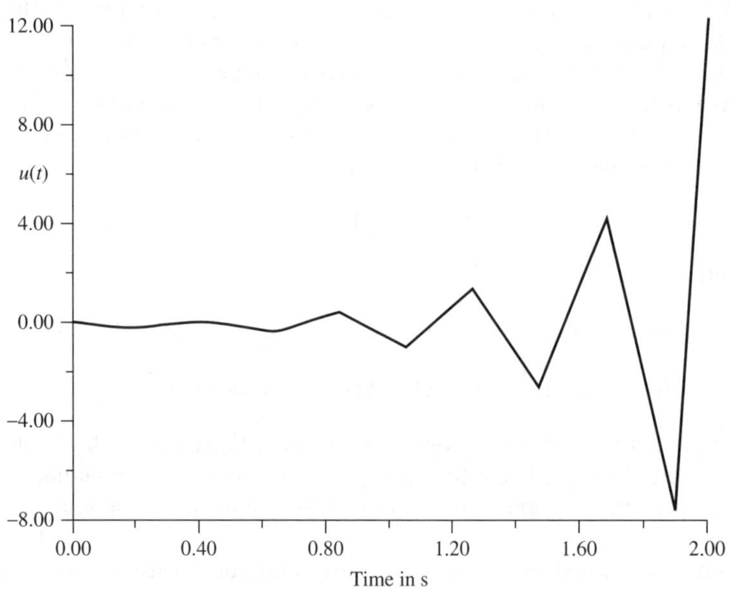

Figure 8.8 SDOF oscillator: unstable (diverging) solution by central difference method; $\omega = 10\,\text{rad/s}$, $\xi = 0.05$ and $u_0 = 0.1$. $\Delta t = 1.05 * \frac{2}{\omega} > \frac{2}{\omega}$

8.4 Newmark-β Method – a Single-Step Implicit Method

A popular numerical technique used for structural dynamic applications is the Newmark β method (Newmark, 1959) that helps achieve better accuracy and stability of integration. In fact, the integration scheme is aimed at achieving unconditional stability for at least linear (time-invariant) problems. The scheme is derivable through truncated Taylor-series expansions for displacement and velocity at t_{n+1} in the following forms:

$$\begin{aligned} \boldsymbol{u}_{n+1} &= \boldsymbol{u}_n + \Delta t\dot{\boldsymbol{u}}_n + \frac{\Delta t^2}{2}\ddot{\boldsymbol{u}}_n + \beta\Delta t^3\dddot{\boldsymbol{u}}_n \\ \dot{\boldsymbol{u}}_{n+1} &= \dot{\boldsymbol{u}}_n + \Delta t\ddot{\boldsymbol{u}}_n + \gamma\Delta t^2\dddot{\boldsymbol{u}}_n \end{aligned} \tag{8.43a,b}$$

β and γ are two parameters introduced to have flexibility in influencing the stability of the method. If linear acceleration is assumed within the time step (t_n, t_{n+1}), we have $\dddot{\boldsymbol{u}}_n = \frac{\ddot{\boldsymbol{u}}_{n+1}-\ddot{\boldsymbol{u}}_n}{\Delta t}$ and Equation 8.43 takes the standard form of the Newmark-β method

$$\begin{aligned} \boldsymbol{u}_{n+1} &= \boldsymbol{u}_n + \Delta t\dot{\boldsymbol{u}}_n + \Delta t^2\left\{\left(\frac{1}{2}-\beta\right)\ddot{\boldsymbol{u}}_n + \beta\ddot{\boldsymbol{u}}_{n+1}\right\} \\ \dot{\boldsymbol{u}}_{n+1} &= \dot{\boldsymbol{u}}_n + \Delta t\{(1-\gamma)\ddot{\boldsymbol{u}}_n + \gamma\ddot{\boldsymbol{u}}_{n+1}\} \end{aligned} \tag{8.44}$$

If $\beta = \frac{1}{4}$ and $\gamma = \frac{1}{2}$, it amounts to assuming an averaged acceleration within the time step (t_n, t_{n+1}). The truncation error is obtained by expanding $\ddot{\boldsymbol{u}}_{n+1}$ (on the RHS of Equation 8.44) in Taylor's series about t_n. Thus, we have:

$$\begin{aligned} \boldsymbol{u}_{n+1} &= \boldsymbol{u}_n + \Delta t\dot{\boldsymbol{u}}_n + \Delta t^2\left\{\left(\frac{1}{2}-\beta\right)\ddot{\boldsymbol{u}}_n + \beta(\ddot{\boldsymbol{u}}_n + \Delta t\dddot{\boldsymbol{u}}_n + ..)\right\} \\ &\Rightarrow \boldsymbol{u}(t_{n+1}) - \boldsymbol{u}_{n+1} = \left(\frac{1}{6}-\beta\right)\Delta t^3\dddot{\boldsymbol{u}}_n(\zeta_1), t_n \le \zeta_1 \le t_{n+1} \\ \dot{\boldsymbol{u}}_{n+1} &= \dot{\boldsymbol{u}}_n + \Delta t\{(1-\gamma)\ddot{\boldsymbol{u}}_n + \gamma(\ddot{\boldsymbol{u}}_n + \Delta t\dddot{\boldsymbol{u}}_n + ..)\} \\ &\Rightarrow \dot{\boldsymbol{u}}(t_{n+1}) - \dot{\boldsymbol{u}}_{n+1} = \left(\frac{1}{2}-\gamma\right)\Delta t^2\dddot{\boldsymbol{u}}_n(\zeta_2), t_n \le \zeta_2 \le t_{n+1} \end{aligned} \tag{8.45a,b}$$

If $\gamma = \frac{1}{2}$, the method is second-order accurate, otherwise it is of first order. Equation 8.44 is utilised to solve the system (semidiscritized) equilibrium equation (8.1). Considering the equilibrium state at t_{n+1}, we obtain the following set of algebraic equations in matrix form:

$$\hat{\boldsymbol{M}}\ddot{\boldsymbol{u}}_{n+1} = \boldsymbol{P}_{n+1} \tag{8.46}$$

Here, $\hat{\boldsymbol{M}} = \boldsymbol{M} + \gamma\Delta t\boldsymbol{C} + \beta\Delta t^2\boldsymbol{K}$ and $\boldsymbol{P}_{n+1} = \boldsymbol{F}_{n+1} - \boldsymbol{C}\{\dot{\boldsymbol{u}}_n + \Delta t(1-\gamma)\ddot{\boldsymbol{u}}_n\} - \boldsymbol{K}\{\boldsymbol{u}_n + \Delta t\dot{\boldsymbol{u}}_n + \Delta t^2\left(\frac{1}{2}-\beta\right)\ddot{\boldsymbol{u}}_n\}$. The system accelerations can be recursively obtained from Equation 8.45. The displacements and velocities then follow from Equations 8.44. The stability of the scheme largely depends on the two parameters β and γ. To discuss this aspect, we consider the test equation $\ddot{u} + 2\xi\omega\dot{u} + \omega^2 u = 0$ and its equilibrium at two instants t_n and t_{n+1}. The expressions $\ddot{u}_n = -2\xi\omega\,\dot{u}_n - \omega^2 u_n$ and

$\ddot{u}_{n+1} = -2\xi\omega\,\dot{u}_{n+1} - \omega^2 u_{n+1}$ corresponding to equilibria of the SDOF oscillator at t_n and t_{n+1} are substituted in Equation 8.44 to obtain the linear mapping:

$$A\begin{pmatrix} u_{n+1} \\ \dot{u}_{n+1} \end{pmatrix} = B\begin{pmatrix} u_n \\ \dot{u}_n \end{pmatrix} \tag{8.47}$$

Here, the matrices A and B are given by

$$A = \begin{bmatrix} 1+\beta\Delta t\omega^2 & 2\xi\omega\beta\Delta t^2 \\ \gamma\omega^2\Delta t & 1+2\xi\omega\gamma\Delta t \end{bmatrix}$$

$$\text{and} \quad B = \begin{bmatrix} 1-\Delta t^2\omega^2\left(\frac{1}{2}-\beta\right) & \Delta t - 2\xi\omega\Delta t^2\left(\frac{1}{2}-\beta\right) \\ -\Delta t\omega^2(1-\gamma) & 1-2\xi\omega\Delta t(1-\gamma) \end{bmatrix}$$

Z-transforming Equation 8.47, we have $z\boldsymbol{A} - \boldsymbol{BI} = 0$ and the characteristic equation as:

$$\begin{aligned} & z^2(1+2\xi\gamma\omega\Delta t+\beta\omega^2\Delta t^2) \\ & - z\left\{(1+\beta\omega^2\Delta t^2)(1-2\xi\omega\Delta t+2\xi\omega\Delta t\gamma)\right. \\ & + \left(1-\frac{\omega^2\Delta t^2}{2}+\omega^2\Delta t^2\beta\right)(1+2\xi\gamma\omega\Delta t)+2\xi\beta\omega^3\Delta t^3(1-\gamma) \\ & \left.+(\gamma\omega^2\Delta t^2-\xi\gamma\omega^3\Delta t^3+2\xi\gamma\beta\omega^3\Delta t^3)\right\} \\ & + \left\{1+2\xi\gamma\omega\Delta t+\beta\omega^2\Delta t^2-\omega^2\Delta t^2\left(\gamma-\frac{1}{2}\right)-2\xi\omega\Delta t\right\} = 0 \end{aligned} \tag{8.48}$$

If $\mathcal{F}(z)(=a_2z^2+a_1z+a_0)$ stands for the quadratic expression on the right-hand side of Equation 8.48, we have $\mathcal{F}(1)=\omega^2\Delta t^2$, $(-1)^n\mathcal{F}(-1)=4+\omega^2\Delta t^2(4\beta-2\gamma)+4\xi\omega\Delta t(2\gamma-1)$. As per Jury's criterion, the necessary conditions for asymptotic stability of the integration scheme are $\mathcal{F}(1)>0$ and $(-1)^n\mathcal{F}(-1)>0$. The first condition is always satisfied and the second one, if $\gamma>\frac{1}{2}$ and $4\beta-2\gamma>0 \Rightarrow \beta>\frac{\gamma}{2}$. The sufficient condition is $a_2>a_0 \Rightarrow \omega^2\Delta t^2\left(\gamma-\frac{1}{2}\right)+2\xi\omega\Delta t>0$, which is also true if $\gamma>\frac{1}{2}$. This is irrespective of whether the eigenvalues of $\boldsymbol{A}$ are real or complex. Thus, we observe that the Newmark-β method is unconditionally stable (i.e. for any choice of Δt) if $\gamma>\frac{1}{2}$ and $\beta>\frac{\gamma}{2}$. When the two parameters violate this requirement of unconditional stability, there is a need to examine if the scheme is conditionally stable or unstable. For example, if $\gamma<\frac{1}{2}$ and $\beta>\frac{\gamma}{2}$, stability is ensured if $(-1)^n\mathcal{F}(-1)>0 \Rightarrow 4+\omega^2\Delta t^2(4\beta-2\gamma)-4\xi\omega\Delta t(1-2\gamma)>0$ and $a_2>a_0$. The latter condition implies that $\omega\Delta t<\frac{2\xi}{\frac{1}{2}-\gamma}$. The former leads to the following inequality in terms of Δt:

$$4+\omega^2\Delta t^2(4\beta-2\gamma)-4\xi\omega\Delta t(1-2\gamma)>0 \tag{8.49}$$

The left-hand side of the above equation is quadratic in $\omega\Delta t$. If the discriminant $\xi^2(1-2\gamma)^2-(4\beta-2\gamma)$ is greater than zero, it admits two real positive roots, r_1 and r_2, given by:

$$r_1, r_2 = \frac{2\left\{\xi(1-2\gamma) \pm \sqrt{\xi^2(1-2\gamma)^2 - (4\beta - 2\gamma)}\right\}}{(2\beta - \gamma)} \tag{8.50}$$

Both the roots are positive and hence, with the assumption that $r_1 < r_2$, condition (8.49) leads to the requirement that $\omega\Delta t < r_1$ or $\omega\Delta t > r_2$. Along with this requirement, Δt has also to satisfy the condition $\omega\Delta t < \frac{2\xi}{\frac{1}{2}-\gamma}$ to ensure the stability of the scheme. If the discriminant is less than zero, the roots are complex and the inequality in Equation 8.49 is satisfied for any choice of Δt. However, the time step has to satisfy the sufficient condition, $\omega\Delta t < \frac{2\xi}{\frac{1}{2}-\gamma}$. In any case, this ($\gamma < \frac{1}{2}$ and $\beta > \frac{\gamma}{2}$) corresponds to conditional stability. If $\gamma > \frac{1}{2}$ and $\beta < \frac{\gamma}{2}$, we observe that the conditions $\mathcal{F}(1) > 0$ and $a_2 > a_0$ are satisfied and the only other condition to be checked for stability is $(-1)^n\mathcal{F}(-1) > 0$. This leads to the following inequality in terms of Δt: $4 - \omega^2\Delta t^2(2\gamma - 4\beta) + 4\xi\omega\Delta t(2\gamma - 1) > 0$. This is similar to the inequality in Equation 8.49 with the difference that the two roots r_1 and r_2 of the quadratic on the LHS are always real and of opposite sign given by:

$$r_1, r_2 = \frac{2\left\{\xi(2\gamma-1) \pm \sqrt{\xi^2(2\gamma-1)^2 + (2\gamma - 4\beta)}\right\}}{(\gamma - 2\beta)} \tag{8.51}$$

Assuming that r_1 is positive, inequality (Equation 8.51) results in the condition for stability as $0 < \omega\Delta t < r_1$. This is also a case of conditional stability. The last possible case of $\gamma < \frac{1}{2}$ and $\beta < \frac{\gamma}{2}$ (in the first quadrant of $\beta - \gamma$ plane) needs both the conditions $a_2 > a_0$ and $(-1)^n\mathcal{F}(-1) > 0$ to be checked for stability. The former requires that $\omega\Delta t < \frac{2\xi}{\frac{1}{2}-\gamma}$. The latter yields $4 - \omega^2\Delta t^2(2\gamma - 4\beta) - 4\xi\omega\Delta t(1 - 2\gamma) > 0$. This inequality is again similar to the one in Equation 8.49. It results in the condition for stability as $0 < \omega\Delta t < r_1$, where r_1 is the positive real root given by

$$r_1 = \frac{2\left\{-\xi(1-2\gamma) + \sqrt{\xi^2(1-2\gamma)^2 + (2\gamma - 4\beta)}\right\}}{(\gamma - 2\beta)} \tag{8.52}$$

Thus, for $\gamma < \frac{1}{2}$ and $\beta < \frac{\gamma}{2}$, the simultaneous requirements $\omega\Delta t < \frac{2\xi}{\frac{1}{2}-\gamma}$ (due to the condition $a_2 > a_0$) and $0 < \omega\Delta t < r_1$ (as given in Equation 8.52) must be ensured for stability of the scheme. Table 8.1 provides a summary of these observations.

Example 8.4: Newmark-β method applied to an SDOF oscillator

The SDOF oscillator ($\Omega = 10$ rad/s and $\xi = 0.05$) in Example 8.1 is again considered to examine the stability of the Newmark-β method and its dependence on the twin parameters β and γ. The stable solutions obtained with $\gamma > \frac{1}{2}$ and $\beta > \frac{\gamma}{2}$ are shown in Figure 8.9 along with the exact solution. The conditional stability of the scheme for the other cases (Table 8.1) is illustrated in Figures 8.10–8.12. In each of these cases, stable solutions are obtained with specific choices of Δt that satisfy the respective Jury's stability criteria. The figures show the effect of fine tuning the step size on the accuracy of the numerical solution (with respect to the exact one). The effect is manifest in reducing the amplitude and periodicity errors that are noticeable in the stable solutions with larger time steps. Unstable solutions obtained with Δt violating the stability criterion are also shown in these figures.

Table 8.1 Newmark-β method; dependence of stability on β and γ

Unconditional stability	Conditional stability		
$\gamma > \frac{1}{2}$ and $\beta > \frac{\gamma}{2}$	$\gamma < \frac{1}{2}$, $\beta > \frac{\gamma}{2}$ If r_1 and r_2 $(r_2 > r_1)$ are real, $\omega\Delta t < r_1$ or $\omega\Delta t > r_2$ (Equation 8.43) and $\omega\Delta t < \frac{2\xi}{\frac{1}{2} - \gamma}$ If r_1 and r_2 are complex, $\omega\Delta t < \frac{2\xi}{\frac{1}{2} - \gamma}$	$\gamma > \frac{1}{2}$, $\beta < \frac{\gamma}{2}$ $0 < \omega\Delta t < r_1$ (Equation 8.44)	$\gamma < \frac{1}{2}$, $\beta < \frac{\gamma}{2}$ $0 < \omega\Delta t < r_1$ (Equation 8.45) and $\omega\Delta t < \frac{2\xi}{\frac{1}{2} - \gamma}$

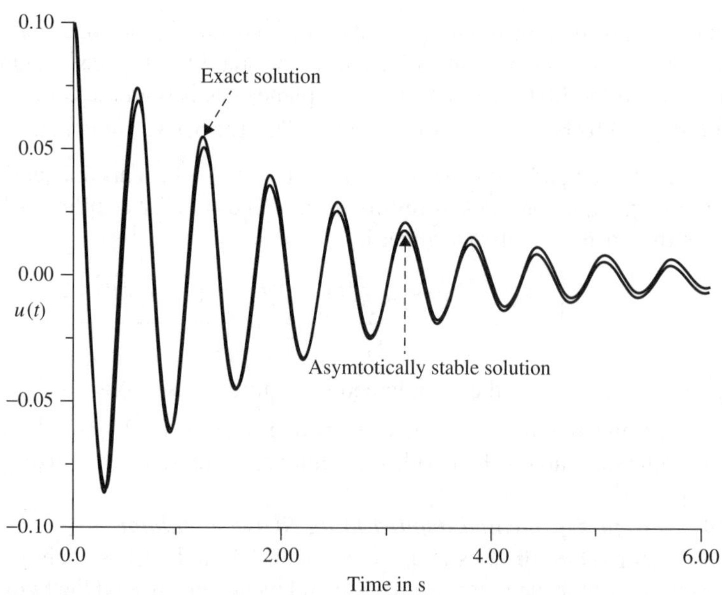

Figure 8.9 SDOF oscillator; unconditionally stable solutions by Newmark-β method; $\omega =$ 10 rad/s, $\xi = 0.05$ and $u_0 = 0.1$; $\gamma = 0.6(> \frac{1}{2})$ and $\beta = 0.4(> \frac{\gamma}{2})$, $\Delta t = 0.01$

8.4.1 Some Degenerate Cases of the Newmark-β Method and Stability

There may arise a few degenerate cases with values of γ and β lying on the boundaries of the stability zone described in Table 8.1. If we consider such cases (with either $\gamma = 1/2$ or $\beta = \gamma/2$ or both), it is not possible to directly infer the stability of the Newmark-β scheme from the conditions listed in Table 8.1. However, for these specific cases, the

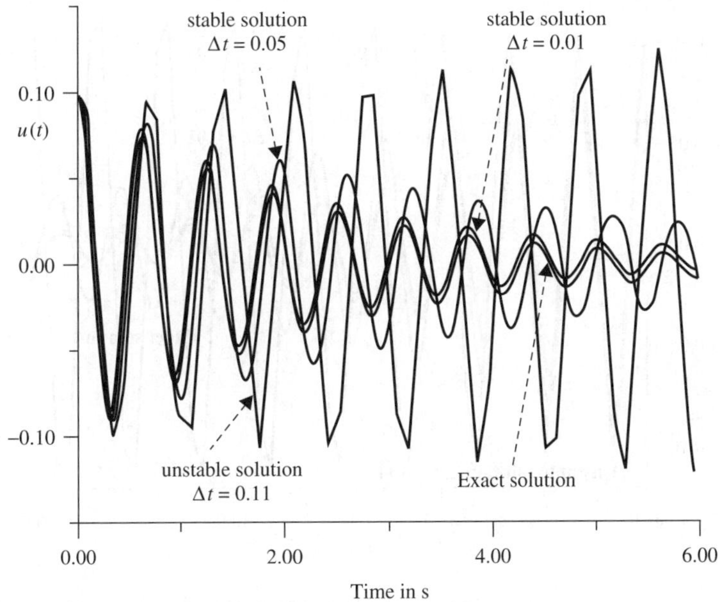

Figure 8.10 SDOF oscillator; conditionally stable and unstable solutions by Newmark-β method; $\omega = 10\,\text{rad/s}$, $\xi = 0.05$ and $u_0 = 0.1$, $\gamma = 0.4\left(< \frac{1}{2}\right)$, $\beta = 0.25(> \frac{\gamma}{2})$, r_1, r_2 complex (Equation 8.43) and $\omega\Delta t < 2\xi/(1/2 - \gamma)$ for stability

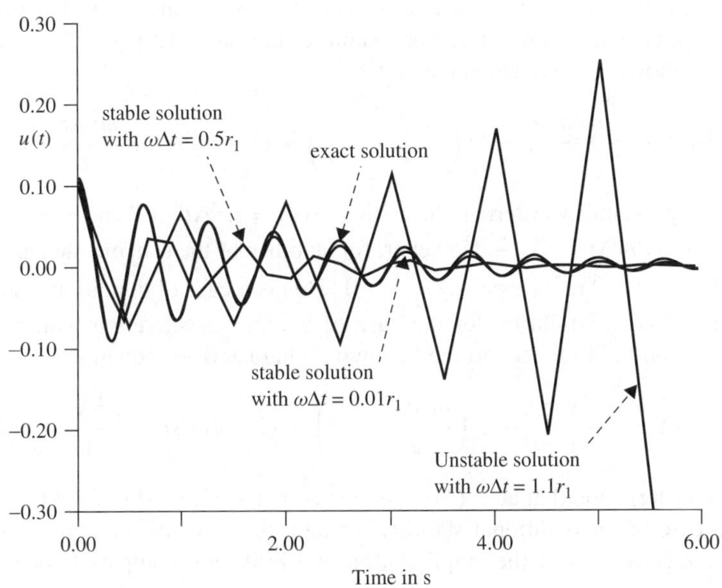

Figure 8.11 SDOF oscillator; conditionally stable and unstable solutions by Newmark-β method; $\omega = 10$ rad/s, $\xi = 0.05$ and $u_0 = 0.1$; $\gamma = 0.6\left(> \frac{1}{2}\right)$, $\beta = 0.25(< \frac{\gamma}{2})$, $r_1 = 4.57$ and r_2 negative (Equation 8.44); $0 < \omega\Delta t < r_1$, for stability

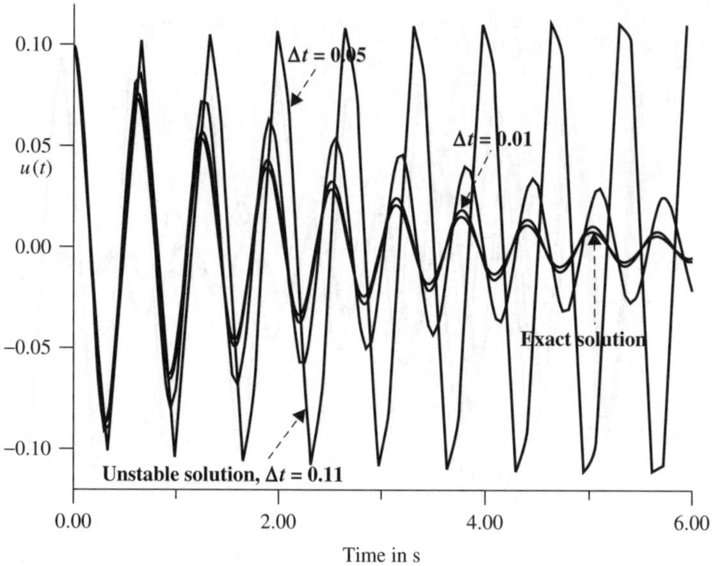

Figure 8.12 SDOF oscillator; conditionally stable and unstable solutions by Newmark-β method; $\omega = 10$ rad/s, $\xi = 0.05$ and $u_0 = 0.1$; $\gamma = 0.4\left(< \frac{1}{2}\right), \beta = 0.15(< \frac{\gamma}{2})$, $r_1 = 4.373$ and r_2 negative (Equation 8.45), $0 < \omega\Delta t < r$, and $\omega\Delta t < \frac{2\xi}{\frac{1}{2}-\gamma}$ for stability

characteristic equation takes a simpler form (Equation 8.48) and the stability criteria can directly be applied with more ease. For example, the case with $\gamma = 1/2$ and $\beta = 1/6$ results in the following characteristic equation:

$$z^2\left(1+\xi\omega\Delta t+\frac{\omega^2\Delta t^2}{6}\right)+2z\left(\frac{\omega^2\Delta t^2}{3}-1\right)+(1-\xi\omega\Delta t+\frac{\omega^2\Delta t^2}{6})=0 \tag{8.53}$$

In this case, Jury's stability criteria indicate that $\mathcal{F}(1) = \omega^2\Delta t^2 > 0$ and $a_2 = 1 + \xi\omega\Delta t + \frac{\omega^2\Delta t^2}{4} > a_0 = 1 - \xi\omega\Delta t + \frac{\omega^2\Delta t^2}{4}$. However, for stability of the scheme, the necessary condition $(-1)^n\mathcal{F}(-1) > 0$ requires $4 - \frac{\omega^2\Delta t^2}{3} > 0$. This is a case of conditional stability with $\omega\Delta t < \sqrt{12} \cong 3.464$. Similarly, for the case of average acceleration with $\gamma = 1/2$ and $\beta = 1/4$, Equation 8.48 reduces to the following characteristic equation:

$$z^2\left(1+\xi\omega\Delta t+\frac{\omega^2\Delta t^2}{4}\right)+z\left(\frac{\omega^2\Delta t^2}{2}-2\right)+(1-\xi\omega\Delta t+\frac{\omega^2\Delta t^2}{4})=0 \tag{8.54}$$

The stability criteria yield that $\mathcal{F}(1) = \omega^2\Delta t^2 > 0$, $(-1)^n\mathcal{F}(-1) = 4 > 0$ and $a_2 > a_0$. Thus, it is a case of unconditional stability. In fact, the roots of the above characteristic equation (the eigenvalues of the amplification matrix $\boldsymbol{A}$) are complex (since $\xi < 1$) and are given by:

$$\lambda_1, \lambda_2 = \frac{1-\dfrac{\omega^2\Delta t^2}{4} \pm j\omega\Delta t\sqrt{1-\xi^2}}{1+\xi\omega\Delta t+\dfrac{\omega^2\Delta t^2}{4}} \tag{8.55}$$

The spectral radius is given by $\rho(\boldsymbol{A}) = \sqrt{\frac{1+\xi^2\omega^2\Delta t^2+\left(\frac{\omega^2\Delta t^2}{4}\right)^2-\frac{3\omega^2\Delta t^2}{2}}{1+\xi\omega\Delta t+\frac{\omega^2\Delta t^2}{4}}}$ and it always remains less than unity (unconditional stability). It is also informative to observe that Equation 8.48 represents the characteristic equation of the explicit Euler method if $\gamma = \beta = 0$ and that of the central difference method if $\gamma = \frac{1}{2}$ and $\beta = 0$.

8.4.2 Undamped Case – Amplitude and Periodicity Errors

For an undamped case with $\xi = 0$, the characteristic equation is obtained from Equation 8.48 as

$$z^2(1+\beta\omega^2\Delta t^2) + z\{\omega^2\Delta t^2(\gamma+1/2) - 2(1+\beta\omega^2\Delta t^2)\} + 1 + \beta\omega^2\Delta t^2 + \omega^2\Delta t^2\left(\frac{1}{2}-\gamma\right) = 0 \tag{8.56}$$

Jury's stability criteria applied to the above equation helps to identify the stability characteristics of the integration scheme for any combination of the parameters β and γ. For example, with $\gamma > \frac{1}{2}$ and $\beta > \gamma/2$, $\mathcal{F}(1) = \omega^2\Delta t^2$, $(-1)^n\mathcal{F}(-1) = 4 + (4\beta - 2\gamma)\omega^2\Delta t^2$ and $a_2 - a_0 = \omega^2\Delta t^2(\gamma - 1/2)$. These conditions are satisfied for any $\Delta t \neq 0$ and thus the scheme is unconditionally stable for $\gamma > \frac{1}{2}$ and $\beta > \gamma/2$ as in the damped case. A closer look at this undamped case ($\ddot{u} + \omega^2 u = 0$) helps to quantify the amplitude and periodicity errors (Hilber and Hughes, 1978). This further highlights the need for a possible tuning and control of Δt to reduce these errors.

8.4.3 Amplitude and Periodicity Errors

For an undamped oscillator, Equation 8.56 admits complex roots (yielding harmonic motion) if:

$$(\gamma+1/2)^2\frac{\omega^2\Delta t^2}{1+\beta\omega^2\Delta t^2} \leq 4 \Rightarrow \omega^2\Delta t^2 \leq \frac{4}{(\gamma+1/2)^2 - 4\beta} \tag{8.57}$$

If the complex roots λ_1, λ_2 are expressed in the amplitude-phase or form, $\rho e^{\pm j\Phi}$ where ρ is the amplitude (and the spectral radius too) and Φ, the phase angle, these are given by

$$\rho = \frac{\left\{1+\beta\omega^2\Delta t^2 + \left(\frac{1}{2}-\gamma\right)\omega^2\Delta t^2\right\}^{1/2}}{(1+\beta\omega^2\Delta t^2)^{1/2}},$$

$$\Phi = \mathrm{Tan}^{-1}\frac{\omega\Delta t\left[1+\left\{\beta - \frac{1}{4}\left(\frac{1}{2}+\gamma\right)^2\right\}\omega^2\Delta t^2\right]^{1/2}}{1+\left\{\beta - \frac{1}{2}\left(\frac{1}{2}+\gamma\right)\right\}\omega^2\Delta t^2} \tag{8.58a,b}$$

If $\gamma \geq 1/2$ and β is so chosen as to satisfy Equation 8.57, we have $\rho \leq 1$ and the scheme is unconditionally stable. With initial condition $u(0) = u_0$, the numerical solution can be written as $u_n = u_0\rho^n e^{in\Phi}$. The exact solution to the undamped test equation is, however, known to be $u(t) = u_0 \cos\omega t \Rightarrow u_n = u_0 e^{in\omega\Delta t}$. The exact spectral radius is unity and the amplitude error is thus $\rho - 1$. The relative periodicity error is given by $\frac{\omega\Delta t}{\Phi} - 1$. With $\gamma = 1/2$ (central difference, average acceleration schemes), the amplitude error is zero

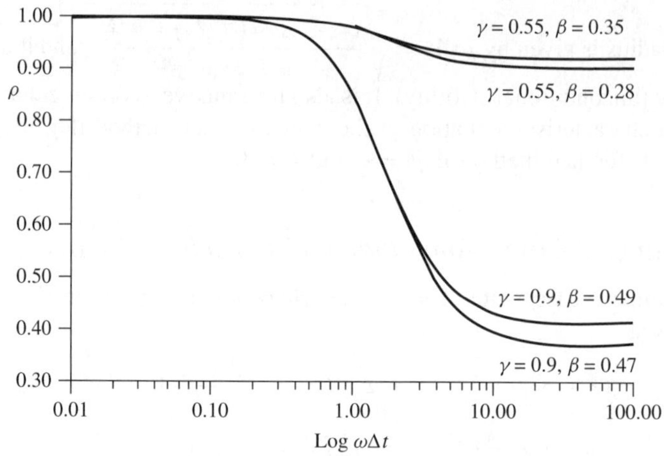

(a)

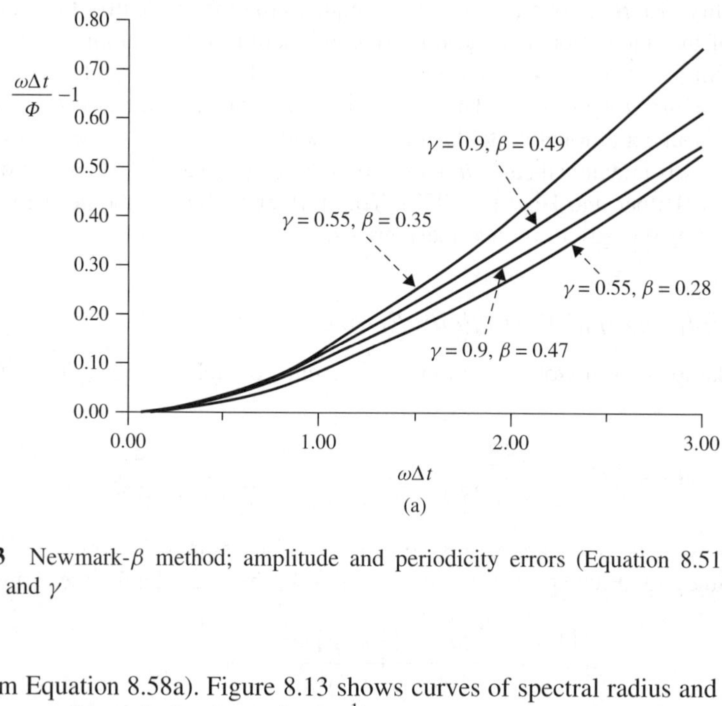

(a)

Figure 8.13 Newmark-β method; amplitude and periodicity errors (Equation 8.51) for typical values of β and γ

($\rho = 1$ from Equation 8.58a). Figure 8.13 shows curves of spectral radius and the relative periodicity error for typical values of $\gamma (> \frac{1}{2})$.

From Figure 8.13, it is inferred that amplitude error remains zero for smaller ω (assuming Δt fixed) and increases for larger frequencies. The amplitude error may be construed as the effect of an external damping induced by the algorithm. This algorithmic damping, in fact, is a desirable feature of an integration scheme in that it helps to numerically dissipate the contributions from spurious high-frequency modes. As stated earlier, these spurious modes invariably arise due to the semidiscretization of a continuous system (wherein the approximated field variable has a limited order of continuity). The Newmark-β method, whilst exhibiting this desirable feature of high-frequency dissipation for values of $\gamma > 1/2$, is, however, only first-order accurate. The HHT-α (Hilber, Hughes and

Taylor, 1977) and generalized-α (Chung and Hulbert, 1993) methods are variants of the Newmark-β method, devised to achieve second-order accuracy along with a control on the high-frequency dissipation.

8.5 HHT-α and Generalized-α Methods

In these methods, the finite difference approximations remain the same as in the Newmark-β method (Equation 8.44). They, however, involve a modification to the equilibrium equation 8.2. Specifically, in HHT-α method, the damping and elastic forces are taken to be a linear combination of those corresponding to the two ends of the time interval (t_n, t_{n+1}). The external force is also assumed to admit a similar combination and the modified equilibrium equations are given by:

$$\begin{aligned} &\boldsymbol{M}\ddot{\boldsymbol{u}}_{n+1} + \{(1-\alpha_f)\boldsymbol{C}\dot{\boldsymbol{u}}_{n+1} + \alpha_f \boldsymbol{C}\dot{\boldsymbol{u}}_n\} + \{(1-\alpha_f)\boldsymbol{K}\boldsymbol{u}_{n+1} + \alpha_f \boldsymbol{K}\boldsymbol{u}_n\} \\ &\quad = \{(1-\alpha_f)\boldsymbol{f}_{n+1} + \alpha_f \boldsymbol{f}_n\} \end{aligned} \tag{8.59}$$

α_f is a parameter defining the relative weights of the damping, elastic and external forces (Figure 8.14) corresponding to the two instants t_n and t_{n+1}. In the generalized-α method, the inertia forces at both these instants are also involved in the time-discrete equilibrium equations. The time-stepping map for this case is given by:

$$\begin{aligned} &\{(1-\alpha_m)\boldsymbol{M}\ddot{\boldsymbol{u}}_{n+1} + \alpha_m \boldsymbol{M}\ddot{\boldsymbol{u}}_n\} + \{(1-\alpha_f)\boldsymbol{C}\dot{\boldsymbol{u}}_{n+1} + \alpha_f \boldsymbol{C}\dot{\boldsymbol{u}}_n\} \\ &\quad + \{(1-\alpha_f)\boldsymbol{K}\boldsymbol{u}_{n+1} + \alpha_f \boldsymbol{K}\boldsymbol{u}_n\} = \{(1-\alpha_f)\boldsymbol{f}_{n+1} + \alpha_f \boldsymbol{f}_n\} \end{aligned} \tag{8.60}$$

α_m is similar to α_f (see Figure 8.14 for a graphical description of how it works). The generalized-α method is more general in that $\alpha_m = 0$ yields the HHT-α method and reduces to the Newmark-β method with $\alpha_m = 0$ and $\alpha_f = 0$. Focusing our attention to the generalized-α method only, we utilise the Newmark-β finite difference equations 8.44 to solve the equilibrium equation 8.60. This leads to the following set of algebraic equations in matrix-vector form:

$$\begin{aligned} &[(1-\alpha_m)\boldsymbol{M} + (1-\alpha_f)\gamma\Delta t\boldsymbol{C} + (1-\alpha_f)\beta\Delta t^2\boldsymbol{K}]\ddot{\boldsymbol{u}}_{n+1} \\ &\quad = (1-\alpha_f)\boldsymbol{f}_{n+1} + \alpha_f \boldsymbol{f}_n - \left[\alpha_m \boldsymbol{M} + (1-\alpha_f)\Delta t(1-\gamma)\boldsymbol{C} + (1-\alpha_f)\Delta t^2\left(\frac{1}{2}-\beta\right)\boldsymbol{K}\right]\ddot{\boldsymbol{u}}_n \\ &\quad - [\boldsymbol{C} + (1-\alpha_f)\Delta t\boldsymbol{K}]\dot{\boldsymbol{u}}_n - \boldsymbol{K}\boldsymbol{u}_n \end{aligned} \tag{8.61}$$

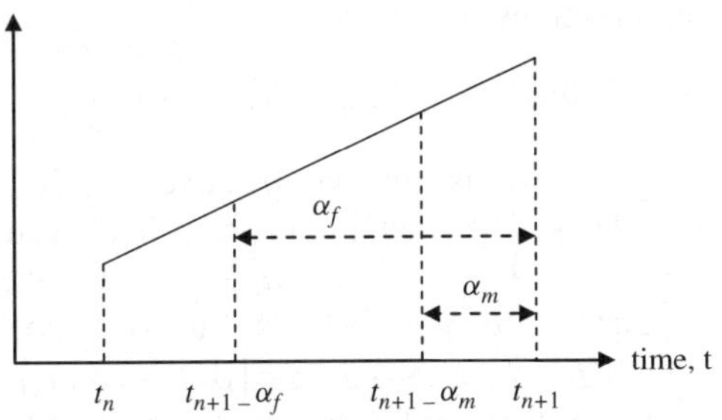

Figure 8.14 HHT-α and generalized-β methods; parameters α_f and α_m defining the relative participation of the damping, elastic, external and inertia forces in the equilibrium equation

The above recursive equation yields $\ddot{\boldsymbol{u}}_{n+1}$ at time t_{n+1}, from which $\dot{\boldsymbol{u}}_{n+1}$ and $\boldsymbol{u}_{n+1}$ can be determined using the finite difference equations 8.44.

If we examine the stability of this method via the test equation $\ddot{u} + 2\xi\omega\dot{u} + \omega^2 u = 0$, we get the following linear mapping of the system states from t_n to t_{n+1}:

$$\begin{bmatrix} 1 & 0 & -\beta\Delta t^2 \\ 0 & 1 & -\gamma\Delta t \\ (1-\alpha_f)\omega^2 & 2\xi\omega(1-\alpha_f) & 1-\alpha_m \end{bmatrix} \begin{pmatrix} u_{n+1} \\ \dot{u}_{n+1} \\ \ddot{u}_{n+1} \end{pmatrix}$$

$$= \begin{bmatrix} 1 & \Delta t & \Delta t^2\left(\dfrac{1}{2} - \beta\right) \\ 0 & 1 & \Delta t(1-\gamma) \\ -\alpha_f\omega^2 & -2\xi\omega\alpha_f & -\alpha_m \end{bmatrix} \begin{pmatrix} u_n \\ \dot{u}_n \\ \ddot{u}_n \end{pmatrix} \tag{8.62}$$

Z-transforming the above equation gives the characteristic equation in the form: $\mathcal{F}(z) = a_3 z^3 + a_2 z^2 + a_1 z + a_0 = 0$, where the coefficients $a_i, i = 0, 1, 2$ and 3 are given by

$$\begin{aligned}
a_3 &= (1-\alpha_m) + 2\xi\omega\Delta t\gamma(1-\alpha_f) + \beta\omega^2\Delta t^2(1-\alpha_f) \\
a_2 &= -2 + 3\alpha_m + 2\xi\omega\Delta t(1-\alpha_f)(1-2\gamma) + 2\xi\omega\Delta t\gamma\alpha_f \\
&\quad + \omega^2\Delta t^2(1-\alpha_f)\left(\frac{1}{2} + \gamma - 2\beta\right) + \omega^2\Delta t^2\alpha_f\beta \\
a_1 &= 1 - 3\alpha_m + 2\xi\omega\Delta t\alpha_f(1-2\gamma) - 2\xi\omega\Delta t(1-\alpha_f)(1-\gamma) \\
&\quad + \omega^2\Delta t^2(1-\alpha_f)\left(\frac{1}{2} - \gamma + \beta\right) + \omega^2\Delta t^2\alpha_f\left(\frac{1}{2} + \gamma - 2\beta\right) \\
a_0 &= \alpha_m + 2\xi\omega\Delta t\alpha_f(\gamma - 1) + \omega^2\Delta t^2\alpha_f\left(\frac{1}{2} + \beta - \gamma\right)
\end{aligned} \tag{8.63}$$

The two necessary conditions for stability are $\mathcal{F}(1) > 0$, $(-1)^n\mathcal{F}(-1) > 0$. Knowing $a_i, i = 0, 1, 2$ and 3 (Equation 8.63), we find that $\mathcal{F}(1) = \Delta t^2\omega^2 > 0$. Further, $(-1)^n\mathcal{F}(-1) = -4(2\alpha_m - 1) - 4\xi\omega\Delta t(2\gamma - 1)(2\alpha_f - 1) - \omega^2\Delta t^2(2\alpha_f - 1)(4\beta - 2\gamma)$. For asymptotic stability, we infer that the condition $(-1)^n\mathcal{F}(-1) > 0$ is satisfied if we have $\alpha_m < \frac{1}{2}$ and $\alpha_f < 1/2$, in addition to having $\frac{1}{2} < \gamma < 1$ and $\beta > \gamma/2$. The latter two requirements on the values of γ and β are the same as those needed for asymptotic stability of the Newmark-β method. Since the characteristic equation in the present case is of degree 3, the number of sufficient conditions for stability is two. Referring to Table 8.2, we have these conditions in the form:

$$|a_3| > |a_0| \quad \text{and} \quad |b_0| > |b_2| \tag{8.64}$$

With $\alpha_m < \frac{1}{2}$ and $\alpha_f < 1/2$, a_3 is always positive. With $\frac{1}{2} < \gamma < 1$, a_0 may be positive or negative. Thus, the condition $|a_3| > |a_0|$ leads to the requirement $a_3 > |a_0| \Rightarrow -a_3 < a_0 < a_3 \Rightarrow a_0 + a_3 > 0$ and $a_0 - a_3 < 0$. We find that $a_0 + a_3 = 1 + 2\xi\omega\Delta t(\gamma - \alpha_f) + \omega^2\Delta t^2(\beta - \alpha_f\gamma + \frac{\alpha_f}{2})$ is always positive and $a_0 - a_3 = (-1 + 2\alpha_m) - 2\xi\omega\Delta t\{\gamma(1 - 2\alpha_{f)} + \alpha_f\} - \omega^2\Delta t^2\{\beta(1 - 2\alpha_f) + \alpha_f(\gamma - \frac{1}{2})\}$ is always negative for any Δt and the first sufficient condition is satisfied. Considering the second sufficient condition, we require $|b_0| > |b_2| \Rightarrow |a_0^2 - a_3^2| > |a_0 a_2 - a_3 a_1|$. The last inequality is derivable by analytical means that is not straightforward. Instead, a

Table 8.2 Jury's stability criteria (Appendix H); table of coefficients of the characteristic equation

Row numbers	z^0	z^1	z^2	z^3
1	a_0	a_1	a_2	a_3
2	a_3	a_2	a_1	a_0
3	$b_0 = \begin{vmatrix} a_0 & a_3 \\ a_3 & a_0 \end{vmatrix}$	$b_1 = \begin{vmatrix} a_0 & a_2 \\ a_3 & a_1 \end{vmatrix}$	$b_2 = \begin{vmatrix} a_0 & a_1 \\ a_3 & a_2 \end{vmatrix}$	–

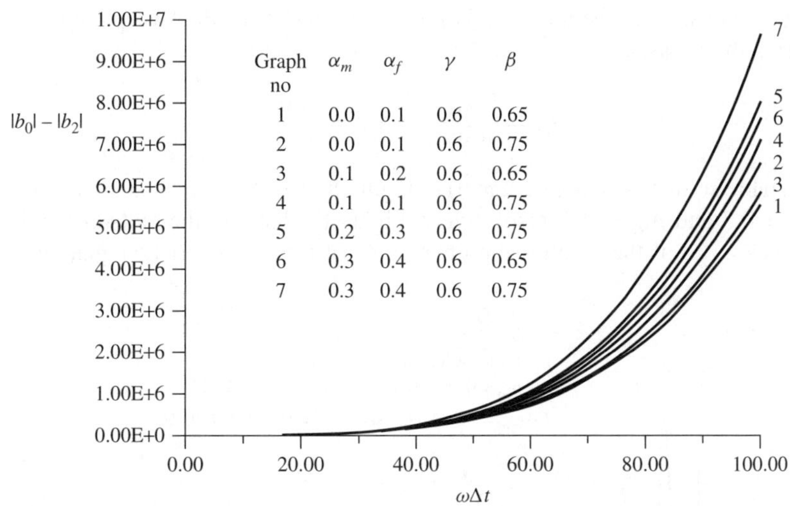

Figure 8.15 Generalized-α method; numerical testing of the sufficient condition $|b_0| > |b_2|$ for unconditional stability

computational procedure is adopted here to examine this condition. Figure 8.15 shows the curves of $|b_0| - |b_2|$ that remains positive for the set of parameters chosen, showing a case of unconditional stability of the integration scheme. In fact, selection of these parameter values is guided by the fact that the generalized-α method is shown (Chung and Hulbert, 1993) to be unconditionally stable if $\alpha_m \le \alpha_f \le 1/2$ and $\gamma = \frac{1}{2} - \alpha_m + \alpha_f$ with $\beta \ge \frac{1}{4} + \frac{1}{2}(\alpha_f - \alpha_m)$. The following Example 8.5 illustrates a few cases of (unconditionally) stable solutions obtained by the generalized-α method as applied to an SDOF oscillator.

Example 8.5: Generalized-α method applied to an SDOF oscillator

Figure 8.16 show the solution of the unforced oscillator $\ddot{u} + 2\xi\omega\dot{u} + \omega^2 u = 0$ with the initial conditions $u(0) = u_0 = 0.1$ and zero initial velocity, using the generalized-α method. The solutions are obtained with such values of the four parameters that they correspond to the unconditionally stable regime. The exact solution of the oscillator is also shown in Figure 8.16. The solution graph with $\alpha_m = 0$ is due to the HHT-α method. The two numerical solutions match well with the exact solution since, with the selected $\Delta t = 0.01$ (with

$\omega\Delta t = 1.0$), the algorithmic damping introduced by the method is enough to reduce the amplitude error that is otherwise present in the normal Newmark-β method (Figure 8.9).

With proper tuning of the parameters α_m and α_f, the generalized-α method provides higher algorithmic damping in the high-frequency range and minimises low-frequency dissipation. The twin requirements depend on the behaviour of spectral radius $\rho(\boldsymbol{A})$ with respect to ω. In the present case, the amplification matrix $\boldsymbol{A}$ possesses three eigenvalues of which one is a spurious root and the other two physically admissible. These roots $\lambda_{1,2}$ are complex conjugates. It may be shown (Chung and Hulbert (1993)) that the generalized-α method has optimal dissipation characteristics if $\beta = \frac{1}{4}(1 - \alpha_m + \alpha_f)^2$. Further, if $\rho_\infty (:= \lim_{\omega\Delta t \to \infty} |\lambda_{1,2}|)$ is the spectral radius at the high-frequency limit, it is given by $\rho_\infty = \frac{\alpha_f - \alpha_m - 1}{\alpha_f - \alpha_m - 1}$ and the minimum low-frequency dissipation is achieved when $\alpha_f = \frac{\alpha_m + 1}{3}$. Thus, if a desirable ρ_∞ is prescribed, we have the two parameters α_m and ρ_f in terms of ρ_∞ as:

$$\alpha_m = \frac{2\rho_\infty - 1}{1 + \rho_\infty}, \rho_f = \frac{\rho_\infty}{1 + \rho_\infty} \tag{8.65}$$

The two stable solutions (graphs 2 and 3) in Figure 8.16 of Example 8.5 are in fact obtained with $\rho_\infty = 0.5$ and $\rho_\infty = 0.8$, respectively. Figure 8.17 shows the plots of the three roots (eigenvalues $\lambda_{1,2,3}$ of the amplification matrix $\boldsymbol{A}$) of the characteristic equation $\mathcal{F}(z)$ with

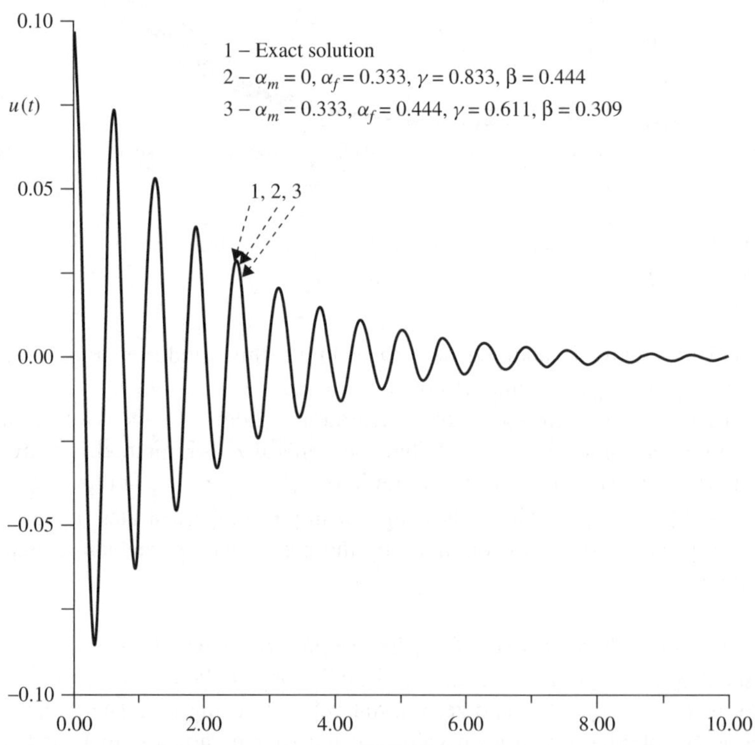

Figure 8.16 Generalized-α method; unconditionally stable solutions of an unforced SDOF oscillator with $\omega = 10, \xi = 0.05$; $\Delta t = 0.01$

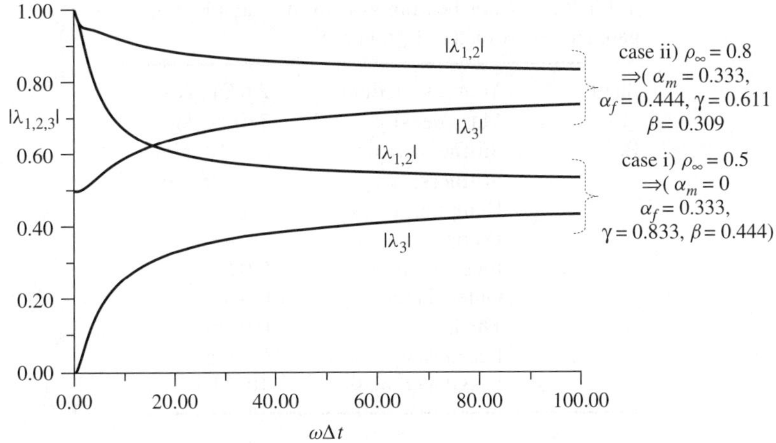

Figure 8.17 Generalized-α method; eigenvalues of $\boldsymbol{A}$, $(\lambda_{1,2,3})$ for case (i) spectral radius $\rho_\infty = 0.5$ and case (ii) spectral radius $\rho_\infty = 0.8$

respect to $\omega\Delta t$. Each set of $|\lambda_{1,2}|$ and $|\lambda_3|$ curves asymptotically approach respective spectral radius ρ_∞ in the limit as $\omega\Delta t \to \infty$.

Example 8.6: An accelerating rotor (Lallane and Ferraris, 1998)
Rotor bearing systems experience transient motions during start-up and shut-down operations and during their passage through a critical speed. These motions are related to the rotor-shaft accelerations. Figure 8.18 shows a rotor bearing system with a symmetric circular shaft of length $l = 0.4$ m with diameter $D = 0.02$ m. The other geometric and material properties are listed in Table 8.3. The operating speed of the rotor is 5000 rpm.

The rotor bearing system is semidiscretized using the FEM, whilst accounting for the rotation effects (Chapter 7). Beam elements are used to model the shaft and a mass element to simulate the disk. The bearings are modelled by springs and damper elements that are characterised by 2×2 stiffness and damping matrices. Dynamic condensation (Chapter 7) is used to reduce the system dimension to arrive at a 2-dof model representing

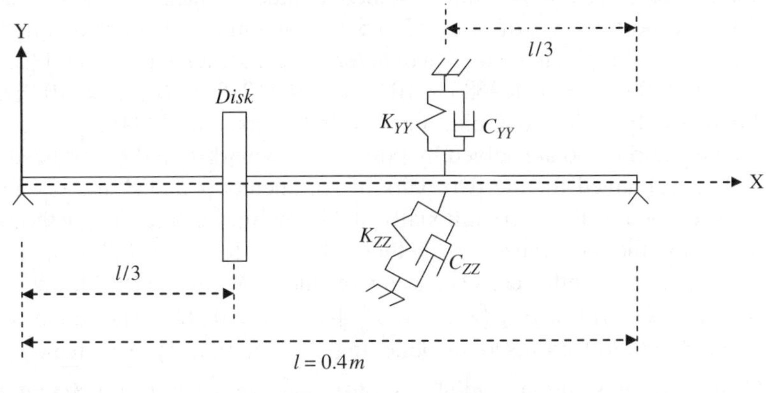

Figure 8.18 Rotor bearing system (accelerating rotor) in Example 8.6

Table 8.3 Rotor bearing system in Example 8.6; geometric and material properties

Shaft	Young's modulus	$2.0E11$ N/m^2
	Mass density	7800 kg/m^3
Bearing	Stiffness, K_{YY}	2×10^5 N/m
	Stiffness, K_{ZZ}	5×10^5 N/m
	Damping, C_{YY}	–
	Damping, C_{ZZ}	–
Disk	Inner diameter	0.02 m
	Outer diameter	0.3 m
	Thickness	0.03 m
	Eccentricity	0.15 m
	Unbalance mass	10^{-4} kg

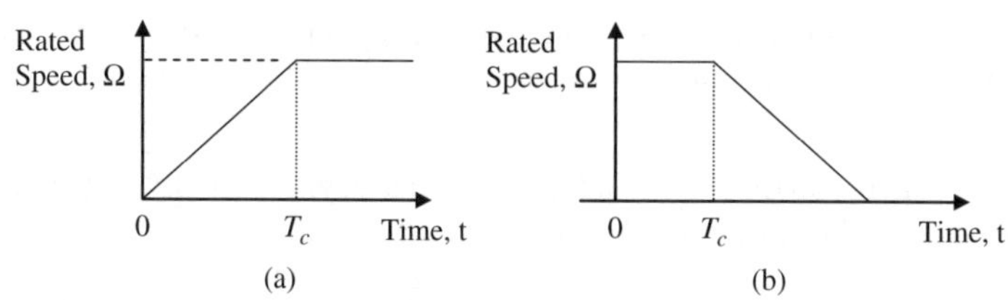

Figure 8.19 Rotor bearing system in Example 8.6; (a) accelerating rotor, linear law: $\dot{\Phi} = a_L t$, with acceleration $=a_L$ and (b) decelerating rotor, linear law: $\dot{\Phi} = -a_L t$, with deceleration $=-a_L$

the two translational degrees of freedom (displacement in the Y and Z directions) at the node where the disk is located. With the parameter values in Table 8.3, the reduced-order model (Lallane and Ferraris, 1998) is given by:

$$\ddot{y}_1 - a_1\dot{\Phi}\dot{y}_2 + ca_2\dot{y}_1 + a_3 y_1 = f\dot{\Phi}^2 \sin\Phi - f\ddot{\Phi}\cos\Phi$$
$$\ddot{y}_2 - a_1\dot{\Phi}\dot{y}_1 + ca_4\dot{y}_2 + a_5 y_2 = f\dot{\Phi}^2 \sin\Phi - f\ddot{\Phi}\cos\Phi \qquad (8.66)$$

$y_1(t)$ and $y_2(t)$ are the displacements (respectively in the Y and Z directions) at the disk node. $\dot{\Phi}$ is the rotation speed of the shaft, assumed to undergo linearly varying acceleration (Figure 8.19) for a short interval of time (12 s) before attaining the targeted uniform speed $\Omega = 5000$ RPM. For the given system parameters, the coefficients $a_i, i = 1, 2, .., 5$ are given by: $a_1 = 0.2009$, $a_2 = 1.049685 \times 10^4$, $a_3 = 9.41218 \times 10^4$, $a_4 = 10.9867 \times 10^4$ and $a_5 = 9.09027 \times 10^{-3}$. The damping factor c is the taken as 0.0002.

The coupled equation 8.66 are solved by both the Newmark-β and generalized-α methods. Solutions are obtained for the two pairs of parameters γ and β and α_f and α_m so chosen that correspond to the unconditionally stable regimes of the two methods.

The solution by the generalized-α method pertains to $\rho_\infty = 0.8$ (enabling high-frequency dissipation) with corresponding values of $\alpha_f = 0.444$, $\alpha_m = 0.333$, $\gamma(=\frac{1}{2} - \alpha_m + \alpha_f) = 0.611$ and $\beta(=\frac{1}{4}(1 - \alpha_m + \alpha_f)^2) = 0.309$. The same values of $\gamma = 0.611$ and $\beta = 0.309$ are used in obtaining the solution by Newmark-β method. Figure 8.20 shows the solution (radial response $R_D(t) = \sqrt{y_1^2(t) + y_2^2(t)}$ at the disk node) obtained by both the methods with $\Delta t = 0.00012$ s.

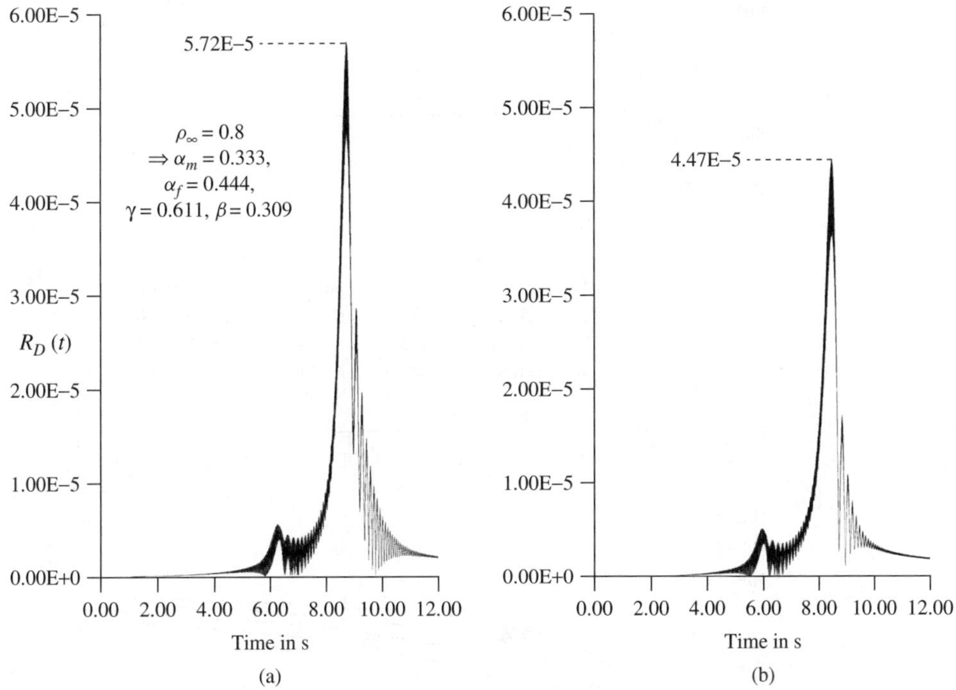

Figure 8.20 Accelerating rotor; solution with $\Delta t = 0.00012$ sec. by (a) generalized-α method with $\rho_\infty = 0.8$ and (b) Newmark-β method with $= 0.611$, $\beta = 0.309$

The relative performance of the two methods with respect to Δt is shown in Figure 8.21. In this figure, the maximum radial displacement (experienced by the accelerating rotor within the integration interval of 12 s) is plotted against the total number of integration steps. The generalized-α method yields accurate solutions even with less integration steps (equivalent to larger Δt).

Further, the generalized-α method, irrespective of the chosen ρ_∞, converges faster to the maximum radial displacement (about 0.057 mm). On the other hand, in the Newmark-β method, not only is the convegence slower (even with Δt as small as 6×10^{-5} s $-2{,}00{,}000$ integration steps), the computed solutions are also shown to be sensitively dependent on variations in the parameters γ and β.

8.6 Conclusions

Direct integration methods provide a universal tool for computing the response histories of semidiscretized dynamical systems and are specifically useful for nonlinear dynamic systems, an account of which is beyond the scope of this book. Even for a class of linear systems, for example stiff systems characterised by high-frequency components in the modal response so that the modal summation may be too slowly convergent to be practicable, appropriately chosen direct integration schemes could be usefully employed by numerically damping out the such components that barely contribute to the response.

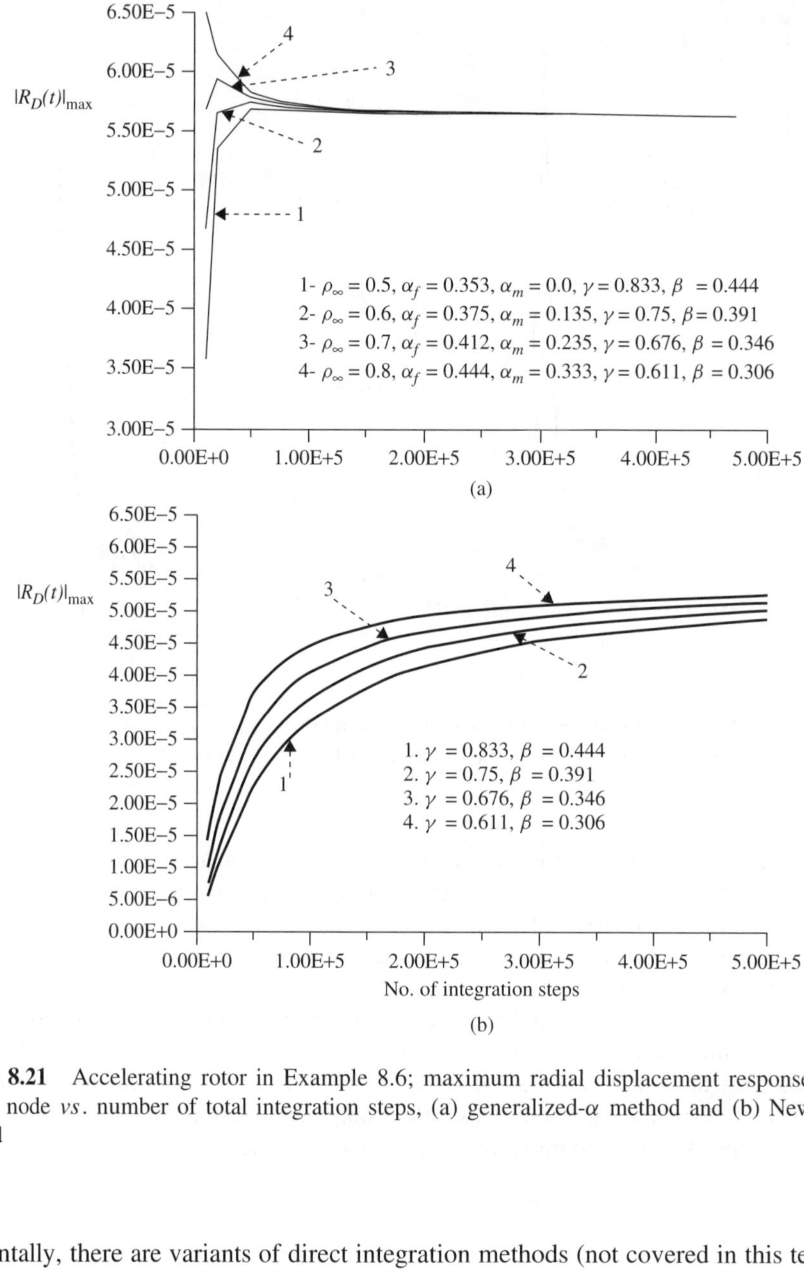

Figure 8.21 Accelerating rotor in Example 8.6; maximum radial displacement response, $R_D(t)$ at disk node *vs*. number of total integration steps, (a) generalized-α method and (b) Newmark-β method

Incidentally, there are variants of direct integration methods (not covered in this text) that use automatic time stepping and thus provide, in some sense, a more optimal utilisation of the available computational resources.

The next chapter is devoted to an elementary exposition of a mathematical model, derived within the broad framework of probability theory, that enables describing and propagating the uncertainty in the input forces to quantify the uncertainty in the structural dynamic response.

Exercises

1. Consider the system of ODEs:

$$\dot{\boldsymbol{x}} = \boldsymbol{A}\boldsymbol{x}, \boldsymbol{x} \in \mathbb{R}^n, \quad \boldsymbol{A} \in \mathbb{R}^{n\times n}, \quad n > 1 \tag{8.67}$$

Let the initial condition vector be given by $\boldsymbol{x}_0 \neq 0$. Suppose that **A** is diagonalisable with the set of eigenvalues $\{\lambda_j | j \in [1, n]\}$ and the matrix of eigenvectors $\boldsymbol{Q} = \{v_j | j \in [1, n]\}$, where v_j denotes the jth eigenvector. Then, show that one can write the solution of the above ODE as:

$$\boldsymbol{x}(t) = \boldsymbol{Q}e^{\Lambda t}\boldsymbol{Q}^{-1}\boldsymbol{x}_0 \tag{8.68}$$

where $\Lambda = diag\{\lambda_j\} \in \mathbb{R}^{n\times n}$.

2. Consider the linear system of ODEs:

$$\dot{\boldsymbol{x}} = \boldsymbol{f}(t, \boldsymbol{x}), \quad \boldsymbol{x} \in \mathbb{R}^n, \quad \boldsymbol{x}(0) = \boldsymbol{x}_0 \tag{8.69}$$

and define the backward difference operators:

$$\nabla^0 \boldsymbol{f}_{j+1} = \boldsymbol{f}_{j+1}, \nabla^i \boldsymbol{f}_{j+1} = \nabla^{i-1}\boldsymbol{f}_{j+1} - \nabla^{i-1}\boldsymbol{f}_j, \text{ for } i \geq 1 \tag{8.70}$$

Now, consider the following generic form of the so-called Adams–Bashforth (AB) method for numerical integration (h being the time step size, assumed to be uniform):

$$\frac{1}{h}(x_{j+1} - x_j) = \left(1 + \frac{1}{2}\nabla^1 + \frac{5}{12}\nabla^2 + \frac{3}{8}\nabla^3 + \frac{251}{720}\nabla^4 + \cdots\right)\boldsymbol{f}_j \tag{8.71}$$

Retaining the first k- terms on the RHS of the above equation results in the kth-order multistep AB method. Discuss the spectral stability of these methods for $k = 1, 2$ and 3.

Notations

a	Real constant
$a_i, i = 0, 1, 2, 3$	Real constants
$\boldsymbol{A}$	Amplification matrix
b	Real constant
$b_i, i = 0, 1, 2$	Real constants
$\boldsymbol{B}$	Matrix
C_{yy}, C_{zz}	Bearing damping coefficients
$d(\cdot)$	Metric
D	Diameter of a shaft
e_n	Local truncation error
f, f_m	Frequency
$\mathcal{F}(z)$	Function of z transform
$\boldsymbol{F}(t)$	External loading function
$\boldsymbol{F}_n$	Discretized forcing at t_n

K_{yy}, K_{zz}	Bearing stiffness coefficients
$\boldsymbol{L}_n$	Load term at t_n
(M, d)	Metric space
r_1, r_2	Roots (real or complex) of stability Equations (8.50, 8.51 and 8.52)
$\boldsymbol{R}(\boldsymbol{u}, \dot{\boldsymbol{u}}, \ddot{\boldsymbol{u}}, t)$	Nonlinear restoring force vector
$\boldsymbol{R}_n$	Load vector
$R_D(t)$	Radial response at the disk node $\left(=\sqrt{y_1^2(t) + y_2^2(t)}\right)$
T	Time period of the oscillator $\left(=\frac{2\pi}{\omega}\right)$
$\boldsymbol{T}$	Linear mapping (transformation)
$\boldsymbol{u}(t)$	Displacement response vector (function of time)
$\boldsymbol{u}_n, \dot{\boldsymbol{u}}_n, \ddot{\boldsymbol{u}}_n$	Approximated displacement, velocity and acceleration (discretized) at t_n
$\boldsymbol{u}_0, \dot{\boldsymbol{u}}_0, \ddot{\boldsymbol{u}}_0$	Initial displacement, velocity and acceleration
$\boldsymbol{x}_n$	System state vector comprising of $u_n, \dot{u}_n, \ddot{u}_n$
$\boldsymbol{y}(t) = (\boldsymbol{u}^T(t), \dot{\boldsymbol{u}}^T(t))^T$	Response vector (state space representation)
$y_1(t), y_2(t)$	Displacements (respectively in the Y and Z directions) at the disk node on the shaft (Example 8.6)
$Z(\cdot)$	z-Transform
α	Constant
α_f	Parameter in HHT-α method defining the relative weights of the damping, elastic and external forces corresponding to the two instants t_n and t_{n+1}
α_m	Parameter in generalized-α method defining the relative weights of the mass corresponding to the two instants t_n and t_{n+1}
β	Parameter in Newmark-β method
γ	Parameter in Newmark-β method
κ	Real constant
λ_j	Complex eigenvalue $(= \sigma_j + j\omega_j)$
λ_1, λ_2	Eigenvalues of the amplification matrix $\boldsymbol{A}$ (Equation 8.55)
$\rho(\cdot)$	Spectral radius (Equation 8.23)
ρ_∞	Spectral radius at the high frequency limit
ς	Time instant in the interval $[t_n, t_{n+1}]$
χ	Real constant
ω_m	Radian frequency $(2\pi f_m)$
Δt	Time step
Φ	Phase angle

References

Atkinson, K. and Han, W. (2004) *Elementary Numerical Analysis*, John Wiley& Sons, Inc., New York.

Bathe, E.L. (1996) *Finite Element Procedures*, Prentice-Hall, Cliffs, NJ.

Brooks, R.M. and Schmitt, K. (2009) The contraction mapping principle and some applications. *Journal of Differential Equations: Monograph*, **9**, 90.

Chung, J. and Hulbert, G.M. (1993) A time integration algorithm for structural dynamics with improved numerical dissipation: the generalized-α method. *ASME Journal of Applied Mechanics*, **60**, 371–375.

Conte, S.D. and Boor, C. (1981) *Elementary Numerical Analysis – An Algorithmic Approach*, McGraw Hill, Singapore.

Hilber, H.M. and Hughes, T.J.R. (1978) Collocation, dissipation and 'overshoot' for time integration schemes in structural dynamics. *Earthquake Engineering and Structural Dynamics*, **6**, 99–117.

Hilber, H.M., Hughes, T.J.R. and Taylor, R.L. (1977) Improved numerical dissipation for time integration algorithms in structural dynamics. *Earthquake Engineering and Structural Dynamics*, **5**, 283–293.

Jury, E.I. (1974) *Inners and Stability of Dynamical Systems*, John Wiley& sons, Inc., London.

Kreyszig, E. (2001) *Introductory Functional Analysis with Applications*, John Wiley & Sons, Inc., Singapore.

Lallane, M. and Ferraris, G. (1998) *Rotordynamics Prediction in Engineering*, John Wiley & Sons, Inc., UK.

Newmark, N.M. (1959) A method dynamics for structural dynamics. *ASCE, Journal of Engineering Mechanics Division*, **85** (EM3), 67–94.

Shannon, C.E. (1949) Communication in the presence of noise. *Proceedings of Institute of Radio Engineers*, **37** (1), 10–21.

Bibliography

Hoff, C. and Pahl, P.J. (1988) Development of an implicit method with numerical dissipation from a generalized single step algorithm for structural dynamics. *Computer Methods in Applied Mechanics and Engineering*, **67**, 367–385.

Macek, R.W. and Aubert, B.H. (1995) A mass penalty technique to control the critical time increment in explicit dynamic finite element analysis. *Earthquake Engineering and Structural Dynamics*, **24**, 1315–1331.

9

Stochastic Structural Dynamics

9.1 Introduction

Discussion in the previous chapters on structural dynamics and the response evaluation methods implicitly assumes that both the system model and the loading function are deterministic (i.e. specifiable without any uncertainty). However, we find uncertainties invariably in all physical phenomena or engineering systems. Sources of uncertainties, in general, are of epistemic or aleatory type (Menezes and Schueller, 1996). The aleatory type of uncertainty is mostly intrinsic in that the variables exhibit randomness characterised by their natural and unpredictable variations when they are observed or measured. The epistemic type arises from modelling and measurement errors and these uncertainties may be reducible by use of additional knowledge/data, if available on the variables involved. Due to this uncertainty or randomness, the 'perfect' predictability of input forces and the structural response in terms of known deterministic functions is infeasible. In the deterministic approach followed so far, one ignores the randomness and models the input variables in some averaged sense in the hope that the system output, so computed, may once more be interpreted as an averaged response and that any possible variations could be accounted for by bringing in a 'factor of safety'. For further illustration, Figure 9.1 shows typical wind velocity wave forms (assumed to be a collection at a location for a short period on different days under 'identical' conditions). These signals show the so-called 'intrinsic' (aleatory) random fluctuations in their histories, extreme values and possibly in its frequency content. It is known (Davenport, 1961; Kaimal *et al.*, 1972) that the wind velocity also exhibits random variations along the height within the boundary layer. Thus, the wind velocity may be viewed as a random function of space and time (i.e. a random field). One may find similar randomness/stochasticity in many other system models, for example road/rail unevenness (Dodds and Robson 1973, Patadia and Crafts 1979), ocean waves (Holmes et al., 1983) and seismic events (Clough and Penzien, 1996). Specifically in the context of structural dynamics, the system properties, either material or geometric – such as elastic moduli, density, cross-sectional dimensions, and so on – are themselves subject to uncertainties (Ang and Tang, 1984; Melchers, 1999). For a continuous structure, the parametric uncertainties constitute a set of random fields, time varying or

Elements of Structural Dynamics: A New Perspective, First Edition. Debasish Roy and G Visweswara Rao.

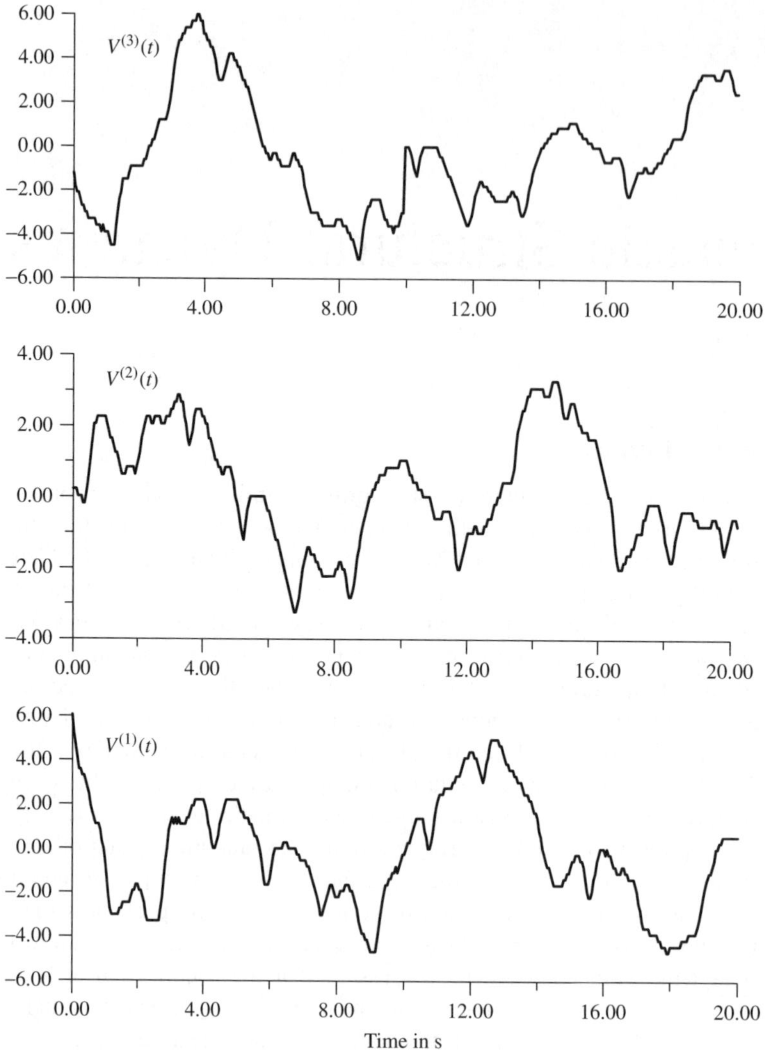

Figure 9.1 Random signals; typical wind velocity (in m/s) wave forms, $V^{(i)}(t)$

otherwise. However, for the following semidiscrete structural dynamic system (involving no space variations):

$$\boldsymbol{M}\ddot{\boldsymbol{u}} + \boldsymbol{C}\dot{\boldsymbol{u}} + \boldsymbol{K}\boldsymbol{u} = \boldsymbol{f}(t) \tag{9.1}$$

the randomness in the input forcing and other system parameters may best described through random functions in time alone. As we shall see, such randomly temporal functions are often described as stochastic processes. In writing Equation 9.1, the following implications must be borne in mind. (i) The system dynamic equilibrium is governed by a stochastic partial differential equation (PDE) (with inputs/parameters in the form of random fields). (ii) Special techniques such as stochastic finite elements (Vanmarcke, 1983; Liu *et al.*, 1986, 1987) are needed to semidescretise these random fields. (iii) The element matrices $\boldsymbol{M}$, $\boldsymbol{C}$, $\boldsymbol{K}$ and the forcing vector function $\boldsymbol{f}(t)$ are matrix/vector-valued stochastic processes. A stochastic finite element method (SFEM) is essentially an extension of the deterministic finite element method (FEM) (Chapter 3) and exhaustive reviews on the method can be found in the literature (Schueller, 2001; Sudret and Der Kiureghian, 2000; Stefanou, 2009). Given the focus of the book to be on introductory structural dynamics, SFEM is not discussed and hence the system parameters are assumed to be strictly deterministic. Accordingly, we only deal with stochastically forced structural dynamic systems, that is $\boldsymbol{f}(t)$ in Equation 9.1 is a vector stochastic process. Moreover, Equation 9.1 is now a stochastic (ordinary) differential equation (SDE). It is obvious that the solution to Equation 9.1 and other derived response quantities (e.g. stresses) are also stochastic and the objective is to express such response unknowns in terms of the input stochastic force. As a precursor to this exercise, it is needed to put in place a probabilistic mathematical framework, wherein the function $\boldsymbol{f}(t)$ can be adequately described, whilst satisfactorily accounting for the underlying physics/mechanics that may likely have a bearing in constructing $\boldsymbol{f}(t)$. Before we embark upon our current aim, which is to solve the SDE (Equation 9.1), a brief exposition of basic concepts on probability spaces, random variables and stochastic processes is, however, be essential.

9.2 Probability Theory and Basic Concepts

Fundamental to the probability theory is the notion of an experiment, its outcomes and the sample space. Simple examples are random experiments on 'tossing a coin' and 'throwing a dice'. The former example has only two outcomes 'head' or 'tail' with the sample space $\Omega = \{\text{head, tail}\}$ and in the case of the latter, $\Omega = \{f1, f2, f3, f4, f5, f6\}$ with only six outcomes representing the six faces of the dice. These two experiments are examples of a discrete sample space, wherein Ω is countable (finite or countably infinite). Consider now the hypothetical experiment of recording wind velocity at a specific location. The velocity V so obtained form a uncountably infinite set of outcomes (in this case, velocity values that are positive real numbers) that define a continuous sample space. In any case, a nonempty set $\mathcal{F}$ consisting of all observable subsets belonging to Ω (discrete or continuous) is known as $\sigma -$ algebra if the set is closed under complementation and countable unions of its members (subsets including Ω). For example, the discrete sample space corresponding to the case of Bernoulli trials (Papoulis, 1991), sample space Ω contains only two outcomes – 'success' or 'failure' (as 'head' or 'tail' in the tossing of a coin). Denoting, for convenience, the word 'success' by the letter 's' and 'failure' by 'f', we have $\Omega = \{\text{s, f}\}$ and $\sigma -$ algebra, $\mathcal{F}$ is the power set of Ω containing $2^{\Omega} = 2^2 = 4$ elements: ϕ, s, f, Ω. Thus, in this simple example, the power set itself is a $\sigma -$ algebra. In general, given Ω, it is possible to have many $\sigma -$ algebras (based on subsets of Ω that themselves can qualify as sample spaces). The intersection of these $\sigma -$ algebras is also a $\sigma -$ algebra. Now, consider a measure $\wp$ on $\mathcal{F}$ such that $\wp : F \to [0, \infty]$ with the following properties:

$$
\begin{aligned}
&\wp(E) \geq 0, \forall E \in \mathcal{F}, \\
&\wp\left(\cup_{i=1}^{\infty} A_i\right) = \sum_{i=1}^{\infty} \wp(A_i) \forall A_1, A_2, \ldots \in \mathcal{F}, \text{ with } A_i \cap A_j = \phi, i \neq j \\
&\wp(\phi) = 0,
\end{aligned}
\tag{9.2}
$$

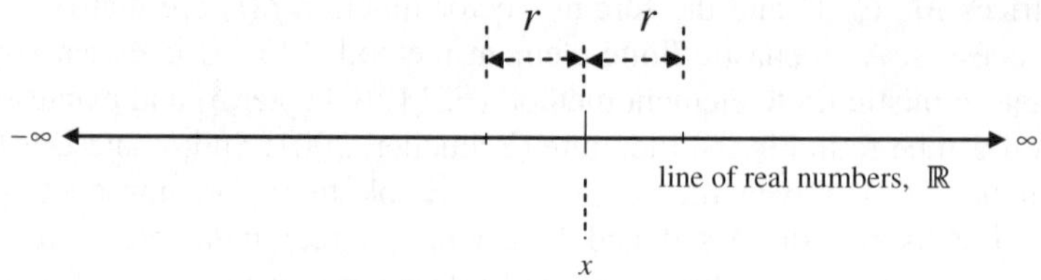

Figure 9.2 Continuous sample space, $\Omega \equiv \mathbb{R}$; probability measure of an event

Here, ϕ is the null set Ω^c. The triplet $(\Omega, \mathcal{F}, \wp)$ is known as the measure space and the elements of $\mathcal{F}$ are known as measurable sets. It follows from the properties of a measure that if $A_i, i = 1, 2..$ are not disjoint, $\wp(\cup_{i=1}^{\infty} A_i) \leq \sum_{i=1}^{\infty} \wp(A_i)$. If $\wp$ is normalised such that $\wp(\Omega) = 1$, then $(\Omega, \mathcal{F}, \wp)$ is called a probability space and $\wp$ the probability measure, with the properties in Equation 9.2 being the axioms of probability. Note that for a nonempty subset $B \subset \Omega$, $(B, \mathcal{F}_B, \wp_B)$ is also a probability space with σ algebra $\mathcal{F}_B = B \cap \mathcal{F}$ and with sample space reduced from Ω to B and the new measure $\wp_B$ satisfying $\wp_B(B) = 1$. Now, for a discrete sample space, if the probability measure (point probability) of each outcome, $\omega \in \Omega$, is known, the probability measure of an event $E \in \mathcal{F}$ is given by $\wp(E) = \sum_{\omega \in E} \wp(\omega)$. For finite Ω, the summation is straightforward. The summation is infinite for countably infinite Ω, and must be convergent. An example is the case of the experiment where one seeks for the probability of having 'success for the first time' during a number of Bernoulli trials. Define the event $E = \{s, fs, ffs, fffs, ...\}$. If $\wp(s) = \wp(f) = 1/2$, then $\wp(E) = \sum_{r=1}^{\infty} \frac{1}{2^r}$. This infinite sum converges to 1. For a continuous sample space, a point measure carries no 'information' in that probability of any individual outcome $\omega \in \Omega$ is zero. This is analogous to the property of a Lebesgue measure defined on any topological space (Rudin, 1976), say $\mathbb{R}$. The measure of an open set (a, b) or the closed set $[a, b]$ in $\mathbb{R}$ is $b - a$ (the same) and the measure of a single point $x \in (a, b)$ is zero. Further, for a continuous sample space, integration replaces the summation in obtaining $\wp(E)$. For example, if $\Omega \subset \mathbb{R}$ and $x, r \in \mathbb{R}$, the probability of the event, $E = (x - r, x + r)$ (Figure 9.2) is given by:

$$\wp(E) = \wp(x - r, x + r) = \int_{x-r}^{x+r} d\wp (\to 0 \text{ as } r \to 0) \tag{9.3}$$

The axiomatic properties of probability spaces and measures in Equation 9.2 form the basis of modern probability theory (Kolmogorov, 1956) that provides a unified treatment for both discrete and continuous Ω. On the other hand, a more classical approach, utilising the so-called relative frequency concept, deals only with discrete Ω. It associates any element (event) E of $\mathcal{F}$ with a probability equal to the ratio of the number of favourable outcomes to the total number of outcomes. The concept derives justification from the law of large numbers (due to Bernoulli; see Appendix I).

9.3 Random Variables

In many cases, we may have readily discernible associations of outcomes ($\omega \in \Omega$) and events ($E \in \mathcal{F}$) with (real) numbers. However, in many others, outcomes may also be descriptive and non-numeric namely, $\Omega = \{\text{failure, no failure}\}$ as in the case, say, of the structural response exceeding a specified critical level or otherwise. However, a numerical/computational treatment of uncertainty requires that the outcomes/events be made to correspond (1:1) to real numbers via an appropriate mapping from Ω to $\mathbb{R}^n$. Thus, we define the random variable to be a (invertible) mapping $X : \Omega \to \mathbb{R}^n$ such that if U is an open set in $\mathbb{R}^n$, then $E = X^{-1}(U) = \{\omega \in \Omega : X(\omega) \in U\} \in \mathcal{F}$. X is thus a function between two measurable spaces $(\Omega, \mathcal{F})$ and $(\mathbb{R}^n, \mathcal{B})$ where $\mathcal{B}$ is the σ algebra (called the Borel σ algebra when referred to $\mathbb{R}^n$) generated by the subsets in $\mathbb{R}^n$. If X is a continuous (and hence bounded) mapping $\Omega \to \mathbb{R}^n$, it is measurable with respect to Borel σ algebra $\mathcal{B}$. If the random variable is discrete (also called simple), it maps outcomes of Ω to a countable set of integers in $\mathbb{R}^n$. For convenience of exposition, assume $n = 1$ (i.e. the case of scalar-valued random variables). Thus, in this case, $X(\omega_i \in \Omega) = x_i \in \mathbb{N} \subset \mathbb{R}$, $P(x_i) = \wp(\omega_i)$, $i = 1, 2..m$, where 'm' is the total number of outcomes in Ω (i.e. the cardinality of Ω). Here, we define a probability mass function $p(x)$ of the discrete random variable in terms of point probabilities $P(X = x_i)$, $i = 1, 2..m$ as:

$$p(X = x_i) = \sum_{k=1}^{m} P(X = x_k)\delta_{ik} \tag{9.4}$$

where δ_{ik} is the Kronecker delta. If the interest is to know the probability of the event $\{X(\omega) \leq x_i\}$, it is given by a probability distribution function denoted by $F_X(x_i)$ in the form:

$$F_X(x_i) = P(-\infty \leq X \leq x_i) = \sum_{k=1}^{i} P(X = x_k)U(X - x_k) \tag{9.5}$$

$U(\cdot)$ denotes a unit-step function. $F_X(x)$ is thus a nondecreasing function with a finite number of discontinuities at points x_i, $i = 1, 2, \ldots, n$ and is piecewise differentiable. Accordingly, $p(x) = F_X'(x)$ everywhere except at the points of discontinuity. One can typically extend the argument set $B = \{x_k : k \in [1, n]\}$ of the function F_X to the entire $\mathbb{R}$ (i.e. B with countably infinite elements) and thus (in the limit) define $F_X(x)$ for a continuous random variable as:

$$F_X(x_i) = \sum_{i=1}^{x_i} P(x_i), \forall x_i \in B \text{ and}$$

$$F_X(x) = F_X(x_i), \quad x_i \leq x < x_{i+1} \tag{9.6}$$

following an appropriate ordering of $B \in \mathcal{B}$, the Borel σ- algebra. Note that, in this simple representation of a continuous random variable (which is fundamental to the theory of continuous random variables), every $x_i \in B$ is essentially representative of an infinitesimally small interval in $\mathbb{R}$. $F_X(x)$ is also known as the cumulative distribution function (*CDF*). For a real-valued continuous random variable, $X : \Omega \to \mathbb{R}$, the probability measure $\wp(\Omega) = 1$ can be transferred to the Borel measure $P(-\infty \leq X \leq \infty) = 1$. Further, if $\mathcal{F} \supset E = X^{-1}(B = \{a \leq x \leq b\} \subset \mathcal{B})$, we have:

$$\wp(E) = P(a \leq X \leq b) = P(X \leq b) - P(X \leq a) = F_X(b) - F_X(a) \tag{9.7}$$

Since point probabilities $P(a)$ and $P(b)$ are zero, $P(a \leq x \leq b) = P(a < x < b)$. If $F_X(x)$ is differentiable, let $f_X(x) = \frac{dF_X(x)}{dx} = F'_X(x)$ denote the probability density function (*pdf*) of the continuous random variable. Thus, we have:

$$P(x \leq X \leq x + dx) = F_X(x + dx) - F_X(x) = \frac{dF_X(x)}{dx} dx = f_X(x)dx \tag{9.8}$$

Thus, $P(a \leq X \leq b) = \int_a^b f_X(x)dx$. Equation 9.8 implies that $\int_{-\infty}^{\infty} f_X(x)dx = P(-\infty \leq x \leq \infty) = 1$. It also shows that the *CDF* $F_X(x)$ satisfies the properties:

i) $F_X(-\infty) = 0$

ii) $F_X(\infty) = 1$

iii) if $x < y$, then $F_X(x) \leq F_X(y)$ for $\forall x, y \in \mathbb{R}$. (9.9)

Property (iii) implies that $F_X(x)$ is a nondecreasing continuous function (unlike the discrete case where it was piecewise continuous). A random variable $X : \Omega \to \mathbb{R}$ being continuous with $P(\{\omega \in \Omega | X(\omega) = x\}) = 0 \ \forall x \in \mathbb{R}$ renders $F_X(x)$ absolutely continuous. Further, $F_X(x)$ is often differentiable so that $f_X(x) = F'_X(x), \forall x \in \mathbb{R}$. In general, we may write $F_X(x) = \int_{-\infty}^{x} f_X(x)dx, \ \forall x \in \mathbb{R}$ (as long as $f_X(x)$ is integrable, which is less restrictive than $F_X(x)$ being strictly differentiable). Thus, taking the case of a discrete random variable for instance, one may still define a (generalized) *pdf*, $f_X(x)$ by using Dirac delta functions $\delta(X - x)$ in the form:

$$f_X(x_i) = \sum_{k=1}^{n} P(X = x_k)\delta(X - x_i) \tag{9.10}$$

such that we have $F_X(x_i) = \int_{-\infty}^{x_i} f_X(x)dx$ as defined in Equation 9.5.

9.3.1 Joint Random Variables, Distributions and Density Functions

Extension of the above definitions for a scalar-valued random variable to the vector-valued case is straightforward. Consider such a random variable (often called a joint random variable) $\boldsymbol{X}(\omega \in \Omega) := (X_1(\omega) = x_1, X_2(\omega) = x_2, \ldots\ldots, X_m(\omega) = x_m)$ defined on $(\Omega, \mathcal{F}, \wp)$. It is understood here that $\boldsymbol{X}$ maps Ω to $\mathbb{R}^m$ with the associated Borel σ algebra on the subsets of $\mathbb{R}^m$. With $:= (x_1, x_2, \ldots., x_m)$, the joint *CDF* $F_X(\boldsymbol{x})$: $\mathbb{R}^m \to [0, 1]$ is defined as:

$$F_X(\boldsymbol{x}) = P(X_1 \leq x_1, X_2 \leq x_2, \ldots., X_m \leq x_m) \tag{9.11}$$

If $F_X(\boldsymbol{x})$ is sufficiently differentiable, we have the *pdf* as $\frac{\partial^m F_X(\boldsymbol{x})}{\partial x_1 \partial x_2 \ldots \partial x_m}$. Then $F_X(\boldsymbol{x})$ is given by the m-dimensional integral in terms of *pdf*:

$$F_X(\boldsymbol{x}) = \int_{-\infty}^{x_1} \int_{-\infty}^{x_2} \cdots \int_{-\infty}^{x_m} f_X(x_1, x_2, \ldots., x_m) dx_1 dx_2 \ldots . dx_m \tag{9.12}$$

If $\boldsymbol{X}$ is a discrete (and joint) random variable, its *pdf* can be expressed in terms of the point probabilities as:

$$f_X(\boldsymbol{x}_i) = \sum_{k=1}^{n} P(X_1 = x_{k1}, X_2 = x_{k2}, \ldots\ldots, X_m = x_{km})\delta_{ik}, \ i = 1, 2 \ldots, m \tag{9.13}$$

Here, the cardinality of each of the random variables is assumed to be same as n. $F_{\boldsymbol{X}}(\boldsymbol{x})$ is a nondecreasing function and has properties analogous to the distribution function of a single random variable (Equation 9.9), namely, $F_{\boldsymbol{X}}(x_1, x_2, \ldots x_{i-1}, -\infty, x_{i+1}, \ldots, x_m) = 0$ for any $i \leq m$ and $F_{\boldsymbol{X}}(\infty, \infty, \ldots, \infty) = 1$. $F_{X_k}(x_k), k = 1, 2 \ldots, m$ are the marginal (i.e. individual) probability distribution functions corresponding to the scalar components of $\boldsymbol{X}$. These are projections (restrictions) of $F_{\boldsymbol{X}}(\boldsymbol{x})$ to one-dimensional (sample) spaces and are obtained as:

$$F_{X_k}(x_k) = \int_{-\infty}^{\infty}\int_{-\infty}^{\infty}\cdots\int_{-\infty}^{x_k}\cdots\int_{-\infty}^{\infty} f_{\boldsymbol{X}}(x_1, x_2, \ldots x_k .., x_m)dx_1 dx_2 \ldots dx_k .. dx_m,$$
$$k = 1, 2 \ldots, m \tag{9.14}$$

The marginal density functions $f_{X_k}(x_k), k = 1, 2 \ldots, m$ follow from the existence of $\frac{\partial F_{X_k}(x_k)}{\partial x_k}$ and are given by the following $(m-1)$-dimensional integral with respect to dx_i, $i = 1, 2, \ldots m, i \neq k$:

$$f_{X_k}(x_k) = \int_{-\infty}^{\infty}\int_{-\infty}^{\infty}\cdots\int_{-\infty}^{\infty} f_{\boldsymbol{X}}(x_1, \ldots x_k .., x_m)dx_1 \ldots dx_{k-1}dx_{k+1} \ldots .. dx_m \tag{9.15}$$

Similar to the definitions in Equations 9.14 and 9.15 above, one also obtains l-dimensional marginal joint distribution/density functions from $F_{\boldsymbol{X}}(\boldsymbol{x})$ by integrating out all variables except the l variables of interest.

9.3.2 Expected (Average) Values of a Random Variable

The distribution or density function may fully describe a random variable X and thus provide the probabilities of outcomes/events of the associated experiment. However, we often seek a few (or rather a sequence of) deterministic quantities (called moments) that, consistent with its distribution, also characterise X. These moments are of practical significance in engineering analysis and design. The first (order) moment, denoted by $E(X)$, is known as the expectation or mean of X. For a scalar-valued random variable X, it is given by the improper integral:

$$E(X) = \int_{\mathcal{B}} xP(dB) = \int_{\Omega} X^{-1}(x)d\wp \tag{9.16}$$

dB is a differential element in $\mathcal{B}$, the Borel σ algebra and the integral is known as the Lebesgue integral (Rudin, 1976). For the integral to exist, $\int_{-\infty}^{\infty} |x|P(dB)$ must be finite $(< \infty)$. If $dB = \{x < X < x + dx\}$, Equation 9.16 takes the form:

$$E(X) = \int_{-\infty}^{\infty} xP(x < X < x + dx) = \int_{-\infty}^{\infty} x\{F_X(x + dx) - F_x(x)\} \tag{9.17}$$

If $F(x)$ is differentiable, $F_X(x + dx) - F_X(x) = dF_X(x)$ and the integral in Equation 9.17 with respect to the nondecreasing real function $F_x(x)$ is the Riemann–Stieltjes integral (Rudin, 1976). Further, writing $dF_X(x) = f_X(x)dx$ and assuming that the integrand $xf_X(x)$ is Reimann integrable, the expectation of X is given by:

$$E(X) = \int_{-\infty}^{\infty} xf_x(x)dx \tag{9.18}$$

The above definition applies to both discrete and continuous random variables except that, for discrete X, summation replaces the integration as:

$$E(X) = \sum_{k=1}^{n} x_k f_x(x_k) \tag{9.19}$$

Generalising the above definition of expectation to include functions of a random variable, consider a function $g(X)$: $\mathbb{R} \rightarrow \mathbb{R}$ such that $\int_{\mathbb{R}} |g(x)|P(dB) \leq \infty$. Then we have:

$$E(g(X)) = \int_{\mathbb{R}} g(x)P(dB) = \int_{-\infty}^{\infty} g(x) f_x(x)dx \tag{9.20}$$

The mth-order moment of X are defined by $E(X^m) = \int_{-\infty}^{\infty} x^m f_x(x)dx$ in the case of continuous random variable and by $E(X^m) = \sum_{k=1}^{n} x_k^m f_x(x_k)$ in the discrete case. Using the above definition of $E(X)$, one can define a L^p 'norm' $\|\boldsymbol{X}\|_p$ of a vector-valued random variable $\boldsymbol{X} : \Omega \rightarrow \mathbb{R}^n$ as:

$$\|\boldsymbol{X}\|_p = \left\{ \int_{\mathcal{B}} |\boldsymbol{x}|^p P(dB) \right\}^{1/p} \tag{9.21}$$

Here, $|\boldsymbol{x}|^p = \sum_{i=1}^{n} |x_i|^p$. Note that the above definition is not a norm in the strictest sense in that $\int_{\mathbb{R}} |\boldsymbol{x}|^p P(dB) = 0$ may have nonzero X as a solution, when the probability of occurrence of such nonzero values of X is zero. Somewhat loosely, the above norm associates with L^p spaces that are Banach spaces. $L^2(P)$, in particular, could be thought of as a Hilbert space with inner product $E(X, Y) = \int_{-\infty}^{\infty} xy f_{XY}(x, y)dxdy$, where X, $Y \in L^2(P)$. Whilst a parallel with Banach spaces of deterministic real analysis could be drawn in this way, notions of such spaces in probability theory are typically not made much use of.

A random variable is also characterised by its central moments – the moments centred about its mean. For instance, consider a vector random variable $\boldsymbol{X}(\omega \in \Omega) := (X_1(\omega) = x_1, X_2(\omega) = x_2, \ldots\ldots, X_m(\omega) = x_m)$ and denote $\boldsymbol{x} := (x_1, x_2, \ldots\ldots, x_m)$. If we denote $E(\boldsymbol{X})$ by $\boldsymbol{\mu}_X$, the mean vector, the central moments are given by:

$$\begin{aligned} E\{(\boldsymbol{X} - \boldsymbol{\mu}_X)^m\} &= \int_{-\infty}^{\infty} (\boldsymbol{x} - \boldsymbol{\mu}_X)^m f_X(\boldsymbol{x})d\boldsymbol{x}, \text{ for continuous } \boldsymbol{X} \\ &= \sum_{k=1}^{n} (\boldsymbol{x}_k - \boldsymbol{\mu}_X)^m f_X(\boldsymbol{x}_k), \text{ for discrete } \boldsymbol{X} \end{aligned} \tag{9.22}$$

Specifically, for $m = 2$, $E(\boldsymbol{X}^2)$ is known as the mean square value and $E\{(\boldsymbol{X} - \boldsymbol{\mu}_X)^2\}$, the variance of $\boldsymbol{X}$ denoted by $\boldsymbol{\sigma}_X^2$. The variance is a measure of dispersion of distribution of $\boldsymbol{X}$ about its mean. Note that $\boldsymbol{\sigma}_X$ is known as the standard deviation of $\boldsymbol{X}$. $\boldsymbol{\sigma}_X = \boldsymbol{0}$ signifies that $\boldsymbol{X}$ is deterministic and no longer random. Expanding $E\{(\boldsymbol{X} - \boldsymbol{\mu}_X)^2\}$, we have $\boldsymbol{\sigma}_X^2 = E(\boldsymbol{X}^2) - 2\boldsymbol{\mu}_X E(\boldsymbol{X}) + \boldsymbol{\mu}_X^2 = E(\boldsymbol{X}^2) - \boldsymbol{\mu}_X^2$. Here, one can define a covariance matrix C_m as:

$$C_m = \begin{bmatrix} C_{11} & C_{12} & \ldots\ldots & C_{1m} \\ C_{21} & C_{22} & \ldots\ldots & C_{2m} \\ \ldots & \ldots & \ldots\ldots & \ldots \\ C_{m1} & C_{m2} & \ldots\ldots & C_{mm} \end{bmatrix} \tag{9.23}$$

where $C_{jk} = E[(X_j - \mu_{X_j})(X_k - \mu_{X_k})]$. If C_m is diagonal with $C_{jk} = 0, j \neq k$, we say that the vector random variable $\boldsymbol{X}$ is uncorrelated and the diagonal elements are merely the mean square values $C_{jj} = E[X_j^2]$.

9.3.3 Characteristic and Moment-Generating Functions

Continuing with the vector random variable $\boldsymbol{X} : \boldsymbol{\Omega} \to \mathbb{R}^n$, a characteristic function of $(\boldsymbol{X})$ is defined by:

$$\begin{aligned}\boldsymbol{\emptyset}_X(\omega_1, \omega_2, \ldots., \omega_n) &= E[\exp(i\omega_1 x_1 + i\omega_2 x_2 + \cdots\cdot + i\omega_n x_n)] \\ &= \int_{\mathbb{R}^n} e^{i(\omega_1 x_1 + i\omega_2 x_2 + \cdots\cdot + \omega_n x_n)} f_X(\boldsymbol{x}) d\boldsymbol{x} \end{aligned} \tag{9.24}$$

The above definition implies that $\boldsymbol{\emptyset}_X(\cdot)$ is the Fourier transform of the *pdf* of $\boldsymbol{X}$. Letting $\boldsymbol{\omega} = (\omega_1, \omega_2, \ldots., \omega_n)$, we note that since $f_X(\boldsymbol{x}) \geq 0$, $\boldsymbol{\emptyset}_X(\boldsymbol{\omega})$ is maximum at $\boldsymbol{\omega} = \boldsymbol{0}$. Thus, with $\boldsymbol{\emptyset}_X(\boldsymbol{0}) = \boldsymbol{1}$, $\boldsymbol{\emptyset}_X(\boldsymbol{\omega}) \leq \boldsymbol{\emptyset}_X(\boldsymbol{0})$.

Replacing $i\boldsymbol{\omega}$ with $\boldsymbol{s} = (s_1, s_2, \ldots., s_n)$ in Equation 9.24, we obtain the moment-generating function $\psi(\boldsymbol{s}) = \int_{\mathbb{R}^n} e^{(\boldsymbol{sx})} f_X(\boldsymbol{x}) d\boldsymbol{x}$ where $(\boldsymbol{sx}) = s_1x_1 + s_2x_2 + \cdots\cdot + s_nx_n$. Differentiation of $\psi(\boldsymbol{s})$ with respect to 's', to different orders yields information on the joint moments of the random variable $\boldsymbol{X}$. For example, with $n = 1$, we find that $\psi'(0) = \int_{-\infty}^{\infty} x e^{(sx)} f_X(\boldsymbol{x}) d\boldsymbol{x} = E(X)$ and thus in general, $\psi^m(0) = E(X^m)$.

9.4 Conditional Probability, Independence and Conditional Expectation

Conditional probability is related to the probability of a subset of events when information on the occurrence of another subset (not necessarily disjoint) is available. Thus, given $(\Omega, \mathcal{F}, P)$ and two subsets $A, B \subset \mathcal{F}$, the probability of A conditioned on the subset B (i.e. with B having occurred being a given hypothesis) is known as the conditional probability $P(A|B)$ or $P_B(A)$. It may be defined as:

$$P_B(A) = P(A|B) = \frac{P(A \cap B)}{P(B)}, \; P(B) > 0 \tag{9.25}$$

From the above definition, we find that $P_B(A)$ is proportional to $P(A \cap B)$. For a given B, $P(B)$ can be considered to be a normaliser so that when A is varied, $0 \leq P_B(A) \leq 1$. Conditioning restricts the sample space to $B \in \Omega$ leading to a new probability space $(B, \mathcal{F}_B, P_B)$ with $\mathcal{F}_B$, the sub-σ algebra associated with B. Thus, the conditional probability is always associated with a reduced sample space and, in this sense, P_B is a restriction to $\mathcal{F}_B$ of P. When the conditional probability $P_B(A)$ equals the unconditional probability $P(A)$, we say that event A is statistically independent of the event B (in that the probability of A is unaffected by that of B). This also implies that B is also independent of A and in either case, the joint probability is given by (Equation 9.25):

$$P(A \cap B) = P(A)P(B) \tag{9.26}$$

Now consider a random variable (discrete or continuous) $X : \Omega \to \mathbb{R}$ with $A, B \in \mathbb{R}$. We define the conditional distribution function as:

$$F_X(x|B) := \frac{P(X \leq x \cap B)}{P(B)} \tag{9.27}$$

The conditional probability distribution function satisfies all the properties of $F_X(x)$ as stated in Equation 9.9 with probabilities replaced by conditional probabilities. The conditional density function $f_X(x|B)$, for instance, is the derivative (if it exists) of $F_X(x|B)$:

$$f_X(x|B) = \frac{dF_X(x|B)}{dx} = \lim_{\Delta x \to 0} \frac{F(x+\Delta x|B) - F(x|B)}{\Delta x}$$
$$= \lim_{\Delta x \to 0} \frac{P(x \le X \le x+\Delta x|B)}{\Delta x} \tag{9.28}$$

As an example, we consider an event A describing the life of a component beyond, say, t units, conditioned to the event B corresponding to the information that it has already survived for a certain period, say, b units. Here, the life of the system, say T, is a continuous random variable $T : \Omega \to (0, \infty) \subset \mathbb{R}$ and we are interested in $P(A = \{T \ge t\})$ given the event $B = \{T > b\}$. Now $P(A|B) = P(T \ge t|T > b)$ and it is given in terms of the conditional distribution function $F_T(t|B)$ as:

$$P(T \ge t|B) = 1 - F_T(t|B) = 1 - \frac{P(T \le t \cap T > b)}{P(T > b)}$$
$$= 1 - \frac{F_T(t) - F_T(b)}{1 - F_T(b)} = \frac{1 - F_T(t)}{1 - F_T(b)}, \quad t > b = 1, t \le b \tag{9.29}$$

$F_T(t)$ and $f_T(t)$ stand for the *CDF* and *pdf* of the random variable T.

Conditional probabilities for vector random variables can be defined in a similar manner. Let $X, Y : \Omega \to \mathbb{R}$ be two random variables defined on the probability space $(\Omega, \mathcal{F}, P)$. With $A = (X \le x)$ and $B = (Y \le y)$, the conditional probability, $P(X \le x|Y \le y)$ is the conditional distribution function $F_X(x|Y \le y)$ given by:

$$F_X(x|Y \le y) = \frac{P(X \le x \cap Y \le y)}{P(Y \le y)} = \frac{F_{XY}(x, y)}{F_Y(y)} = \frac{\int_{-\infty}^{y} \int_{-\infty}^{x} f_{XY}(x, y)dxdy}{\int_{-\infty}^{\infty} \int_{-\infty}^{x} f_{XY}(x, y)dxdy} \tag{9.30}$$

It follows from Equation 9.30 that if $F_{XY}(x, y)$ is differentiable, one can define the conditional density function $f_X(x|Y \le y) = \frac{\partial F_{XY}(x,y)/\partial x}{F_Y(y)}$. To find $f_X(x|Y = y)$, we first consider $F_X(x|y \le Y \le y + \Delta y)$, which is given by:

$$F_X(x|y \le Y \le y+\Delta y) = \frac{F_{XY}(x, y+\Delta y) - F_{XY}(x, y)}{F_Y(y+\Delta y) - F_Y(y)}$$
$$= \frac{\int_{-\infty}^{\infty} \int_{y}^{y+\Delta y} f_{XY}(x, y)dydx}{f_Y(y)\Delta y} \tag{9.31}$$

Differentiating with respect to x, we have:

$$f_X(x|y \le Y \le y+\Delta y) = \frac{\int_{y}^{y+\Delta y} f_{XY}(x, z)dz}{f_Y(y)\Delta y} = \frac{f_{XY}(x, y)\Delta y}{f_Y(y)\Delta y}$$

Hence, in the limit as $\Delta y \to 0$, we obtain $f_X(x|y)$ as:

$$f_X(x|y) = \lim_{\Delta y \to 0} f_X(x|y \le Y \le y+\Delta y) = \frac{f_{XY}(x, y)}{f_Y(y)} \tag{9.32}$$

Extending the definition of independent events to random variables, we say that X and Y are independent if:

$$f_X(x|y) = f_X(x) \Rightarrow f_{XY}(x, y) = f_X(x) f_Y(y) \tag{9.33}$$

Following Equation 9.33 we have that if two random variables are independent, they are uncorrelated. That is, since $E(XY) = \int_{-\infty}^{\infty} \int_{-\infty}^{\infty} xy f_{XY}(x, y) dx dy = \int_{-\infty}^{\infty} x f_X(x) dx \int_{-\infty}^{\infty} y f_Y(y) dy = E(X)E(Y) = \mu_X \mu_y$ we have:

$$C_{XY} = E[(X - \mu_X)(Y - \mu_Y)] = E(XY) - \mu_X \mu_Y = 0 \tag{9.34}$$

Here, note that if two random variables are uncorrelated they are not necessarily independent (left as an exercise to the reader).

9.4.1 Conditional Expectation

Consider a random variable X defined on $(\Omega, \mathcal{F}, P)$ such that $E(\|X\|) < \infty$ (i.e. X is integrable). Let $B \subset \mathcal{F}$ be a sub-σ algebra, then the conditional expectation $E(X|B)$ is defined by:

$$E(X|B) = \int_B x P(dB) \tag{9.35}$$

The existence and uniqueness of $E(X|B)$ follow from Radon–Nikodym theorem (Williams, 1991). Following Equations 9.16–9.18, the conditional expectation takes the form:

$$E(X|B) = \int_{-\infty}^{\infty} x f_X(x|B) dx \tag{9.36}$$

We note that $E(X|B)$ is a function (a new random variable via appropriate functional restriction as noted earlier), mapping $\Omega \to \mathbb{R}$ and is B-measurable. This is more evident if we consider the expectation of X conditioned on another random variable Y associated with sub-σ algebra, say $\mathcal{F}_Y \subset \mathcal{F}$. This is given by $E(X|Y) = \int_{-\infty}^{\infty} x f_X(x|Y = y) dx$, which is a function of Y, say $g(Y)$. Utilising Equation 9.33, we have, (If $g(Y)$ is integrable):

$$\begin{aligned} E(g(Y)) = E\{E(X|Y = y)\} &= \int_{-\infty}^{\infty} \left(\int_{-\infty}^{\infty} x f_X(x|y) dx \right) f_Y(y) dy \\ &= \int_{-\infty}^{\infty} \int_{-\infty}^{\infty} x f_{XY}(x, y) dx dy = \int_{-\infty}^{\infty} x f_X(x) dx = E(X) \end{aligned} \tag{9.37}$$

Analogous to the definition of conditional expectation, conditional variance can be defined. That is, if we denote $E(X|Y)$ by $\mu_{X/Y}$, then, the conditional variance, $\sigma_{X/Y}^2 = E[(x - \mu_{X/Y})^2 |Y)] = \int_{-\infty}^{\infty} (x - \mu_{X/Y})^2 f_X(x|Y = y) dx$.

9.5 Some oft-Used Probability Distributions

Several probability distribution and density functions, associated with a random variable X, are popularly used (probably more often than the others) in the literature. Choice of a distribution may, however, be facilitated through an understanding of the nature of uncertainties in the mathematical model itself or some physical phenomena associated with the model (some of which may not have been incorporated within the model band therefore qualify as additional forms of uncertainty). Some of these distributions, such as binomial, Poisson (which are discrete), uniform, normal and Rayleigh (which are continuous) are described in the following sections.

9.5.1 Binomial Distribution

For ease of exposition, consider X to be a scalar-valued random variable. A binomial distribution is related to a random experiment consisting of a fixed n number of Bernoulli trials where the random variable X (which is X_n, to be exact) is the number of 'successes' in these independent n trials. Introduce a parameter θ (a constant) representing the probability of a success in each Bernoulli trial. Here, the sample space is the set $\{1, 2, \ldots, n\} \subset \mathbb{N} \subset \mathbb{R}$ and the Borel σ algebra, $\mathcal{B}$ containing 2^n subsets (power set) of the sample space. Since $(1-\theta)$ denotes the probability of failure, the probability $P(X = r)$ is given by $\binom{n}{r}\theta^r(1-\theta)^{n-r}$ which is incidentally the $r + 1^{th}$ term in the binomial expansion of $\{\theta + (1-\theta)\}^n = 1$. Moreover, we also have $\sum_{r=1}^{n}\binom{n}{r}\theta^r(1-\theta)^{n-r} = 1$. Thus, the pdf of the binomial distribution symbolically expressed as $b(r; n, \theta)$ can be written in the form:

$$f_X(r) = \sum_{k=0}^{n}\binom{n}{k}\theta^k(1-\theta)^{n-k}\delta(X-r) \tag{9.38}$$

It follows from the above equation that the CDF, $F_X(r) = \sum_{k=0}^{r} p(X = k)U(X-k)$. The mean of X is given by $E(X) = \mu_X = \sum_{k=0}^{n}\binom{n}{k}\theta^k(1-\theta)^{n-k} = n\theta$ and the variance by $\sigma_X^2 = E(X^2) - \mu_X^2 = n\theta(1-\theta)$.

9.5.2 Poisson Distribution

Computation of probabilities from binomial distribution as $n \to \infty$ and $\theta \to 0$ is cumbersome. In this limiting case, together with the assumption of $n\theta(= \lambda$, say) remaining finite, it can be shown that (Feller, 1968):

$$P(X = k) = \binom{n}{k}\theta^k(1-\theta)^{n-k} \approx \frac{e^{-\lambda}\lambda^k}{k!} \tag{9.39}$$

The above limiting approximation to $b(r; n \cdot \theta)$ is known as the Poisson distribution represented by a single parameter λ. The Poisson distribution finds wide application as a useful model for random experiments with rare events (θ being small). As an example, let us consider the case of a possible structural failure when the response (such as displacement or stress) histories (similar to those in Figure 9.1) exceed a critical level c. Here, we take the number of critical level crossings in a time interval $(0, T)$ as the random variable X_T. If the interval is divided into n equal parts of Δt such that $n = \frac{T}{\Delta t}$, X_T follows a binomial distribution. Let us now assume that (i) the probability, θ of crossing the level c exactly once in a sufficiently small Δt is proportional to Δt, (ii) the level-crossing events in any two subintervals are independent of each other and (iii) probability of critical level crossing more than once in Δt is negligible. In the limiting case of large n (i.e. large T)

and small θ (e.g. if the level c is high), the above assumptions leads to a Poisson distribution (Equation 9.39). As such, the probability of no failure within $(0, T)$ is equivalent to knowing the probability of the number of critical level crossings being zero. We can compute this probability as $P(X_T = 0) = e^{-\lambda}$, provided, of course, $\lambda(= n\theta)$ is known. Note that for a Poisson distribution, $E(X_T) = \sum_{k=0}^{n} k \frac{e^{-\lambda}\lambda^k}{k!} = \lambda$ and $\sigma_X^2 = \lambda$.

9.5.3 Normal Distribution

A scalar-valued random variable $X : \Omega \to \mathbb{R}$ is normal (or Gaussian), if the pdf of X is given by:

$$f_X(x) = \frac{1}{\sigma\sqrt{2\pi}} e^{-\frac{1}{2}\left(\frac{x-\mu}{\sigma}\right)^2} \tag{9.40}$$

The normal distribution is thus represented by two parameters $\mu \in \mathbb{R}$ and $\sigma \in (0, \infty)$ and is usually written as $N(\mu, \sigma)$. μ and σ may indeed be shown to be the mean and standard deviation of X with the density being symmetric about μ. The distribution function $F_X(x)$ is given by:

$$F_X(x) = \frac{1}{\sigma\sqrt{2\pi}} \int_{-\infty}^{X} e^{-\frac{1}{2}\left(\frac{x-\mu}{\sigma}\right)^2} dx \tag{9.41}$$

$F_X(x)$ needs to be evaluated numerically as no antiderivative of $f_X(x)$ in Equation 9.40 is analytically known. A scalar-valued standard normal variable $X \sim N(0, 1)$ is one with zero mean and unit variance. Thus, the pdf and CDF of $N(0, 1)$ are, respectively, given by $f_X(x) = \frac{1}{\sqrt{2\pi}} e^{-\frac{x^2}{2}}$ and $F_X(x) = \frac{1}{\sqrt{2\pi}} \int_{-\infty}^{x} e^{-\frac{x^2}{2}} dx = \frac{1}{2} + \mathrm{erf}(x)$, where $\mathrm{erf}(x) = \frac{1}{\sqrt{2\pi}} \int_0^x e^{-\frac{x^2}{2}} dx$ is the error function, whose values are typically available in standard tables. The characteristic function for $X \sim N(\mu, \sigma)$ is given by $\exp(i\mu\omega - \frac{\sigma\omega^2}{2})$ (Equation 9.24 with $n = 1$).

Normal random variables are commonly encountered in engineering applications. For instance, measurement errors in experiments are often modelled by a normal distribution. Even a discrete random variable like binomial, $b(r; n, \theta)$ is approximated by a normal distribution when n is large and $n\theta(1 - \theta) \gg 1$. This result is due to the DeMoivre–Laplace limit theorem (Feller, 1968). With n large, a binomial random variable may be well approximated by a Poisson variable for $\lambda(= n\theta)$ small and by a normal variable for λ large. Moreover, normal distribution also derives its importance from the central limit theorem (Appendix I) that implies that sum of a large number of independent and identically distributed random variables is approximately normal. This justifies a normal approximation to a binomial random variable, if the latter is considered to be a sum of n independent Bernoulli random variables. Finally, a normal random variable remains normal under linear transformation. Specifically, with X being a normal random variable and given $Y = aX + b$ such that $a, b \in \mathbb{R}$, Y is also normal with mean $\mu_Y = E(Y) = aE(X) = a\mu_X + b$ and variance $\sigma_Y^2 = E(Y - \mu_Y)^2 = a^2\sigma_X^2$. More generally, the vector random variable $\boldsymbol{X} = (X_1, X_2, \ldots. X_n)$ is jointly normal if only if $Y = a_1X_1 + a_2X_2 + \cdots + a_nX_n$ is normal for any $a_j, j = 1, 2..n \in \mathbb{R}$ (Appendix I). With $E(\boldsymbol{X}) = \boldsymbol{\mu}$ and $\boldsymbol{C}$ denoting the covariance matrix, the joint density function is given by:

$$f_X(\boldsymbol{x}) = \frac{1}{(2\pi)^{n/2}|\boldsymbol{C}|} exp\left\{-\left(\frac{1}{2}\boldsymbol{X}\boldsymbol{C}^{-1}\boldsymbol{X}^T - \boldsymbol{\mu}\boldsymbol{X}\right)\right\} \tag{9.42}$$

Here, $\boldsymbol{C} \in \mathbb{R}^{nxn}$ is a symmetric positive definite matrix. This property is made use of in simulation studies to transform $\boldsymbol{X}$ to a vector $\boldsymbol{Y}$ of uncorrelated random variables $(Y_1, Y_2, \ldots Y_n)$. Here, one can use the transformation $\boldsymbol{Y} = \boldsymbol{V}^T\boldsymbol{X}$ where $\boldsymbol{V}$ is the orthonormalised eigenvector matrix of $\boldsymbol{C}$ with $\boldsymbol{V}^T\boldsymbol{V} = \boldsymbol{I}$. Diagonalisation of $\boldsymbol{C}$ leads to $\boldsymbol{V}^T\boldsymbol{C}\boldsymbol{V}=\boldsymbol{\Lambda}$ with $\boldsymbol{\Lambda}$ being a diagonal matrix containing the eigenvalues of $\boldsymbol{C}$. $\boldsymbol{\Lambda}$ is also the covariance matrix of $\boldsymbol{Y}$. The diagonal elements give the correlations $E\left[Y_j^2\right]$ of the random variables of $\boldsymbol{Y}$.

In structural analysis involving random external loading, a response quantity could be regarded as a transformation of the input random parameters (defining the loading) through the governing equations of motion and hence the response (at a given time) would admit characterisation as a normal random variable if the input parameters are themselves normal and the governing equations of motion linear.

9.5.4 Uniform Distribution

Let a random variable X take values in the real interval $[a, b]$, $b > a$ such that its CDF is:

$$\begin{aligned} F_X(x) &= 0, x < a \\ &= \frac{x-a}{b-a}, \quad a \le x \le b \\ &= 0, \quad x > b \end{aligned} \tag{9.43}$$

$F_X(x)$ is differentiable everywhere except at $x = a$ and $x = b$. This gives the pdf of X as:

$$\begin{aligned} f_X(x) &= \frac{1}{b-a}, \quad x \in [a, b] \\ &= 0, \quad x \notin [a, b] \end{aligned} \tag{9.44}$$

Such a random variable is said to be uniformly distributed in $[a, b]$ and denoted by $U(a, b)$. A standard uniform variate is $U(0, 1)$ with uniform *pdf* in [0, 1]. A significant application of the uniform distribution is in simulation studies where it is extensively used to generate random numbers with other (specified) distributions. Generation of random numbers uniformly distributed in [0, 1] is relatively easy with present-day computing machines and random numbers with a specified probability distribution (other than uniform) are obtained via an appropriate transformation (Appendix I).

9.5.5 Rayleigh Distribution

A scalar-valued random variable $X : \Omega \to \mathbb{R}_+$ is Rayleigh if the pdf and CDF of X are given by:

$$f_X(x) = \frac{x}{\sigma^2} e^{-\frac{x^2}{2\sigma^2}}$$

$$F_X(x) = 1 - e^{-\frac{x^2}{2\sigma^2}}, \quad x \ge 0 = 0, \quad x < 0 \tag{9.45a,b}$$

For this distribution, we have the mean $= \sigma\sqrt{\frac{\pi}{2}}$ and variance $= \frac{(4-\pi)\sigma^2}{2}$. The distribution is functionally related to the normal random variable in that if $X$and Y are independent zero mean normal random variables with variance σ^2, then $Z = \sqrt{X^2 + Y^2}$ follows a Rayleigh distribution (Appendix I).

9.6 Stochastic Processes

A vector-valued stochastic process (also known as a random process) is a mapping $X(\omega, t) : \Omega \times \mathbb{R}^+ \to \mathbb{R}^n$ $(n \geq 1)$ with t being an indexing parameter. If the parameter is time $t \in T \subset \mathbb{R}^+$, then for each $\omega \in \Omega$ the stochastic process $X(t, \omega)$ is a function of t alone (often called a path), and for each t, it is a random variable defined on a probability space $(\Omega, \mathcal{F}_t, P)$. If $T = \mathbb{N} = \{1, 2 \ldots\}$, $X(t)$ (dropping the second argument ω for the sake of brevity) is a discrete process. Whilst the indexing parameter may, in principle, also be any other independent variable (including a vector, e.g. the space variables, in which case the process is often referred to as a random field), we presently restrict the parameter to denote the scalar time variable only. Many natural/engineering phenomena such as wind, seismic events or road/rail undulations are examples of stochastic processes (Figure 9.1). Structural responses under these loads are also stochastic. For these loading or response processes, one often attempts at qualifying the associated stochastic process by seeking the probabilities of the form $P(X(t_1) \in \mathcal{F}_1, X(t_2) \in \mathcal{F}_2, \ldots., X(t_k) \in \mathcal{F}_k)$ for any choices of $0 \leq t_1 \leq t_2 \leq \cdots\cdots t_k \leq T$, where $\mathcal{F}_i, i = 1, 2.., k$ are Borel sets in $\mathbb{R}^n$. In other words, a stochastic process is typically characterised by all its finite-dimensional joint distribution of the form:

$$F_X(x_1, x_2, \ldots. x_m; t_1, t_2, \ldots., t_m) = P(X(t_1) \leq x_1, X(t_2) \leq x_2, \ldots, X(t_m) \leq x_m) \tag{9.46}$$

for any choices of $0 \leq t_1 \leq t_2 \leq \cdots\cdots t_m \leq T$. If this distribution function is differentiable, we have the joint density function $f_X(x_1, x_2, \ldots. x_m; t_1, t_2, \ldots., t_m) = \frac{\partial^n F_X(x_1, x_2, \ldots. x_m; t_1, t_2, \ldots., t_m)}{\partial x_1 \partial x_2, \ldots. \partial x_m} = P(x_1 \leq X(t_1) \leq x_1 + dx_1, x_2 \leq X(t_2) \leq x_2 + dx_2, \ldots, x_m \leq X(t_n) \leq x_n + dx_m)$. Thus, for a given stochastic process, we can look for all such finite-dimensional joint distributions associated with the random variables that correspond to the 'snapshots' of the stochastic process at different time instants. Kolmogorov's extension theorem provides the basis for proving the reverse – given a family of distributions of finite-dimensional random variables that is 'consistent' (see below), there exists a stochastic process satisfying the distributions. The consistency requirement (Øksendal, 2003) states that for all $t_1, t_2, \ldots., t_k \in T$, if $\gamma_{t_1, t_2, \ldots., t_k}$ is a probability measure on $\mathbb{R}^{nk}$ such that (i) $\gamma_{\sigma(1), \sigma(2), \ldots., \sigma(k)}(\mathcal{F}_1 \times \mathcal{F}_1 \times \cdots \times \mathcal{F}_k) = \gamma_{t_1, t_2, \ldots., t_k} \quad (\mathcal{F}_{\sigma^{-1}(1)} \times \mathcal{F}_{\sigma^{-1}(2)} \times \cdots \times \mathcal{F}_{\sigma^{-1}(k)})$ for all permutations of σ (a permutation parameter) on $(1, 2 \ldots k)$ and (ii) $\gamma_{t_1, t_2, \ldots., t_k (\mathcal{F}_1 \times \mathcal{F}_1 \times \cdots \times \mathcal{F}_k)} = \gamma_{t_1, t_2, \ldots., t_k, t_{k+1}, \ldots t_{k+m}, (\mathcal{F}_1 \times \mathcal{F}_1 \times \cdots \times \mathcal{F}_k \times \mathbb{R}^n \times \cdots. \times \mathbb{R}^n)}$ for all $m \in \mathbb{N}$, then there exists a probability space $(\Omega, \mathcal{F}, P)$ and a stochastic process $X(\omega, t)$ on Ω; $X_t : \Omega \to \mathbb{R}^n$ such that $\gamma_{t_1, t_2, \ldots., t_k (\mathcal{F}_1 \times \mathcal{F}_1 \times \cdots \times \mathcal{F}_k)} = P(X_{t_1} \in \mathcal{F}_1, X_{t_2} \in \mathcal{F}_2, \ldots, X_{t_k} \in \mathcal{F}_k)$ for all $t \in T, k \in \mathbb{N}$ and for all Borel sets $\mathcal{F}_i, i = 1, 2..k$. The first consistency condition implies that the joint probability information is invariant with respect to the permutations of its arguments and the second one that higher-dimensional distributions contain the lower-dimensional (marginal) ones.

9.6.1 Stationarity of a Stochastic Process

A stochastic process $X(t), t \in [0, T]$ (considered to be a scalar-valued process for expositional clarity), is called stationary if for any m time instants

$t_1, t_2, \ldots, t_m$ and any time shift τ, the random vectors $\{X(t_1), X(t_2), \ldots, X(t_m)\}$ and $\{X(t_1+\tau), X(t_2+\tau), \ldots, X(t_m+\tau)\}$ have the same probability distribution. Specifically, this property is referred to as stationarity of order m. $X(t)$ is covariance (weakly or wide-sense) stationary if it is stationary of order 2. From this definition it follows that, for a stationary process, the pdf is invariant in time translations, that is for any $\tau > 0$, we have:

$$f(X_1, X_2, \ldots, X_n; t_1, t_2, \ldots, t_m) = f(X_1, X_2, \ldots, X_m; t_1+\tau, t_2+\tau, \ldots, t_m+\tau) \tag{9.47}$$

Thus, for a weakly stationary process (i.e. a stationary process with $m = 2$), the covariance $C_X(s,t)$ (also known as the autocovariance function given by $E[(X(s) - \mu_{X(s)})(X(t) - \mu_{X(t)})]$) is a function of time difference $t - s$. It follows that the autocorrelation function, $R_{XX}(s,t) (= E[X(s)X(t)])$ is also a function of the time shift only. Note that stationarity of order m implies stationarity of all lower orders. Thus, weak stationarity implies that $E[X(t)]$ is constant (since, for $m = 1$, the probability distribution of $X(t)$ is invariant of t). Also observe that $C_{XX}(s,t) = E[X(s)X(t)] - \mu_{X(s)}\mu_{X(t)} = R_{XX}(s,t) - \mu_{X(s)}\mu_{X(t)}$.

If we consider the process:

$$X(t) = \sum_{j=1}^{n\in\mathbb{N}} (A_j \cos\omega_j t + B_j \sin\omega_j t) \tag{9.48}$$

where A_j and B_j are independent random variables, we have $E[X(t)] = \sum_{j=1}^{n} (E(A_j)\cos\omega_j t + E(B_j)\sin\omega_j t)$. For (weak) stationarity of $X(t)$, the mean is t-invariant, which requires $E(A_j) = E(B_j) = 0$ as necessary conditions. The autocorrelation function of $X(t)$ is given by:

$$\begin{aligned} E[X(s)X(t)] &= E\left[\sum_j^n\sum_k^n (A_jA_k\cos\omega_j s\cos\omega_k t + A_jB_k\cos\omega_j s\sin\omega_k t \right. \\ &\qquad \left. + B_jA_k\sin\omega_j s\cos\omega_k t + B_jB_k\sin\omega_j s\sin\omega_k t)\right] \\ &= \sum_j^n\sum_k^n (E[A_jA_k]\cos\omega_j s\cos\omega_k t + E[A_jB_k]\cos\omega_j s\sin\omega_k t \\ &\qquad + E[B_jA_k]\sin\omega_j s\cos\omega_k t + E[B_jB_k]\sin\omega_j s\sin\omega_k t) \end{aligned} \tag{9.49a}$$

If the random variables A_j and B_j, $j = 1, 2..n \in \mathbb{N}$ are uncorrelated and with equal variances, then $X(t)$ is (weakly) stationary. To examine this, first assume $E[A_jB_k] = E[B_jA_k] = 0$ and $E[A_jA_k] = E[B_jB_k] = \sigma_j^2\delta_{jk}$ $\forall j,k$ in Equation 9.44a. This results in:

$$\begin{aligned} E[X(s)X(t)] &= \sum_{j=1}^{n\in\mathbb{N}} E\left[A_j^2\right]\left(\cos\omega_j s\cos\omega_k t + E\left[B_j^2\right]\sin\omega_j s\sin\omega_j t\right) \\ &= \sum_j^{n\in\mathbb{N}} \sigma_j^2\cos\omega_j(t-s) \end{aligned} \tag{9.49b}$$

Thus, $E[X(s)X(t)]$ is a function of $(t-s)$ only and hence $X(t)$ is at least weakly stationary.

9.6.2 Properties of Autocovariance/Autocorrelation Functions of Stationary Processes

The autocorrelation function $R_{XX}(t, t+\tau)$ is maximum at $\tau = 0$, which is derived from the following inequality.

$$\begin{aligned} &E[(X(t) \pm X(t+\tau))^2] \geq 0 \\ &\Rightarrow E[X^2(t) \pm 2X(t)X(t+\tau) + X^2(t+\tau)] \geq 0 \\ &= R_{XX}(0) \pm 2R_{XX}(\tau) + R_{XX}(0) = 2(R_{XX}(0) \pm R_{XX}(\tau)) \geq 0 \\ &\Rightarrow R_{XX}(0) \geq R_{XX}(\tau) \geq 0 \ \forall\ \tau \end{aligned} \tag{9.50}$$

$R_{XX}(\tau)$ is an even function of τ as $E[X(t)X(t+\tau)] = E[X(t)X(t-\tau)]$ and it follows that $R'_{XX}(0) = \left.\frac{dR_X(\tau)}{d\tau}\right|_{\tau=0} = 0$. A typical autocorrelation function is shown in Figure 9.3.

For the derivative process $\dot{X}(t)$, the autocorrelation function $R_{\dot{X}\dot{X}}(t, t+\tau)$ is derived as follows:

$$\begin{aligned} &\frac{dR_{XX}(\tau)}{d\tau} = \frac{d\{E[X(t)E(X(t+\tau)]\}}{d\tau} = E[X(t)\dot{X}(t+\tau)] = E[X(t-\tau)\dot{X}(t)] \\ &\Rightarrow \frac{d^2R_{XX}(\tau)}{d\tau^2} = \frac{dE[X(t-\tau)\dot{X}(t)]}{d\tau} = -E[\dot{X}(t-\tau)\dot{X}(t)] = -E[\dot{X}(t)\dot{X}(t+\tau)] \end{aligned} \tag{9.51}$$

Thus, the autocorrelation function of a derivative process is given by $R_{\dot{X}\dot{X}}(t, t+\tau) = -\frac{\partial^2 R_{XX}(\tau)}{\partial t^2}$. For example, if $X(t)$ is a displacement process, the second derivative of its autocorrelation function is also the autocorrelation of the velocity but for a change of sign. Incidentally, observe that $E[\dot{X}(t)] = \frac{dE[X(t)]}{dt} = 0$ since $E[X(t)]$ is a constant for a (weakly) stationary process. For convenience, a 'weakly stationary' process will henceforth be described as a 'stationary' process.

9.6.3 Spectral Representation of a Stochastic Process

In general, sample paths of a stochastic process $X(t)$ need not even be integrable owing to their (possibly) unbounded variations over finite intervals. Hence, they do not necessarily

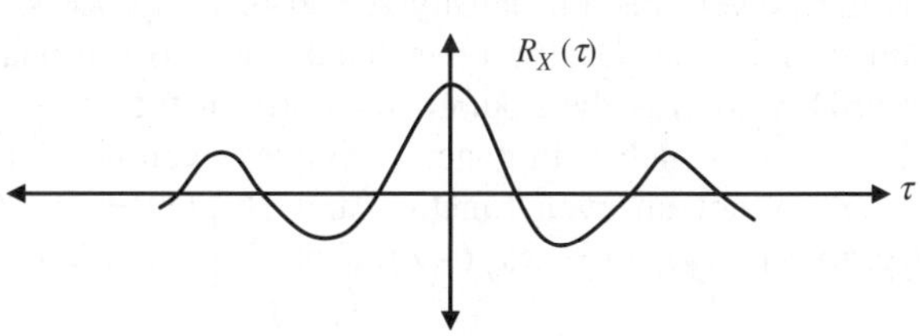

Figure 9.3 Autocorrelation function $R_X(\tau)$ of a stationary process – an even function

belong to the class of functions (e.g. square integrable ones) usually treated in Fourier analysis. However, an integral representation for $X(t)$ is possible via Cramer's spectral representation:

$$X(t) = \int_{\mathbb{R}} e^{i\lambda t} d\Phi(\lambda) \tag{9.52}$$

where $\Phi(\lambda)$ is an amplitude function on the frequency axis $\mathbb{R}$ with the above representation being identified as the quadratic mean limit of appropriate summands (Cramer and Leadbetter, 1967; see Lippi, 2004 for some issues regarding the convergence of the above representation through a finite summand). Assume that the infinitesimal increments $d\Phi(\lambda)$ are uncorrelated random variables (complex) with zero mean and a measure of variance given by the function $S_{XX}(\lambda)$. More precisely, we may write:

$$E[d\Phi(\lambda_1)d\Phi^*(\lambda_2)] = S_{XX}(\lambda_1)\delta(\lambda_1 - \lambda_2)d\lambda_1 d\lambda_2 \tag{9.53}$$

The superscript '$*$' denotes complex conjugation and $\delta(\cdot)$ is the Dirac delta function. From Equation 9.52, we have:

$$\begin{aligned} E[X(t_1)X(t_2)] &= E\int_{\mathbb{R}} e^{-i\lambda_1 t_1} d\Phi^*(\lambda_1) \int_{\mathbb{R}} e^{i\lambda_2 t_2} d\Phi(\lambda_2) \\ &= \iint_{\mathbb{R}} e^{i(\lambda_2 t_2 - \lambda_1 t_1)} E[d\Phi^*(\lambda_1)d\Phi(\lambda_2)] \\ &= \iint_{\mathbb{R}} e^{i(\lambda_2 t_2 - \lambda_1 t_1)} S_{XX}(\lambda_1)\delta(\lambda_1 - \lambda_2)d\lambda_1 d\lambda_2 \\ &= \int_{\mathbb{R}} e^{i\lambda(t_2 - t_1)} S_{XX}(\lambda)d\lambda \end{aligned} \tag{9.54}$$

Since $X(t)$ is stationary, the extreme left-hand side of the above equation is the autocorrelation function $R_{XX}(t_2 - t_1)$. With $\tau = t_2 - t_1$, we have:

$$R_{XX}(\tau) = \int_{\mathbb{R}} e^{i\lambda\tau} S_{XX}(\lambda)d\lambda \tag{9.55}$$

Equation 9.55 shows the Fourier transform relationship between the autocorrelation function $R_{XX}(\tau)$of a stationary stochastic process and the spectral function $S_{XX}(\lambda)$. Taking the inverse Fourier transform of Equation 9.55, we have:

$$S_{XX}(\lambda) = \int_{\mathbb{R}} R_{XX}(\tau)e^{-i\lambda\tau} d\lambda \tag{9.56}$$

Equations 9.55 and 9.56 are known as Wiener–Kienchine relations and $S_{XX}(\lambda)$ is generally referred to as the power spectral density (PSD) of the process. Both $R_{XX}(\tau)$ and $S_{XX}(\lambda)$ are real-valued and even. On the other hand, for two stationary processes $X(t)$ and $Y(t)$, the cross-PSD $S_{XY}(\lambda)$ is the Fourier transform of the crosscorrelation function $R_{XY}(\tau) = E[X(t)Y(t+\tau)]$, which is in general complex even if $X(t)$ and $Y(t)$ are real. This is true as $R_{XY}(\tau)$ is not an even function and $R_{XY}(\tau) = R_{YX}(-\tau)$. $S_{XY}(\lambda)$ has Hermitian symmetry, that is $S_{XY}(\lambda) = S^*_{XY}(-\lambda)$. From Equation 9.55, we have:

$$R_{XX}(0) = E[X^2(t)] = \int_{\mathbb{R}} S_{XX}(\lambda)d\lambda \tag{9.57}$$

which indicates that the area under a PSD curve yields the mean square value of a stationary stochastic process. Thus, $\sqrt{\int_{\mathbb{R}} S_{XX}(\lambda)d\lambda}\left(=\sqrt{R_{XX}(0)}\right)$ is the root mean square (RMS) value that is often used in structural analysis and design whilst quantifying the response process.

9.6.4 $S_{XX}(\lambda)$ as the Mean Energy Density of $X(t)$

Consider, as before, a stationary stochastic process $X(t)$ and let a segment of this process, denoted by $X_T(t)$, be defined as:

$$\begin{aligned} X_T(t) &= X(t), \quad |t| \le T \\ &= 0, \quad |t| > T \end{aligned} \tag{9.58}$$

A measure of 'energy' of this segment (for each path of $X(t)$) is given by $\int_{-T}^{T} X_T^2(t)dt$ and the time-averaged power is:

$$P_{X_T} = \frac{1}{2T}\int_{-T}^{T} X_T^2(t)dt \tag{9.59}$$

The definition of $X_T(t)$ enables the integration limits in the above equation extended to $-\infty$ to ∞. Now, by Parseval's theorem, $\lim_{T\to\infty} \frac{1}{2T}\int_{-T}^{T} X_T^2(t)dt = \lim_{T\to\infty} \frac{1}{2T}\int_{-T}^{T} |X_{TF}(\lambda)|^2 d\lambda$, where $X_{TF}(\lambda)$ is the Fourier transform of $X_T(t)$. Here, we assume that the Fourier transform exists for the restricted process $X_T(t)$. P_{X_T} in Equation 9.59 is a random variable whose expectation is obtained as:

$$E[P_{X_T}] = \frac{1}{2T}\int_{-\infty}^{\infty} E[|X_{TF}(\lambda)|^2]d\lambda \tag{9.60}$$

Taking the limit as $T \to \infty$ (if it exists), the average power (ensemble average) of P_X of the process $X(t)$ is given by:

$$E[P_X] = \lim_{T\to\infty} \frac{1}{2T}\int_{-\infty}^{\infty} E[|X_{TF}(\lambda)|^2]d\lambda = \int_{-\infty}^{\infty} \lim_{T\to\infty} \frac{E[|X_{TF}(\lambda)|^2]}{2T}d\lambda \tag{9.61}$$

The integrand $\lim_{T\to\infty} \frac{E[|X_{TF}(\lambda)|^2]}{2T}$ in Equation 9.61 is a measure of the average power contained in $X(t)$ corresponding to the frequency λ and has the units of power per unit frequency. Thus, from Equations 9.57 and 9.59 it must be $S_{XX}(\lambda)$, the PSD function. In fact, the interpretation of PSD via this limiting strategy forms the basis for the experimental procedure (Harris and Crede, 1961) in obtaining the PSD of a signal interpreted as a stochastic process. Further, the above definition of PSD based on the notion of average power (a squared quantity) implies that $S_{XX}(\lambda) \ge 0, \forall\ \lambda$.

A stochastic process $X(t)$ is called a 'white' noise (name derived after white light that contains all visible-light frequencies in its spectrum) if its PSD is constant over all frequencies $\lambda \in (-\infty, \infty)$ (Figure 9.4a). Thus, if we define $S_{XX}(\lambda) = S_0 \in \mathbb{R}$, the inverse Fourier transform of $S_{XX}(\lambda)$ gives the autocorrelation function $R_X(\tau)$ as $S_0\delta(\tau)$ (Figure 9.4b). Such a white-noise process of constant PSD (i.e. with infinite power content) is practically unrealisable as a graph of its path as a function of time does not exist (see the section on Wiener processes for an alternative definition of a white-noise process). A more realistic form is a so-called 'band-limited white noise' process (as shown in Figure 9.5a along with its autocorrelation in Figure 9.5b).

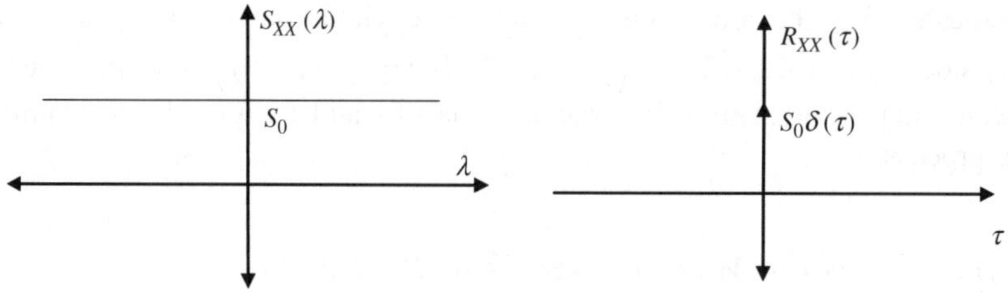

Figure 9.4 (a) PSD and (b) autocorrelation of white-noise process

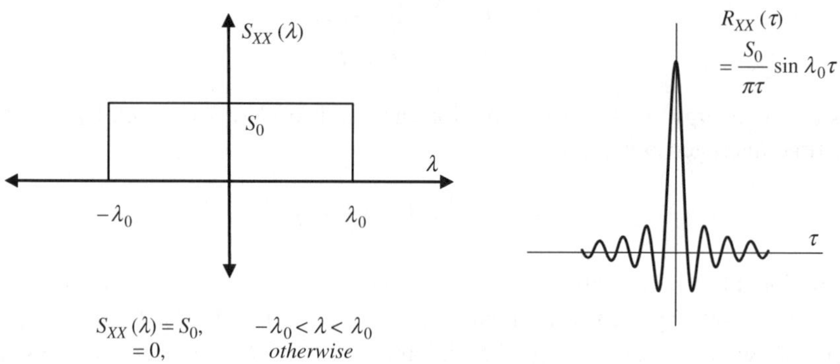

Figure 9.5 (a) PSD and (b) autocorrelation of a band-limited white noise

9.6.5 Some Basic Stochastic Processes

Stochastic processes such as Markov, Weiner and Poisson provide useful mathematical models to represent a broad class of random signals encountered in engineering and physics. The notion of conditional probability and expectation is central to the development of a Markov process. Weiner and Poisson processes are those with independent increments in that, for all choices of $0 < t_1 < \cdots < t_m$, the random increments $X(t_1) - X(0), X(t_2) - X(t_1), \ldots, X(t_m - t_{m-1})$ are independent. Whilst a Poisson process has increments with Poisson probability distributions, those corresponding to the Weiner process are Gaussian.

9.6.5.1 Markov Process

A stochastic process $X(t)$ is Markov if for all sequence of times $t_1 \leq t_1 \leq \cdots \leq t_m$ with $m \in N$, we have the identity:

$$P\{X(t_m) \leq X(t_{m-1}) | X(t_{n-1}), X(t_{n-1}), \ldots, X(t_1)\} = P\{X(t_n) \leq X(t_{n-1}) | X(t_{n-1})\} \tag{9.62}$$

Thus, for a Markov process, a future state is dependent only on the present and not on the past. A Markov chain is a discrete time and hence a discrete state version of a Markov process.

9.6.5.2 Poisson Process

A Poisson process is a discrete state continuous time stochastic process. It is a counting process that is typical of knowing the failure of a system component via, say, the number of crossings of a stochastic process $X(t)$ of a level $c \in \mathbb{R}$ in an interval $[0, T] \subset (0, \infty)$. The number of occurrences of the event in any finite collection of nonoverlapping intervals $\in T$ is an independent random variables typical of an independent increment process.

9.6.5.3 Wiener Process

A Weiner process provides the mathematical model for Brownian motion that concerns the random movement of a particle in a fluid due to its bombardment with the fluid molecules. In fact, random walk is a discrete-time analogue of the continuous-time Brownian motion. The statistical characteristics of the Weiner process are thus derivable through appropriate limiting operations on random walk models (Papoulis, 1991). If $X(t)$ is a Weiner process, it is a zero-mean Gaussian process with correlation function $E[X(t)X(s)] = \min(t, s)$, so that $E[X^2(t)] = t$. It follows that $E[X^2(0)] = 0$ and thus $X(0) = 0$. As a limit of random walk model, the Weiner process is known to have the following properties:

1. It is a Gaussian process, that is for all $0 < t_1 < \cdots < t_m$, the vector random variable $\boldsymbol{Y} = \{X(t_1), X(t_2), \ldots, X(t_m)\} \in \mathbb{R}^m$ has a multinormal distribution. Thus, there exist a vector $\boldsymbol{\mu} = (\mu_1, \mu_2, \ldots, \mu_m) \in \mathbb{R}^m$ and a non-negative definite matrix $\boldsymbol{C} = [C_{ij}] \in \mathbb{R}^{m\times m}$ such that $E\left[e^{i\sum_{j=1}^{m}\lambda_j Y_j}\right] = \exp\left(-\frac{1}{2}\sum_{k=1}^{m}\sum_{j=1}^{m}\lambda_j\left[C_{jk}\right]\lambda_k + i\sum_{j=1}^{m}\lambda_j\mu_j\right)$ for all $\boldsymbol{\lambda} = (\lambda_1, \lambda_2, \ldots, \lambda_m) \in \mathbb{R}^m$. If this holds, $\boldsymbol{\mu}$ is the mean vector and $\boldsymbol{C}$ is the covariance matrix with $C_{jk} = E[(Y_j - \mu_j)(Y_k - \mu_k)]$.
2. It is a stochastic process with independent increments. To prove this, we use the fact that $X(t)$ is normal and normal random variables are independent iff (if and only if) they are uncorrelated. Hence, it is enough to show that:

$$E[(X(t_i) - X(t_{i-1}))(X(t_j) - X(t_{j-1}))] = 0, \quad t_i < t_j \tag{9.63}$$

   The left-hand side of the above equation can be expanded and evaluated as follows:

$$\begin{aligned} &E[(X(t_i) - X(t_{i-1}))(X(t_j) - X(t_{j-1}))] \\ &= E[X(t_i)X(t_j) - X(t_i)X(t_{j-1}) - X(t_{i-1})X(t_j) + X(t_{i-1})X(t_{j-1})] \\ &= t_i - t_i - t_{i-1} + t_{i-1} = 0 \end{aligned} \tag{9.64}$$

   The characteristics of the Weiner process, $E[X^2(t)] = t$ and $E[X(t)X(s)] = \min(t, s)$ are utilised in simplifying Equation 9.64.
3. The paths of the Wiener process can be considered to be continuous. Here, we refer to Kolmogorov's continuity condition: If for all $T > 0$, there exist $\gamma > 0, \beta > 0, D > 0$ such that $E[|X(t) - X(s)|^\gamma] \leq D|t - s|^{1+\beta}$, $0 \leq t, s \leq T$, then there exists a continuous version of $X(t)$ (i.e. $X(t)$ as a process is not defined uniquely and a discontinuous version of $X(t)$ satisfying all these requirements may also exist). However, it suffices here to know that two stochastic processes may have same finite-dimensional probability distributions and considered to be same, but their path properties may be different.

In the present case, since $X(t)$ is normal, we know that the fourth-order moment, $E[X^4(t)] = 3E[X^2(t)] = 3t^2$ and it follows that:

$$E[|X(t) - X(s)|^4] = 3|t - s|^2 \tag{9.65}$$

The above equation shows that the Wiener process satisfies the Kolmogorov's continuity condition with $\gamma = 4$, $D = 3$ and $\beta = 1$ and hence it is a continuous version.

4. A Weiner process is not differentiable. To show this, consider $\frac{\Delta X}{\Delta h} = \frac{X(t+\Delta h)-X(t)}{\Delta h}$ with Δh being an infinitesimal interval in $\mathbb{R}$. $\frac{\Delta X}{\Delta h}$ is a random variable with $E\left[\left(\frac{\Delta X}{\Delta h}\right)^2\right] = \frac{1}{\Delta h}$. If $\lim_{\Delta h \to 0} \frac{\Delta X}{\Delta h}$ converges in some sense (see Appendix I on convergence of a sequence of random variables) to a limit, then the sequence of the characteristic functions $E\left[e^{i\lambda \frac{\Delta X}{\Delta h}}\right]$ converges to a limit that must be a continuous function in λ. We know that for a zero-mean normal random variable Z, the characteristic function $\Phi_Z(\lambda) = E\left[e^{i\lambda z}\right] = e^{-i\lambda^2 \alpha_z^2}$, where σ_z^2 is the variance of Z. Hence, we have:

$$\lim_{\Delta h \to 0} E\left[e^{i\lambda \frac{\Delta X}{\Delta h}}\right] = \lim_{\Delta h \to 0} e^{\frac{-\lambda^2}{2\Delta h^2}} = 1, \quad \lambda = 0 \text{ and } = 0, \text{ otherwise} \tag{9.66}$$

We find from the above equation that the characteristic function in the limit is not continuous and hence the assumed convergence of the derivative of the Weiner process is lost. However, we define a generalized derivative, $\dot{X}(t)$ and show it to be the white-noise process $W(t)$ with $E[W(t)] = 0$ and $E[W(t)W(s)] = \delta(t - s)$. Here, $\delta(\cdot)$ is the Dirac delta function. To this end, consider a deterministic function $g(t)$, a smooth function having derivatives of all orders and write:

$$\int_0^t g(t)\dot{X}(t)dt = g(t)X(t) - \int_0^t \dot{g}(t)X(t)dt \tag{9.67}$$

Note that the above equation is the result of integrating $\int_0^t g(t)\dot{X}(t)dt$ by parts. Now taking expectation on both sides, we have:

$$E\left[\int_0^t g(t)\dot{X}(t)dt\right] = 0 \Rightarrow E[\dot{X}(t)] = 0 \tag{9.68}$$

Further, let $p(t) = E\left[\int_0^t \int_0^t g(s)g(s')\dot{X}(s)\dot{X}(s')dsds'\right]$ and we have:

$$\begin{aligned}
p(t) &= \int_0^t \int_0^t g(s)g(s')E\left[\dot{X}(s)\dot{X}(s')\right]dsds' \\
&= E\left[\int_0^t g(s)\dot{X}(s)ds \int_0^t g(s')\dot{X}(s')ds'\right] \\
&= E\left(g(t)X(t) - \int_0^t \dot{g}(s)X(s)ds\right)^2, \text{ (from Equation 9.67)} \\
&= g^2(t)E\left[X^2(t)\right] - 2g(t)\int_0^t \dot{g}(s)E\left[X(t)X(s)\right]ds \\
&\quad + \int_0^t \int_0^t \dot{g}(s)\dot{g}(s')E\left[X(s)X(s')\right]dsds' \\
&= tg^2(t) - 2g(t)\int_0^t \dot{g}(s)s\,ds + \int_0^t \int_0^t \dot{g}(s)\dot{g}(s')\min(s, s')dsds'
\end{aligned}$$

$$\Rightarrow \frac{dp(t)}{dt} = \{g^2(t) + 2tg(t)\dot{g}(t)\} - 2\left\{\dot{g}(t)\int_0^t \dot{g}(s)sds + tg(t)\dot{g}(t)\right\}$$

$$+ 2\dot{g}(t)\int_0^t g(s)s\,ds$$

$$= g^2(t)$$

$$\Rightarrow p(t) = \int_0^t g^2(t)dt = \left[\int_0^t \int_0^t g(s)g(s')E\left[\dot{X}(s)\dot{X}(s')\right]dsds'\right] \tag{9.69}$$

However,

$$\int_0^t g^2(t)dt = \int_0^t \int_0^t g(s)g(s')\delta(s-s')dsds' \tag{9.70}$$

Comparing Equations 9.69 and 9.70, we infer:

$$E[\dot{X}(s)\dot{X}(s')] = \delta(s-s') \tag{9.71}$$

Thus, the autocorrelation function of $\dot{X}(t)$ is a Dirac delta function. Combining the two results in Equations 9.68 and 9.71, we have that the (generalized/formal) derivative of a Wiener process is a white-noise process. The autocorrelation function of the white-noise process shows that no two different points on a sample path of the process are correlated, that is it is completely memoryless. As such, the sample histories of a white-noise process are characterised by a very violent oscillatory behaviour. It is easily generated on a computer and through an appropriate filtering of such a process, other stationary stochastic processes with any specified PSD function can be readily generated. This is extremely useful in time-series analysis and stochastic structural dynamics.

9.7 Stochastic Dynamics of Linear Structural Systems

Having described so far, though in brief, a stochastic process and its characteristics by use of probability theory, we now consider the effect of such a random loading on linear time-invariant (LTI) dynamical systems (Crandall, 1963; Lin, 1967; Nigam, 1983; Yang, 1986; Nigam and Narayanan, 1994). We assume that the external loading, considered to be a stochastic process, is specified in terms of its characteristics such as the mean, autocorrelation/autocovariance. The objective here is to determine how these first- and second-order characteristics of the input process are transformed by the linear system in both the time and frequency domains.

9.7.1 Continuous Systems under Stochastic Input

For example, for a linear continuous system (Chapters 2 and 3) the governing equation of motion is:

$$\boldsymbol{L}\{Y(z,t)\} = F(z,t) \tag{9.72}$$

Here, $\boldsymbol{L}$ is the partial differential operator that can be expressed as $\boldsymbol{L}_s + m\frac{\partial^2}{\partial t^2} + c\frac{\partial}{\partial t}$ where $\boldsymbol{L}_s$ is a self-adjoint linear operator (Chapter 2) in a spatial coordinate vector z

($\in \boldsymbol{\Sigma}$, the domain of the continuum). For instance, for a continuous beam of length l, $\boldsymbol{L}_s = EI\frac{\partial^4 Y}{\partial z^4}$ with $z \in [0, l]$, E and I being the Young's modulus and the moment of inertia, respectively. $Y(z, t)$ is the field variable representing the system response, that is the deflection of the continuous system. $F(z, t)$ is the external loading function, which is in reality a random field. Suppose that the input characteristics of $F(z, t)$ are known in terms of its mean $\mu_F(z, t)$ and autocorrelation $R_{FF}(z_1, z_2, t_1, t_2)$. At this stage, stationarity is not assumed for $F(z, t)$. Following the method of separation of variables described in Chapter 2 and then using the orthonormal eigenfunctions $\Phi_j(z)$, $j \in \mathbb{N}$ of the resulting Sturm–Liouville boundary-value problem, we express the system response as:

$$Y(z, t) = \sum_{j=1}^{\infty} q_j(t)\Phi_j(z) \tag{9.73}$$

$q_j(t)$, $j \in \mathbb{N}$ are the modal responses each of which is governed by the single degree of freedom (SDOF) oscillator equation (modal equation):

$$\ddot{q}_j(t) + 2\xi_j\omega_j\dot{q}_j(t) + \omega_j^2 q_j(t) = F_j(t), \;\; F_j(t) = \int_{\Sigma} \Phi_j(z)\frac{F(z, t)}{m}dz \tag{9.74}$$

Here, $F_j(t)$ is the generalized force (modal force). ω_j, $j \in \mathbb{N}$ are the natural frequencies of the system. ξ_j, $j \in \mathbb{N}$ are the modal damping ratios. With zero initial conditions, the solution to Equation 9.58 is given by the convolution integral:

$$q_j(\tau) = \int_0^t F_j(\tau)h_j(t-\tau)d\tau = \int_{-\infty}^{\infty} F_j(\tau)h_j(t-\tau)d\tau \tag{9.75}$$

$h_j(t)$, $j \in \mathbb{N}$ is the impulse response function of each modal equation (see Chapter 5):

$$h_j(t) = \frac{e^{-\xi\omega_j t}}{m\omega_{dj}}\sin\omega_{dj}t \tag{9.76}$$

$\omega_{dj} = \omega_j\sqrt{1-\xi_j^2}$. Whilst $\Phi_j(z)$ and $h_j(t)$ are deterministic functions, $F_j(t)$ and $q_j(t)$ are stochastic input and the resulting output. In view of Equation 9.74, the input $F_j(t)$ to these modal equations is related to the input process $F(z, t)$ by the relationships:

$$\mu_{F_j}(t) = E[F_j(t)] = \int_{\Sigma} \Phi_j(z)\frac{E[F(z, t)]}{m}dz = \int_{\Sigma} \frac{\Phi_j(z)}{m}\mu_F(z, t)dz \text{ and}$$

$$\begin{aligned} R_{F_jF_k}(t_1, t_2) &= \int_{\Sigma} \Phi_j(z_1)\Phi_k(z_2)\frac{E[F(z_1, t_1)F(z_2, t_2)]}{m^2}dz_1dz_2 \\ &= \int_{\Sigma}\int_{\Sigma} \frac{\Phi_j(z_1)\Phi_k(z_2)}{m^2}R_{FF}(z_1, z_2, t_1, t_2)dz_1dz_2 \end{aligned} \tag{9.77a,b}$$

From Equation 9.73, the first- and second-order moments of the output process $Y(z, t)$ are given in terms of moments of modal responses as:

$$\begin{aligned} \mu_Y(z, t) = E[Y(z, t)] &= \sum_{j=1}^{\infty} \Phi_j(z)E[q_j(t)] \\ &= \sum_{j=1}^{\infty} \Phi_j(z)\int_{-\infty}^{\infty} E[F_j(\tau)]h_j(t-\tau)d\tau \quad \text{and} \end{aligned}$$

$$R_{YY}(z_1, z_2, t_1, t_2) = E[Y(z_1, t_1)Y(z_2, t_2)] = \sum_{j=1}^{\infty}\sum_{k=1}^{\infty} E[q_j(t_1)q_k(t_2)]\Phi_j(z_1)\Phi_k(z_2)$$

(9.78a,b)

The above equations indicate that the system response moments depend on those of the modal responses, $q_j(\tau)$, $j \in \mathbb{N}$ and in turn of the input forcing function. Thus, knowing these input characteristics to each of the modal equations, the task finally reduces to determining the response of an SDOF oscillator under a stochastic input.

9.7.1.1 Response of a SDOF Oscillator under a Stochastic Input

For this purpose, let us take a typical oscillator:

$$\ddot{q}(t) + 2\xi\omega\dot{q}(t) + \omega^2 q(t) = G(t) \Rightarrow \boldsymbol{L}q(t) = G(t) \tag{9.79}$$

Since for linear systems, $E[\boldsymbol{L}q(t)] = \boldsymbol{L}(E[q(t)])$, the output mean, $E[q(t)]$ is the response of the system with $E[G(t)]$ as the input. Thus, using Equation 9.75, we obtain:

$$\mu_q(t) = E[q(t)] = \int_{-\infty}^{\infty} E[G(\tau)]h(t-\tau)d\tau = \int_{-\infty}^{\infty} \mu_G(t)h(t-\tau)d\tau \tag{9.80}$$

To obtain the second-order moments, we first multiply Equation 9.79 by $G(t_1)$ to give $\boldsymbol{L}\{G(t_1)q(t)\} = G(t_1)G(t)$ from which we obtain:

$$\boldsymbol{L}\{E[G(t_1)q(t)]\} = E[G(t_1)G(t)] \Rightarrow R_{Gq}(t_1, t) = \int_{-\infty}^{\infty} R_{GG}(t_1, t)h(t-\tau_2)d\tau_2$$

$$\Rightarrow R_{Gq}(t_1, t_2) = \int_{-\infty}^{\infty} R_{GG}(t_1, t_2)h(t_2-\tau_2)d\tau_2 \tag{9.81}$$

Similarly, by multiplying Equation 9.79 by $q(t_2)$ and taking the expectation operation, we get:

$$\boldsymbol{L}\{E[q(t_2)q(t)]\} = E[q(t_2)G(t)] \Rightarrow R_{qq}(t, t_2) = \int_{-\infty}^{\infty} R_{Gq}(t, t_2)h(t-\tau_1)d\tau_1$$

$$\Rightarrow R_{qq}(t_1, t_2) = \int_{-\infty}^{\infty} R_{Gq}(t_1, t_2)h(t_1-\tau_1)d\tau_1 \tag{9.82}$$

Substituting the expression for $R_{Gq}(t_1, t_2)$ from Equation 9.81 in Equation 9.82 gives:

$$R_{qq}(t_1, t_2) = \int_{-\infty}^{\infty}\int_{-\infty}^{\infty} R_{GG}(\tau_1, \tau_2)h(t_1-\tau_1)h(t_2-\tau_2)d\tau_1 d\tau_2 \tag{9.83}$$

Equations 9.80 and 9.83 together describe the relationships between the input and output mean and autocorrelation functions for an SDOF oscillator.

Knowing these output characteristics of an SDOF oscillator, we can determine the continuous system characteristics $\mu_Y(z, t)$ and $R_{YY}(z_1, z_2, t_1, t_2)$ from Equation 9.78a,b. Specifically, Equation 9.78b requires crosscorrelation $R_{q_jq_k}(t_1, t_2)$ between $q_j(t)$ and $q_k(t)$ of two oscillators driven by the generalized forces $G_j(t)$ and $G_k(t)$ and whose impulse response functions are $h_j(t)$ and $h_k(t)$, respectively. We can extend the result in Equation 9.83 to obtain

$$R_{q_jq_k}(t_1,t_2)=\int_{-\infty}^{\infty}\int_{-\infty}^{\infty}R_{G_jG_k}(\tau_1,\tau_2)h_j(t_1-\tau_1)h_k(t_2-\tau_2)d\tau_1 d\tau_2 \quad (9.84)$$

9.7.1.2 Response of an SDOF Oscillator to a Stationary Input

If the input $G(t)$ is stationary, that is $\mu_G(t)=\mu_{(G)}=$ constant and $R_{GG}(t_1,t_2)=R_{GG}(t_2-t_1)$, the SDOF oscillator output characteristics in Equations 9.80, 9.83 and 9.84 take the form:

$$\mu_q(t)=\mu_{(G)}\int_{-\infty}^{\infty}h(t-\tau)d\tau \text{ and}$$

$$R_{qq}(t_1,t_2)=\int_{-\infty}^{\infty}\int_{-\infty}^{\infty}R_{GG}(\tau_2-\tau_1)h(t_1-\tau_1)h(t_2-\tau_2)d\tau_1 d\tau_2 \quad (9.85a,b)$$

From the result in the above equations, it is inferred that the oscillator output (and so the output of each mode in Equation 9.79) is nonstationary even if the input is stationary or otherwise. The case with the response of the original continuous system is the same as it is a linear superposition of the modal responses (Equation 9.73).

In the specific case of a white-noise input with $R_{GG}(\tau_2-\tau_1)=\mathrm{I}\delta(\tau_2-\tau_1)$, Equation 9.83 takes the form:

$$R_{qq}(t_1,t_2)=\mathrm{I}\int_{-\infty}^{\infty}\int_{-\infty}^{\infty}\delta(\tau_2-\tau_1)h(t_1-\tau_1)h(t_2-\tau_2)d\tau_1 d\tau_2$$

$$=I\int_{-\infty}^{\infty}h(t_1-\tau_1)h(t_2-\tau_1)d\tau_1 \quad (9.86a)$$

Substituting for the impulse response functions (Equation 9.76), we evaluate the above integral (Lin, 1967) and finally obtain:

$$R_{qq}(t_1,t_2)=\frac{I}{4\xi\omega^3}e^{-\xi\omega(t_2-t_1)}\left[\frac{e^{-2\xi\omega t_1}}{1-\xi^2}\left\{\xi^2\cos\omega_d(t_1+t_2)-\xi\sqrt{1-\xi^2}\sin\omega_d(t_1+t_2)\right.\right.$$
$$\left.\left.-\cos\omega_d(t_2-t_1)\right\}+\left\{\cos\omega_d(t_2-t_1)+\frac{\xi}{\sqrt{1-\xi^2}}\sin\omega_d(t_2-t_1)\right\}\right] \quad (9.86b)$$

For large times ($t_1,t_2\to\infty$) but with finite $t_2-t_1=\tau$, observe that $R_{qq}(t_1,t_2)$ attains stationarity, that is $\lim_{\substack{t_1,t_2\to\infty \\ t_2-t_1\to \text{finite}}} R_{qq}(t_1,t_2)\to R_{qq}(\tau)$ and this so-called (stochastic) steady-state correlation is given by:

$$R_{qq}(\tau)=\frac{I}{4\xi\omega^3}e^{-\xi\omega|\tau|}\{\cos\omega_d\tau+\frac{\xi}{\sqrt{1-\xi^2}}\sin\omega_d|\tau|\},\ -\infty\le\tau\le\infty \quad (9.87)$$

The absolute sign in the above equation is required since $R_{qq}(\tau)$ is an even function of τ. Note that Equations 9.85 and 9.87 correspond to the stochastic response of the SDOF oscillator in the time domain, similar to the transient response of the oscillator under a deterministic time-varying input (Chapter 5).

9.7.1.3 Response of an SDOF Oscillator in the Frequency Domain

Knowing the spectral characteristics of the input $G(t)$ in terms of its PSD $S_{GG}(\lambda)$, we can obtain the PSD of the response $q(t)$. This is again analogous to the frequency-response analysis (in steady state) of an SDOF oscillator. Referring to Equation 9.83 and assuming that the input process $G(t)$ is stationary with $R_{GG}(\tau_1, \tau_2) = R_{GG}(\tau_2 - \tau_1) = \int_{-\infty}^{\infty} S_{GG}(\lambda)e^{i\lambda(\tau_2-\tau_1)}d\lambda$, we have output correlation:

$$\begin{aligned} R_{qq}(t_1, t_2) &= \int_{-\infty}^{\infty}\int_{-\infty}^{\infty} R_{GG}(\tau_2 - \tau_1)h(t_1 - \tau_1)h(t_2 - \tau_2)d\tau_1 d\tau_2 \\ &= \int_{-\infty}^{\infty}\int_{-\infty}^{\infty}\left(\int_{-\infty}^{\infty} S_{GG}(\lambda)e^{i\lambda(\tau_2-\tau_1)}d\lambda\right)h(t_1 - \tau_1)h(t_2 - \tau_2)d\tau_1 d\tau_2 \end{aligned} \tag{9.88}$$

The transformations $t_1 - \tau_1 = \alpha_1$ and $t_2 - \tau_2 = \alpha_2$ modify the above integral as:

$$\begin{aligned} R_{qq}(t_1, t_2) &= \int_{-\infty}^{\infty}\int_{-\infty}^{\infty}\left(\int_{-\infty}^{\infty} S_{GG}(\lambda)e^{i\lambda(\tau_2-\tau_1+\alpha_1-\alpha_2)}d\lambda\right)h(\alpha_1)h(\alpha_2)d\alpha_1 d\alpha_2 \\ &= \int_{-\infty}^{\infty}\int_{-\infty}^{\infty}\left(\int_{-\infty}^{\infty} S_{GG}(\lambda)e^{i\lambda(t_2-t_1)}d\lambda\right)h(\alpha_1)e^{i\lambda\alpha_1}h(\alpha_2)e^{-i\lambda\alpha_2}d\alpha_1 d\alpha_2 \\ &= \int_{-\infty}^{\infty} S_{GG}(\lambda)e^{i\lambda(t_2-t_1)}\left(\int_{-\infty}^{\infty} h(\alpha_1)e^{i\lambda\alpha_1}d\alpha_1\right)\left(\int_{-\infty}^{\infty} h(\alpha_2)e^{-i\lambda\alpha_2}d\alpha_2\right)d\lambda \\ &= \int_{-\infty}^{\infty} [S_{GG}(\lambda)H^*(\lambda)H(\lambda)]e^{i\lambda(t_2-t_1)}d\lambda \end{aligned} \tag{9.89}$$

$H(\lambda)$ is the complex frequency function of the SDOF oscillator (Chapter 5) and is the Fourier transform of its impulse response function, $h(t)$. That is:

$$H(\lambda) = \mathcal{F}(h(t)) = \int_{-\infty}^{\infty} h(\alpha)e^{-i\lambda\alpha}d\alpha = \frac{1}{(\lambda^2 - \omega^2) + i2\xi\lambda\omega} \tag{9.90}$$

$H^*(\lambda)$ in Equation 9.89 is the complex conjugate of $H(\lambda)$. Equation 9.89 shows that (i) $R_{qq}(t_1, t_2)$ is a function of $(t_2 - t_1)$ and hence the oscillator output process, $q(t)$ is stationary and (ii) it can be recognised to be an inverse Fourier transform of $S_{GG}(\lambda)H^*(\lambda)H(\lambda)$, yielding $R_{qq}(t_2 - t_1)$. Hence, we deduce that the output PSD of an SDOF oscillator under a stochastic input of PSD, $S_{GG}(\lambda)$ is given by:

$$S_{qq}(\lambda) = S_{GG}(\lambda)H^*(\lambda)H(\lambda) \Rightarrow S_{qq}(\lambda) = S_{GG}(\lambda)|H(\lambda)|^2 \tag{9.91}$$

where $H(\lambda)$ is the frequency response function of the oscillator.

Example 9.1: Response of a continuous uniform beam under moving random load [Fryba 1972]

We refer to the Example 2.3 and assume that the moving load $F(z, t)$ is a concentrated load moving with uniform velocity V_s on the beam (Figure 9.6) with properties: mass density, $l = 10\,\text{m}$, $EI == 1.67 \times 10^6\,\text{Nm}^2$ and $\text{m} = 78.5\,\text{kg/m}$. $F(z, t) = \delta(z - V_s t)G(t)$ where $G(t)$ is given to be a stationary stochastic process with mean, $E([G(t)] = G_0$ and auto-correlation function $R_{GG}(t_1, t_2) = 2\pi S_0\delta(t_2 - t_1)$. Here, $G_0 = Mg$ and is a constant with units of Newtons. Here, M is the mass of the moving load. 'g' stands for the acceleration

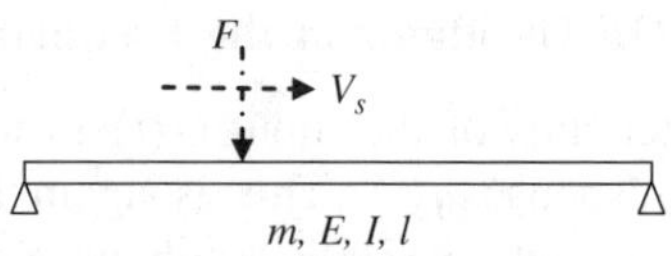

Figure 9.6 A simply supported beam under moving random load

due to gravity. $G(t)$ is recognised as white noise with a constant spectral density S_0. For the simply supported beam, the eigenfunctions, $\Phi_j(z) = \sqrt{2}\sin j\pi z$ (Chapter 2), which are orthonormal.

The eigenfrequencies (natural frequencies) of the beam are given by $\omega_j^2 = \frac{EI}{m}\frac{j^4\pi^4}{L^4}$. The modal force in Equation 9.74 is $G_j(t) = \int_\Sigma \Phi_j(z)\frac{F(z,t)}{m}dz = \frac{\sqrt{2}}{m}\sin j\pi V_s t G(t)$. Hence, we have:

$$R_{G_jG_k}(t_1,t_2) = \frac{2}{m^2}\sin j\pi V_s t_1 \sin k\pi V_s t_2 R_{GG}(t_1,t_2) \tag{9.92}$$

Equation 9.84 gives:

$$\begin{aligned} R_{q_jq_k}(t_1,t_2) &= \frac{2}{m^2}\int_{-\infty}^{\infty}\int_{-\infty}^{\infty} \begin{matrix}\sin j\pi V_s\tau_1 \sin k\pi V_s\tau_2\ h_j(t_1-\tau_1)h_k(t_2-\tau_2)\\ \cdot 2\pi S_0\delta(\tau_2-\tau_1)\cdot d\tau_1 d\tau_2\end{matrix} \\ &= \frac{4\pi S_0}{m^2}\int_{-\infty}^{\infty}\sin j\pi V_s\tau_1 \sin k\pi V_s\tau_1\ h_j(t_1-\tau_1)h_k(t_2-\tau_1)d\tau_1 \end{aligned} \tag{9.93}$$

The first-order and second-order moments of the output process $Y(z,t)$ can now be obtained from Equation 9.78 Thus, we have:

$$\begin{aligned} \mu_Y(z,t) &= \sum_{j=1}^{\infty} E[q_j(t)]\Phi_j(z) = \sum_{j=1}^{\infty}\left(\int_{-\infty}^{\infty} E[G_j(\tau)]h_j(t-\tau)d\tau\right)\Phi_j(z) \\ &= \frac{2}{m}\sum_{j=1}^{\infty}\sin j\pi z\left(\int_{-\infty}^{\infty}\sin j\pi V_s\tau E[G(\tau)]\frac{e^{-\xi\omega_j(t-\tau)}}{m\omega_{dj}}\sin\omega_{dj}(t-\tau)d\tau\right) \end{aligned} \tag{9.94}$$

Considering only the first mode in the above summation gives the mean deflection as $\frac{2G_0}{m^2\omega_{dj}}e^{-\xi\omega_j t}\sin\pi z\left(\int_{-\infty}^{\infty} e^{\xi\omega_j\tau}\sin\pi V_s\tau\sin\omega_{d1}(t-\tau)d\tau\right)$. From Equation 9.78b, the autocorrelation of the beam deflection is obtained as $R_{YY}(z_1,z_2,t_1,t_2) = \sum_{j=1}^{\infty}\sum_{k=1}^{\infty} R_{q_jq_k}(t_1,t_2)\Phi_j(z_1)\Phi_k(z_2)$. Ignoring the modal crosscorrelations, the mean square value $E[Y^2(z,t)]$ of the beam deflection is given by $R_{YY}(z,z,t,t)$ and it is obtained by using Equation 9.93 as:

$$E[Y^2(z,t)] = \frac{8\pi S_0}{m^2}\sum_{j=1}^{\infty}\sin^2 j\pi z\int_{-\infty}^{\infty}\sin^2 j\pi V_s\tau\ h_j^2(t-\tau)d\tau \tag{9.95}$$

The integration involved in obtaining $E[Y^2(z,t)]$ from the above equation can be evaluated in closed form (see Fryba, 1972 for details of the derivation). It can be seen that even though the input moving load is stationary, the response moments are functions of time, implying that the beam response is a nonstationary process.

9.7.2 *Discrete Systems under Stochastic Input – Modal Superposition Method*

Semidiscretization of a linear continuous system leads to an multidegree-of-freedom (MDOF) system (Chapters 3 and 4) governed by matrix vector differential equations ordinary differential equations (ODEs) rewritten hereunder:

$$\boldsymbol{M}\ddot{\boldsymbol{x}} + \boldsymbol{C}\dot{\boldsymbol{x}} + \boldsymbol{K}\boldsymbol{x} = \boldsymbol{F}(t) \tag{9.96}$$

Let N be the number of *degrees of freedom* in the discrete model. In the case of FEM, $\boldsymbol{x}(t)$ is the response vector ($N \times 1$) that corresponds to the unknown displacements at the nodal *degrees of freedom*. $\boldsymbol{M}, \boldsymbol{C}$ and $\boldsymbol{K}$ are $N\ x\ N$ matrices that are generally symmetric. $\boldsymbol{F}(t)$ is the vector of nodal forces that are assumed to be stochastic. Equation 9.96 is usually referred to as a stochastic differential equation. Similar to the deterministic case, the 'classical' approach of modal superposition can be utilised to solve the SDEs to obtain the system response in terms of its statistical moments, given the similar description of the input forcing. A more general approach is to solve the SDE (Equation 9.96) by application of stochastic calculus for the response moments or by a direct method of time integration to obtain the moments or the pathwise (sample wise) solutions. Whilst the methods based on the stochastic calculus and numerical integration of SDEs are dealt with in the subsequent sections, the modal dynamics (Chapter 5), which is primarily suitable for linear (LTI) systems, is described here. This method, as already observed, requires solving a set of uncoupled SDOF oscillators each driven by a stochastic modal input as in Equation 9.74 If $m(\ll N)$ is the number of eigenmodes considered to be adequate for modal summation, the system response is given by:

$$\boldsymbol{x}(t) = \boldsymbol{\Phi}\boldsymbol{q}(\boldsymbol{t}) \tag{9.97}$$

$\boldsymbol{\Phi}$ is the $N \times m$ modal vector matrix and $\boldsymbol{q}(t)$, the $m \times 1$ vector of modal responses. Suppose that $\boldsymbol{F}(t)$ is a stationary stochastic process with mean $\boldsymbol{\mu}_F \in \mathbb{R}^N$ and correlation matrix $\boldsymbol{R}_{FF}(\tau) \in \mathbb{R}^{N\times N}$. From Equation 5.91, we obtain the corresponding characteristics (i.e. mean and correlation) of the modal force vector $\boldsymbol{G}(t)$ as:

$$\boldsymbol{\mu}_G = \boldsymbol{\Phi}^{\mathrm{T}}\boldsymbol{\mu}_F \in \mathbb{R}^m \text{ and } \boldsymbol{R}_{GG}(\tau) = \boldsymbol{\Phi}^{\mathrm{T}}\boldsymbol{R}_{FF}(\tau)\boldsymbol{\Phi} \in \mathrm{R}^{m\times N} \tag{9.98}$$

Using $\boldsymbol{\mu}_G$ and $\boldsymbol{R}_{GG}(\tau)$, we obtain the mean vector $\boldsymbol{\mu}_q$ and the correlation matrix $\boldsymbol{R}_{qq}(\tau)$ of $\boldsymbol{q}(t)$. Assuming the existence of a steady-state (time invariance) for these modal responses as $t \to \infty$, one finally arrives at the system haracteristics:

$$\boldsymbol{\mu}_x = \boldsymbol{\Phi}\boldsymbol{\mu}_q \in \mathbb{R}^N \text{ and } \boldsymbol{R}_{xx}(\tau) = \boldsymbol{\Phi}\boldsymbol{R}_{qq}(\tau)\boldsymbol{\Phi}^T \in \mathbb{R}^{N\times N} \tag{9.99}$$

For response in the frequency domain, assume that the input PSD matrix $\boldsymbol{S}_{FF}(\lambda)$ is given. Then, the PSD matrix of the modal forces is obtained as:

$$\boldsymbol{S}_{GG}(\lambda) = \boldsymbol{\Phi}^T\boldsymbol{S}_{FF}(\lambda)\boldsymbol{\Phi} \tag{9.100}$$

Using $\boldsymbol{S}_{GG}(\lambda)$, we obtain the modal response PSD matrix:

$$\boldsymbol{S}_{qq}(\lambda) = \boldsymbol{H}^*(\boldsymbol{\lambda})\boldsymbol{S}_{GG}(\lambda)\boldsymbol{H}(\lambda) \tag{9.101}$$

$\boldsymbol{H}(\lambda)$ is the $m \times m$ diagonal matrix of frequency response functions of the m modal equations. Finally, the PSD of the stationary system response is obtained via Equation 9.97:

$$\boldsymbol{S}_{xx}(\lambda) = \boldsymbol{\Phi} \boldsymbol{S}_{qq}(\lambda) \boldsymbol{\Phi}^T \tag{9.102}$$

We find system mean square responses from either $R_{x_j x_j}(0)$, or by evaluating $\int_0^\infty S_{x_j x_j}(\lambda) d\lambda$, $j = 1, 2.., N$.

9.8 An Introduction to Ito Calculus

A far more general and insightful treatment of structural dynamic systems (linear, LTI or even nonlinear) under stochastic loading conditions is possible through a systematic application of stochastic calculus (Klebaner, 1998), of which the Ito calculus is the most widely used and the one adopted in the rest of this chapter. Towards a simple exposition of the basic ideas, consider a 1D (possibly hypothetical) SDE given by:

$$\frac{dX}{dt} = a(t, X) + \sigma(t, X) W(t) \tag{9.103}$$

Here, $\sigma(t, X) W(t)$ is the modelled noise term, which may be system dependent, that is dependent on $X(t)$. In such a case, the noise is called multiplicative. In the special case, where $\sigma(t, X) = \sigma(t)$, the noise term is referred to as additive. Note that $W(t)$ is a 1D white-noise process and we should have:

i) $W(t_1)$ is independent of $W(t_2)$ for any $t_1 \neq t_2$

ii) $W(t)$ is stationary, and

iii) $E[W(t)] = 0 \; \forall t$ (9.104)

Whilst such a stochastic process $W(t)$ is formally referred to as the white noise, its physical existence is hardly justifiable. For instance, $W(t)$ cannot have continuous paths. To see this, consider the process $W(t)$ over a finite interval $[0, T]$ of interest and discretize this interval into $N > 0$ subintervals using the $N + 1$ points $0 = t_0 < t_1 < t_2 < \cdots < t_N = T$. One may now consider the following approximation, $W^N(t)$ for $W(t)$:

$$W^N(t) = \max(-N, \min(N, W(t)) \; \forall \; t \in [0, T] \tag{9.105}$$

For any $0 \le t \le T$, we have:

$$\begin{aligned} E\left[(W^N(t) - W^N(s))^2\right] &= E\left[(W^N(t))^2\right] + E\left[(W^N(s))^2\right] \text{ (using property i)} \\ &= 2E\left[(W^N(t))^2\right] \text{ (using property ii)} \\ &> 0, \text{ even as } t \to s \end{aligned} \tag{9.106}$$

In other words, $E\left[(W^N(t) - W^N(s))^2\right]$ cannot be bounded by $C(t-s)^{1+\alpha}$, $\alpha > 0$ and hence by Kolmogorov's continuity theorem, $W^N(t)$ and so $W(t)$ cannot be continuous. Accordingly, one must look for an alternative representation of the SDE in Equation 9.103, wherein $W(t)$ is replaced by a more meaningful stochastic process. Towards this, with $0 = t_0 < t_1 < t_2 < \cdots < t_N = T$, we write the following discrete version of the SDE:

$$X_{j+1} - X_j = a(t_j, X_j)h_j + \sigma(t_j X_j)W_j h_j \tag{9.107}$$

where $h_j = t_{j+1} - t_j$ is the time step $X_j := X(t_j)$, $W_j := W(t_j)$, and so on. Writing $Q_j = W_j h_j$ and passing to the limit as $h_j \to 0$, we wish to see if the process $Q(t)$ makes sense. It follows from properties (i–iii) of $W(t)$ that $Q(t)$ should have stationary independent increments with zero mean. But we know that a version of such a process with continuous paths is the Brownian motion $B(t)$, that is we have $Q(t) = B(t)$. Hence, Equation 9.107 may be written as:

$$X_{j+1} = X_0 + \sum_{l=0}^{j} a(t_l, X_l)h_l + \sum_{l=0}^{j} \sigma(t_l, X_l)\Delta B_l \tag{9.108}$$

where $\Delta B_l = B(t_{l+1}) - B(t_l)$ is the Brownian increments. Passing to the limit as $h_l \to 0$, we can then write a continuous form of Equation 9.108 as:

$$X(t) = X_0 + \int_0^t a(s, X(s))ds + \int_0^t \sigma(s, X(s))dB(s) \tag{9.109}$$

Obtaining the solution $X(t)$ above clearly requires an appropriate interpretation and evaluation of $\int_0^t \sigma(s, X(s))dB(s)$, called the 'stochastic integral'. Moreover, one must also identify a suitable class of functions $\sigma(t, X)$ for which the stochastic integral is well defined. First, let σ be additive and a simple function, that is it is expressible in the form:

$$\sigma(t, w) = \sum_{j \geq 0} \boldsymbol{\Gamma}_j(w) \cdot \boldsymbol{I}_{\left[\frac{j}{2^n}, \frac{j}{2^{n+1}}\right)}(t), \quad w \in \Omega \tag{9.110}$$

where $\boldsymbol{I}$ is the indicator function. Here, the stochastic integral may be written as:

$$\int_{t_m}^{t_p} \sigma(t, w)dB(t, w) = \sum_{j \geq 0} \boldsymbol{\Gamma}_j(w)[B_{j+1}(w) - B_j(w)] \tag{9.111}$$

where

$$\begin{aligned} t_j &= \frac{j}{2^n} \text{ if } t_m \leq \frac{j}{2^n} \leq t_p \\ &= t_m \ if \ \frac{j}{2^n} < t_m \\ &= t_p \ if \ \frac{j}{2^n} > t_p \end{aligned}$$

It now remains to choose $\boldsymbol{\Gamma}_j(w)$ in the variables-separated right-hand side of Equation 9.110. In the Ito calculus, we choose $\boldsymbol{\Gamma}_j(w) = \sigma(t_j, w)$. Whilst it is tempting to choose a more general form $\boldsymbol{\Gamma}_j(w) = \sigma(\tilde{t}_j, w)$ with $\tilde{t}_j \in [t_j, t_{j+1}]$, the stochastic integral will admit different evaluations depending on the precise location of $\tilde{t}_j$, as shown below. Let

$\sigma(t, w) = B(t, w)$, the standard Brownian motion and consider the following couple of approximations summands for $B(t, w)$:

$$A_1(t, w) = \sum_{j \geq 0} B\left(\frac{j}{2^n}, w\right) \boldsymbol{I}_{\left[\frac{j}{2^n}, \frac{j}{2^{n+1}}\right)}(\boldsymbol{t})$$

$$A_2(t, w) = \sum_{j \geq 0} B\left(\frac{j+1}{2^n}, w\right) \boldsymbol{I}_{\left[\frac{j}{2^n}, \frac{j}{2^{n+1}}\right)}(\boldsymbol{t}) \tag{9.112a,b}$$

Then we have:

$$E\left[\int_0^T A_1(t, w) dB(t, w)\right] = \sum_{j \geq 0} E\left[B\left(t_j\right)\left(B\left(t_{j+1}\right) - B\left(t_j\right)\right)\right]$$
$$= 0 \tag{9.113a}$$

$$E\left[\int_0^T A_2(t, w) dB(t, w)\right] = \sum_{j \geq 0} E\left[B\left(t_{j+1}\right)\left(B\left(t_{j+1}\right) - B\left(t_j\right)\right)\right]$$
$$= \sum_{j \geq 0} E\left[\left(B\left(t_{j+1}\right) - B\left(t_j\right)\right)^2\right]$$
$$= T \tag{9.113b}$$

The results in Equation 9.113a,b are quite different, even as $A_1(t, w)$ and $A_2(t, w)$ appear to be reasonable approximants for $B(t, w)$ (at least from our background of deterministic continuous functions). Note that the result in Equation 9.113a corresponds to the Ito interpretation. Another popular branch of stochastic calculus, referred to as the Stratonovich calculus, takes $\tilde{t}_j = (t_{j+1} + t_j)/2$. In the rest of this chapter, we mainly adhere to an exposition based on Ito's theory.

9.8.1 *Brownian Filtration*

Let $B(t, w)$ be a 1D Brownian motion as above and define F_t to be the σ algebra generated by the set of random variables $\{B(t_i, w) | t_i \in [0, T]\}$. The set $\mathcal{N}_t = \{F_s | s \leq t\}$ is referred to as the Brownian filtration and may be interpreted as containing the 'history' of $B(s, \omega)$ for $s \leq t$. Generalising this definition to other stochastic processes (including the higher-dimensional ones) is straightforward. Note that $\mathcal{N}_s \subset \mathcal{N}_t$ for all $s < t$ and hence the filtration is ever increasing.

9.8.2 *Measurability*

A random variable $\gamma(\omega)$ is F_t-measurable if $\sigma(\gamma(\omega)) \subset F_t$. For instance, $B\left(\frac{t}{3}, \omega\right)$ is F_t-measurable, but $B(3t, \omega)$ is not.

9.8.3 *An Adapted Stochastic Process*

The stochastic process $X(t, \omega)$ is said to be adapted to the filtration $\mathcal{N}_t$ if for any $t \geq 0$, the function $\omega \rightarrow X(t, \omega)$ is $\mathcal{N}_t$-measurable.

9.8.4 Ito Integral

Let $\sigma(t, w)$ be F_t- measurable and $E\int_0^T \sigma^2(t,\omega)dt < \infty$. Then $\int_0^T \sigma(t,\omega)dB(t,\omega)$ is referred to as the Ito integral. Note that $\sigma(t,\omega)$ is much more general than just a simple function. However, the approximation of $\sigma(t,\omega)$ through simple functions (as in Equation 9.110) plays a crucial role in demonstrating the convergence of the Ito integral (see Øksendal, 2003 for a proof). Observe that $E\left[\int_0^T \sigma(t,\omega)dB(t,\omega)\right] = 0$. This is obvious if $\sigma(t,\omega)$ admits a simple representation; otherwise one can show that there exists a sequence $\{\sigma_n(t,\omega)\}$ of simple representations such that $E\int_0^T (\sigma(t,\omega) - \sigma_n(t,\omega))^2 dt \to 0$ as $n \to \infty$. The second moment of the Ito integral is given by the so-called Ito isometry, which is the following identity:

$$E\left[\left(\int_0^T \sigma(t,\omega)dB(t,\omega)\right)^2\right] = E\int \sigma^2(t,\omega)dt \tag{9.114}$$

As before, it suffices to prove this identity when $\sigma(t,\omega)$ is simple (as the general case follows from a limiting argument involving the mean square error). Thus, using Equation 9.109, we may write:

$$E\left[\left(\int_0^T \sigma(t,\omega)dB(t,\omega)\right)^2\right] = \sum_{i,j} E[\Gamma_i \Gamma_j \Delta B_i \Delta B_j] \tag{9.115}$$

where $\Delta B_i = B(t_{i+1}) - B(t_i)$, $\Gamma_i = \sigma(t_i,\omega)$, and so on. Thus, the right-hand side of Equation 9.115 reduces to:

$$\begin{aligned} E\left[\left(\int_0^T \sigma(t,\omega)dB(t,\omega)\right)^2\right] &= \sum_j E\left[\Gamma_j^2\right](t_{j+1} - t_j) \\ &= E\left[\int_0^t \sigma^2 dt\right] \end{aligned} \tag{9.116}$$

which provides the desired identity.

Example 9.2: Ito integral $\int_0^t B(s)dB(s) = \frac{1}{2}B^2(t) - \frac{1}{2}t$

Let $B(0) = 0$ and use the following sequence of simple functions to approximate $B(t,\omega)$:

$$B_n(t,\omega) = \sum_{j=0}^{n} B(t_j,\omega) I_{t_j,t_{j+1}}(t),\ n > 0 \tag{9.117}$$

Now, we may write:

$$\begin{aligned} E\left[\int_0^t (B_n(s) - B(s))^2 ds\right] &= E\left[\sum_j \int_{t_j}^{t_{j+1}} (B(t_j) - B(s))^2 ds\right] \\ &= \int_{t_j}^{t_{j+1}} (s - t_j)ds = \sum_j \frac{1}{2}(t_{j+1} - t_j)^2 \to 0 \text{ as } \Delta t_j \to 0 \end{aligned} \tag{9.118}$$

Hence, we can say:

$$\int_0^t B(s)dB(s) = \lim_{\Delta t_j \to 0} \int_0^t B_n(s)dB(s) = \lim_{\Delta t_j \to 0} \sum_j B(t_j)\Delta B_j \tag{9.119}$$

where $\Delta B_j = B(t_{j+1}) - B(t_j)$. Now:

$$\begin{aligned}\Delta\left(B^2(t_j)\right) &= B^2(t_{j+1}) - B^2(t_j)\\ &= \left(B(t_{j+1}) - B(t_j)\right)^2 + 2B(t_j)\left(B(t_{j+1}) - B(t_j)\right)\\ &= (\Delta B_j)^2 + 2B_j\Delta B_j\end{aligned} \tag{9.120}$$

Thus, using $B(0) = 0$, we get:

$$B^2(t) = \sum_j \Delta\left(B^2(t_j)\right) = \sum_j (\Delta B_j)^2 + 2\sum_j B(t_j)\Delta B_j$$

$$\text{Or, } \sum_j B(t_j)\Delta B_j = \frac{1}{2}B^2(t) - \frac{1}{2}\sum_j (\Delta B_j)^2 \tag{9.121}$$

Since $\sum_j (\Delta B_j)^2 \to t$ as $\Delta t_j \to 0$, we finally have:

$$\int_0^t B(s)dB(s) = \frac{1}{2}B^2(t) - \frac{1}{2}t \tag{9.122}$$

Interestingly, unlike the deterministic case wherein $\int_0^t x(t)dx(t) = \frac{1}{2}x^2(t)$ we have an extra term $-\frac{1}{2}t$ in the evaluation of the Ito integral. This is one of the distinguishing feature of Ito's calculus.

9.8.5 *Martingale*

A stochastic process $X(t), t \geq 0$ is called a martingale if $E[X(t)] < \infty$ (i.e. the process is integrable) and for any $s > 0$, we have the following identity:

$$E[X(t+s)|F(t)] = X(t) \quad \text{a.s.} \tag{9.123}$$

Moreover, $X(t)$ is a supermartingale if:

$$E[X(t+s)|F(t)] < X(t) \quad \text{a.s.} \tag{9.124}$$

And a submartingale if:

$$E[X(t+s)|F(t)] > X(t) \quad \text{a.s.} \tag{9.125}$$

The martingale property of $X(t)$, which owes its origin from betting in casinos, implies that if the process $X(t) = x$ is known at a time t, then its expected value at any future time remains the same.

Example 9.3: $B(t)$ ***is a martingale***
To see this, first note that $E[B(t)]= 0$, which shows that it is integrable. Then one can write for $s > 0$:

$$\begin{aligned}E[B(t+s)|F(t)] &= E[(B(t)+(B(t+s)-B(t)))|F(t)]\\ &= E[B(t)|F(t)]+E[B(t+s)-B(t)|F(t)]\\ &= B(t)+E[B(t+s)-B(t)] \text{ (since } B(t) \text{ is } F_t\text{-measurable and the}\\ &\quad \text{increment } B(t+s)-B(t) \text{ is independent of } F_t)\\ &= B(t) \end{aligned} \tag{9.126}$$

which shows that $B(t)$ is martingale.

Example 9.4: The Ito integral $I(t)=\int_0^t \sigma(s,\omega)dB(s\omega)$ ***is a martingale***
We know that $I(t)$ is integrable as $E[I(t)]= 0$. Now let $u < t$ and observe that $I(t)$ is F_t-measurable. Hence:

$$\begin{aligned}E[I(t)|F_u] &= E\left[\int_0^t \sigma(s)dB(s)\bigg|\, F_u\right]\\ &\quad \text{(here the second argument } \omega \text{ is dropped for convenience)}\\ &= \int_0^u \sigma(u)dB(u)+\int_u^t \sigma(s)dB(s)\bigg|\, F_u\\ &= \int_0^u \sigma(u)dB(u)+E\left[\int_u^t \sigma(s)dB(s)\right]\\ &= \int_0^u \sigma(u)dB(u)\\ &= I(u)\end{aligned} \tag{9.127}$$

Thus, $I(t)$ is a martingale.

In the above, we have again made use of the simple representation to see that $\int_u^t \sigma(s)dB(s)$ is independent of F_t. Also note that if $E\left[\int_0^t \sigma^2(s)ds\right]$ is unbounded (not integrable), then Ito's integral may fail to be a martingale. However, even in this case, $I(t)$ would be a 'local' martingale (i.e. an appropriately stopped version of $I(t)$ would still be a martingale).

9.8.6 Ito Process

Note that the Ito integral $I(t)=\int_0^t \sigma(s)dB(s)$ defines an adapted stochastic process, which is called an Ito process. One may also write down a differential version of the process as follows:

$$dI(t)=\sigma(t,x)dB(t) \tag{9.128}$$

Indeed, the above equation may also be readily interpreted as an SDE, for which a more general form is given by Equation 9.103, wherein $a(t, X)$ is referred to as the 'drift' term and $\sigma(t, X)dB(t)$ as the 'diffusion' (or 'noise') term. $\sigma(t, x)$ is sometimes called the diffusion coefficient. A cornerstone of Ito's calculus is the Ito formula that is stated below for 1D (scalar-valued) functions of the 1D Brownian motion $B(t)$.

9.8.6.1 Ito's Formula

Let $f(B(t))$ be at least twice differentiable in the argument. Then, for any t, we have:

$$f(B(t)) = f(0) + f'(B(s))dB(s) + \frac{1}{2}f''(B(s))d(s) + \cdots. \tag{9.129a}$$

where $f'(x) = \frac{df}{ds}$ and so on. Alternatively, we can write:

$$df(B(t)) = f'(B(t))dB(t) + \frac{1}{2}f''(B(t))dt + \cdots. \tag{9.129b}$$

In order to prove this formula, we need the following important result on $B(t)$. Consider a partition $\mathcal{T}^N = \{t_j | j \in [0, N]\}$ of the interval $[0, t]$ into N subintervals such that $\delta_N = \max_{j\in[0,N-1]}\{(t_{j+1} - t_j)\}$. Then, one can show that:

$$\sum_{j=0}^{N-1}(B(t_{j+1}) - B(t_j))^2 = t, \text{ in } L^2(P) \tag{9.130}$$

$L^2(P)$ is the space of all square integrable random variables with respect to P, the associated probability measure. First, observe that $\lim_{\delta N\to 0}\sum_{j=0}^{N-1} h_j = \int_0^t ds = t$ (where $h_j = t_{j+1} - t_j$). Now, in order to prove the required equality in $L^2(P)$, we take the expectation of the difference (between the LHS and RHS of the equality) as:

$$\begin{aligned} E\left[\sum_{j=0}^{N-1}\left((\Delta B_j)^2 - h_j\right)\right]^2 &= E\left[\left(\sum_{j=0}^{N-1}\left((\Delta B_j)^2 - h_j\right)\right)^2 \middle| F_{t_j}\right] \\ &= 2\sum_{j=0}^{N-1} h_j^2 \le 2\delta_N \sum_{j=0}^{N-1} h_j \\ &= 2\delta_N t \to 0 \;\; as \;\; \delta_N \to 0 \end{aligned} \tag{9.131}$$

Thus, we get the remarkable result that:

$$\left(\sum_{j=0}^{N-1}(\Delta B_j)^2\right) - t \xrightarrow{L^2(P)} 0 \tag{9.132}$$

More general variants of this identity, namely

$$\lim_{\delta N\to 0}\sum_{j=0}^{N-1}\psi_{t_j}(\Delta B_j)^2 = \int_0^t \psi_{t_j} ds \tag{9.133a}$$

$$\text{Or, } \lim_{\delta N \to 0} \sum_{j=0}^{N-1} \psi\left(B\left(t_j\right)\right)\left(\Delta B_j\right)^2 = \int_0^t \psi(B(s))ds \tag{9.133b}$$

may also be readily established in the $L^2(P)$ setting. The proof of the formula now starts with the identity:

$$f(B(t)) = f(0) + \sum_{j=0}^{N-1} \left(f\left(B(t_{j+1})\right) - f\left(B(t_j)\right)\right) \tag{9.134}$$

Applying Taylor's expansion to $f\left(B(t_{j+1})\right) - f\left(B(t_j)\right)$, we get:

$$f\left(B(t_{j+1})\right) = f\left(B(t_j)\right) + f'\left(B(t_j)\right)\Delta B_j + \frac{1}{2} f''(\theta_j)(\Delta B_j)^2 \tag{9.135}$$

where $\theta_j \in (B(t_j), B(t_{j+1}))$. Substituting the above in the identity (Equation 9.134) and using the identity in Equation 9.133a, we get:

$$f(B(t)) = f(0) + \sum_{j=0}^{N-1} f'(B(t_j))\Delta B_j + \frac{1}{2}\sum_{j=0}^{N-1} f''(\theta_j)h_j \tag{9.136}$$

Passing to the limit as $\delta_N \to 0$ immediately yields Ito's formula. The SDE form of Ito's formula also immediately follows.

Example 9.5: $\int_0^t B(s)dB(s) = \frac{1}{2}B^2(t) - \frac{1}{2}t$
Let $f(B(t)) = B^2(t)$. The integral form of Ito's formula yields, with $B(0) = 0$:

$$B^2(t) = 2\int_0^t B(s)dB(s) + t, \text{ that is} \tag{9.137}$$

$\int_0^t B(s)dB(s) = \frac{1}{2}B^2(t) - t$ (as obtained earlier in Example 9.2).

Example 9.6: SDE form of Ito's formula for $f(B(t)) = e^{B(t)}$
By using the SDE form of Ito's formula we have:

$$d\left(e^{B(t)}\right) = e^{B(t)}dB(t) + \frac{1}{2}e^{B(t)}dt \tag{9.138}$$

where $\frac{1}{2}e^{B(t)}$ is the drift coefficient and $e^{B(t)}$ is the diffusion coefficient. In other words, the stochastic process $X(t) = e^{B(t)}$ satisfies the SDE:

$$dX(t) = X(t)dB(t) + \frac{1}{2}X(t)dt \tag{9.139}$$

A more useful form of Ito's formula arises in the context of the following scenario. Let the process $X(t)$ be the solution of the following SDE:

$$dX(t) = a(t)dt + \sigma(t)dB(t) \tag{9.140}$$

If $f(B(t))$ is at least twice differentiable in $X(t)$ then what would be the SDE for $(X(t))$? Before writing down Ito's formula for this case, it is convenient to introduce the notion of quadratic variation (QV) of a stochastic process $X(t)$.

9.8.6.2 Quadratic Variation (QV) of $X(t)$

Given the partition $\mathcal{T}^N$ of $[0, t]$ into N intervals as before, QV of $X(t)$ is defined as:

$$[X, X](t) = \lim_{\delta N \to 0} \sum_{j=0}^{N-1} \left(X(t_{j+1}) - X(t_j)\right)^2 \tag{9.141}$$

Example 9.7: QV of Brownian motion $B(t)$

From Example 9.2, it is readily observed that the QV of $B(t)$ is given by:

$$[B, B](t) = t \tag{9.142}$$

Example 9.8: QV of a deterministic process

Let $X(t)$ be a continuous and at least once-differentiable deterministic process. Then, using a deterministic Taylor expansion in the definition of QV, it is not difficult to see that:

$$[X, X](t) = 0 \tag{9.143}$$

This may be contrasted with the behaviour of $B(t)$, for which QV is a characteristic shared by the diffusion terms in the SDEs, as made precise in the next example.

Example 9.9: QV of the Ito integral process

Consider the Ito integral process $I(t) = \int_0^t \sigma(s)dB(s)$. Through a direct application of the definition of QV (alternatively from the proof of Ito's isometry), it follows that;

$$[I, I](t) = \int_0^t \sigma^2(s)ds \tag{9.144}$$

Example 9.10: QV of the SDE (Equation 9.140)

Finally, consider the SDE for $X(t)$ Since the QVs of any terms containing the drift coefficient vanish, we readily get:

$$[X, X](t) = \int_0^t \sigma^2(s)dB(s) \tag{9.145}$$

A nonzero QV $[X, X](t)$ implies that the process $X(t)$ is not of bounded variation within any finite time interval. We are now ready to state Ito's formula for the scalar function $f(X(t))$, where $X(t)$ is the solution of the SDE (Equation 9.139). Specifically, we have:

$$df(X(t)) = f'(X(t))dX(t) + \frac{1}{2}f''(X(t))d[X, X](t)$$

$$\text{Or, } df(X(t)) = f'(X(t))dX(t) + \frac{1}{2}f''(X(t))\sigma^2(t)dt$$

$$\text{Or, } df(X(t)) = \left(f'(X(t))\,a(t) + \frac{1}{2}f''(X(t))\,\sigma^2(t)\right)dt + f'(X(t))\sigma(t)dB(t) \tag{9.146}$$

The above is then the SDE form of the Ito's formula for $f(X(t))$ Its integral form is:

$$f(X(t)) = f(X(0)) + \int_0^t f'(X(s))dX(s) + \int_0^t \frac{1}{2} f''(X(s))\sigma^2(s)ds \tag{9.147}$$

Example 9.11: Ito's formula for a scalar function $f(X(t))$
Let $f(X(t)) = \log X(t)$, where $X(t)$ is the solution of the SDE $dX(t) = X(t)dB(t) + \frac{1}{2}X(t)dt$. Noting that $(\log x)' = 1/x$ and $(\log x)'' = -\frac{1}{x^2}$, we get via Ito's formula:

$$\begin{aligned} d\log X(t) &= \frac{1}{X(t)} dX(t) - \frac{1}{2X^2(t)} X^2(t)dt \text{ (by using } \sigma(t) = X(t)) \\ &= dB(t) + \frac{1}{2}dt - \frac{1}{2}dt = dB(t) \\ &\Rightarrow \log(X(t)) = \log(X(0) + B(t)) \\ &\Rightarrow X(t) = X(0)e^{B(t)} \end{aligned} \tag{9.148}$$

9.8.6.3 Quadratic Covariation

Consider the following two 1D SDEs:

$$\begin{aligned} dX(t) &= a_X(t)dt + \sigma_X(t)dB(t) \\ dY(t) &= a_Y(t)dt + \sigma_Y(t)dB(t) \end{aligned} \tag{9.149a,b}$$

Then the quadratic covariation process $[X, Y](t)$ is defined in terms of the following limit over decreasing partitions $\mathcal{T}^N$ of $[0, t]$:

$$[X, Y](t) = \lim_{\delta N \to 0} \sum_{j=0}^{N-1} \left(X\left(t_{j+1}\right) - X(t_j)\right)\left(Y\left(t_{j+1}\right) - Y(t_j)\right) \tag{9.150}$$

Example 9.12: Quadratic covariation process for the SDEs in Equation 9.149
From Equation 9.149, we have:

$$d[X, Y](t) = dX(t)dY(t) = \sigma_X(t)\sigma_Y(t)(dB(t))^2 = \sigma_X(t)\sigma_Y(t)dt$$

The above result implies that $[X, Y](t) = \int_0^t \sigma_X(t)\sigma_Y(t)dt$.

9.8.6.4 Integration by Parts

Expanding the RHS in Equation 9.150, we get:

$$[X,Y](t) = \lim_{\delta N\to 0}\left[\sum_{j=0}^{N-1}\left(X(t_{j+1})Y(t_{j+1}) - X(t_j)Y(t_j)\right) - \sum_{j=0}^{N-1} X(t_j)\left(Y(t_{j+1}) - Y(t_j)\right)\right.$$
$$\left. - \sum_{j=0}^{N-1} Y(t_j)\left(X(t_{j+1}) - X(t_j)\right)\right]$$
$$= X(t)Y(t) - X(0)Y(0)$$
$$- \lim_{\delta N\to 0}\left[\sum_{j=0}^{N-1} X(t_j)\left(Y(t_{j+1}) - Y(t_j)\right) - \sum_{j=0}^{N-1} Y(t_j)\left(X(t_{j+1}) - X(t_j)\right)\right] \quad (9.151)$$

The last two terms on the RHS of the above equation converge *a.s* to the Ito integrals $\int_0^t X(s)dY(s)$ and $\int_0^t Y(s)dX(s)$, respectively. Thus, we have the following expression for the quadratic covariation process:

$$[X,Y](t) = X(t)Y(t) - X(0)Y(0) - \int_0^t X(t)dY(s) - \int_0^t Y(s)dX(s) \quad (9.152)$$

Rearranging the terms in the above identity, we arrive at the following formula for Ito integration by parts:

$$X(t)Y(t) - X(0)Y(0) = \int_0^t X(s)dY(s) + \int_0^t Y(s)dX(s) + [X,Y](t) \quad (9.153a)$$

Or, in the differential form, we have:

$$d(X(t)Y(t)) = X(t)dY(t) + Y(t)dX(t) + d[X,Y](t) \quad (9.153b)$$

Example 9.13: Integration by parts – an alternative form of QV process
The above formula (Equation 9.153) gives the following alternative expression for the QV process:

$$[X,X](t) = X^2(t) - X^2(0) - 2\int_0^t X(s)dX(s) \quad (9.154)$$

Example 9.14: Quadratic covariation $[f(B),B](t)$
Let $f(B(t))$ be twice differentiable in the argument and we intend to find $[f(B(t)), B(t)](t)$ via Ito's formula, we have:

$$df(B) = f'(B)dB - \frac{1}{2}f''(B)dt \quad (9.155)$$

Thus, we may obtain:

$$\begin{aligned} d[f(B), B](t) &= df(B(t))dB(t) \\ &= f'(B(t))(dB(t))^2 + \frac{1}{2}f''(B(t))dB(t)dt \\ &= f'(B(t))dt \text{ (since } (dB)^2 = dt \text{ and } dBdt = 0) \end{aligned} \quad (9.156)$$

Thus,

$$[f(B), B](t) = \int_0^t f'(B(t))dt \quad (9.157)$$

9.8.6.5 Higher-Dimensional Ito's Formula

Modal decomposition of LTI structural dynamical systems allows us to work with a collection of SDOF oscillators, each of which is a 2D dynamical system, whilst treating MDOF oscillators. Accordingly, it would be worthwhile to consider the 2D version of Ito's formula before discussing the more general case. Towards this, consider two stochastic processes $X(t)$ and $Y(t)$ (both being Ito diffusion processes and hence solutions of a couple of 1D SDOFs) and a scalar function $f(X(t), Y(t))$ so that it is at least twice differentiable in its arguments. Here, Ito's formula takes the form:

$$\begin{aligned} df(X(t), Y(t)) &= \frac{\partial f}{\partial x}(X(t), Y(t))dX(t) + \frac{\partial f}{\partial y}(X(t), Y(t))\,dY(t) \\ &+ \frac{1}{2}\frac{\partial^2 f}{\partial x^2}(X(t), Y(t))\,\sigma_x^2 dt + \frac{1}{2}\frac{\partial^2 f}{\partial y^2}(X(t), Y(t))\,\sigma_y^2 dt \\ &+ \frac{\partial^2 f}{\partial x \partial y}(X(t), Y(t))\,\sigma_x(t)\sigma_y(t) \end{aligned} \quad (9.158)$$

where $\sigma_x(t)$ and $\sigma_y(t)$ are the diffusion coefficients in the SDEs for $X(t)$ and $Y(t)$ respectively. Note that the above formula is valid even for multiplicative diffusion coefficients, that is when $\sigma_x = \sigma_x(X(t))$ and $\sigma_y = \sigma_y(Y(t))$. It is clear that the derivation of this formula has required both the QVs $[X, X](t)$ and $[Y, Y](t)$ and the quadratic covariation $[X, Y](t)$. The details of the proof are straightforward and are left as an additional exercise for the reader.

Example 9.15: Solution of SDEs corresponding to an SDOF system by higher-dimensional Ito's formula

Let us now consider the additively excited 2D SDE corresponding to an SDOF oscillator as follows:

$$\begin{aligned} dX(t) &= Y(t)dt \\ dY(t) &= -2\xi\omega Y(t) - \omega^2 X(t) + \sigma_Y(t)dB(t) \end{aligned} \quad (9.159a,b)$$

Where, the equations have been written in the incremental form (as $B(t)$ is not differentiable) and $Y(t) = \dot{X}(t)$ is the velocity. Now, let us use Ito's formula to obtain an SDE for $f(X(t)Y(t)) = X^2(t)$. Using Equation 9.153, we may write:

$$dX^2(t) = 2X(t)Y(t)dt \quad (9.160)$$

and we actually have the random ODE:

$$\frac{d}{dt}X^2(t) = 2X(t)Y(t) \text{ (no diffusion term)} \tag{9.161}$$

Now let $f = X(t)Y(t)$. We have the associated SDE:

$$\begin{aligned} d(X(t)Y(t)) = Y^2(t)dt + X(t)\left(-2\xi\omega Y(t) - \omega^2 X(t)\right)dt \\ + X(t)\sigma_Y(t)dB(t) + dX(t)dY(t) \end{aligned} \tag{9.162}$$

Since the last term $dX(t)dY(t) = d[X,Y](t)$ in the above RHS is zero (as the SDE for $X(t)$ has no diffusion term), we finally arrive at the following SDE for $X(t)Y(t)$:

$$d(X(t)Y(t)) = (Y^2(t) - 2\xi\omega Y(t)X(t) - \omega^2 X^2(t))dt + \sigma_Y(t)X(t)dB(t) \tag{9.163}$$

Finally, let $f = Y^2(t)$, so that we may write:

$$\begin{aligned} dY^2(t) &= 2Y(t)\left(-2\xi\omega Y(t) - \omega^2 X(t)\right)dt + 2Y(t)\sigma_Y(t)dB(t) + {\sigma^2}_Y(t)dt \\ \Rightarrow dY^2(t) &= \left(-4\xi\omega Y^2(t) - 2\omega^2 X(t)Y(t) + \sigma_Y^2(t)\right)dt + 2Y(t)\sigma_Y(t)dB(t) \end{aligned} \tag{9.164}$$

Note that the SDEs for $X(t)Y(t)$ and $Y^2(t)$ have Ito diffusion processes, which evolve as zero-mean martingales. Hence, supposing that we wish to find the second-moment statistics for the SDOF oscillator, we can take expectations of Equations 9.160, 9.163 and 9.164 to arrive at the following system of ODEs:

$$\begin{aligned} \frac{d}{dt}[EX^2(t)] &= 2E[X(t)Y(t)] \\ \frac{d}{dt}E[X(t)Y(t)] &= -2\xi\omega E[X(t)Y(t)] - \omega^2 E\left[X^2(t)\right] + E\left[Y^2(t)\right] \\ \frac{d}{dt}E[Y^2(t)] &= -4\xi\omega E\left[Y^2(t)\right] - 2\omega^2 E[X(t)Y(t)] + \sigma_Y^2(t) \end{aligned} \tag{9.165a–c}$$

Equation 9.165 constitute a closed system of linear ODEs in the second moments $E[X^2(t)]$, $E[X(t)Y(t)]$ and $E[Y^2(t)]$ that may be solved (analytically or through numerical integration), provided known initial conditions are supplied. In the special case of the SDOF system response being stationary, we have $\frac{d}{dt}E[X^2] = \frac{d}{dt}E[XY] = \frac{d}{dt}E[Y^2] = 0$ and $\sigma_Y^2(t) = \sigma_Y^2$, so that the above set of ODEs reduces to one of three linear algebraic equations in as many unknowns. Solving this system yields:

$$E[X^2] = E[Y^2]/\omega^2, \quad E[Y^2] = \sigma_Y^2/4\xi\omega, \quad E[XY] = 0 \tag{9.166}$$

The principles of deriving Ito's formula for still higher dimensions with the diffusion vector being contributed by a set of independently evolving scalar (and standard) Brownian motion processes, $\vec{B}(t) = \{B_1(t), B_1(t), \ldots, B_1(t)\}^T$ remain precisely the same. Hence, we state the formula directly in what follows. Thus, consider a $p \geq 2$ dimensional dynamical system driven by the above q-dimensional Brownian vector $\vec{B}(t)$ with the SDEs in the following indicial form:

$$dX_i(t) = a_i(\vec{X}, t)dt + \sum_{j=1}^{q} \sigma_{ij}(t)dB_j(t), \quad i = 1, 2 \ldots, p \tag{9.167}$$

where $\vec{X} = \{X_1(t), X_2(t), \ldots., X_p(t)\}^T$ denotes the response vector and $\sigma = [\sigma_{ij}]$ is a $p \times q$-dimensional (generally rectangular) diffusion matrices. Also note that $\vec{a} = \{a_1(\vec{X},t), a_1(\vec{X},t), \ldots., a_1(\vec{X},t)\}^T$ defines the drift vector for the above system of SDEs and that, in the context of MDOF systems, p is usually an even number with $\frac{p}{2}$ equations corresponding to $\frac{p}{2}$ displacement components) having no diffusion terms. Here, F_t denotes the σ-field generated by the Brownian vector $\vec{B}(s)$, $s \le t$ and $\sigma_{ij}(t)$ (if random) are adapted to F_t. For linear dynamical systems of concern in this book, $a_i(\vec{X},t)$ must be linear in $\vec{X}$, even though Ito's formula has the same form for nonlinear drift terms and for multiplicative diffusion coefficients $\sigma_{ij}(t) = \sigma_{ij}(\vec{X},t)$ as well. An additional fact that needs to be used in deriving the formula in this case is the quadratic covariation of the scalar components of $\vec{B}(t)$, which we state below.

9.8.6.6 Quadratic Covariation of Brownian Components

For any two Brownian motion components $B_k(t)$ and $B_l(t)$, $k, l \in [1,q]$ and $k \ne l$, we have $[B_k, B_l](t) = 0$ in $L^2(P)$. To see this, once more consider a partition $\mathcal{T}^N$ of $[0,t]$ and, using the definition of quadratic covariation, construct the sum:

$$Q_N = \sum_{j=0}^{N-1} \left(B_k(t_{j+1}) - B_k(t_j)\right)\left(B_l(t_{j+1}) - B_l(t_j)\right) \tag{9.168}$$

The independence of $B_k(t)$ and $B_l(t)$ implies that $E[Q_N] = 0$. Moreover, using the fact that variance of Q_N is the sum of variances of the individual terms (which follows via the independence of Brownian increments), we get:

$$\begin{aligned}\text{Var}\, Q_N &= \sum_{j=0}^{N-1} E\left[\left(B_k(t_{j+1}) - B_k(t_j)\right)^2\right] E\left[\left(B_l(t_{j+1}) - B_l(t_j)\right)^2\right] \\ &= \sum_{j=0}^{N-1} (t_{j+1} - t_j)^2 \le \delta_N t_j;\ \ \delta_N := \max_j (t_{j+1} - t_j)\end{aligned} \tag{9.169}$$

Hence, $\text{Var}\, Q_N = E[Q_N^2] \to 0$ as $\delta_N \to 0$, showing that the claim is valid in $L^2(P)$. Now the statement of Ito's formula goes as follows: Let $f(\vec{X})$ be a scalar C^2 function of its arguments so that $f(\vec{X})$ is an Ito diffusion process. Then the SDE for $f(\vec{X})$ is given by:

$$df(X_1, X_2, \ldots X_p) = \sum_{j=1}^{p} \frac{\partial f(\vec{X})}{\partial X_i} d\vec{X}(t) + \sum_{j,k=1}^{p} \frac{1}{2} \frac{\partial^2 f(\vec{X})}{\partial X_i \partial X_j} d[X_i, X_j](t) \tag{9.170}$$

Note that the quadratic covariation of Brownian components must be used whilst evaluating $d[X_i, X_j](t)$.

Example 9.16: Integration by parts

Let $p = 2$ and the component processes X_1, X_2 be driven by independent Brownian motion components $B_1(t)$ and $B_2(t)$. If $f(X_1, X_2) = X_1 X_2$, then we get:

$$d(X_1, X_2)(t) = X(t)dY(t) + Y(t)dX(t) \tag{9.171}$$

The above is true as $d(X_1, X_2)(t) = 0$, which in turn is implied by the fact that $d(B_1, B_2)(t) = 0$. Remark: If f is an explicit function of t, then the SDE of f remains the same as in the RHS of Equation 9.170 except that the drift coefficient contains the additional term $\frac{\partial f}{\partial t}$. This readily follows from a Taylor expansion of f.

9.8.7 Computing the Response Moments

In Example 9.15, response moments of the second order for an SDOF oscillator under white-noise excitation were derived directly using Ito's formula. We now formalise this procedure in the context of the p-dimensional SDE given by Equation 9.167. Thus, considering a scalar function $f(\vec{X}t)$ that is at least twice differentiable in the first argument, we recast Ito's formula as:

$$df(\vec{X}(t), t) = \left(\frac{\partial f(\vec{X}(t), t)}{\partial t} + \boldsymbol{L}_t(f(\vec{X}(t), t)) \right) dt + \frac{\partial f(\vec{X}(t), t)}{\partial \vec{X}} \sigma(t) d\vec{B}(t) \quad (9.172a)$$

where the operator $\boldsymbol{L}_t$ is defined as:

$$\boldsymbol{L}_t f = \sum_{i=1}^{p} a_i(\vec{X}(t), t) \frac{\partial f}{\partial X_i} + \frac{1}{2} \sum_{i,j=1}^{p} \sum_{l=1}^{q} \sigma_{il} \sigma_{jl} \frac{\partial^2 f}{\partial X_i \partial X_j} \quad (9.172b)$$

Note that, in Equation 9.172a, $\frac{\partial f(\vec{X}(t),t)}{\partial \vec{X}} = \left\{ \frac{\partial f}{\partial X_1}, \frac{\partial f}{\partial X_1}, \ldots, \frac{\partial f}{\partial X_p} \right\}$, $\sigma(t)$ is a $p \times q$-dimensional diffusion matrix and $\vec{B}(t)$ is a q-dimensional Brownian motion vector. The integral form of Equation 9.172 may be written as:

$$f(\vec{X}(t), t) = f(\vec{X}(0), 0) + \int_0^t \left(\frac{\partial f(\vec{X}(u), u)}{\partial u} + \boldsymbol{L}_u(f(\vec{X}(u), u)) \right) du$$
$$+ \int_0^t \frac{\partial f(\vec{X}(u), u)}{\partial \vec{X}} \sigma(u) d\vec{B}(u) \quad (9.173)$$

where the last term (Ito's integral) on the RHS of the above equation is a zero-mean martingale (as $\left| \frac{\partial f}{\partial \vec{X}} \right|$ is bounded owing to the stipulated differentials of f with respect to the components of $\vec{X}$). By taking expectations on both sides of Equation 9.173, we arrive at what is referred to as Dynkin's formula:

$$E[f(\vec{X}(t), t)] = f(\vec{X}(0), 0) + E\left[\int_0^t (\boldsymbol{L}_u f(X(u), u) + \frac{\partial f(X(u), u)}{\partial u}) du \right] \quad (9.174)$$

Example 9.17: Solution of the SDE corresponding to an MDOF system in terms of the response moments

Consider an MDOF system as shown in Figure 9.7. It is a 2-dof system represented by the response vector $(x_1, x_2)^T$ and excited by zero mean white-noise processes $W_1(t)$ and $W_2(t)$, as shown in the figure.

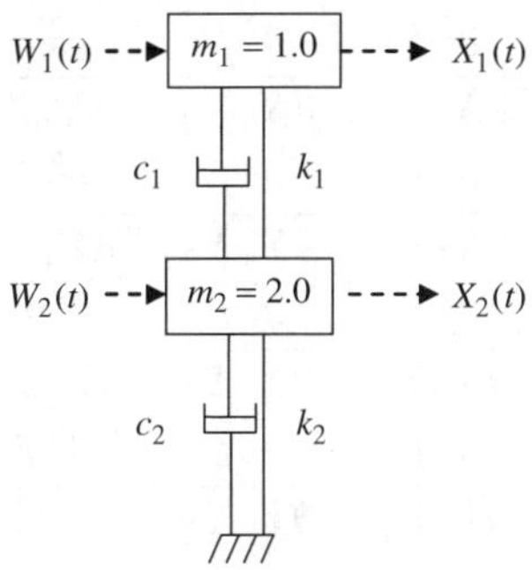

Figure 9.7 MDOF system under stochastic input; stiffnesses, $k_1 = 1200$, $k_2 = 1800$, damping values, $c_1 = 5.0$, $c_2 = 2.0$

Defining the two velocities $y_1(t) := \dot{x}_1(t)$ and $y_2(t) := \dot{x}_2(t)$, we can write the discrete form of SDEs for the MDOF system as follows:

$$\begin{aligned} dX_1(t) &= Y_1(t)dt \\ dY_1(t) &= (-C_1Y_1(t) + C_2Y_2(t) - K_1X_1(t) + K_2X_2(t))dt + \sigma_2(t)dB_1(t)/m_1 \\ dX_2(t) &= Y_2(t)dt \\ dY_2(t) &= (C_3Y_1(t) - C_4Y_2(t) + K_3X_1(t) - K_4X_2(t))dt + \sigma_4(t)dB_2(t)/m_2 \end{aligned} \quad (9.175\text{a–d})$$

Here, $K_1 = 1200.0$, $K_2 = -1200.0$, $K_3 = -600.0$ and $K_4 = 1500.0$. And, $C_1 = 5.0$, $C_2 = -5.0$, $C_3 = -2.5$ and $C_4 = 3.5$. The above system of SDEs is four-dimensional ($p = 4$) and the Brownian motion vector $(B_1, B_2)^T$is two-dimensional ($q = 2$). Thus, formally, $W_1(t)dt \equiv dB_1(t)$ and $W_2(t)dt \equiv dB_2(t)$. $\sigma(t)$, the $p \times q$ matrix of diffusion coefficients is $\begin{bmatrix} 0 & \sigma_2 & 0 & 0 \\ 0 & 0 & 0 & \sigma_4 \end{bmatrix}^T$, where $\sigma_2 = 1.0$ and $\sigma_4 = 0.5$ are the assumed intensities of the two white-noise processes $W_1(t)$ and $W_2(t)$, respectively. The vector of drift coefficients is:

$$\begin{aligned} a(t) = \{&Y_1(t), -C_1Y_1(t) + C_2Y_2(t) + K_1X_1(t) - K_2X_1(t), Y_2(t), C_3Y_1(t) \\ &-C_4Y_2(t) - K_3X_1(t) + K_4X_2(t)\}^T \end{aligned} \quad (9.176)$$

Here, we may utilise either Dynkin's formula in Equation 9.174 (which directly yields the ODEs for the moments) or Ito's formula in Equation 9.158 (which yield the associated SDE-s and thus requires the additional step of taking expectations on both sides) to solve for the second-order moments. Towards this, we have to choose appropriate scalar functions $f(\vec{X}:=(X_1, Y_1, X_2, Y_2)^T, t)$, which in this case are the second-order monomials in terms of the response variables, that is $f = X_1^2, Y_1^2, X_1Y_1, X_2^2, Y_2^2, X_2Y_2, X_1X_2, Y_1Y_2, X_1Y_2$ and X_2Y_1 and write Ito's (or Dynkin's) formula for each such f in turn. First, let $f = X_1^2$ and noting that f is at least twice differentiable in $\vec{X}$, we have:

$$\frac{\partial f}{\partial \vec{X}} = \left(\frac{\partial f}{\partial X_1}, \frac{\partial f}{\partial Y_1}, \frac{\partial f}{\partial X_2}, \frac{\partial f}{\partial Y_2}\right) = (2X_1, 0, 0, 0) \text{ and}$$

$$\left[\frac{\partial^2 f}{\partial \boldsymbol{X}_i \boldsymbol{X}_j}\right] = \begin{bmatrix} \frac{\partial^2 f}{\partial \boldsymbol{X}_1^2} & \frac{\partial^2 f}{\partial \boldsymbol{X}_1 \boldsymbol{Y}_1} & \frac{\partial^2 f}{\partial \boldsymbol{X}_1 \boldsymbol{X}_2} & \frac{\partial^2 f}{\partial \boldsymbol{X}_1 \boldsymbol{Y}_2} \\ \frac{\partial^2 f}{\partial \boldsymbol{Y}_1 \boldsymbol{X}_1} & \frac{\partial^2 f}{\partial \boldsymbol{Y}_1^2} & \frac{\partial^2 f}{\partial \boldsymbol{Y}_1 \boldsymbol{X}_2} & \frac{\partial^2 f}{\partial \boldsymbol{Y}_1 \boldsymbol{Y}_2} \\ \frac{\partial^2 f}{\partial \boldsymbol{X}_2 \boldsymbol{X}_1} & \frac{\partial^2 f}{\partial \boldsymbol{X}_2 \boldsymbol{Y}_1} & \frac{\partial^2 f}{\partial \boldsymbol{X}_i \boldsymbol{X}_2} & \frac{\partial^2 f}{\partial \boldsymbol{X}_2 \boldsymbol{Y}_2} \\ \frac{\partial^2 f}{\partial \boldsymbol{Y}_2 \boldsymbol{X}_1} & \frac{\partial^2 f}{\partial \boldsymbol{Y}_2 \boldsymbol{Y}_1} & \frac{\partial^2 f}{\partial \boldsymbol{Y}_2 \boldsymbol{X}_2} & \frac{\partial^2 f}{\partial \boldsymbol{Y}_2^2} \end{bmatrix} = \begin{bmatrix} 2 & 0 & 0 & 0 \\ 0 & 0 & 0 & 0 \\ 0 & 0 & 0 & 0 \\ 0 & 0 & 0 & 0 \end{bmatrix} \tag{9.177}$$

Ito's formula in Equation 9.172 gives SDE for $X_1^2(t)$ as:

$$d(X_1^2(t)) = 2X_1 Y_1 dt \tag{9.178}$$

If $f = Y_1^2$, then:

$$\frac{\partial f}{\partial \vec{X}} = \left(\frac{\partial f}{\partial X_1}, \frac{\partial f}{\partial Y_1}, \frac{\partial f}{\partial X_2}, \frac{\partial f}{\partial Y_2}\right) = (0, 2Y_1, 0, 0) \text{ and } \left[\frac{\partial^2 f}{\partial \boldsymbol{X}_i \boldsymbol{X}_j}\right] = \begin{bmatrix} 0 & 0 & 0 & 0 \\ 0 & 2 & 0 & 0 \\ 0 & 0 & 0 & 0 \\ 0 & 0 & 0 & 0 \end{bmatrix} \tag{9.179}$$

The SDE for Y_1^2 is:

$$\begin{aligned} d(Y_1^2(t)) &= 2Y_1\left(-C_1 Y_1(t) + C_2 Y_2(t) - K_1 X_1(t) + K_2 X_2(t)\right) dt + \frac{1}{2}\left(2\sigma_2^2(t)\right) dt \\ &\quad + 2Y_1(t)\sigma_2(t) dB_1(t) \\ &= 2Y_1\left(-C_1 Y_1(t) + C_2 Y_2(t) - K_1 X_1(t) + K_2 X_2(t)\right) dt + \sigma_2^2(t) dt \\ &\quad + 2Y_1(t)\sigma_2(t) dB_1(t) \end{aligned} \tag{9.180}$$

If $f = X_1 Y_1$, we have:

$$\frac{\partial f}{\partial \vec{X}} = \left(\frac{\partial f}{\partial X_1}, \frac{\partial f}{\partial Y_1}, \frac{\partial f}{\partial X_2}, \frac{\partial f}{\partial Y_2}\right) = (Y_1, X_1, 0, 0) \text{ and } \left[\frac{\partial^2 f}{\partial \boldsymbol{X}_i \boldsymbol{X}_j}\right] = \begin{bmatrix} 0 & 1 & 0 & 0 \\ 1 & 0 & 0 & 0 \\ 0 & 0 & 0 & 0 \\ 0 & 0 & 0 & 0 \end{bmatrix} \tag{9.181}$$

The SDE for $X_1 Y_1$ is:

$$d(X_1 Y_1) = \left[Y_1^2 + X_1\left(-C_1 Y_1(t) + C_2 Y_2(t) - K_1 X_1(t) + K_2 X_2(t)\right)\right] dt \tag{9.182}$$

Similar steps yield the SDEs for the other response variables as:

$$\begin{aligned} d\left(X_2^2(t)\right) &= 2X_2 Y_2 dt, \\ d\left(Y_2^2(t)\right) &= 2Y_2\left(C_3 Y_1(t) - C_4 Y_2(t) + K_3 X_1(t) - K_4 X_2(t)\right) dt + \sigma_4^2(t) dt \\ &\quad + 2Y_2(t)\sigma_4(t) dB_2(t), \end{aligned}$$

$$
\begin{aligned}
d\left(X_2Y_2\right) &= \left[Y_2^2 + X_2\left(C_3Y_1(t) - C_4Y_2(t) + K_3X_1(t) - K_4X_2(t)\right)\right]dt,\\
d\left(X_1X_2\right) &= \left(X_2Y_1 + X_1Y_2\right)dt,\\
d\left(Y_1Y_2\right) &= \begin{bmatrix} Y_2\left(-C_1Y_1(t) + C_2Y_2(t) - K_1X_1(t) + K_2X_2(t)\right) \\ \text{[4pt]} + Y_1\left(C_3Y_1(t) - C_4Y_2(t) + K_3X_1(t) - K_4X_2(t)\right)\end{bmatrix}dt,\\
d\left(X_1Y_2\right) &= \left[Y_1Y_2 + X_1\left(C_3Y_1(t) - C_4Y_2(t) + K_3X_1(t) - K_4X_2(t)\right)\right]dt \text{ and}\\
d\left(X_2Y_1\right) &= \left[Y_1Y_2 + X_2\left(C_3Y_1(t) - C_4Y_2(t) + K_3X_1(t) - K_4X_2(t)\right)\right]dt
\end{aligned}
$$

(9.183–9.189)

By taking expectation on both sides of SDEs (Equations 9.178, 9.180, 9.182–9.189), we obtain the following system of coupled ODEs for the second-order statistics of the MDOF system (as an exercise, the reader may try obtaining them directly using Dynkin's formula):

$$
\begin{aligned}
\frac{dE\left[X_1^2(t)\right]}{dt} &= 2E\left[X_1Y_1\right],\\
\frac{dE\left[Y_1^2(t)\right]}{dt} &= -2C_1E\left[Y_1^2\right] - 2K_1E\left[X_1Y_1\right] + 2C_2E\left[Y_1Y_2\right] + 2K_2E\left[Y_1X_2\right] + \sigma_2^2,\\
\frac{dE\left[X_1Y_1\right]}{dt} &= E\left[Y_1^2\right] - C_1E\left[X_1Y_1\right] - K_1E\left[X_1^2\right] + C_2E\left[X_1Y_2\right] + K_2E\left[X_1X_2\right],\\
\frac{dE\left[X_2^2(t)\right]}{dt} &= 2E\left[X_2Y_2\right],\\
\frac{dE\left[Y_2^2(t)\right]}{dt} &= -2C_4E\left[Y_2^2\right] - 2K_4E\left[X_2Y_2\right] + 2C_3E\left[Y_1Y_2\right] + 2K_3E\left[X_1Y_2\right] + \sigma_4^2,\\
\frac{dE\left[X_2Y_2\right]}{dt} &= E\left[Y_2^2\right] - C_4E\left[X_2Y_2\right] - K_4E\left[X_2^2\right] + C_3E\left[X_2Y_1\right] + K_3E\left[X_1X_2\right],\\
\frac{dE\left[X_1X_2\right]}{dt} &= E\left[X_2Y_1\right] + E\left[X_1Y_2\right],\\
\frac{dE\left[Y_1Y_2\right]}{dt} &= -C_1E\left[Y_1Y_2\right] - K_1E\left[X_1Y_2\right] + C_2E\left[Y_2^2\right] + K_2E\left[X_2Y_2\right]\\
&\quad - C_4E\left[Y_1Y_2\right] - K_4E\left[X_2Y_1\right] + C_3E\left[Y_1^2\right] + K_3E\left[X_1Y_1\right],\\
\frac{dE\left[X_1Y_2\right]}{dt} &= E\left[Y_1Y_2\right] + C_3E\left[X_1Y_1\right] - C_4E\left[X_1Y_2\right]\\
&\quad + K_3E\left[X_1^2\right] - K_4E\left[X_1X_2\right] \quad \text{and}\\
\frac{dE\left[X_2Y_1\right]}{dt} &= E\left[Y_1Y_2\right] + C_3E\left[X_2Y_1\right] - C_4E\left[X_2Y_2\right] + K_3E\left[X_1X_2\right] - K_4E\left[X_2^2\right]
\end{aligned}
$$

(9.190–9.199)

The above ODEs are solved (subject to identically zero initial conditions) to get the histories of the second-order moments, some of which are shown in Figures 9.8 and 9.9.

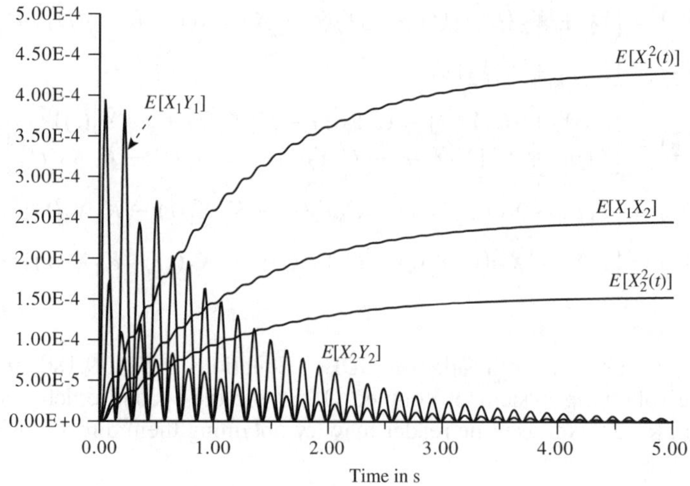

Figure 9.8 Time histories of response moments of an MDOF system by Ito's formula (Equation 9.172); $E[X_1^2(t)]$, $E[X_2^2(t)]$, $E[X_1X_2]$, $E[X_1Y_1]$ and $E[X_2Y_2]$

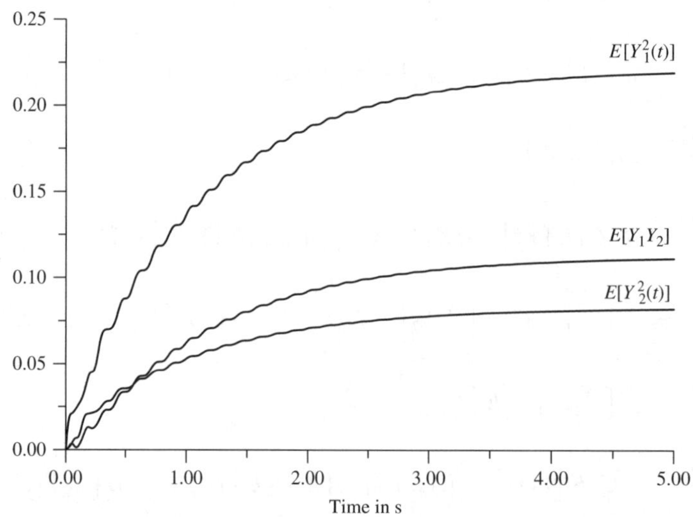

Figure 9.9 Time histories of response moments of an MDOF system by Ito's formula (Equation 9.172); $E[Y_1^2(t)]$, $E[Y_2^2(t)]$, $E[Y_1Y_2]$

It is interesting to note from Equation 9.173 that if $\boldsymbol{L}_u f(X(u), u) + \frac{\partial f(X(u),u)}{\partial u} = 0$ identically for all $u \in [0, t]$, then the second term on the RHS of the equation vanishes so that $f(\vec{X}(t), t)$ becomes a martingale with mean $E[f(\vec{X}(0), 0)]$. This observation yields yet another useful way of computing the expectation, namely Kolmogorov's backward equation approach as stated below:

Let $f(\vec{X}(t), t)$ solve the backward PDE:

$$L_t f(\vec{X}(t), t) + \frac{\partial f(\vec{X}(t), t)}{\partial t} = 0 \quad \text{with} \quad f(\vec{X}(t), T)) = g(\vec{X}) \tag{9.200}$$

Then, subject to an adequate smoothness of f as prescribed earlier, we have the following identity:

$$f(\vec{X}(t), t) = E[g(\vec{X}(T))|\vec{X}(t) = \vec{X}] \tag{9.201}$$

To see that the above assertion is true, we first note via the martingale property of $f(\vec{X}(t), t)$ that:

$$E[f(\vec{X}(T), T)|F_t] = f(\vec{X}(t), t) \tag{9.202}$$

To proceed further, we need what is referred to as the Markov property of $\vec{X}(t)$. Consider a stochastic process $\vec{X}(t)$ with the associated σ-field F_t. The process is called (strongly) Markov if, for $s < t$, we have:

$$E[\vec{X}(t)|F_s] = E[\vec{X}(t)|\vec{X}(s)]$$

This means that the conditional expectation given some past history is the same as that given the most recent past. It can be shown (Øksendal, 2003) that the solution $\vec{X}(t)$ to the SDE (Equation 9.167) is indeed Markovian.

Now, by the Markov property of $\vec{X}(t)$ and Equation 9.56, we get $E[f(\vec{X}(T), T)| X(t) = X] = f(Xt)$. However since $f(X(T), T) = g(X(T))$, the assertion follows.

9.8.8 Time Integration of SDEs

Here, we are concerned only with linear SDEs corresponding to linear structural dynamical systems. Referring once more to the p-dimensional SDE in Equation 9.167, we see that if $a_i(\vec{X}t)$ is linear in $\vec{X}$, then the vector $\{\boldsymbol{a}\} = \{a_i\}$ admits the decomposition $\{\boldsymbol{a}\} = [\boldsymbol{A}]\{\vec{X}\}$, where $\boldsymbol{A}(t)$ is a $p \times p$-dimensional (and possibly time-invariant) matrix. We have the following linear system of SDEs (with only additive noises):

$$d\vec{X}(t) = \boldsymbol{A}(t)\vec{X}(t)dt + \sigma(t)d\vec{B}(t) \tag{9.203}$$

Suppose we can find a so-called fundamental solution matrix (FSM) $Q(t, t_0)$ satisfying the matrix-valued ODE:

$$\frac{dQ(t, t_0)}{dt} = \boldsymbol{A}(t)Q(t, t_0) \tag{9.204}$$

where t_0 is the time at which the initial conditions are specified with $Q(t, t_0) = I$, the $p \times p$-dimensional identity matrix. In the special case of the drift coefficient matrix $\boldsymbol{A}(t) = \boldsymbol{A}$ being time-invariant, we have:

$$Q(t, t_0) = \exp(\boldsymbol{A}(t - t_0)) \tag{9.205}$$

where the above matrix exponential may be evaluated using, say, an eigenvalue decomposition or Taylor expansion. In the latter case, we have:

$$\exp(\boldsymbol{A}(t-t_0)) = I + \boldsymbol{A}(t-t_0) + \frac{1}{2}\boldsymbol{A}^2(t-t_0)^2 + \frac{1}{6}\boldsymbol{A}^3(t-t_0)^3 + \cdots\cdots\cdots \quad (9.206)$$

Once we find $Q(t, t_0)$, the solution to the SDE (Equation 9.203) may be written as:

$$\vec{X}(t) = Q(t, t_0)\left[\vec{X}(t_0) + \int_{t_0}^{t} Q^{-1}(s, t_0)\sigma(s)d\vec{B}(s)\right] \quad (9.207)$$

In addition, if Equation 9.203 has an additional deterministic forcing vector $\vec{G}(t)dt$ appearing in the drift terms, then the solution $\vec{X}(t)$ takes the form:

$$\vec{X}(t) = Q(t, t_0)\left[\vec{X}(t_0) + \int_{t_0}^{t} Q^{-1}(s, t_0)\vec{G}(s)ds + \int_{t_0}^{t} Q^{-1}(s, t_0)\sigma(s)d\vec{B}(s)\right] \quad (9.208)$$

The derivation of $Q(t, t_0)$ for the case of $\boldsymbol{A}(t)$ being an explicit function of t requires a Lie-group-based approach and is beyond the scope of this book.

If one adopts a Taylor expansion in Equation 9.206 to evaluate $Q(t, t_0)$ for the case of $\boldsymbol{A}$ being time invariant, the expansion may converge slowly as $(t - t_0)$ increases. Accordingly, it would be more convenient to use expansion in Equation 9.206 recursively over a partition $\mathcal{T}^N = \{t_j | j \in [0, N]\}$ of $[0, t]$ such that we have:

$$\vec{X}(t_{j+1}) = Q(t_{j+1}, t_j)\left[\vec{X}(t_j) + \int_{t_j}^{t_{j+1}} Q^{-1}(t, t_j)\sigma(t)d\vec{B}(t)\right] \quad (9.209a)$$

where j^{th} FSM $Q(t_{j+1}, t_j)$ is given by:

$$Q(t_{j+1}, t_j) = I + Ah_j + \frac{1}{2}A^2h_j^2 + \frac{1}{6}A^3h_j^3 + \cdots \quad (9.209b)$$

Here, $h_j = t_{j+1} - t_j$. Thus, Equation 9.209 provides a numerical recursive strategy for time integration of a class of linear SDEs.

In what follows, we briefly describe yet another popularly used recursive time integration scheme for linear SDEs (Equation 9.203), what is referred to as the Euler–Maruyama (EM) method (or simply Euler's method). As with other such time-integration methods, we use the partition $\mathcal{T}^N$ of $[0, t]$ and write the EM map for time stepping as:

$$\vec{X}(t_{j+1}) = \vec{X}(t_j) + A(t_j)\vec{X}(t_j)h_j + \sigma(t_j)\{\vec{B}(t_{j+1}) - \vec{B}(t_j)\},\ j = 0, 1, ..N-1 \quad (9.210)$$

where $\vec{X}(0)$ is specified as the initial condition. Note that the RHS of the above mapping contains only $\vec{X}(t_j)$ (which is known within the recursion framework) and not $\vec{X}(t_{j+1})$ (which needs to be computed). Hence, the above method is explicit (see Chapter 8). A drift-implicit version of the EM method may be written as:

$$\vec{X}(t_{j+1}) = \vec{X}(t_j) + A(t_{j+1})\vec{X}(t_{j+1})h_j + \sigma(t_j)\left\{\vec{B}(t_{j+1}) - \vec{B}(t_j)\right\} \quad (9.211a)$$

A straightforward rearrangement of terms yields:

$$\vec{X}(t_{j+1}) = \left[I + A(t_{j+1})h_j\right]^{-1}\left[\vec{X}(t_j) + \sigma(t_j)\left\{\vec{B}(t_{j+1}) - \vec{B}(t_j)\right\}\right] \tag{9.211b}$$

How good is the quality of the approximation generated by the explicit or implicit EM method? In order to answer this question, we need to understand the stochastic Taylor expansion (STE) that is derivable from Ito's formula. The STE forms the basis for derivation of most of the numerical integration schemes for SDEs. Even though an exposition of the STE is beyond our current scope, it may be worthwhile to make a brief beginning, mainly to motivate further reading on these lines. Towards this, we go back to the example containing the SDOF system:

$$dX(t) = Y(t)dt, dY(t)(= -2\eta\omega Y(t) - \omega^2 X(t))dt + \sigma(t)dB(t) \tag{9.212}$$

Suppose that, given a partition $\mathcal{T}^N$ of the time axis of interest, we expand $X(t)$, $t \in [t_j, t_{j+1}]$ about $X(t_j)$. Then, we can write:

$$X(t_{j+1}) = X(t_j) + \int_0^t Y(t)dt \tag{9.213}$$

Now, using the integral form of Taylor's formula on $Y(t)$ based at time t_j, we can write:

$$X(t_{j+1}) = X(t_j) + \int_{t_j}^{t_{j+1}}\left[Y(t_j) + \int_{t_j}^{t}(-2\eta\omega Y(u) - \omega^2 X(u))du + \int_{t_j}^{t}\sigma(u)dB(u)\right]dt$$

or,

$$X(t_{j+1}) = X(t_j) + Y(t_j)h_j - 2\eta\omega\int_{t_j}^{t_{j+1}}\int_{t_j}^{t} Y(u)dudt - \omega^2\int_{t_j}^{t_{j+1}}\int_{t_j}^{t}(X(u))dudt$$
$$+\int_{t_j}^{t_{j+1}}\int_{t_j}^{t}\sigma(u)dB(u)dt \tag{9.214}$$

The last double integral on the RHS above, involving integration with respect to both $B(t)$ and t, is generally referred to as a multiple stochastic integral (MSI). Numerical evaluations of MSIs are generally a tough task, and this mainly explains why higher order numerical schemes for SDEs are difficult to derive. As with the explicit EM scheme, if we truncate the RHS by retaining only the first two terms and treating all the double integrals as part of the remainder, then it is important to assess the order of these terms in terms of the step size h_j. For instance, if $X(t)$, $Y(t)$ and $\sigma(t)$ are sufficiently smooth in t, (ensuring the boundedness of their appropriate derivatives), then it is readily seen that $\int_{t_j}^{t_{j+1}}\int_{t_j}^{t} Y(u)dudt$ and $\int_{t_j}^{t_{j+1}}\int_{t_j}^{t}(X(u))dudt$ are $O(h_j^2)$. For the third integral $\int_{t_j}^{t_{j+1}}\int_{t_j}^{t}\sigma(u)dB(u)dt$, we must recall the identity $(\Delta B)^2 = \Delta t$ in $L^2(P)$ (i.e. $(dB)^2 = dt$, or, $dB = \sqrt{dt}$, in a pathwise sense). It thus follows that the last MSI is of $O\left(\sqrt{h_j}\cdot h_j\right)$, that is $O\left(h_j^{3/2}\right)$, which provides the formal error order (local) for the explicit EM-based approximation of the displacement component $X(t)$. We can perform a precisely similar exercise with the velocity $Y(t)$ to yield:

$$
\begin{aligned}
Y(t_{j+1}) = {} & Y(t_j) + Y(t_j)h_j + \left(-2\eta\omega Y(t_j) - \omega^2 X(t_j)\right) h_j + \sigma(t_j)\left\{B(t_{j+1}) - B(t_j)\right\} \\
& - 2\eta\omega \int_{t_j}^{t_{j+1}} \int_{t_j}^{t} \{-2\eta\omega Y(u) - \omega^2 X(u)\} du dt \\
& - 2\eta\omega \int_{t_j}^{t_{j+1}} \int_{t_j}^{t} \sigma(u) dB(u) dt - \omega^2 \int_{t_j}^{t_{j+1}} \int_{t_j}^{t} Y(u) du dt \\
& + \int_{t_j}^{t_{j+1}} \int_{t_j}^{t} \sigma(u) dB(u) dt
\end{aligned} \tag{9.215}
$$

The explicit EM method corresponds to retaining only the first three terms on the RHS above and treating all the double integrals (two of them being MSIs) as constituents of the remainder. Following the same arguments as for $X(t)$, we see that the remainder is of $O\left(h_j^{3/2}\right)$, which determines the local error order for the approximated velocity.

If a specific realisation of the stochastic excitation $F(t, \omega) := \sigma(t, \omega) dB(t, \omega)$, $\omega \in \Omega$, is given as the input force, then one of the time-integration schemes as discussed above may be directly used to obtain the corresponding realisation, say $X(t, \omega)$, of the approximated response. Such a solution, computed pathwise, is referred to as 'strong'. As an example, such a strong solution is appropriate to describe the motion of a shear frame under a prescribed earthquake acceleration history. Engineering interest is, however, often more focused in obtaining 'weak' solutions, which are only required to provide accurate information on the statistical moments (and possibly the transitional *pdf*) of a response functional and not a pathwise accurate description. In the context of time integration schemes, such weak solutions may be obtained via a Monte Carlo integration approach. Here, an ensemble of N_{samp} realisations $\{X(t, \omega_j); j \in [1, N_{\text{samp}}]\}$ are computed corresponding an ensemble of associated forcing $\{F(t, \omega_j); j \in [1, N_{\text{samp}}]\}$. Now, for a given function $f(X(t))$, we approximate its expectation as $E(f(X(t))) \cong \frac{1}{N_{\text{samp}}} \sum_{j=1}^{N_{\text{samp}}} f(X(t, \omega_j))$, which converges to $E(f(X(t)))$ as $N_{\text{samp}} \to \infty$ by the strong law of large numbers (see Appendix I). Interestingly, the jth element $F(t, \omega_j)$ of the forcing ensemble need not at all be pathwise correspondent to the jth realisation $\sigma(t, \omega_j) dB(t, \omega_j)$ of the true forcing. Rather, it suffices to ensure that the two ensembles $\{F(t, \omega_j)\}$ and $\{\sigma(t, \omega_j) dB(t, \omega_j)\}$ have the same statistical moments up to a certain order (for any t). The existing literature on weak integration approaches is indeed quite rich (see, for instance, Kloeden and Platen, 1999 for a detailed exposition).

9.9 Conclusions

In this chapter, an attempt has been made to lay down the basic framework, based on both 'classical' and a more rigorous stochastic calculus, approaches, to compute the response (either in terms of the statistical moments or pathwise) of linear structural dynamic systems under additive stochastic noises/excitations. The true power of the approach based on stochastic calculus lies in the ease it offers in treating nonlinear dynamic systems and a host of other problems in system identification or stochastic structural control applications. Whilst these aspects are not within the current scope, it is hoped that the material presented may prepare the young reader for a more insightful understanding of such relatively advanced topics.

Exercises

1. Given that x_j, $j = 1, 2, \ldots, n$ are normal random variables, we know that $\boldsymbol{X} = a_1 x_1 + a_2 x_2 + \cdots\cdots + a_n x_n$ is jointly normal for any $a_i \in \mathbb{R}$, i=1,2,...n. If the random variables x_j have zero mean and covariance matrix C, then show that
   (a) the joint characteristic function $\emptyset_X(\Omega) = \exp\left(-\frac{1}{2}\Omega C \Omega^T\right)$
   (b) $E[x_1 x_2 x_3 x_4] = C_{12}C_{34} + C_{13}C_{24} + C_{14}C_{23}$, where $C_{jk} = E[x_j x_k]$.

2. Random variables x_j, $j = 1, 2, \ldots, n$ are independent and uniformly distributed in the interval [0, 1] with means m_j and variances σ_j^2, $j = 1, 2, \ldots, n$. Compare the density of $\boldsymbol{X} = x_1 + x_2 + \cdots\cdots + x_n$ for $n = 2$, 3 and 4 with the normal approximation $f_X(\boldsymbol{x}) = \frac{1}{\sigma\sqrt{2\pi}} \exp\left(-\frac{1}{2}\left(\frac{x-m}{\sigma}\right)^2\right)$ with $m = m_1 + m_2 + \cdots + m_n$ and $\sigma^2 = \sigma_1^2 + \sigma_2^2 + \cdots + \sigma_n^2$.
3. Find the mean square velocity of a continuous uniform beam shown in Figure 9.10 when it is subjected to a stochastic forcing at $x = a$ with its mean zero and its PSD given by

$$\begin{aligned} S_F(\lambda) &= S_0, \quad -\lambda_c \le \lambda \le \lambda_c \\ &= 0, \ \lambda > |\lambda_c| \end{aligned}$$

4. For the SDOF system shown in Figure 9.11 below, find the mean square response of the relative displacement when the base acceleration $\ddot{g}(t)$ is idealised white noise with autocorrelation $R(\tau) = 2\pi S_0 \delta(\tau)$.
5. An N *dof* MDOF system under support excitation is represented by equations (6.26) and (6.34) of Chapter 6 in their original and modal form respectively. These equations in terms of the N-dimensional relative displacement vector z(t) are rewritten hereunder:

$$\boldsymbol{M}\ddot{z} + \boldsymbol{C}\dot{z} + \boldsymbol{K}z = -\boldsymbol{MT}\ddot{\boldsymbol{g}}(t) \qquad \text{(equations in the original } dofs\text{)} \qquad \text{(a)}$$

$$\ddot{q}_j + 2\xi_j\omega_j\dot{q}_j + \omega_j^2 q_j = p_j(t), \quad j = 1, 2, \ldots, m \ \text{(modal equations)} \qquad \text{(b)}$$

The reader would be familiar with the notations in the above equations. In particular, $p_j(t) = -\boldsymbol{\Phi}_j^T \boldsymbol{MT}\ddot{g}(t)$ Assume that $\ddot{g}(t)$ is a 1-dimensional and stationary stochastic

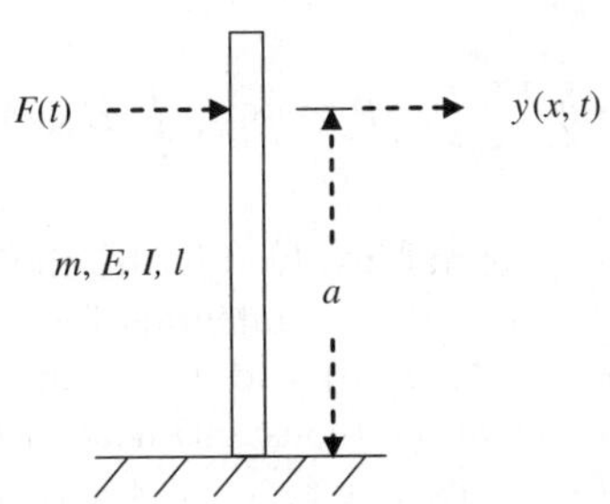

Figure 9.10 Continuous beam under stochastic excitation, $F(t)$

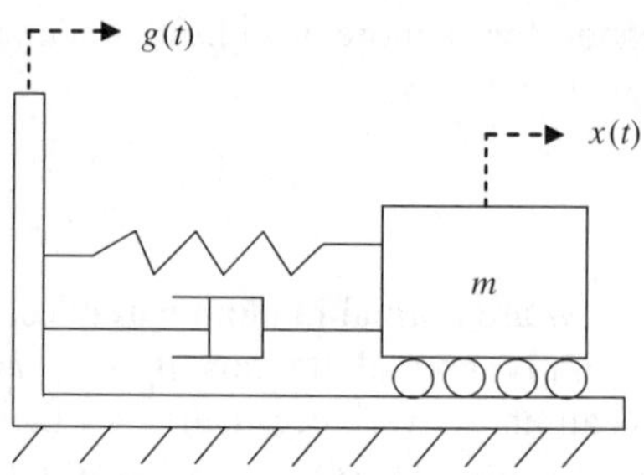

Figure 9.11 SDOF oscillator under random base excitation

process with a prescribed acceleration PSD function $S_{\ddot{g}\ddot{g}}(\lambda)$. Derive an expression for the response correlations, $R_{z_i z_i}(\tau), i = 1, 2, \ldots N$. (Refer to the notes below on SRSS and CQC modal summation rules that are derivable from a random vibration approach applied to MDOF systems).

Hint: For this example, $\boldsymbol{T} = \{1, 1, \ldots., 1\}_{N\times 1}$ take $P_j = -\boldsymbol{\Phi}_j^T \boldsymbol{M}\boldsymbol{T}$, which is the participation factor for the jth mode. Note that the PSD matrix $S_{qq}(\lambda)$ is given by equation (9.101). Specifically, $S_{q_j q_k}(\lambda) = P_j P_k H_j(\lambda) H_k^*(\lambda) S_{\ddot{g}\ddot{g}}(\lambda)$ from which the cross-correlation $R_{q_j q_k}(\tau)$ is obtained as:

$$R_{q_j q_k}(\tau) = \int_{\mathbb{R}} S_{q_j q_k}(\lambda) e^{i\lambda\tau} d\lambda = P_j P_k \int_{\mathbb{R}} H_j(\lambda) H_k^*(\lambda) S_{\ddot{g}\ddot{g}}(\lambda) e^{i\lambda\tau} d\lambda \qquad \text{(c)}$$

Finally, equation (9.99) gives the required result on the correlation matrix $R_{zz}(\tau)$ in terms of $R_{q_j q_k}(\tau), j, k = 1, 2.., m$.

Notes on SRSS and CQC modal summations: For designers, it may be necessary to have a quick (i.e. computationally cheap) access to such information as the upper bounds (or peak values) of system responses, e.g. displacements or derived quantities like stresses, reactions forces etc. For instance, the response spectrum method (see Chapter 6) of structural dynamic analysis uses the peak estimates of such response quantities for design purposes. These peak estimates are generally related to the associated mean square values (see Bibliography), derivable from the elementary random vibration theory. In the present case, the mean square values of the response *dof*s $z_i(t), i = 1, 2 \ldots, N$, are obtainable from equation (c) with $\tau = 0$ as:

$$\langle z_i(t) z_i(t) \rangle = R_{z_i z_i}(0) = \sum_{j=1}^{m} \sum_{k=1}^{m} \Phi_{ij} R_{q_j q_k}(0) \Phi_{ik}$$

$$= \sum_{j=1}^{m} \sum_{k=1}^{m} P_j P_k \Phi_{ij} \Phi_{ik} \int_{\mathbb{R}} H_j(\lambda) H_k^*(\lambda) S_{\ddot{g}\ddot{g}}(\lambda) d\lambda \qquad \text{(d)}$$

While the evaluations of the integral involved in the above equation are cumbersome, it is nevertheless possible to simplify the integrals for cases involving low damping. Specifically, for low values of ξ, $H_j(\lambda)$ and $H_k(\lambda)$ are narrowly peaked and hence $S_{\ddot{g}\ddot{g}}(\lambda)$ may be assumed to remain flat (constant) near these peaks. Thus, for the terms in equation (d) with $j = k$, $R_{q_j q_j}(0) = S_{\ddot{g}\ddot{g}}(\omega_j) \int_{\mathbb{R}} \left| H_j(\lambda) \right|^2 d\lambda$. By contour integration, we have $\int_{\mathbb{R}} \left| H_j(\lambda) \right|^2 d\lambda = \frac{\pi \omega_j}{2\omega_j}$. For the cross terms with $j \neq k$, two possibilities arise.

For well-separated modes, $H_j(\lambda)$ and $H_k(\lambda)$ are also separated implying that their contributions to the integral in equation (d) are negligible. On the other hand, for closely spaced modes, $H_j(\lambda)$ and $H_k(\lambda)$ typically have significant overlaps so that $S_{\ddot{g}\ddot{g}}(\omega_j)$ and $S_{\ddot{g}\ddot{g}}(\omega_k)$ may not differ much from each other. Hence the integral in equation (d) can be approximated as $S_{\ddot{g}\ddot{g}}(\omega_j)\int_{\mathbb{R}} H_j(\lambda)H_k^*(\lambda)d\lambda$. With $H_j(\lambda)$ given in equation (9.90), the integral $\int_{\mathbb{R}} H_j(\lambda)H_k^*(\lambda)d\lambda$ can be evaluated by contour integration [Der Kiureghian 1980] as:

$$\int_{\mathbb{R}} H_j(\lambda)H_k^*(\lambda)d\lambda = \frac{\pi}{2}\sqrt{\frac{\omega_j\omega_k}{\xi_j\xi_k}}\frac{8\sqrt{\xi_j\xi_k}\left(\xi_j+r\xi_k\right)r^{3/2}}{\left(1-r^2\right)^2+4\xi_j\xi_k r\left(1+r^2\right)+4\left(\xi_j^2+\xi_j^2\right)r^2} \tag{e}$$

Here $r=\omega_k/\omega_j$ with $\omega_j>\omega_k$. From the above discussion, we note that: (i) for a system with well separated modes, cross terms in the modal summation (d) may be ignored and the mean square values are approximately given by $R_{z_iz_i}(0)=\sum_{j=1}^m P_j^2\Phi_{ij}^2 S_{\ddot{g}\ddot{g}}\left(\omega_j\right)\int_{\mathbb{R}}\left|H_j(\lambda)\right|^2 d\lambda, i=1,2,\ldots,N$; and (ii) for a system with closely spaced modes, cross modal terms need to be retained in the summation. In either case, if the mean square values of the modal quantities can be related to their peak values [Clough and Penzien 1996], the modal summation corresponds to the SRSS (square root of the sum of the squares) rule in case (i) and the CQC (complete quadratic combination) rule in case (ii) to arrive at the peak response estimates for the system. These combination rules, whose developments predate the appearance of fast computing machines, have nevertheless been used extensively in the response spectrum method of structural analysis and design.

6. Show that
   (a) $B^2(t)$ is a martingale
   (b) $\exp\left(-uB(t)-\frac{u^2}{2}t\right)$ is a martingale where u is real.

7. For the 2-dof shear frame model in Example 9.17, numerically integrate the SDE by the explicit EM method with a uniform step size $h=0.001$ s. Towards generating Brownian increments $B(t_{j+1})-B(t_j)$ for the EM based velocity updates (see Equation 9.72), use an ensemble size with $N_{\text{samp}}=1000$. Compute the second-order moments using the Monte Carlo approach and compare the results obtained using Dynkin's formula.

Notations

a.e	Almost everywhere
a.s	Almost surely
$a(t,X)$	Drift term in SDE (Equation 9.103)
$\vec{a}$	Drift vector $\left(=\left\{a_1(\vec{X},t),a_1(\vec{X},t),\ldots,a_1(\vec{X},t)\right\}^T\right)$
$\boldsymbol{a}(t)$	Vector of drift coefficients (Equation 9.176)
$A_1A_2,\ldots$	Events ($\in\mathcal{F}$)
$A_1(t,w), A_2(t,w)$	Approximating summands for Brownian motion (Equation 9.112)
$\boldsymbol{A}(t)$	$p\times p$ dimensional matrix

$\mathcal{B}$	Borel σ algebra
$B_t, B(t), B(t, \omega)$	Brownian motion
c_1, c_2	Damping coefficients
C_1, C_2, C_3, C_4	Damping coefficients
C_{XY}	Covariance, $E[(X - \mu_X)(Y - \mu_Y)]$
C, C_m	Covariance matrix with $C_{jk} = E\left[\left(X_j - \mu_{X_j}\right)\left(X_k - \mu_{X_k}\right)\right]$
$C_{XX}(s, t)$	Autocovariance function of the random process, X
E	Event
$E(X)$	Expectation of a random variable, X
$E(X\|B)$	Conditional expectation
f	Probability density function (*pdf*) $\left(= \frac{dF_X(x)}{dx}\right)$
$f(X(t))$	Scalar-valued (at least twice differentiable) function of the random process, $X(t)$
$f\ (X(t), Y(t))$	Scalar-valued (at least twice differentiable) function of the random processes $X(t)$ and $Y(t)$
$f(\vec{X}, t)$	Scalar-valued (at least twice differentiable) function of vector random processes $\vec{X}$)
F	Cumulative (probability) distribution function (CDF)
$\boldsymbol{F}(t)$	Vector of nodal forces in a discrete system
$\mathcal{F}$	$\sigma-$algebra
$\mathcal{F}_i, i = 1, 2 .., k$	Borel sets in $\mathbb{R}^n$
F_t	The σ-algebra generated by the set of random variables $\{B(t_i, \omega)\|t_i \in [0, T]\}$
$F_j(t)$	Generalized force (modal force) in j^{th} mode
$F(t, \omega)$	Stochastic excitation
$g(\cdot)$	Function of a random variable
$\vec{G}(t)$	Deterministic forcing vector
h_j	Time step $(= t_{j+1} - t_j)$
$\boldsymbol{I}(t)$	Indicator function (Equation 9.110)
$G(t)$	Forcing function of an SDOF oscillator (Equation 9.79)
k_1, k_2	Stiffness coefficients
K_1, K_2, K_3, K_4	Stiffness coefficients
L_t	Differential operator (see Equation 9.172b)
$N(\mu, \sigma)$	Normal random variable
$\mathcal{N}_t$	Brownian filtration $(= \{F_s\|s \leq t\})$
N_{samp}	Number of realisations (Monte Carlo integration approach) p an integer
$P, \wp, \wp_B$	Probability measures
$P(A\|B)$ or $P_B(A)$	Conditional probability (probability of A conditioned on the subset B with $A, B \subset \mathcal{F}$)
q	An integer
$\boldsymbol{Q}(t, t_0)$	Fundamental solution matrix (Equation 9.204)
$R_X(\tau)$	Autocorrelation function of the stationary random process X
$R_{XX}(t, s)$	Autocorrelation function of the random process X
$S_{XX}(\lambda)$	Power spectral density of the stationary random process X
t	Time variable

$\mathcal{T}^N$	Partition of the interval $[0, t]$ $(= \{t_j \mid j \in [0, N]\})$
V_s	Velocity of moving load
$\boldsymbol{x}(t)$	Response vector in a discrete system
$\vec{X}$	Vector response process $(= \{X_1(t), X_2(t), \ldots, X_p(t)\}^T)$
X, Y	Random variables
$\boldsymbol{X}, \boldsymbol{Y}$	Vector (joint) random variable
$X(\omega, t)$	Random process
$[X, X](t)$	Quadratic variation of the random process, $X(t)$
$[X, Y](t)$	Quadratic covariation process of $X(t)$ and $Y(t)$
$W(t), W_1(t), W_2(t)$	1D white-noise process
δ_N	$:= \max_{j(t_{j+1}-t_j)}$
θ	Parameter in binomial distribution (Equation 9.38)
λ	Parameter in Poisson distribution (Equation 9.39)
μ_X	Mean of a random variable, X
ϕ	Null space
$\boldsymbol{\emptyset}_{\boldsymbol{X}}(\omega_1, \omega_2, \ldots, \omega_n)$	Characteristic function of vector random variable, $\boldsymbol{X}$
σ_X^2	Variance of a random variable, X
$\sigma_{X/Y}^2$	Conditional variance
$\sigma(t, X)$	Noise (diffusion) term in SDE (9.103)
$\boldsymbol{\sigma}$	$p \times q$-dimensional (generally rectangular) diffusion matrices $(= [\sigma_{ij}])$
τ	Parameter denoting a time shift
$\psi(\cdot)$	Moment generating function
ω	Outcome of a random experiment $(\in \Omega)$
ΔB_l	Brownian increments $(= B(t_{l+1}) - B(t_l))$
Ω	Sample space (discrete or continuous)

References

Ang, A. H-S. and Tang, W. H. (1984) *Probability Concepts in Engineering Planning and Design, Volume II Decision, Risk and Reliability*, John Wiley & Sons, New York, USA.

Clough, R.W. and Penzien, J. (1996) *Dynamics of Structures*, McGraw Hill, Singapore.

Crandall, S.H. and Mark, W.D. (1963) *Random Vibration in Mechanical Systems*, Academic Press, New York.

Davenport, A.F. (1961) Application of statistical concepts to the wind loading of structures. *Proceedings of the Institution of Civil Engineers*, **19**, 449–472.

Dodds, C.J. and Robson, J.D. (1973) The description of road surface roughness, *Journal of Sound and Vibration*, Vol. 31(2), pp. 175–183.

Feller, W. (1968) *An Introduction to Probability Theory and Its Applications*, Vol. 1, 3rd Edition, John Wiley & Sons, New York, USA.

Fryba, L. (1972) *Vibration of solids and structures under moving loads*, Noordhoff International, Groningen, Netherlands.

Harris, C.M. and Crede, C.E. (1961) *Shock and Vibration Handbook*, McGraw-Hill, New York, USA.

Holmes, P., Chaplin. J.R. and Tickell, R.G., (1983) Wave loading and structure response, *Design of Offshore Structures*, *Proceedings of a conference organized by Institution of Civil Engineers*, London, pp. 3–12.

Kaimal, J.C., Wyngaard, J.C., Izumi, Y. and Cote, O.R., (1972) Spectral characteristics of surface layer turbulence. *Quarterly Journal of the Royal Meteorological Society*, **98**, 563–589.

Klebaner, F.C. (1998) *Introduction to Stochastic Calculus with Applications*, Imperial College Press.

Kloeden, P.E. and Platen, E. (1999) *Numerical Solution of Stochastic Differential Equations*, Springer-Verlag, Berlin.

Kolmogorov, A.N. (1956) *Foundations of the Theory of Probability*, 2nd English Edition, Chelsea Publishing Company, New York, USA

Øksendal, B. (2003) *Stochastic Differential Equations, An Introduction with Applications*, Springer-Verlag, Heidelberg.

Lin, Y.K. (1967) *Probabilistic Theory of Structural Dynamics*, 2nd edn, McGraw-Hill, New York.

Liu, W.K., Belytschko, T. and Mani, A. (1986) Random fields. Finite element. *International Journal for Numerical Methods in Engineering*, Vol. 23, pp. 1831–1845.

Liu, W.K., Mani, A. and Belytschko, T. (1987) Finite elements in probabilistic mechanics, *Probabilistic Engineering Mechanics*, Vol. 2(4), pp. 201–213.

Menezes, R.C.R. and Schueller, G.I. (1996) On structural reliability assessment considering mechanical model uncertainties. *Proceeding of the International Workshop on Uncertainty: Models and Measures*, Lambrecht, Germany (eds H.G. Natke and Y. Ben Halm), pp. 173–186.

Melchers, R. E. (1999) *Structural Reliability and Prediction*, 2nd Edition, John Wiley & Sons, Chichester, UK.

Nigam, N.C. (1983) *Introduction to Random Vibrations*, MIT Press, Massachusetts.

Nigam, N.C. and Narayanan, S. (1994) *Applications of Random Vibrations*, Narosa Publishing House, New Delhi, India.

Papoulis, A. (1991) *Probability, Random Variables and Stochastic Processes*, 3rd edn, McGraw Hill.

Patadia, S. and Crafts, W.H. (1979) The investigation of locomotive dynamics via a large degree of freedom modelling, *Transactions of ASME, Journal of Engineering for Industry*, Vol. 101, pp. 397–402.1

Rudin, W. (1976) *Principles of Mathematical Analysis*, McGraw-Hill.

Schüeller, G.I. (2001) Computational stochastic mechanics – recent advances, *Computers and Structures*, Vol. 79, pp. 2225–2234.

Stefanou, G. (2009) The stochastic finite element method: Past, present and future, *Comput. Methods Appl. Mech. Eng*, Vol. 198, pp. 1031–1051.

Sudret, B. and Der Kiureghian, A. (2000) *Stochastic Finite Element Methods and Reliability – A State-of-the-Art Report*, Report no. UCB/SEMM-2000/8, University of California, Berkeley, USA.

Vanmarcke, E.H. (1983) *Random Fields. Analysis and Synthesis*, The MIT Press, Cambridge, MA, USA.

Williams, D. (1991) *Probability with Martingales*, Cambridge University Press.

Yang, C.Y. (1986) *Random Vibration of Structures*, John Wiley & Sons, Inc., New York.

Bibliography

On Monte Carlo Approach

Hammerseley, J.M. and Handscomb, D.C. (1964) *Monte Carlo Methods*, Chapman & Hall, London.

On SRSS and CQC Rules for Modal Summation

Der Kiureghian, A. (1980) A response spectrum method for random vibrations in *Report UCB/EERC - 80/15*, University of California, Berkeley, CA.

Appendix A

1. **Lipschitz domain**
   A Lipschitz domain (or a domain with a Lipschitz boundary) is a domain in Euclidean space with sufficiently regular boundary that can be thought of locally as being the graph of a Lipschitz continuous function. Many partial differential equations (PDEs) and variational problems are defined on the Lipschitz domain.
   Lipschitz continuous function: Given metric spaces $(X, d_X)$and(Y, d_Y), a function $f{:}X \to Y$ is Lipschitz continuous if there exists $K \geq 0$ such that, for all x_1 and x_2 in X, $d_Y\left(f(x_1)f(x_2)\right) \leq K d_X(x_1, x_2)$.
2. **To prove $\frac{1}{J}\frac{dJ}{dt} = \nabla \cdot v$**
   Proof:
   Here, $J = \det F$, where F is the deformation gradient given by:

$$F = \begin{bmatrix} \frac{\partial x_1}{\partial X_1} & \frac{\partial x_2}{\partial X_1} & \frac{\partial x_3}{\partial X_1} \\ \frac{\partial x_1}{\partial X_2} & \frac{\partial x_2}{\partial X_2} & \frac{\partial x_3}{\partial X_2} \\ \frac{\partial x_1}{\partial X_3} & \frac{\partial x_2}{\partial X_3} & \frac{\partial x_3}{\partial X_3} \end{bmatrix} \tag{A.1}$$

   Thus, we have:

$$J = e_{ijk}\frac{\partial x_1}{\partial X_i}\frac{\partial x_2}{\partial X_j}\frac{\partial x_3}{\partial X_k} \tag{A.2}$$

   Here, e_{ijk} is a permutation symbol given by:

$$\begin{aligned} e_{ijk} &= 1, \quad \textbf{if } i, j, k \textbf{ take values in the cyclic order}, \\ &= -1, \quad \textbf{if } i, j, k \textbf{ take values in acyclic order}, \\ &= 0, \quad \textbf{two or all of } i, j, k \textbf{ take the same value} \end{aligned} \tag{A.3}$$

   Now we can express $\frac{dJ}{dt}$ as:

$$\frac{dJ}{dt} = \sum_{m=1}^{3} e_{ijk}\frac{\partial v_m}{\partial X_i}\frac{\partial x_2}{\partial X_j}\frac{\partial x_3}{\partial X_k} \text{ with } \frac{\partial v_m}{\partial X_i} = \frac{d}{dt}\left(\frac{\partial x_m}{\partial X_i}\right)$$

Elements of Structural Dynamics: A New Perspective, First Edition. Debasish Roy and G Visweswara Rao.

$$= \sum_{m=1}^{3} e_{ijk} \frac{\partial v_m}{\partial x_1} \frac{\partial x_1}{\partial X_i} \frac{\partial x_2}{\partial X_j} \frac{\partial x_3}{\partial X_k} = e_{ijk} \frac{\partial x_1}{\partial X_i} \frac{\partial x_2}{\partial X_j} \frac{\partial x_3}{\partial X_k} \sum_{m=1}^{3} \frac{\partial v_m}{\partial x_1}$$

$$= J\,(\nabla.v)$$

$$\Rightarrow \frac{1}{J}\frac{dJ}{dt} = \nabla \cdot v \tag{A.4}$$

3. **Localisation Theorem**
Consider Φ to be a continuous field (scalar, vector or tensor) on V. If, for all closed sets $B \in V$, $\int_B \Phi(u) dV = \mathbf{0}$, then $\Phi = \mathbf{0}$ for all $u \in \boldsymbol{V}$.
Proof:
Let us define:

$$I = \left| \Phi\left(u_0\right) - \frac{1}{(B_r)_{vol}} \int_{B_r} \Phi\,(u)\,dV \right| = \left| \frac{1}{(B_r)_{vol}} \int_{B_r} \left(\Phi\left(u_0\right) - \Phi\,(u)\right) dV \right| \tag{A.5}$$

where B_r is the closed ball of radius $r > \mathbf{0}$ centred at u_0 and $(B_r)_{vol}$ is the volume of the ball. According to a theorem in real analysis (Rudin, 1976), we have:

$$\begin{aligned} I &\leq \frac{1}{(B_r)_{vol}} \int_{B_r} |\Phi(u_0) - \Phi(u)| dV \\ &\leq \frac{1}{(B_r)_{vol}} \int_{B_r} Sup_{u \in B_r} |\Phi(u_0) - \Phi(u)| dV \\ &= \max_{u \in B_r} |\Phi(u_0) - \Phi(u)| \end{aligned} \tag{A.6}$$

max replaces sup in the above equation due to the continuity and compactness of B_r. Since $\Phi(u)$ is continuous, $I \to \mathbf{0}$ as $r \to \mathbf{0}$. Hence, from Equation A.5 and the given hypothesis, we have:

$$\Phi\left(u_0\right) = \lim_{r \to \mathbf{0}} \left(\frac{1}{(B_r)_{vol}} \int_{B_r} \Phi\,(u)\,dV \right) = \mathbf{0} \tag{A.7}$$

Since u_0 is arbitrary, we have the result that $\Phi\,(u) = 0\ \forall u \in V$.

References

Rudin, W. (1976) *Principles of Mathematical Analysis*, McGraw-Hill, New York.

Appendix B

1. **Homogeneous and nonhomogeneous** partial differential equations (PDEs)
   If every term of the PDE involves the dependent variable or any of its derivatives, the PDE is homogeneous and otherwise it is nonhomogeneous. As an example, let us consider the wave equation either of a vibrating string or a membrane given by $-\Delta u + m\frac{\partial^2 u}{\partial t^2} = f(\boldsymbol{x}, t)$. If the forcing term is zero the PDE is said to be homogeneous; otherwise it is said to be nonhomogeneous.
2. **St. Venant–Kirchhoff type material**
   In the context of 3D elasticity theory, if σ and ε, respectively, denote the symmetric stress and strain tensors for a linear isotropic elastic material, the constitutive relation is given by

   $$\sigma(\varepsilon) = \lambda(trace\ \varepsilon)I_{3\times 3} + 2\mu\varepsilon \tag{B.1}$$

   where λ and μ are material constants, called Lame parameters. Such a material is often referred to as the St. Venant–Kirchhoff type. Also note that the equations relating the Lame parameters to the more popularly used elasticity parameters E and ν (Young's modulus and Poisson's ratio, respectively) are given by:

   $$E = \frac{\mu(3\lambda + 2\mu)}{\lambda + \mu} \text{ and } \nu = \frac{\lambda}{2(\lambda + \mu)} \tag{B.2}$$

3. **Cauchy-type boundary condition**
   Boundary conditions (BCs) imposed as a weighted average of Dirichlet and Neumann BCs are sometimes called the Cauchy-type boundary conditions. For example, for a second-order ordinary differential equation (ODE) with dependent variable y, the Cauchy boundary conditions may be specified in terms of (combinations of) y and/or y' at a given initial or boundary point. Note that if we specify a value to y on the boundary it is a Dirichlet boundary condition and a value to the normal derivative at the boundary is known as a Neumann boundary condition.
4. **Norm and normed linear spaces**
   A norm is an operation $\|.\|$ on a function space or vector (linear) space V. It is a real-valued function on a space V satisfying the following conditions for all $x, y \in V$ and $a \in F$, whether the field F is often the set of reals $\mathbb{R}$:

   $$\|x\| \geq 0, \quad \text{and } \|x\| = 0 \quad iff \quad x = 0$$

   $$\|ax\| = |a|\ \|x\|$$

   $$\|x + y\| \leq \|x\| + \|y\| \text{ (triangle inequality)} \tag{B.3}$$

Elements of Structural Dynamics: A New Perspective, First Edition. Debasish Roy and G Visweswara Rao.

An example of a vector norm is its length (Euclidean norm) given by $\|x\|_2 = \left(\sum_{i=1}^{n} |x_i|^2\right)^{\frac{1}{2}}$, where n is the vector size. Similarly, if V is a space of all pointwise bounded continuous functions $f(x)$ over the domain Ω, then one can define a functional norm as: $\|f\|_\infty = \{\max |f(\boldsymbol{x})|$ for all points x in $\Omega\}$. Norms are useful in quantifying error estimates and rates of convergence of approximations.
A *normed space* is one with a norm defined on it. Euclidean space is a normed space. L^p space is defined with a norm: $\|x\|_p = \left(\sum_{i=1}^{n} |x_i|^p\right)^{\frac{1}{p}}$. For $p = 2$ it is the $\boldsymbol{L}^2$ space and the $\boldsymbol{L}^2$ norm is given by:

$$\|x\| = \left(\sum_{i=1}^{n} |x_i|^2\right)^{\frac{1}{2}} \tag{B.4}$$

5. **Inner product space**
An inner product space is a space with an inner product defined on it. An inner product on V is a mapping of $V \times V \to \mathbb{K}$, the scalar field of V; that is, for every pair $x, y \in V$, there is an associated scalar. It is generally denoted by $\langle \cdot \rangle$ and has the properties for all $x, y \in V$ and $\alpha \in \mathbb{K}$:

$$\begin{aligned}
&\langle x, y\rangle = \overline{\langle y, x\rangle} \\
&\langle x, x\rangle \geq 0 \\
&\langle x, x\rangle = 0 \to x = 0 \\
&\langle \alpha x, y\rangle = \alpha \langle x, y\rangle \\
&\langle x + y, z\rangle = \langle x, z\rangle + \langle y, z\rangle
\end{aligned} \tag{B.5}$$

The overbar indicates conjugation in case of complex spaces. For a real space V, the inner product exhibits symmetry, that is, $\langle x, y\rangle = \langle y, x\rangle$. The standard inner product for a vector space V is the familiar *dot* product $x \cdot y$. If V is a function space and we consider two functions $f, g \in V$ over the domain Ω, then the standard inner product is given by:

$$\langle fg\rangle = \int_\Omega fg d\boldsymbol{x}$$

If the inner product is zero, we say that the functions f and g are orthogonal over Ω. An inner product on V defines a norm on V given by:

$$\|f\| = \sqrt{\langle f, f\rangle} \tag{B.6}$$

6. **Cauchy sequence**
A sequence $\{f_n\}$ in a subset U of a normed space V is called a Cauchy sequence if:

$$\lim_{m,n\to\infty} \|f_m - f_n\| = \boldsymbol{0} \tag{B.7}$$

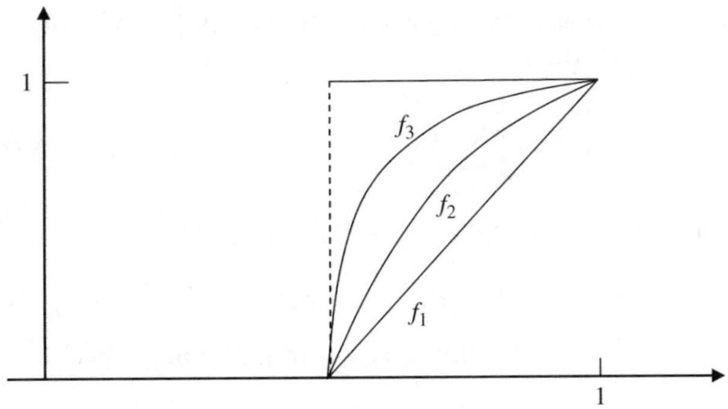

Figure B.1 Cauchy sequence and convergence

The above definition means that if for any given $\varepsilon > \mathbf{0}$, there exists a number $N(\varepsilon)$ such that:

$$\left\| f_m - f_n \right\| < \varepsilon \quad \text{for} \quad m, n > N(\varepsilon) \tag{B.8}$$

Every convergent sequence is a Cauchy sequence but not every Cauchy sequence is convergent. The reason for this is that the members in $\{f_n\}$ may converge to a limit, but the limit may not be the part of the space. As an example, consider the space $V = C[0, 1]$ with the integral norm, $\|f\|^2 = \int_0^1 f^2 dx$. Let the sequence $\{f_n\}$ in V be defined by (Figure B.1):

$$\begin{aligned} f_n &= 0, \qquad 0 \le x \le \frac{1}{2} \\ &= \left(x - \frac{1}{2}\right)^{\frac{1}{n}}, \qquad \tfrac{1}{2} \le x \le 1 \end{aligned} \tag{B.9}$$

It can be verified that $\{f_n\}$ is a Cauchy sequence. Now, the limit of the sequence is:

$$\begin{aligned} \lim_{n \to \infty} f_n &= 0, \qquad 0 \le x \le \tfrac{1}{2} \\ &= 1, \qquad \frac{1}{2} \le x \le 1 \end{aligned} \tag{B.10}$$

The above limit is a discontinuous function and $\notin C[0, 1]$. Hence, the Cauchy sequence is not convergent in V.

7. **Complete space**
   A general definition for a complete space is given by:
   'A subset U of a normed space V is complete if every Cauchy sequence in U converges to an element in U'.
   For example, a finite-dimensional normed space V is complete. To prove this, we consider a Cauchy sequence $\{u_j\}$ $j = 1, 2, \ldots$ in V. Let $B = \{v_1, v_2, \ldots, v_n\}$ be any basis for V. Then any member $u_j \in V$ can be expressed as a linear combination of the basis functions:

$$u_j = \alpha_{j1} v_1 + \alpha_{j2} v_2 + \ldots + \alpha_{jn} \mathbf{v}_n \tag{B.11}$$

Here, $\alpha_{ji} i = 1, 2, \ldots n$ are real constants. Since $\{u_j\}$ is Cauchy, we have for any $\varepsilon > \mathbf{0}$, an integer $N(\varepsilon)$ such that:

$$\left\| u_j - u_k \right\| = \left\| \sum_{i=1}^{n} (\alpha_{ji} - \alpha_{ki}) v_i \right\| < \varepsilon \text{ for } j, k > N(\varepsilon) \tag{B.12}$$

It is also true that for every choice of scalars $\beta_1, \beta_2 \ldots, \beta_n$:

$$\left\| \beta_1 v_1 + \beta_2 v_2 + \ldots + \beta_n v_n \right\| \geq c \left(|\beta_1| + |\beta_2| + \ldots + |\beta_n| \right)$$

(by repeated triangle inequality) (B.13)

In view of the above Equation B.12 gives:

$$\left| \alpha_{ji} - \alpha_{ki} \right| \leq \sum_{i=1}^{n} \left| \alpha_{ji} - \alpha_{ki} \right| < \frac{\varepsilon}{c}, c \text{ real.} \tag{B.14}$$

The above equation implies that $\{\alpha_{ji}\}$ $j = 1, 2, \ldots$ is Cauchy for each fixed i. Therefore, α_{ji} converges to a scalar α_i, say. Now let us define:

$$u = \alpha_1 v_1 + \alpha_2 v_2 + \ldots + \alpha_n v_n \tag{B.15}$$

Then, we have:

$$\left\| u_j - u \right\| = \left\| \sum_{i=1}^{n} (\alpha_{ji} - \alpha_i) v_i \right\| \leq \sum_{i=1}^{n} \left| \alpha_{ji} - \alpha_i \right| \left| v_i \right| \text{ (by the Schwarz inequality)} \tag{B.16}$$

Since $\alpha_{ji} \to \alpha_i$ for each i, $u_j \to u$ in V. Hence, V is a complete normed space.

8. **Hilbert space ($\mathcal{H}$)**
   A complete inner product space is a Hilbert space $\mathcal{H}$. An example for $\mathcal{H}$ is $\boldsymbol{L}^2(a, b)$ equipped with the $\boldsymbol{L}^2$-norm $\left(= \|x\| = \left(\sum_{i=1}^{n} |x_i|^2 \right)^{\frac{1}{2}} \right)$.
9. **Dense space**
   If X and Y are two subsets of a normed space, then Y is said to be dense in X if and only if there are points in X arbitrarily close to points in Y or given any point $X_0 \in X$ and any number $\varepsilon > 0$, then it is possible to find $Y_0 \in Y$ such that $\left\| X_0 - Y_0 \right\| < \varepsilon$.
10. **Separable space**
   A normed space V is said to be separable if it contains a countable subset that is dense in V. As an example, **R** is separable since the set of rationals **Q** is dense in **R** and is countable. A set **A** is *countable* if it is in one-to-one correspondence with **N**. For instance, any function in a separable function space may be represented via a linear combination of the elements of a countable set of basis functions.
11. **Square integrable functions**
   Functions having the property:

$$\int_{\Omega} ||f(x)||^2 dx < \infty \tag{B.17}$$

are known as square integrable. For example, functions in $\boldsymbol{L}^2(\Omega)$ possess this property.

12. **Closed and convex spaces**

A subset X of a linear space V is said to be convex if $u, v \in X$ implies:

$$M = \{w \in V | w = \alpha u + (1 - \alpha) v, \quad 0 \leq \alpha \leq 1\} \subset X \tag{B.18}$$

M is called a closed segment with boundary points u and v; and any other $w \in M$ is called an interior point. As an example, the closed ball $B = \{u \in V : \|u\| \leq 1\}$ in a normed space V is convex. Every subspace of V is convex.

The convexity also implies that in a normed space V, the set M of best approximation to a given point w out of a subspace U of V is convex.

13. **Null space**

For a linear operator $\boldsymbol{T}$ defined by:

$$\boldsymbol{T} : V \rightarrow U, \boldsymbol{T}v = u, v \in V \quad \text{and} \quad u \in U \tag{B.19}$$

the null space $N(\boldsymbol{T})$ of $\boldsymbol{T}$ is the set of all elements in the domain of $\boldsymbol{T}$ whose image is zero, that is

$$N(\boldsymbol{T}) = \{v \in V, \quad \boldsymbol{T}v = 0\} \tag{B.20}$$

The dimension of $N(\boldsymbol{T})$ is known as the nullity. If $\boldsymbol{T}$ is a differential operator (like $\boldsymbol{L} - \lambda \boldsymbol{I}$ with $\boldsymbol{L}$ being the Sturm–Liouville operator – Equation 2.83), and nullity is one, it shows that the eigenvalues are distinct and are termed as simple. If $\boldsymbol{T}$ is a matrix operator (like $\mathbf{A} - \lambda \boldsymbol{I}$ with $\mathbf{A}$ being a matrix, the solutions to $\boldsymbol{T}v = \mathbf{0}$ yield the null space of $\boldsymbol{T}$.

Appendix C

1. **Comparison functions and admissible functions**
In seeking an approximate solution to the governing partial differential equation (PDE) (see Equation 3.16) in the form $\hat{u}\,(x,t) = \sum_{j=1}^{N} U_j(x)q_j(t)$, we need to choose appropriate trial functions $U_j(x)$. In general, if the associated operator $\boldsymbol{L}$ is of order $2p$ ($p \in \mathbb{N}$), then the boundary condition operator $\mathcal{B}$ is at most of the order $2p-1$. Also, the solution to the PDE will have four ($2p$) constants of integration along with $2p$ BCs. For example, for the PDE (Equation 3.5) corresponds to a uniform cantilever beam, we have $p=2$ and $\mathcal{B} \equiv \frac{\partial}{\partial x}\left(EI\frac{\partial^2}{\partial x^2}\right)$ corresponding to the natural BC at the free end and is of order three ($=2p-1$). The BCs are four ($=2p$)–two essential (geometric) BCs: $u(0)=0$ and $\frac{\partial u(0)}{\partial x}=0$ and two natural BCs: $EI\frac{\partial^2 u(L)}{\partial x^2}=0$ and $EI\frac{\partial^3 u(L)}{\partial x^3}=0$. Comparison functions are the ones that are $2p$ times differentiable and satisfy all the $2p$ BCs. On the other hand, the admissible functions are those that are p times differentiable and satisfy the essential BCs only.
2. **Proposition: Equation 3.10–$\boldsymbol{KQ} = \omega^2\,\boldsymbol{MQ}$–is obtained by the Rayleigh–Ritz method when applied to continuous system. The eigenvalues of the eigenvalue problem remain an upper bound to the corresponding exact eigenvalue of the continuous system**
Proof: The matrices $\boldsymbol{K}$ and $\boldsymbol{M}$ represent a discrete model for the original continuous system. If the PDE operator $\boldsymbol{L}$ is self-adjoint, the continuous system is self-adjoint. See Chapter 2 for the definition of the self-adjoint system and its properties. Analogously, for a self-adjoint system, $\boldsymbol{K}$ and $\boldsymbol{M}$ are real and symmetric and the eigenvalues are real. If N trial functions $\{W_j(x), j=1,2..,N\}$ are used in the assumed solution, the $\boldsymbol{K}$ and $\boldsymbol{M}$ matrices are of size $N \times N$ and the eigenvalue problem in Equation 3.10 yields estimates $\omega_j^{(N)}$ for the first N eigenvalues of the original continuous system. In principle, since the set of trial functions $\left\{W_j(x)\right\}$ is complete (see *item 3* below for definition of completeness of a set), the exact eigenvalues ω_j^e, $j=1,2,\ldots$ can be obtained by letting $N \to \infty$. Let the eiegnvalues be ordered such that:

$$\omega_1^{(N)} \le \omega_2^{(N)} \le \ldots.. \le \omega_N^{(N)} \quad \text{and} \quad \omega_1^e \le \omega_1^e \le \ldots.. \tag{C.1}$$

Let $\boldsymbol{K}^{(N+1)}$ and $\boldsymbol{M}^{(N+1)}$ be matrices obtained by the Rayleigh–Ritz method using $N+1$ trial functions. Since these matrices result from an addition of an extra trial function $W_{N+1}(x)$ to the set $\left\{W_j(x), j=1,2..,N\right\}$, the original matrices $\boldsymbol{K}^{(N)}$ and $\boldsymbol{M}^{(N)}$ will be contained within $\boldsymbol{K}^{(N+1)}$ and $\boldsymbol{M}^{(N+1)}$, that is:

$$\boldsymbol{K}^{(N+1)} = \begin{bmatrix} \boldsymbol{K}^{(N)} & \times \\ \times & \times \end{bmatrix}, \; \boldsymbol{M}^{(N+1)} = \begin{bmatrix} \boldsymbol{M}^{(N)} & \times \\ \times & \times \end{bmatrix} \tag{C.2}$$

Elements of Structural Dynamics: A New Perspective, First Edition. Debasish Roy and G Visweswara Rao.

where $\times$ stands for elements in the extra row and column of $\boldsymbol{K}^{(N+1)}$. If the eigenvalues of $\boldsymbol{K}^{(N+1)}\boldsymbol{Q} = \omega^2\boldsymbol{M}^{(N+1)}\boldsymbol{Q}$ are denoted by $\omega_j^{(N+1)}$ $j = 1, 2, ..N, N+1$, then the eigenvalue separation property (see *item 4* below for the definition and proof) states that:

$$\omega_1^{(N+1)} \leq \omega_1^{(N)} \leq \omega_2^{(N+1)} \leq \omega_2^{(N)} \leq \ldots.. \leq \omega_N^{(N+1)} \leq \omega_N^{(N)} \leq \omega_{N+1}^{(N+1)} \tag{C.3}$$

By the above property, it means that the eigenvalues of $\boldsymbol{K}^{(N)}\boldsymbol{Q} = \omega^2\boldsymbol{M}^{(N)}\boldsymbol{Q}$ are bracketed by those of $\boldsymbol{K}^{(N+1)}\boldsymbol{Q} = \omega^2\boldsymbol{M}^{(N+1)}\boldsymbol{Q}$. This also means that for increasing N, any j^{th} eigenvalue of successive approximations follows the inequality:

$$\omega_j^{(N+1)} \leq \omega_j^{(N)} \leq \omega_j^{(N-1)} \leq \ldots\ldots\ldots \tag{C.4}$$

Thus, since as $N \to \infty$, $\omega_j^{(N)} \to \omega_j^e$, Equations C.3 and C.4 point to the fact that ω_j^e remains as a lower bound to the estimated eigenvalues for any j, thus completing the proof.

3. **Eigenvalue separation property**
Consider the standard eigenvalue problem $\boldsymbol{A}\boldsymbol{x} = \lambda\boldsymbol{x}$. The eigenvalue separation property states that the eigenvalues of a symmetric matrix $\boldsymbol{A}$ of dimension N bracket those of the matrix of dimension $N-1$ with the last row and column omitted from $\boldsymbol{A}$. This property is based on the minimax characterisation of eigenvalues of $\boldsymbol{A}$ and is explained below.
Rayleigh quotient in general is defined as: $R(\boldsymbol{x}) = \frac{\boldsymbol{x}^T\boldsymbol{A}\boldsymbol{x}}{\boldsymbol{x}^T\boldsymbol{x}}$. If $\lambda_i, i = 1, 2\ldots N$ are the eigenvalues of $\boldsymbol{A}$, the minimax characterisation of eigenvalues is given by:

$$\lambda_r = \max\left(\min\frac{\boldsymbol{x}^T\boldsymbol{A}\boldsymbol{x}}{\boldsymbol{x}^T\boldsymbol{x}}\right), r = 1, 2, ..., N \tag{C.5}$$

with the constraint $\boldsymbol{x}^T\boldsymbol{v}_j = 0$, for $j = 1, 2, \ldots r-1$ and $r \geq 2$. This principle involves choosing vectors $\boldsymbol{v}_j$, $j = 1, 2, \ldots r-1$ and evaluate the minimum of $R(\boldsymbol{x})$ such that $\boldsymbol{x}$ is orthogonal to each of $\boldsymbol{v}_j$, $j = 1, 2, \ldots r-1$. Now, with different sets of vectors $\boldsymbol{v}_j$, we obtain different minima and the maximum of all these minima is the rth eigenvalue, λ_r of A.
Now consider the eigenvalue problems $\boldsymbol{A}\boldsymbol{x} = \lambda\boldsymbol{x}$ and $\boldsymbol{A}^{(m)}\boldsymbol{y}^{(m)} = \lambda^{(m)}\boldsymbol{y}^{(m)}$, where $\boldsymbol{A}^{(m)}$ is obtained by omitting the last m rows and columns from $\mathbf{A}$. By the minimax characterisation of eigenvalues of $\boldsymbol{A}$, we have:

$$\lambda_r = \max\left(\min\frac{\boldsymbol{x}^T\boldsymbol{A}\boldsymbol{x}}{\boldsymbol{x}^T\boldsymbol{x}}\right) \text{ with } \boldsymbol{x}^T\boldsymbol{v}_j = 0,\ j = 1, 2, \ldots, r-1 \text{ and } \boldsymbol{v}_j \text{ arbitrary} \tag{C.6}$$

$$\lambda_{r+1} = \max\left(\min\frac{\boldsymbol{x}^T\boldsymbol{A}\boldsymbol{x}}{\boldsymbol{x}^T\boldsymbol{x}}\right) \text{ with } \boldsymbol{x}^T\boldsymbol{v}_j = 0,\ j = 1, 2, \ldots, r \text{ and } \boldsymbol{v}_j \text{ arbitrary} \tag{C.7}$$

With $m = 1$, we have for $\boldsymbol{A}^{(\mathbf{1})}$:

$$\lambda_r^{(1)} = \max\left(\min\frac{\boldsymbol{y}^T\boldsymbol{A}^{(\mathbf{1})}\boldsymbol{y}}{\boldsymbol{y}^T\boldsymbol{y}}\right) \text{ with } \boldsymbol{y}^T\boldsymbol{w}_j = 0,\ j = 1, 2, \ldots, r-1 \text{ and } \boldsymbol{w}_j \text{ arbitrary} \tag{C.8}$$

Now, in finding λ_r and $\lambda_r^{(1)}$, the same number of constraints might have been used. However, omitting the last row and column in $\boldsymbol{A}$ to obtain $\boldsymbol{A}^{(1)}$ is equivalent to an additional constraint and thus:

$$\lambda_r^{(1)} \geq \lambda_r \tag{C.9}$$

Further, the constraints in Equation C.7 are more restrictive and contain those in Equation C.8 and thus, we have:

$$\lambda_{r+1} \geq \lambda_r^{(1)} \tag{C.10}$$

Combining the results in Equations C.9 and C.10, we finally get:

$$\lambda_r \leq \lambda_r^{(1)} \leq \lambda_{r+1} \tag{C.11}$$

Thus, the first eigenvalue of $\boldsymbol{A}^{(1)}$ is bracketed by the first two eigenvalues of $\boldsymbol{A}$. Continuing the argument further for $m \geq 2$, we obtain $\lambda_r^{(1)} \leq \lambda_r^{(2)} \leq \lambda_{r+1}^{(1)}, \lambda_r^{(2)} \leq \lambda_r^{(3)} \leq \lambda_{r+1}^{(2)}$, and so on. Combining these results yields:

$$\lambda_r^m \leq \lambda_r^{(m+1)} \leq \lambda_{r+1}^m \tag{C.12}$$

4. **Complete set of functions**
   Consider a set of linearly independent functions $\Phi_1, \Phi_2, \ldots$ belonging to a separable function space V (see Appendix B for definition of a separable space). Define a function $f \in V$ and approximate it as the linear combination:

$$f_n = \sum_{j=1}^{N} \alpha_j \Phi_j \tag{C.13}$$

   Here, $\alpha_j\, j = 1, 2, \ldots, N$, are real constants. As N is increased, the norm $\|f - f_n\|^2$ can be made less than any arbitrary small number $\varepsilon > 0$, the set of functions $\Phi_1, \Phi_2, \ldots$ is said to be complete. Alternatively, one may also argue that the limit of every Cauchy sequence in V exists and belongs to V (also see Appendix B for definition of a complete space).

Appendix D

1. **Convex functional**

   A functional, $I(.)$ is defined as a function $I{:}V \to \mathbb{R}$, where V is a linear function space. $I(.)$ is a convex functional if, for $0 < \theta < 1$ and $u, v \in V$,

   $$I(u + \theta(v - u)) \le I(u) + \theta(I(v) - I(u))$$
   $$\Rightarrow I(\theta v + (1 - \theta) u) \le \theta I(v) + (1 - \theta) I(u) \tag{D.1}$$

   A strictly convex functional satisfies:

   $$I(\theta v + (1 - \theta) u) < \theta I(v) + (1 - \theta) I(u) \tag{D.2}$$

   The above definition is similar to that of a convex set and a convex function. Figure D.1 shows convex and strictly convex functions of a single variable.
   The concept of convexity is useful in minimization problems. For example, to locate the extremum (a minimum, maximum or an inflection point) of a C^1 function $f{:}\mathbb{R} \to \mathbb{R}$, we know from elementary calculus that the necessary condition is $\frac{df}{dx} = 0$. If the function is convex (Figure D.1), the condition gives a minimum. However, for a strictly convex function (Figure D.1b) the minimum is unique, while for just convex functions (Figure D.1a) the minimum is given by all points between a and b. These basic ideas are as well applicable to the extremization of a functional $I(.)$, wherein the aim is to identify a function u such that $I(u) \le I(v) \forall v \in V$.
2. $\boldsymbol{C^m(a, b)}$

   $C^m(a, b)$ is the set of all functions that, together with their first m derivatives, are continuous on (a, b).
3. $\boldsymbol{L^2}$ ***spaces***

   If $I = (a, b)$, $L^2(I)$ is a space of square integrable functions on I defined as:

   $$L^2(I) = \left\{ v{:}v \text{ is defined on } I \text{ such that } \int_I v^2 dx < \infty \right\} \tag{D.3}$$

   $L^2(I)$ is a Hilbert space with the inner product $(u, v) = \int_I uv dx$ and with the norm:

   $$\|u\|_{L_2(I)} = \sqrt{(u, u)} = \sqrt{\int_I u^2 dx}, \forall u \in L_2(I) \tag{D.4}$$

   i) *L^p space*

   Let p be a real number ≥ 1. A function u defined on Ω of $\mathbb{R}^n$ is said to belong to $L^p(\Omega)$, if u is measurable and if the integral $\int_\Omega |u|^p dx < \infty$.

Elements of Structural Dynamics: A New Perspective, First Edition. Debasish Roy and G Visweswara Rao.

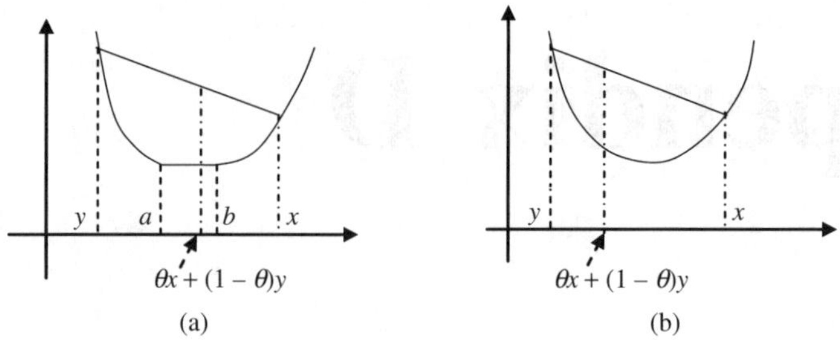

Figure D.1 (a) A convex function and (b) a strictly convex function

4. **Inner product space (also see Appendix B for additional information on inner product spaces)**
   V is an inner product space with the inner product denoted by:

$$(u, v) = \int_0^l uv dx, \forall u, v \in V \tag{D.5}$$

An inner product space is a normed space with the norm defined as $\|u\| = \sqrt{(u, u)}$. Every inner product space satisfies the Cauchy–Schwartz inequality:

$$|(u, v)| \leq \|u\| \, \|v\| \tag{D.6}$$

5. **Sobolev space $H^m(\Omega)$, $H_0^m(\Omega)$) and $W^{m,p}(\Omega)$) with $m = 0, 1, 2, \ldots$ and $p = 1, 2, \ldots$**
   i) $H^m(\Omega)$ spaces

A Sobolev space of order m, denoted by $H^m(\Omega)$, is defined to be the space that consists of functions in $L^2(\Omega)$ that together with all their 'weak' derivatives up to and including those of order m, belong to $L^2(\Omega)$. See item 8 of this appendix for the definition of a weak derivative. Thus,

$$H^m(\Omega) = \left\{v{:}D^\alpha v \in L^2(\Omega), \text{ for all } \alpha \text{ such that } |\alpha| \leq m\right\} \tag{D.7}$$

where $\alpha = \{\alpha_1, \alpha_2, \ldots\}$ is called the multi-index and $D^\alpha v = \frac{\partial^\alpha v_1}{\partial x}$ denotes the weak derivative of v. Note that $H^0(\Omega) = L^2(\Omega)$. $H^m(\Omega)$ is also an inner product space with the Sobolev inner product defined by:

$$(u, v)_{H^m} = \int_\Omega \sum\nolimits_{|\alpha| \leq m} D^\alpha u D^\alpha v \, dx \text{ for } u, v \in H^m(\Omega) \tag{D.8}$$

The Sobolev norm associated with the above inner product is given by:

$$\|u\|_{H^m}^2 = (u, u)_{H^m} = \int_\Omega \sum\nolimits_{\alpha| \leq m} \left(D^\alpha u\right)^2 dx \tag{D.9}$$

We also write $(u, v)_{H^m}$ as:

$$(u, v)_{H^m} = \sum\nolimits_{|\alpha| \leq m} \left(D^\alpha u D^\alpha v\right)_{L^2} \tag{D.10}$$

Hence:

$$\|u\|_{H^m}^2 = \sum_{|\alpha| \le m} \|D^\alpha u\|_{L^2}^2 \tag{D.11}$$

An example for $H^m(\Omega)$

Let us find to which space $H^m(\Omega)$ the function given below (Figure D.2) belongs to.

$$\begin{aligned} u(x) &= x^2, \quad 0 < x \le 1 \\ &= 2x^2 - 2x + 1, \quad 1 < x < 2 \end{aligned} \tag{D.12}$$

To find the solution, let us have the derivatives of $u(x)$:

$$\begin{aligned} u'(x) &= 2x, \quad 0 < x \le 1 \\ &= 4x - 2, \quad 1 < x < 2 \end{aligned} \tag{D.13}$$

Clearly, $u'(x) \in C\,(0, 2)$ and belongs to $L^2(0, 2)$. $u'(x)$ has a weak derivative (see Figure D.2) given by:

$$\begin{aligned} u''(x) &= 2, 0 < x \le 1 \\ &= 4, 1 < x < 2 \end{aligned} \tag{D.14}$$

We find that $u'''(x)$, the derivative of $u''(x)$ is $2\delta(x-1)$, which does not belong to $L^2(0, 2)$. Hence, $u(x)$ is $H^2(0, 2)$. Incidentally, we note that $u'(x)$ is $H^1(0, 2)$.and $u''(x)$ belongs to $H^0\,(0, 2) \equiv L^2(0, 2)$. The norms of these functions are given by:

$$\begin{aligned} \|u(x)\|_{H^2} &= \int_0^2 \left\{u^2 + (u')^2 + (u'')^2\right\} dx = 71.37 \\ \|u'(x)\|_{H^1} &= \int_0^2 \left\{(u')^2 + (u'')^2\right\} dx = 39 \\ \|u(x)\|_{H^0} &= \int_0^2 \left\{(u'')^2\right\} dx = 20 \end{aligned} \tag{D.15a,b,c}$$

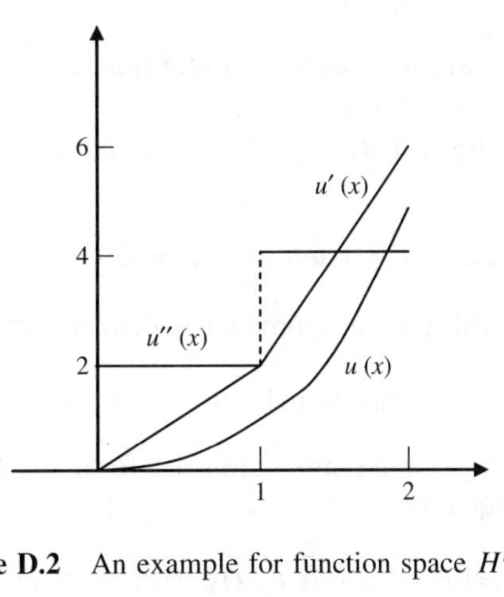

Figure D.2 An example for function space $H^m(\Omega)$

ii) $H_0^m(\Omega)$ spaces $m = 0, 1, 2, \ldots$

$H_0^m(\Omega)$ is a subspace of $H^m(\Omega)$ and certain derivatives of its members vanish at the boundary Γ. Consider a function $u \in H^m(\Omega)$. $H_0^m(\Omega)$ is defined by:

$$H_0^m(\Omega) = \left\{u \in H^m(\Omega) : D^\alpha u = 0 \text{ on } \Gamma, |\alpha| \leq m - 1\right\} \tag{D.16}$$

The function $u(x)$ in Figure D.2 belongs to $H_0^2(\Omega)$ since u and $D^1 u$ are zero at the boundary $x = 0$.

iii) $W^{m,p}$ spaces $m = 0, 1, 2, \ldots$ *and* $p = 1, 2, \ldots$

The Sobolev space $W^{m,p}$ is the space of functions that together with all their weak derivatives up to and including order m belong to $L^p(\Omega)$:

$$W^{m,p}(\Omega) = \left\{u : D^\alpha u \in L^p(\Omega), |\alpha| \leq m\right\} \tag{D.17}$$

6. **Linear form**

$l(.)$ is a linear form on V if $l: V \to \mathbb{R}$, that is $l(u) \in \mathbb{R}$ for $u \in V$ and for all $u, v \in V$, we have:

$$l\,(au + bv) = al(u) + bl(v) \;\; a, b \in \mathbb{R} \tag{D.18}$$

Further, the linear form is **continuous** if there exists a $K \in \mathbb{R}$ such that

$$|l(u)| \leq K \, \|u\|_V \, \forall v \in V \tag{D.19}$$

Here, $\|u\|_V$ is the norm induced by V.

7. **Bilinear form**

$\alpha\,(.,.)$ is a bilinear form on $V \times V$ if $\alpha: V \times V \to \mathbb{R}$, that is $\alpha(u, v) \in \mathbb{R}$ for all $u, v \in V$ and it is linear in both the arguments:

$$\begin{aligned} &\alpha\,(au + bw, v) = a\alpha(u, v) + b\alpha(w, v) \\ &\alpha\,(u, aw + bv) = a\alpha(u, w) + b\alpha(u, v), \forall u, v, w \in V \text{ and } a, b \in \mathbb{R} \end{aligned} \tag{D.20}$$

A bilinear form $\alpha\,(.,.)$ on $V \times V$ is **symmetric** if:

$$\alpha(u, v) = \alpha(v, u), \forall u, v \in V \tag{D.21}$$

A symmetric bilinear form is **positive semidefinite** if $\alpha(u, u) \geq 0, \forall u \in V$ and **positive definite** if $\alpha(u, u) > 0, \forall u \in V$.

A symmetric bilinear form $\alpha(u, v): V \times V \to \mathbb{R}$ is an inner product on V, iff, for $\forall u, v \in V$:

$$\alpha(u, u) \geq 0 \text{ and } \alpha(u, u) = 0 \Longleftrightarrow u = 0. \tag{D.22}$$

The norm associated with the inner product is defined by:

$$\|u\|_V = (\alpha(u, u))^{1/2}, \forall u \in V \tag{D.23}$$

A symmetric positive semidefinite bilinear form $\alpha: V \times V \to \mathbb{R}$ satisfies the Cauchy–Schwartz inequality:

$$|\alpha(u, v)| \leq \sqrt{\alpha(u, u)}\sqrt{\alpha(v, v)}, \forall u, v \in V \tag{D.24}$$

Proof: If either u or v is zero or both are zero functions in V, the result in Equation D.24 is trivially satisfied. In the nontrivial case, consider the following bilinear form with $a \in \mathbb{R}$:

$$\begin{aligned}\alpha(u-av, u-av) &= \alpha(u, u-av) - a\alpha(v, u-av) \\ &\Rightarrow \alpha(u,u) - 2a\alpha(u,v) + a^2\alpha(v,v)\end{aligned} \tag{D.25}$$

Equation D.25 is a quadratic expression in a and for at most one real root, we must have:

$$\alpha(u,v)^2 - \alpha(u,u)\alpha(v,v) \le 0 \Rightarrow |\alpha(u,v)| \le \sqrt{\alpha(u,u)}\sqrt{\alpha(v,v)} \tag{D.26}$$

8. **A continuous function on $H^1(a, b)$**
   A function v is continuous at x iff, for every $\varepsilon > 0$, there exists a $\beta > 0$ such that

$$|x-y| < \beta \Rightarrow |v(x) - v(y)| < \varepsilon \tag{D.27}$$

   We shall test this continuity condition for $u \in H^1(a,b)$. Let $x, y \in (a,b)$ and define $(w, v) := \int_x^y wv dz$. Then:

$$u(y) - u(x) = \int_x^y u'(z)dz = (1, u') \tag{D.28}$$

   $(1, u')$ is a bilinear form. By the Cauchy–Schwartz inequality, we have:

$$\begin{aligned}|u(y) - u(x)| &= |(1,u')| \le \sqrt{(1,1)}\sqrt{(u',u')} \\ &= \left|\int_x^y 1^2 dz\right|^{1/2} \left|\int_x^y u'^2 dz\right|^{1/2} \\ &\le |y-x|^{\frac{1}{2}} \|u'\|_{L^2(x,y)} \quad \text{since } u \in L^2(ab)\end{aligned} \tag{D.29}$$

   If we choose $\beta < \frac{\varepsilon^2}{\|u'\|^2_{L_2(x,y)}}$, then it follows that $|u(y) - u(x)| < \varepsilon$ and $u \in H^1(a,b)$ is continuous.
9. **Compactly supported functions**
   A function has compact support if it is zero outside of a compact set. This means that a function has a compact support if its support is a compact set. Suppose that a function $u(x)$ is defined on a domain Ω and is nonzero only for points in a proper subset A. If $\bar{A}$ is a closure of A, then $\bar{A}$ is the support of u. Now we say that u has a compact support on Ω if $\bar{A}$ is compact, that is if it is a closed and bounded subset of Ω.
10. **Weak derivative of a function**
   Let Ω be a domain in $\mathbb{R}^n$ and $D(\Omega)$ the set of $C^\infty(\Omega)$ functions with compact supports in Ω. Let us also define the set of locally integrable functions in the compact $K \subset \Omega$:

$$L^1_{loc}(\Omega) := \left\{u : u \in L^1_K \forall K \subset \Omega \text{ compact}\right\} \tag{D.30}$$

   Now, we say a given function $u \in L^1_{loc}(\Omega)$ has a weak derivative $D^\alpha u$, provided there exists a function $D^\alpha u \in L^1_{loc}(\Omega)$ such that:

$$\int_\Omega D^\alpha u(x)\phi(x)dx = (-1)^{|\alpha|}\int_\Omega u(x)D^\alpha\phi(x)dx, \ \forall\phi \in D(\Omega) \tag{D.31}$$

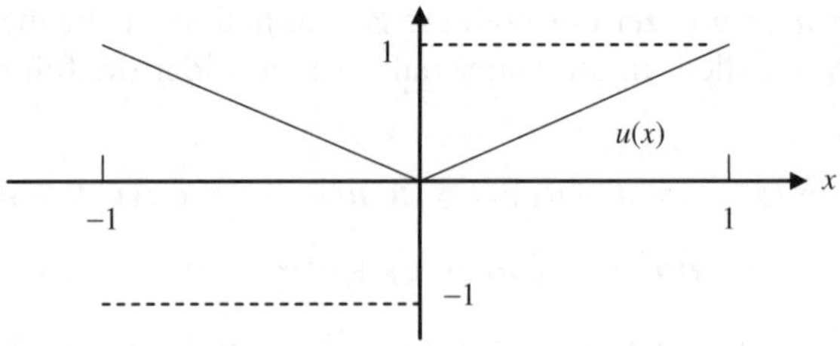

Figure D.3 Function $u(x)$ and its weak derivative

If such a function exists we say that $D^\alpha u$ is the weak derivative of $u(x)$. Here, α is a multi-index and is an ordered n-tuples of non-negative integers. $|\alpha|$ is a notation for the sum $\alpha_1 + \alpha_1 + \ldots + \alpha_n$ with α_i being a non-negative integer. $D^\alpha u$ denotes the weak partial derivative:

$$D^\alpha u = \frac{\partial^{|\alpha|} u}{\partial x_1^{\alpha_1} \partial x_2^{\alpha_2} \ldots \partial x_n^{\alpha_n}} \tag{D.32}$$

Thus, if $|\alpha| = m$, $D^\alpha u$ is one of the mth partial derivatives of u. If $n = 3$, we can take, for example $\alpha = (1, 0, 3)$. With $|\alpha| = 1 + 0 + 3 = 4$, $D^\alpha u$ is the fourth-order partial derivative given by:

$$D^\alpha u = \frac{\partial^4 u}{\partial x_1^1 \partial x_2^0 \partial x_n^3} = \frac{\partial^4 u}{\partial x_1 \partial x_3^3} \tag{D.33}$$

Example for a weak derivative
Consider a function $u(x)$ (Figure D.3) defined by:

$$u(x) = |x| \tag{D.34}$$

The given function $u(x) \in C[-1, 1]$. The classical derivative $u'(x)$ does not exist, since it is not defined at $x = 0$. The weak derivative is the function:

$$\begin{aligned} D^1 u &= -1 \text{ for } -1 \leq x < 0 \\ &= +1 \text{ for } 0 \leq x \leq 1 \end{aligned} \tag{D.35}$$

Proof: We consider the integral $\int_{-1}^{1} u(x)\phi'(x)dx$ where $\phi(x)$ is infinitely differentiable and compactly supported over $[-1, 1]$. Split the interval $[-1, 1]$ into two parts and integrate by parts to obtain:

$$\int_{-1}^{1} u(x)\phi'(x)dx = \int_{-1}^{0} u(x)\phi'(x)dx + \int_{0}^{1} u(x)\phi'(x)dx$$

$$= u(x)\phi(x)|_{-1}^{0} - \int_{-1}^{0} u'(x)\phi(x)dx + u(x)\phi(x)|_{0}^{1} - \int_{0}^{1} u'(x)\phi(x)dx$$

$$= -\left(\int_{-1}^{0} (-1)\phi(x)dx + \int_{0}^{1} (+1)\phi(x)dx\right),$$

since $u(x)$ is continuous at $x = 0$ and $u(0) = 0$

$$= -\int_{-1}^{1} D^1 u\phi(x)dx$$

$$\Rightarrow \int_{-1}^{1} D^1 u\phi(x)dx = (-1)^1 \int_{-1}^{1} u(x)\phi'(x)dx \tag{D.36}$$

Thus, $D^1 u$ is given by Equation D.34. Note that if $u(x)$ is sufficiently smooth such that it belongs to $C^m(\Omega)$, then its weak derivative $D^\alpha u$ coincides with its classical derivative for $|\alpha| \leq m$.

11. **Bound for the interpolation error $u - \pi u$ by use of Taylor's theorem**
We here prove a general form of an interpolation estimate for $y(x)$ when its interpolant is a polynomial $Y(x)$ of degree $m-1$. Y interpolates y, say, at equispaced points x_i, $i = 1, 2, \ldots k$ in the interval $[a, b]$ such that $a = x_1 < x_2 < \ldots\ldots x_{k-1} < x_k = b$ With $l = b - a$ and $h = x_i - x_{i-1}$, the general form of the estimate is:

$$\left\|(y - Y)^{(n)}\right\|_{L^2(a,b)} \leq Ch^{m-n} \left\|u^{\{m\}}\right\|_{L^2(a,b)},\quad 0 \leq n \leq m \tag{D.37}$$

Proof: To prove Equation D.37 we first arrive at the following error estimates using Taylor's theorem:

$$\begin{aligned}
\|y - Y\|_{L_2(a,b)} &\leq C_1 l^2 \left\|y^{;;}\right\|_{L^2(a,b)}\\
\|y - Y\|_{L_2(a,b)} &\leq C_2 l \left\|y'\right\|_{L^2(a,b)}\\
\left\|(y - Y)'\right\|_{L_2(a,b)} &\leq C_3 l \left\|y''\right\|_{L^2(a,b)}
\end{aligned} \tag{D.38a-c}$$

To derive the above three estimates, we use Taylor's theorem:

$$y(x) = \sum_{m=0}^{k} \frac{(x - x_0)^m}{m!} y^m(x_0) + R_k(y) \tag{D.39}$$

where $R_k(y)$ is the remainder given by:

$$R_k(y) := \frac{1}{k!} \int_{x_0}^{x} (x - \xi)^k\, y^{k+1}(\xi)\, d\xi \tag{D.40}$$

If $Y(x)$ is a linear interpolant of $y(x)$, $m = 2$ and we have $Y(x) = \frac{(x-a)}{l} y(b) + \frac{(b-x)}{l} y(a)$. If we use Taylor's theorem and write approximations for $y(a)$ and $y(b)$ in terms of $y(x)$ as:

$$\begin{aligned}
y(a) &= y(x) + (a - x)\, y'(x) + \int_{x}^{a} (a - \xi)\, y''(\xi)d\xi\\
y(b) &= y(x) + (b - x)\, y'(x) + \int_{x}^{b} (b - \xi)\, y''(\xi)d\xi
\end{aligned} \tag{D.41a,b}$$

Using the above expressions, we get the error equation:

$$y(x) - Y(x) = -\frac{(b - x)}{l} \int_{x}^{a} (a - \xi)\, y''(\xi)d\xi - \frac{(x - a)}{l} \int_{x}^{b} (b - \xi)\, y''(\xi)d\xi \tag{D.42}$$

Squaring both sides and using the Cauchy–Schwartz inequality gives:

$$|y(x) - Y(x)|^2 \leq 2\left(\frac{(b-x)}{l}\right)^2 \left(\int_x^a (a-\xi)\, y''(\xi) d\xi\right)^2$$
$$+2\left(\frac{(x-a)}{l}\right)^2 \left(\int_x^b (b-\xi)\, y''(\xi) d\xi\right)^2$$
$$\leq 2\left(\frac{(b-x)}{l}\right)^2 \frac{(x-a)^3}{3} \int_x^a \left|y''\right|^2 d\xi$$
$$+2\left(\frac{(x-a)}{l}\right)^2 \frac{(b-x)^3}{3} \int_x^b \left|y''\right|^2 d\xi \tag{D.43}$$

Since $\max_{a \leq x \leq b}(b-x)^2 (x-a)^3 = \max_{a \leq x \leq b}(x-a)^2 (b-x)^3 = \frac{2^2 3^2}{5^4} l^5$, Equation D.43 simplifies to:

$$|y(x) - Y(x)|^2 \leq \frac{2^3 3^2}{5^4} l^3 \int_a^b \left|y''\right|^2 d\xi \tag{D.44}$$

Integrating both sides over (a, b) and taking square roots, we get the result in Equation D.38a with $C_{12} = \frac{2^3 3^2}{5^4}$. Thus if $y'' \in L^2(a, b)$, the linear interpolation error is of order $O(l^2)$.

If $y' \in L^2(a, b)$ only, then we obtain the result in Equation D.38b as follows. By Taylor's theorem, we have:

$$y(x) = y(a) + \int_a^x y'(\xi)\, d\xi \text{ and } y(x) = y(b) + \int_b^x y'(\xi)\, d\xi \tag{D.45}$$

It follows that:

$$y(x) - Y(x) = \frac{(b-x)}{l} \int_a^x y'(\xi)\, d\xi + \frac{(x-a)}{l} \int_b^x y'(\xi)\, d\xi \tag{D.46}$$

Further simplification yields the error estimate in Equation D.37b with $C_{22} = \frac{8}{27}$. Finally, to get the result in Equation D.37c, we first note that $Y'(x) = \frac{y(b)-y(a)}{l}$. Then, using the same Taylor's expansions for $y(a)$ and $y(b)$ as in Equation D.38, we obtain:

$$y'(x) - Y'(x) = -\frac{1}{l} \int_x^a (a-\xi)\, y''(\xi) d\xi - \frac{1}{l} \int_x^b (b-\xi)\, y''(\xi) d\xi$$

Squaring both sides and using the Cauchy–Schwartz inequality gives:

$$\left|y'(x) - Y'(x)\right|^2 \leq \frac{2}{3l^2} \int_a^b \left|y''(\xi)\right|^2 d\xi \left((b-x)^3 - (x-a)^3\right) \tag{D.47}$$

Integrating the above equation on both sides and taking square root gives the error estimate in Equation D.38c with $C_{32} = 1/3$. From the above three error estimates, we find the pattern that gives the general formula in Equation D.37.

Appendix E

1. **Linearly independent vectors**
   A finite subset A of vectors $v_1 v_2, \ldots, v_n$ $(n \in \mathbb{N})$ in a vector space V is called linearly independent if for any set of scalars $\alpha_1, \alpha_2, \ldots, \alpha_n$, we have:

$$\alpha_1 v_1 + \alpha_2 v_2 +, , , +\alpha_n v_n = 0 \Leftrightarrow \alpha_1 = \alpha_2 = \ldots = \alpha_n = 0 \tag{E.1}$$

2. **Span (.)**
   If $A = \{v_1 v_2, \ldots, v_n\}$ is a finite nonempty subset of a vector space V, A is said to span V if every $u \in V$ can be expressed in the from $u = \alpha_1 v_1 + \alpha_2 v_2 + \ldots + \alpha_n v_n$, where α_i are scalars. We also say $\text{span}(A) = V$ since the vectors of A generate (or span) V.
3. **Basis**
   A basis B of a linear vector (or function) space V is a linearly independent subset of V that generates V. Thus, we have $span(B) = V$ meaning that if $B = \{v_1, v_2, \ldots, v_n\}$, then every vector in V can be *uniquely* expressed in the form:

$$u = \sum_{j=1}^{n} \alpha_j v_j \tag{E.2}$$

   The uniqueness property implied in the above definition is explained as follows. Suppose that it is possible to have:

$$u = \sum_{j=1}^{n} \alpha_j v_j \text{ and } u = \sum_{j=1}^{n} \beta_j v_j \tag{E.3}$$

   Subtracting the second from the first expansion gives:

$$0 = \sum_{j=1}^{n} (\alpha_j - \beta_j) v_j \tag{E.4}$$

   Since B is linearly independent, we have $\alpha_j - \beta_j = 0$, $j = 1, 2, .., n$ which gives $\alpha_j = \beta_j$, $j = 1, 2, .., n$. V is finite-dimensional if it has a basis consisting of a finite number of vectors/functions and the number of vectors/functions in the basis is called the dimension of V and denoted by $\dim(V)$.
4. **Linear transformation**
   Consider two vector spaces V and W over a field F. We call a function $T{:}V \to W$ a linear transformation from V to W if, for all $x, y \in V$ and $\alpha \in F$, we have:

$$T\,(x + \alpha y) = Tx + \alpha T y \tag{E.5}$$

Elements of Structural Dynamics: A New Perspective, First Edition. Debasish Roy and G Visweswara Rao.

5. **Null space of a linear transformation**
If $T{:}V \to W$ is a linear transformation, the null space (or Kernel) $N(T)$ of T is defined as the set of all vectors v in V such that $T(v) = 0$; that is:

$$N(T) = \{v \in V{:}T(v) = 0\} \tag{E.6}$$

For a standard eigenvalue problem $Av = \lambda I v$, the eigenspace E_{λ_j} for a specific eigenvalue λ_j is given by the null space, $N\left(A - \lambda_j I\right) = \left\{v \in V{:}\left[A - \lambda_j I\right]v = 0\right\}$.
6. **Null space of a differential operator $(D - \alpha I)$ has $e^{\alpha t}$ as a basis**
With $D \equiv \frac{d}{dt}$, the null space of the differential operator $(D - \alpha I)$ is given by $N(D - \alpha I) = \{y{:}\,(D - \alpha I)\,y = 0\}$. Consider the first-order homogeneous differential equation (DE):

$$(D - \alpha)\,y = 0 \tag{E.7}$$

$e^{\alpha t}$ is a solution to the DE. Suppose $x(t)$ is any solution to the above DE, then:

$$\frac{dx}{dt} = \alpha x(t) \forall t \in \mathbb{R}^+ \tag{E.8}$$

Now define:

$$z(t) = e^{-\alpha t} x(t) \tag{E.9}$$

and we have $\frac{dz}{dt} = -\alpha e^{-\alpha t} x(t) + e^{-\alpha t}\left(\alpha x(t)\right) = 0 \Rightarrow z(t) = k$, aconstant. Substituting this result in Equation E.9 gives:

$$x(t) = k e^{\alpha t} \tag{E.10}$$

Thus, any solution to the first-order DE is a linear combination of $e^{\alpha t}$ and hence, null space of $(D - \alpha I)$ has $e^{\alpha t}$ as a basis. As a corollary to the above result, we have that the set of all solutions to a homogeneous linear DE with constant coefficients is the null space of $P(D)$, the auxiliary (characteristic) polynomial associated with the DE. Further, if $P(D) = \left(D - \alpha_1\right)\left(D - \alpha_2\right) \ldots . \left(D - \alpha_n\right)$, then $\left\{e^{\alpha_1 t},\ e^{\alpha_2 t}, \ldots ., e^{\alpha_n t}\right\}$ is a basis for the solution space of the DE.
7. **Invertibility of a matrix (or a linear transformation)**
A matrix A is *invertible* if it is nonsingular ($\det A \neq 0$). Since the product of eigenvalues of a matrix A is equal to its determinant, A is invertible only if none of the eigenvalues is zero. This will be clear if we consider the characteristic polynomial $P(\lambda)$ corresponding to the eigenvalue problem $Ax = \lambda x$:

$$P(\lambda) = \det(A - \lambda I) = \lambda^n + \alpha_{n-1}\lambda^{n-1} + \ldots + \alpha_0 \tag{E.11}$$

Here n is the size of the matrix A. It is clear that the constant term α_0 in the above equation is equivalent to $\det A$. It is also true that the product of the eigenvalues $\prod_{j=1}^{n} \lambda_j = \alpha_0$. Hence, we have the result that A is *invertible* if and only if all the eigenvalues are nonzero.
8. **Diagonalisability of a matrix**
A square matrix A is diagonisable if all its eigenvectors corresponding to the eigenvalue problem $Ax = \lambda x$ are linearly independent. Let the eigenvector matrix be denoted by Q with each column representing an eigenvector. Since the eigenvectors are linearly

independent, it follows that Q is invertible and A can be diagonalised by a similarity transformation:

$$Q^{-1} A Q = \Lambda \tag{E.12}$$

Here, Λ is a diagonal matrix and the diagonal elements are the eigenvalues of A. Further, in view of the orthogonality of the eigenvectors, $Q^{-1} = Q^T$, Equation E.12 takes the form:

$$Q^T A Q = \Lambda \tag{E.13}$$

When all the eigenvalues are distinct, the corresponding eigenvectors are linearly independent. Diagonalisability of A in this case is straightforward. In the case of nondistinct or repeated eigenvalues, if the algebraic multiplicity is equal to the geometric multiplicity of an eigenvalue λ, then the eigenspace E_λ of this eigenvalue contains linearly independent eigenvectors and in this case also, the matrix A is diagonisable.

9. **Symmetric eigenvalue problem yields real eigenvalues and orthogonal eigenvectors**
Consider the eigenvalue problem $Ax = \lambda x$, where A is a real symmetric matrix. A is a linear operator on a vector space V and a nonzero vector v is called an eigenvector of A if there exists a scalar λ such that $Av = \lambda v$. The scalar λ is called the eigenvalue corresponding to the eigenvector v. A has at most n distinct eigenvalues. The eigenspace of A is given by the direct sum of the eigenspaces of all its n eigenvalues.
   i) To prove that the eigenvalues of a real symmetric matrix are real
   Assume that λ_j and corresponding eigenvector v_j are complex. By premultiplying $Av_j = \lambda_j v_j$ on both sides by $\bar{v}_j^T$, where $\bar{v}_j$ is the complex conjugate of v_j, we have:

$$\bar{v}_j^T A v_j = \lambda_j \bar{v}_j^T v_j \tag{E.14}$$

   By transposing $Av_j = \lambda_j v_j$, we also have:

$$\bar{v}_j^T A = \bar{\lambda}_j \bar{v}_j^T \tag{E.15}$$

   Postmultiplying the above equation by v_j and subtracting the result from Equation E.14 gives:

$$\left(\lambda_j - \bar{\lambda}_j\right) \bar{v}_j^T v_j = 0 \tag{E.16}$$

   Since v_j is nontrivial, $\lambda_j - \bar{\lambda}_j = 0 \Rightarrow \lambda_j = \bar{\lambda}_j \Rightarrow \lambda_j$ real.
   ii) To prove that the eigenvectors of a real symmetric matrix are orthogonal
   Assume that the eigenvalues are distinct and consider two distinct eigenpairs $\left(\lambda_j, v_j\right)$ and $\left(\lambda_j, v_j\right)$. This gives:

$$Av_j = \lambda_j v_j \tag{E.17}$$

$$Av_k = \lambda_k v_k \tag{E.18}$$

   Premultiplying Equation E.17 by v_k^T and Equation E.18 by v_j^T on both sides, we get:

$$v_k^T A v_j = \lambda_j v_k^T v_j \text{ and } v_j^T A v_k = \lambda_k v_j^T v_k \tag{E.19}$$

Taking transpose of the second equation and subtracting from the first gives:

$$0 = \left(\lambda_j - \lambda_k\right) v_k^T v_j \Rightarrow v_k^T v_j = 0 \text{ since } \lambda_j \neq \lambda_k \tag{E.20}$$

Equation E.20 shows that the eigenvectors are orthogonal, which is true for any $j \neq k$. If we normalise the eigenvectors such that their dot product is unity, we say that the eigenvectors are orthonormal. That is, let the orthonormalised eigenvectors be denoted by $\hat{v}_j$, $j = 1, 2 \ldots, n$, then $\hat{v}_j = \frac{v_j}{\sqrt{v_j^T v_j}}$.

Appendix F

1. **Orders of approximation**
   Assume that a function $f(h)$ is approximated by the function $p(h)$ and that there exists a real constant $M \geq 0$ and a positive integer n so that: $\frac{f(h)-p(h)}{|h^n|} \leq M$ for sufficiently small h, then we say that $p(h)$ approximates $f(h)$ with order of approximation $O(h^n)$ and write $f(h) = p(h) + O(h^n)$. On the other hand, we say that $p(h)$ approximates $f(h)$ with order of approximation $o(h^n)$ if $\lim_{n \to 0} \frac{f(h)-p(h)}{|h^n|} = 0$ and we write $f(h) = p(h) + o(h^n)$. That is saying an error is $o(h^n)$ is stronger than saying the error is $O(h^n)$ (as $h \to 0$).
2. **Numerical integration of the convolution integral by incremental summation**
   The convolution integral is:

$$Y(t) = \frac{1}{\omega_D} \int_0^t p(\tau)\, e^{-\xi\omega(t-\tau)} \sin \omega_D (t-\tau)\, d\tau \tag{F.1}$$

With $\sin \omega_D (t-\tau) = \sin \omega_D t \cos \omega_D \tau - \cos \omega_D t \sin \omega_D \tau$, $Y(t)$ in Equation F.1 can be written as:

$$Y(t) = A(t) \sin \omega_D t - B(t) \cos \omega_D t \tag{F.2}$$

where

$$A(t) = \frac{e^{-\xi\omega t}}{\omega_D} \int_0^t p(\tau)\, e^{\xi\omega\tau} \cos \omega_D \tau d\tau \text{ and}$$
$$B(t) = \frac{e^{-\xi\omega t}}{\omega_D} \int_0^t p(\tau)\, e^{\xi\omega\tau} \sin \omega_D \tau d\tau \tag{F.3a,b}$$

The integrals in Equation F.3 can be numerically evaluated by using, say, Simpson's rule. Towards this, the domain $(0, t)$ is divided into an even number $(2m)$ of equal subintervals (t_i, t_{i+1}) with $t_0 = 0, \Delta t, 2\Delta t, \ldots., (2m-1)\Delta t, 2m\Delta t = t_{2m}$. Each of the integrands in the two integrals of Equation F.3 are evaluated at each of these time instants. If $y(t)$ represents any of these integrands, we have by Simpson's rule (see Figure F.1):

$$\int_0^t y(t)\, dt = \frac{\Delta t}{3}\left(y_0 + 4y_1 + 2y_2 + 4y_3 + \ldots\ldots + 2y_{2m-2} + 4y_{2m-1} + y_{2m}\right) \tag{F.4}$$

Elements of Structural Dynamics: A New Perspective, First Edition. Debasish Roy and G Visweswara Rao.

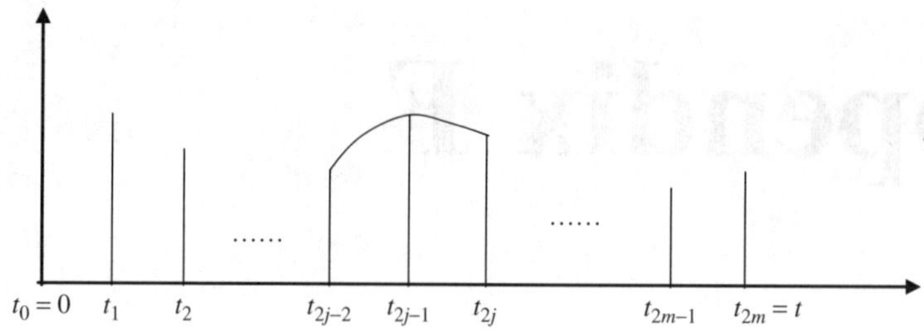

Figure F.1 Evaluation of integrals in Equation F.3 by an incremental summation procedure

Here, $y_j = y(j\Delta t)$. For computational ease, an incremental summation procedure (Clough and Penzien, 1996) can be adopted. To this end, the evaluation of the integrals is accomplished incrementally. That is, if $I(t_{2j-2})$ is the value of the integral in Equation F.4 up to $t_{2j-2}(= t_{2j} - 2\Delta t)$, then $I(t_{2j})$ is given by:

$$I(t_{2j}) = \int_0^{t_{2j}} y(\tau)\,d\tau = I\left(t_{2j-2}\right) + \frac{\Delta t}{3}\left\{y\left(t_{2j-2}\right) + 4y\left(t_{2j-1}\right) + y(t_{2j})\right\} \quad \text{(F.5)}$$

Thus, with $I\left(t_0\right) = 0$, the above incremental summation can be continued to finally get $I(t)$. $A(t)$ and $B(t)$ in Equation F.3 are similarly obtained as:

$$A(t) = \frac{e^{-\xi\omega t}}{\omega_D}\sum_{j=0}^{m} I_1(t_{2j}) \text{ and } B(t) = \frac{e^{-\xi\omega t}}{\omega_D}\sum_{j=0}^{m} I_2(t_{2j}) \quad \text{(F.6)}$$

Here, $I_1(t_{2j})$ and $I_2(t_{2j})$ are:

$$\begin{aligned} I_1\left(t_{2j}\right) &= I_1\left(t_{2j-2}\right) + \frac{\Delta t}{3}\Big\{p\left(t_{2j-2}\right)e^{\xi\omega(t_{2j-2})}\cos\omega_D\left(t_{2j-2}\right) \\ &\quad + 4p\left(t_{2j-1}\right)e^{\xi\omega(t_{2j-1})}\cos\omega_D\left(t_{2j-1}\right) + p\left(t_{2j}\right)e^{\xi\omega t_{2j}}\cos\omega_D t_{2j}\Big\} \\ I_2\left(t_{2j}\right) &= I_2\left(t_{2j-2}\right) + \frac{\Delta t}{3}\Big\{p\left(t_{2j-2}\right)e^{\xi\omega(t_{2j-2})}\sin\omega_D\left(t_{2j-2}\right) \\ &\quad + 4p\left(t_{2j-1}\right)e^{\xi\omega(t_{2j-1})}\sin\omega_D\left(t_{2j-1}\right) + p\left(t_{2j}\right)e^{\xi\omega t_{2j}}\sin\omega_D t_{2j}\Big\} \end{aligned} \quad \text{(F.7a,b)}$$

References

Clough, R.W. and Penzien, J. (1996) *Dynamics of Structures*, McGraw-Hill, Singapore.

Appendix G

1. **Invariant subspace**
   Let V be a vector space and $T:V \rightarrow V$ a linear transformation, A subspace $W \subseteq V$ is T invariant if $T(x) \in W, \forall x \in W$, that is $T(W) \subseteq W$. For example, V, $N(T)$, E_λ are invariant subspaces.
2. **For a positive definite matrix A, all the eigenvalues are positive**
   Consider any vector $x \in V$, a vector space. For a positive definite matrix A (with real elements), we have:

$$< x, Ax> = x^T Ax > 0 \tag{G.1}$$

   If λ is an eigenvalue of A, then:

$$Ax = \lambda x \tag{G.2}$$

   Substituting the above equation in Equation G.1 results in:

$$x^T \lambda x > 0 \Rightarrow \lambda \, \|x\|^2 > 0 \Rightarrow \lambda > 0 \tag{G.3}$$

   The above result shows that the eigenvalues of a positive definite matrix are positive.
3. **For a positive definite matrix A, expressed as LDL^T, the diagonal elements of the diagonal matrix D are greater than zero**
   Consider a matrix $A \in \mathbb{R}^{N\times N}$. The matrix A is decomposed into the form:

$$A = LDL^T \tag{G.4}$$

   Here, L is a lower triangular matrix with $L_{jj} = 1$ and $L_{jk} = 0$ for $k > j, \forall k, j = 1, 2, ..N$. D is a diagonal matrix with diagonal elements $d_{jj}, j = 1, 2, .., N$. It follows that:

$$\det A = \det L \det D \det L^T = 1. \det D.1 = \det D = \prod_{j=1}^{N} d_{jj} \tag{G.5}$$

   Since $\det A = \prod_{j=1}^{N} \lambda_j$, we have:

$$\prod_{j=1}^{N} \lambda_j = \prod_{j=1}^{N} d_{jj} \tag{G.6}$$

   For a positive definite matrix, $\prod_{j=1}^{N} \lambda_j > 0$ (see Item 1 of this appendix) and this leads to $\prod_{j=1}^{N} d_{jj} > 0$. However, this does not ensure that all $d_{jj} > 0, j = 1, 2, \ldots, N$.

Elements of Structural Dynamics: A New Perspective, First Edition. Debasish Roy and G Visweswara Rao.

Now, consider the sequence of matrices $\{A^{(k)}\}$ formed from the original matrix A by deleting the last k rows and columns. Let each of these matrices also be decomposed as:

$$A^{(k)} = L^{(k)} D^{(k)} L^{(k)T} \tag{G.7}$$

Here, $L^{(k)}$ and $D^{(k)}$ are contained in L and D, respectively, in that these can be obtained by omitting the last k rows and k columns from L and D. Consider first the case with $k = N - 1$ when $\lambda_1 = d_{11}$. Since λ_1 of $A^{(N-1)}$ is positive, d_{11} is positive. For $k = N - 2$, $\lambda_1\lambda_2 > 0 \Rightarrow d_{11}d_{22} > 0$. Since $d_{11} > 0$, it follows that $d_{22} > 0$. Continuing this argument, we obtain the result that $d_{jj}, j = 1, 2, .., N$ are all positive.

4. $\nabla_{x^*} \cdot v^* = \text{tr } L^* = \text{tr } L = \nabla_x \cdot v$ **(Equation 7.117)**

Proof:

The velocity gradient L is given by:

$$L = \nabla_x v \tag{G.8}$$

Thus, L is a second-order tensor since the gradient of a vector is a second-order tensor. From Equation G.8, we have:

$$tr L = \nabla \cdot v \tag{G.9}$$

L is related to the deformation gradient F by the relation:

$$\frac{dF}{dt} = \left(\frac{\partial F}{\partial t}\right)_x = LF \tag{G.10}$$

Similarly for Frame O^*, it follows:

$$\text{tr } L^* = \nabla \cdot v^* \text{ and } \frac{dF^*}{dt} = L^* F^* \tag{G.11}$$

Since $F^* = QF$ (frame indifferent), Equation G.11 results in:

$$\frac{d(QF)}{dt} = L^* QF \Rightarrow Q\frac{dF}{dt} + \dot{Q}\, F = L^* QF \tag{G.12}$$

From the above equation, we have:

$$L^* = \left(Q\frac{dF}{dt} + \dot{Q}F\right) F^{-1} Q^{-1} \tag{G.13}$$

Since Q is orthogonal:

$$L^* = QLQ^T + \dot{Q}Q^T \tag{G.14}$$

Since $\dot{Q}Q^T$ is skew-symmetric (see Item 4 of this appendix), $\text{tr}\left(\dot{Q}Q^T\right) = 0$. Further, QLQ^T is an orthogonal similarity transformation, traces of L and QLQ^T remain the same (since trace of a matrix is its first principal invariant). Therefore,

$$\text{tr } L^* = \text{tr } L \Rightarrow \nabla \cdot v^* = \nabla \cdot v \tag{G.15}$$

5. **$W = \dot{Q} Q^T$ is skew-symmetric** ($\mathbb{R}^{3\times3}$)
Proof:
Since Q is orthogonal, we have:

$$Q Q^T = I \tag{G.16}$$

Differentiating both sides of the above equation with respect to time t gives:

$$\frac{dQ}{dt}Q^T + Q\frac{dQ^T}{dt} = 0$$

$$\Rightarrow \frac{dQ}{dt}Q^T = -Q\frac{dQ^T}{dt} = -Q\left(\frac{dQ}{dt}\right)^T$$

$$\Rightarrow \left(\frac{dQ}{dt}Q^T\right)^T = \left\{-Q\left(\frac{dQ}{dt}\right)^T\right\}^T$$

$$\Rightarrow \left(\frac{dQ}{dt}Q^T\right)^T = -\frac{dQ}{dt}Q^T \tag{G.17}$$

The above equation shows that $\dot{Q} Q^T$ is skew-symmetric.

6. **Axial vector of a skew symmetric matrix in $\mathbb{R}^{3\times3}$**
Proof:
Let $W = \dot{Q} Q^T$ be skew-symmetric (see Item 5 of this appendix). An important property of a skew tensor is the existence of an axial vector (or dual vector). This follows from the fact that given a skew tensor W, there exists a unique vector u such that:

$$Wv = u \times v \text{ for } u \in \mathbb{R}^3 \text{ and any vector } v \in \mathbb{R}^3 \tag{G.18}$$

Conversely, given a vector u, there exists a unique skew tensor W such that Equation G.18 holds for every vector v. This is proved in the following:
Let $W = w_{ij}$. Since W is skew, $w_{ij} = -w_{ji}$. So, we have:

$$w_{ij} = \frac{1}{2}\left(w_{ij} - w_{ji}\right) \tag{G.19}$$

In terms of the Kronecker delta, δ_{ij}, and the third-order permutation tensor, ε_{ijk} (see Equation 2.167 for definition of Kronecker delta and Equation A.3 of Appendix A for definition of the permutation tensor), w_{ij} can be written using $\varepsilon - \delta$ identity (see Bibliography of Chapter 1 for definitions on tensors) as:

$$w_{ij} = \frac{1}{2}\left(\delta_{ip}\delta_{jq} - \delta_{jp}\delta_{iq}\right) = \frac{1}{2}\varepsilon_{ijk}\,\varepsilon_{kpq}\,w_{pq} \tag{G.20}$$

Since $\frac{1}{2}\varepsilon_{kpq} w_{pq}$ results in a vector, say $\{u_k\}$, we find that this vector is uniquely generated by W. Thus, Equation G.20 can be written as:

$$w_{ij} = -\varepsilon_{ijk}\,u_k \tag{G.21}$$

Now, choosing any vector v with components v_j, we have the result from Equation G.21:

$$w_{ij}v_j = -\varepsilon_{ijk}\,u_k v_j = \varepsilon_{ijk}\,u_j v_k \Rightarrow Wv = u \times v \tag{G.22}$$

u is called the axial (or dual) vector of $\boldsymbol{W}$. To prove the converse is straightforward. Let a vector u be given. We obtain from Equation G.21 the components w_{ij} of a skew tensor W, which is uniquely generated by u. Equation G.22 readily leads to the result in Equation G.18 and thus W is the unique skew tensor for which u is the axial vector.

7. **Relationship between the angular velocity vector $(\omega_x, \omega_y, \omega_{z,})^T$ of rotating frame O with respect to the fixed frame of reference O^***
As shown in Figure G.1, the (XYZ) system represents the fixed reference frame and (xyz) the rotating frame.
The rotating coordinate system can be constructed (Meirovitch, 1970) from the fixed frame via a set of three angles of rotation (Euler angles) ψ, θ and ϕ. The first rotation is about the Z-axis by the angle ψ resulting in $x'y'Z$ system of axes. The second is about the y'-axis by an angle θ giving $xy'z'$. Now, by a rotation of angle ϕ about the x-axis, we finally obtain the (xyz) system. If ω_x, ω_y and ω_z are the angular velocities in the (xyz) system, they are related to the velocities in the (XYZ) system as:

$$\omega_x = \dot{\phi} - \dot{\psi} \sin\theta$$
$$\omega_y = \dot{\psi} \cos\theta \sin\phi + \dot{\phi} \cos\phi$$
$$\omega_z = \dot{\psi} \cos\theta \cos\phi - \dot{\theta} \sin\phi \tag{G.23a-c}$$

In matrix form, the relationship is given by:

$$\begin{pmatrix} \omega_x \\ \omega_y \\ \omega_z \end{pmatrix} = \begin{bmatrix} -\sin\theta & 1 & 0 \\ \cos\theta \sin\phi & 0 & \cos\phi \\ \cos\theta \cos\phi & 0 & -\sin\phi \end{bmatrix} \begin{pmatrix} \dot{\psi} \\ \dot{\phi} \\ \dot{\phi} \end{pmatrix} \tag{G.24}$$

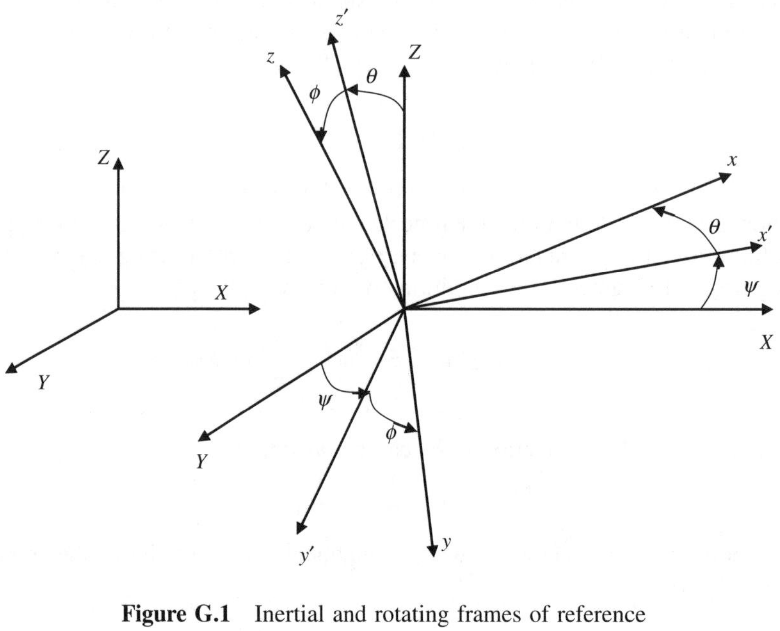

Figure G.1 Inertial and rotating frames of reference

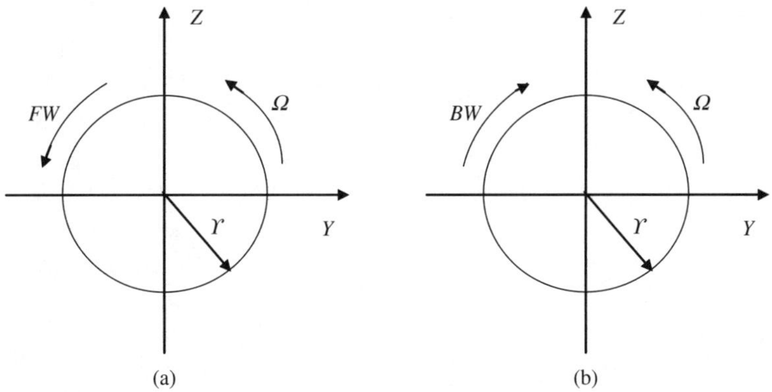

Figure G.2 (a) Forward and (b) backward whirl for a rotating shaft; FW–forward whirl, BW–backward whirl

8. **Forward whirl and backward whirl of a rotating shaft**
   The effect of rotation on the natural frequencies of a rotating shaft is to split each one of those corresponding to the nonrotating shaft into what are usually referred to as backward and forward whirl frequencies (Figure G.2). **Critical speeds** are those rotation speeds that coincide with the shaft whirl speeds (natural frequencies). These speeds correspond to the resonances in rotating shafts and may cause large-amplitude whirls to take place. For further reading on the subject, refer to Rao (1983) and Lalanne and Ferraris (1998).

References

Lalanne, M. and Ferraris, G. (1998) *Rotordynamics Prediction in Engineering*, 2nd edn, John Wiley & Sons Ltd., UK.

Meirovitch. L. (1970) *Methods of Analytical Dynamics*, McGraw-Hill, New York, USA.

Rao, J. S. (1983) Rotordynamics, Wiley Eastern Ltd., New Delhi, India.

Appendix H

1. **Orders of approximation**
   Assume that a function $f(h)$ is approximated by the function $p(h)$ and that there exists a real constant $M \geq 0$ and a positive integer n so that: $\frac{f(h)-p(h)}{|h^n|} \leq M$ for sufficiently small h, then we say that $p(h)$ approximates $f(h)$ with order of approximation $O(h^n)$ and write $f(h) = p(h) + O(h^n)$. On the other hand, we say that $p(h)$ approximates $f(h)$ with order of approximation $o(h^n)$ if $\lim_{h\to 0} \frac{f(h)-p(h)}{|h^n|} = 0$ and we write $f(h) = p(h) + o(h^n)$. That is saying an error is $o(h^n)$ is stronger than saying the error is $O(h^n)$ (as $h \to 0$).
2. **Metric space**
   A set M of elements is termed a *metric* space if for every pair of elements $x, y \in M$ there is defined a real number $d(x, y)$ called the 'distance', also termed as a metric between x and y satisfying the following axiomatic properties:

$$d(x, y) > 0 \text{ if } x \neq y,\ d(x, y) = 0 \text{ iff } x = y$$

$$d(x, y) = d(y, x) \text{ for every } x \text{ and } y \text{ (symmetry)}$$

$$d(x, y) \leq d(x, z) + d(y, z) \text{ for every } z \in M \text{ (triangle inequality)} \qquad \text{(H.1a-c)}$$

   A metric space is usually denoted by (M, d). Sets in $\mathbb{R}$, $\mathbb{R}^2$ and Euclidean space $\mathbb{R}^3$ are examples of metric spaces with the usual concept of distance between any two points. Many sets of functions can be regarded as metric spaces when equipped with a metric suitably defined on these spaces. For example, any set M of continuous functions defined on an interval $[a, b]$ constitutes a metric space if it is associated with a metric defined, say, as the distance between any two member functions $x(t)$ and $y(t)$:

$$d(x, y) := ||x - y||_\infty = \max_{a<t<b} |x(t) - y(t)| \qquad \text{(H.2)}$$

   The above metric satisfies all the properties in Equation H.1. Another choice for the metric may be:

$$d(x, y) = \int_a^b |x(t) - y(t)|\, dt \ (L^1 \text{ norm}) \qquad \text{(H.3)}$$

   Normed and inner product spaces (See Appendix B) are thus metric spaces. In the case of former, we may, for instance, define the metric as:

$$d(x, y) = \|x - y\|_p,\ 1 \leq p \leq \infty \qquad \text{(H.4)}$$

Elements of Structural Dynamics: A New Perspective, First Edition. Debasish Roy and G Visweswara Rao.

Thus, $d(\cdot,\cdot)$ is the metric generated by the norm $\|\cdot\|$ and indeed satisfies the axioms in Equation H.1 for a metric. For the inner product spaces ($p = 2$), the inner product (x, y) associated with these spaces is chosen as the metric.

3. ***z*-transform**

As a Laplace transform is to a continuous signal, so is a z-transform to a sampled (digitised) signal. Indeed, a z-transform is evolved from the Laplace transform and this is explained in the following. The sampling process is a modulation operation in which a pulse train (Figure H.1) with magnitude $\frac{1}{\gamma}$ and period T multiplies a continuous function $f(t)$ and outputs the sampled function $f^*(t)$. If the pulse duration γ is small compared with the sampling period, the pulse train can be represented by an ideal sampler. In this ideal case, since the area of each pulse is unity, the ideal sampler produces the impulse train (Figure H.1e) of unit magnitude that is given by:

$$\delta_T(t) = \sum_{n=-\infty}^{+\infty} \delta(t - nT) \tag{H.5}$$

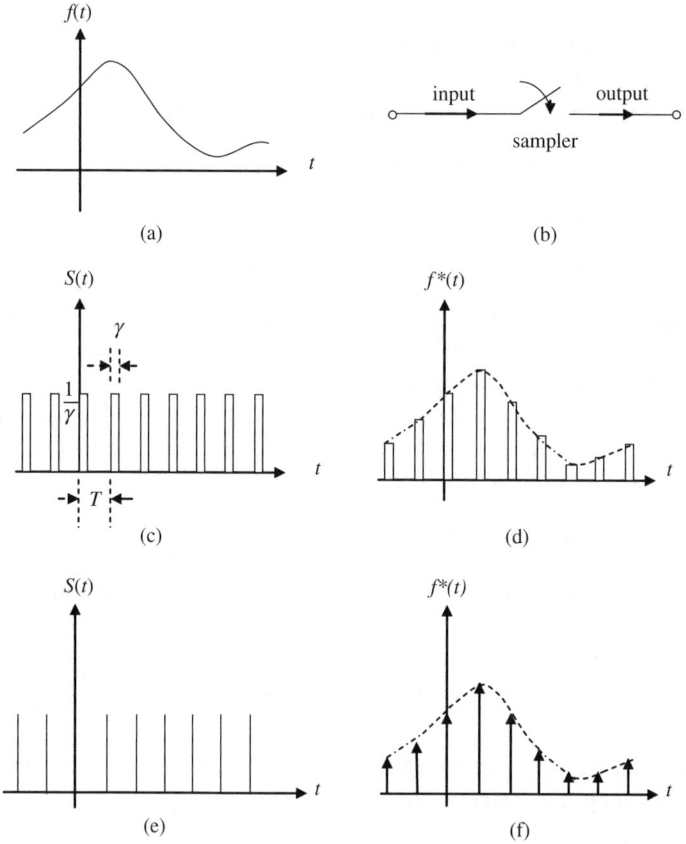

Figure H.1 Sampled signal data; (a) continuous signal, (b) a sampler–periodic sampling, (c) pulse train, (d) pulse sampling approximation, (e) impulse train and (f) impulse sampling approximation

When $f(t) = 0$ for $t < 0$, the Laplace transform of the output impulse sequence $f^*(t)$ (Figure H.1f) is:

$$Lf^*(t) = F^*(s) = L\left(\sum_{n=-\infty}^{+\infty} f(t)\delta(t-nT)\right) = \sum_{n=0}^{\infty} f(nT)e^{-nTs} \quad \text{(H.6)}$$

$f(nT)$ is the function $f(t)$ sampled at the sampling times nT. With the change of variable, $z = e^{Ts}$ in Equation H.6, we get:

$$F^*(s)\Big|_{s=\frac{1}{T}\ln z} = F^*(z) = \sum_{n=0}^{\infty} f(nT)z^{-n} \quad \text{(H.7)}$$

Thus, if a continuous signal $f(t)$, which is zero for $t < 0$, is represented by sampled values $f(nT), n \geq 0$, its one-sided z-transform is given by Equation H.7. Note that z is an arbitrary complex number and $F^*(z)$, a sum of complex numbers, is also complex. As an example of a z-transform, if $f(nT) = a^n$ for $n \geq 0$, and $f(nT) = 0$ for $n < 0$, the z-transform of this geometric sequence is:

$$F^*(z) = \sum_{n=0}^{\infty} a^n z^{-n} = \sum_{n-0}^{\infty} (az^{-1})^n \quad \text{(H.8)}$$

The infinite sum in Equation H.8 is convergent if $|az^{-1}| < 1$ or $|z| > |a|$. Thus, the z-transform for the geometric sequence takes the form:

$$F^*(z) = \frac{1}{1-az^{-1}} = \frac{z}{z-a} \quad \text{for } |z| > |a| \quad \text{(H.9)}$$

In general, the z-transform of any sequence $f(nT)$ will have a region of convergence (ROC) as shown in Figure H.2. Table H.1 gives some of the commonly used z-transforms.

A z-transform is used in the analysis of sampled data systems. The mathematical model of a sampled-data system is in the form of difference equation:

$$\begin{aligned} &c(n) + a_1c(n-1) + \cdots + a_nc(n-k) \\ &= b_0r(n) + b_1r(n-1) + \cdots + b_kr(n-k) \end{aligned} \quad \text{(H.10)}$$

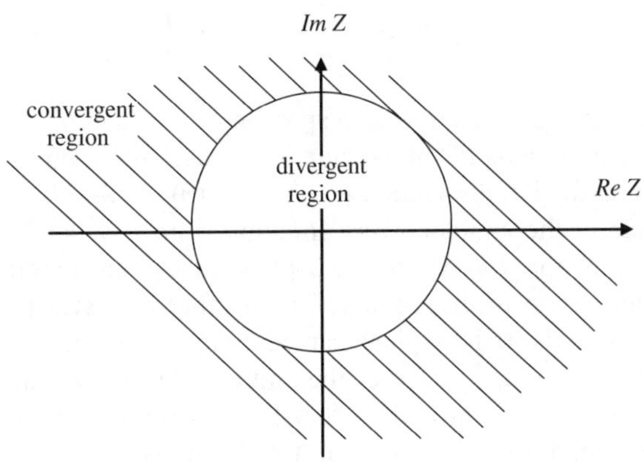

Figure H.2 ROC for the z-transform of a geometric sequence, $f(nT) = a^n$, $n \geq 0$ and $= 0, n < 0$

Table H.1 List of some commonly used z-transforms along with corresponding Laplace transforms

$f(t)$	$f(nt)$	z-transform	L-transform
$\delta(t)$	$\delta(nT)$	1	1
$\delta(t-kT)$	$\delta\{(n-k)T\}$	z^{-k}	e^{-kTs}
$U(t)$	$U(nT)$	$\frac{z}{z-1}$	$\frac{1}{s}$
e^{at}	e^{anT}	$\frac{z}{z-e^{aT}}$	$\frac{1}{s-a}$
$\sin\omega t$	$\sin\omega nT$	$\frac{z\sin\omega T}{z^2-2z\cos\omega T+1}$	$\frac{\omega}{s^2+\omega^2}$
$\cos\omega t$	$\cos\omega nT$	$\frac{z(z-cos\omega T)}{z^2-2z\cos\omega T+1}$	$\frac{s}{s^2+\omega^2}$
t	nT	$\frac{zT}{(z-1)^2}$	$\frac{1}{s^2}$

z-transforming both sides of the above equation and assuming zero initial conditions and casual input $(r(t)=0, t<0)$ gives the z-transfer function as:

$$\frac{C(z)}{R(z)}=\frac{b_0+b_1z^{-1}+\cdots+b_kz^{-k}}{1+a_1z^{-1}+\cdots+a_kz^{-k}} \tag{H.11}$$

Further information on z-transforms, their properties and applications are available, for example in (D'Azzo and Houpis, 1988; Haykin and Veen, 1999). However, a brief outline of the s-plane and z-plane relationship with particular reference to stability aspects of ordinary differential equations (ODEs) is provided below.

s-plane and z-plane relationship

The transformation $z=e^{Ts}$ or $s=\frac{1}{T}\ln z$ maps the s-plane into the z-plane. Specifically, the mapping of the imaginary axis ($j\omega$-axis) of the s-plane (Figure H.3a) is:

$$z=e^{Tj\omega}=e^{\frac{j2\pi\omega}{\omega_s}}=1\angle 2\pi\left(\frac{\omega}{\omega_s}\right) \tag{H.12}$$

Here, $\frac{2\pi}{\omega_s}$ is the sampling frequency corresponding to the sampling period T. Equation H.12 shows that the $j\omega$-axis of the s-plane is mapped into a unit circle in the z-plane (Figure H.3b). In particular, the section $\left(-j\frac{\omega_s}{2}\right)\rightarrow(0)\rightarrow\left(+j\frac{\omega_s}{2}\right)$ of the $j\omega$-axis maps into the unit circle in the anticlockwise direction from $\pi\rightarrow-\frac{\pi}{2}\rightarrow 0\rightarrow\frac{\pi}{2}\rightarrow\pi$. In fact, every section of the $j\omega$-axis in the s-plane that is an integral multiple of ω_s is mapped into a unit circle in the z-plane. This further indicates that the region to the left of the imaginary axis in the s-plane can be divided into strips of width equal to ω_s, each of which maps into the interior of the unit circle in the z-plane. This relationship between the Laplace transform and $z-$transform is relevant if we also consider the stability of, for example the solution to an ODE. Let us consider the single degree of freedom (SDOF) oscillator equation:

$$\ddot{u}+2\xi\,\omega\dot{u}+\omega^2u=0 \tag{H.13}$$

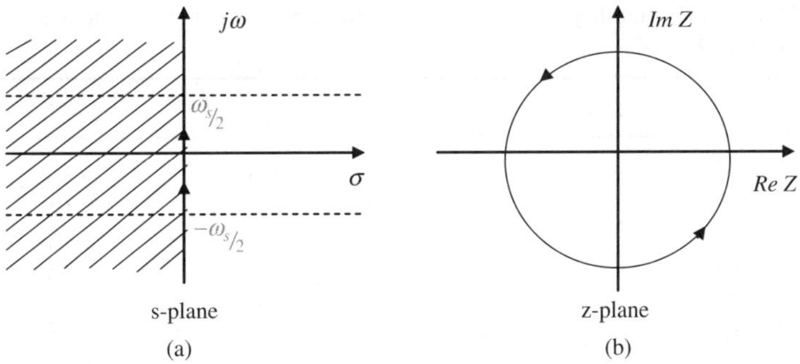

Figure H.3 s-plane and z-plane relationship for a system stability

and take $\boldsymbol{x}(t) = \{u(t)\ \dot{u}(t)\}^T$. The ODE (Equation H.13) can be written in state vector form:

$$\dot{x}(t) = \boldsymbol{A}\boldsymbol{x}(t) \text{ with } \boldsymbol{A} = \begin{bmatrix} 0 & 1 \\ -\omega^2 & -2\xi\omega \end{bmatrix} \tag{H.14}$$

By Laplace transforming the above equation we get:

$$\boldsymbol{X}(s) = [s\boldsymbol{I} - \boldsymbol{A}]^{-1}\boldsymbol{x}(0) \tag{H.15}$$

$\boldsymbol{x}(0)$ is the initial condition vector. The stability of the solution $\boldsymbol{x}(t)$, which is usually known as the complementary solution, depends on the nature of the eigenvalues λ_j, $j = 1, 2$ of the matrix $\boldsymbol{A}$ and hence on the roots of the characteristic equation given by:

$$\det[s\boldsymbol{I} - \boldsymbol{A}] = 0 \Rightarrow s^2 + 2\xi\omega s + \omega^2 = 0 \tag{H.16}$$

The complementary solution is of the form: $\boldsymbol{x}(t) = \boldsymbol{c}_1 e^{\lambda_1 t} + \boldsymbol{c}_2 e^{\lambda_2 t}$. The unknown constant vectors $\boldsymbol{c}_j$, $j = 1, 2$ are to be evaluated from the initial conditions. The roots of Equation H.16 are $s_{1,2} = -\xi\omega \pm i\omega\sqrt{1-\xi^2}$. With $\xi, \omega > 0$, the eigenvalues always possess negative real parts and hence lie in the left-half of the $s-$plane. Thus, the system is asymptotically stable in that the complementary solution $\boldsymbol{x}(t)$ (under no external excitation) decays to zero due to any initial condition disturbance. In fact, the stability of all linear time-invariant (LTI) systems is ensured if their eigenvalues (in general complex) lie in the left-half of the $s-$plane.
Now, if we substitute $s = \frac{1}{T}\ln z$ in the roots of Equation H.16, we get:

$$\ln z = -\xi\omega T \pm i\omega T\sqrt{1-\xi^2} \Rightarrow z_{1,2} = e^{-\xi\omega T} e^{\pm i\omega T\sqrt{1-\xi^2}} \tag{H.17}$$

Equation H.17 shows that with $\xi, \omega T > 0$, $|z_{1,2}| < 1$ and thus the roots lie within an unit circle in the $z-$plane. Thus, the equivalence of the criteria of stability in the s- and z-planes (Figure H.3) is evident.

4. **Jury's stability test** (Jury, 1974)
   Jury's test is a simple means of determining whether or not the roots of a characteristic polynomial lie within a unit circle in the z-transform plane so that one can decide the

Table H.2 Jury's stability criterion; coefficients for testing the sufficient conditions in Equation H.21

Row	z^0	z^1	z^2	....	z^{n-k}	....	z^{n-1}	z^n
1	a_0	a_1	a_2	....	a_{n-k}	....	a_{n-1}	a_n
2	a_n	a_{n-1}	a_{n-2}	....	a_k	....	a_1	a_0
3	b_0	b_1	b_2	....	....	....	b_{n-1}	
4	b_{n-1}	b_{n-2}	....	....	....	....	b_0	
5	c_0	c_1	c_2	....	....	c_{n-2}		
6	c_{n-2}	c_{n-3}	....	....	....	c_0		
⋮	⋮	⋮	⋮	⋮	⋮	⋮		
$2n-5$	r_0	r_1	r_2	r_3				
$2n-4$	r_3	r_2	r_1	r_0				
$2n-3$	s_0	s_1	s_2					

associated system stability (see Item 3 of this Appendix also). Like the Ruth–Hurwitz criterion (D'Azzo and Houpis, 1988), this test also provides both necessary and sufficient conditions for this purpose. Let us consider the nth-degree polynomial:

$$P(z) = a_n z^n + a_{n-1} z^{n-1} + \cdots + a_1 z + a_0, \ a_n > 0 \tag{H.18}$$

The necessary conditions for stability are:

$$P(1) > 0 \ \ and \ \ (-1)^n P(-1) > 0 \tag{H.19}$$

One simple method to assess the sufficient conditions for stability consists of preparing the Table H.2.

In Table H.2, b_k, c_k, d_k, and so on are determinants given by:

$$b_k = \begin{vmatrix} a_0 & a_{n-k} \\ a_n & a_k \end{vmatrix},$$

$$c_k = \begin{vmatrix} b_0 & b_{n-1-k} \\ b_{n-1} & b_k \end{vmatrix}$$

$$d_k = \begin{vmatrix} c_0 & c_{n-2-k} \\ c_{n-2} & c_k \end{vmatrix} \text{ and so on}\ldots. \tag{H.20}$$

Now, the sufficient conditions for stability are:

$$|a_0| < |a_n|$$

$$|b_0| > |b_{n-1}|$$

$$|c_0| > |c_{n-2}|$$

$$|s_0| > |s_2| \tag{H.21}$$

Equation H.21 above contains $n - 1$ conditions for stability (for $n \geq 2$).

References

D'Azzo, J. J. and Houpis, C. H. (1988) Linear Control System Analysis and Design: Conventional and Modern, McGraw-Hill, Singapore.

Haykin, S. and Veen. B. V. (1999) *Signals and Systems*, John Wiley & Sons, Singapore.

Jury, E. I. (1974) *Inners and stability of dynamical systems*, John Wiley & sons, London.

Appendix I

1. **Open and closed sets**
   A subset $A \subseteq \mathbb{R}$ is said to be open if $\exists\, \varepsilon > 0$ such that $(x - \varepsilon, x + \varepsilon) \subseteq A$, $\forall\, x \in A$. In higher dimensions, we may consider an open ball, $\mathcal{B}_r\left(\boldsymbol{x}_0\right) = \boldsymbol{x}, \left|\boldsymbol{x} - \boldsymbol{x}_0\right| < r$ with centre $\boldsymbol{x}_0$ and radius $r > 0$ in $\mathbb{R}^n$. These open sets are measurable. For $n = 1$, the measure is merely the distance between x and x_0 on the real line $\mathbb{R}$.
   A subset $B \subseteq \mathbb{R}$ is said to be closed if $A = B^c$ is open.
2. **Simulation of a random variable with a specified distribution $F_X(x)$**
   Assume that the cumulative distribution function (CDF) $F_X(x)$ is strictly increasing with $0 \le F_X(x) \le 1$. For standard uniform variate $U(0, 1)$, we have (Figure I.1):
   $$F_U(u) = P\left(U \le u\right) = u \tag{I.1}$$
   Using the transformation $F_X(x) = u$ and utilising Equation I.1, we obtain the cumulative probability:
   $$P\left(X \le x\right) = P\left(F_X^{-1}(U) \le x\right) = P\left(U \le F_X(x)\right) = F_U\left(F_X(x)\right) = F_X(x) \tag{I.2}$$
   Thus, with a set of numbers $u_1, u_2, \ldots.$ (generated as realisations of the random variable U), the corresponding set $x_1, x_2, \ldots.$ for the random variable X is obtained via the inverse transformation:
   $$x_i = F_X^{-1}\left(u_i\right), i = 1, 2 \ldots \tag{I.3}$$
3. **Rayleigh distribution from two normal random variables**
   Let X and Y be independent and identically distributed (i.i.d) random variables distributed as $N(0.\sigma^2)$. To find the distribution of the random variable $Z = \sqrt{X^2 + Y^2}$, we proceed as follows:
   $$F_Z(z) = P\left(Z \le z\right) = P\left(\sqrt{X^2 + Y^2} \le z\right) = \int_\Omega f_{XY}\left(x, y\right) dxdy \tag{I.4}$$
   Here, $d\Omega = dxdy$ is the region in the x-y plane that satisfies $\sqrt{X^2 + Y^2} \le z$. Hence, the requirement is to find the point set in the X-Y plane such that the events $\{\omega{:}Z(\omega) \le z\}$ and $\{\omega{:}\,\{X(\omega), Y(\omega)\} \in d\Omega\}$ are equal (Figure I.2).
   Using polar coordinates $x = r\cos\theta$ and $y = r\sin\theta$, we have $dx\,dy = r dr d\theta$ and $\sqrt{X^2 + Y^2} \le z \Rightarrow r \le z$. Noting that $f_{XY}\left(x, y\right) = \frac{1}{2\pi\sigma^2} e^{-\frac{x^2+y^2}{2\sigma^2}}$, we have:
   $$F_Z(z) = \frac{1}{2\pi\sigma^2}\int_0^{2\pi} d\theta \int_0^z r e^{-\frac{r^2}{2\sigma^2}} dr = \left(1 - e^{-\frac{z^2}{2\sigma^2}}\right) U(z) \tag{I.5}$$

Elements of Structural Dynamics: A New Perspective, First Edition. Debasish Roy and G Visweswara Rao.

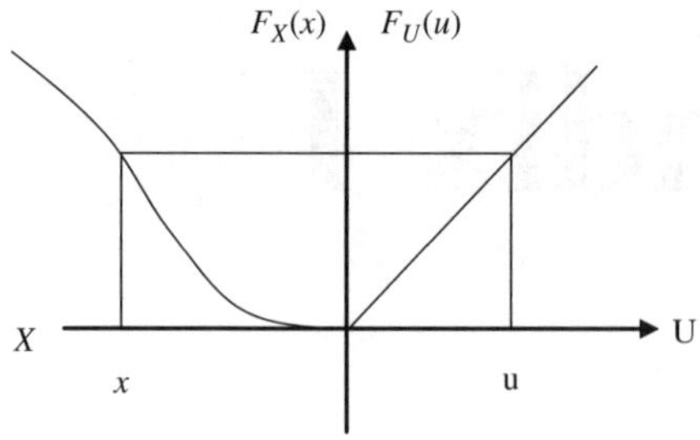

Figure I.1 Generation of realisations for X of specified $F_X(x)$ via a transformation using $U(0, 1)$

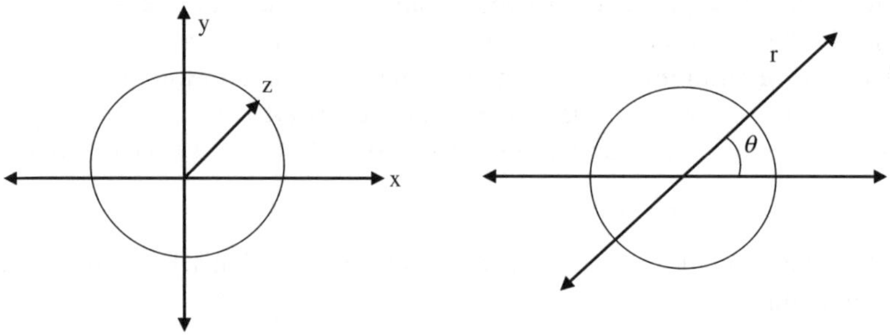

Figure I.2 Rayleigh distribution from two normal random variables (RVs)

The pdf given by:

$$f_Z(z) = \frac{z}{\sigma^2} e^{-\frac{z^2}{2\sigma^2}} U(z) \tag{I.6}$$

U(z) is the Heavyside unit step function. $f_Z(z)$ is the pdf of a Rayliegh distribution.

4. **Convergence of a sequence of random variables**
   Consider a sequence of random variables $\{X_n\} := \{X_1, X_2, \ldots..X_n\}$. Different criteria for the convergence of the sequence are given in the following.
   a) Convergence everywhere
      If X is a random variable such that $\lim_{n\to\infty} \{X_n\} = X$, then the sequence $\{X_n\}$ is said to converge to X everywhere.
   b) Convergence almost surely $(a.s.)$
      If $P(X_n \to X) = 1$ as $n \to \infty$, then the sequence $\{X_n\}$ is said to converge to X $a.\ s.$.
   c) Convergence in probability
      If $P(|X_n - X| > \varepsilon) \to 0$ for every $\varepsilon > 0$ as $n \to \infty$, then the sequence $\{X_n\}$ is said to converge to X in probability.
   d) Convergence in the mean square sense
      If $E[|X_n - X|^2] \to 0$ as $n \to \infty$, then the sequence $\{X_n\}$ is said to converge to X in the mean square sense.
   e) Convergence in distribution (or weak convergence)

If $F_n(x)$ and $F(x)$ are the distribution functions of X_n and X, respectively, such that $F_n(x) \to F(x)$ as $n \to \infty$ for every point of continuity of $F(x)$, then the sequence $\{X_n\}$ is said to converge to X in distribution.

5. **Markov and Chebychev inequalities**

Markov inequality: For a random variable X and any $\alpha > 0$:

$$P(|X| \geq \alpha) \leq \frac{E[|X|]}{\alpha} \tag{I.7}$$

Proof: Define $A = \{\omega \in \Omega |\, |X(\omega)| > \alpha\}$. Then, if $I_A(\omega)$ denotes the indicator function of the set A, it follows that $|X(\omega)| \geq \alpha I_A(\omega)$. Thus, we have $E(|X|) \geq \alpha E(I_A) = \alpha P(A) = \alpha P(|X| \geq \alpha)$.

Chebychev inequality: Let $E[X] = \mu$. For any $\varepsilon > 0$:

$$P\left(|X - \mu| \geq \varepsilon\right) \leq \frac{\sigma^2}{\varepsilon^2} \tag{I.8}$$

Proof:

$$P\left(|X - \mu| \geq \varepsilon\right) = \int_{-\infty}^{-\mu-\varepsilon} f(x)dx + \int_{\mu+\varepsilon}^{\infty} f(x)dx = \int_{|x-\mu|\geq\varepsilon} f(x)dx \tag{I.9}$$

where $f(x)$ denotes the *pdf* of X. Further,

$$\sigma^2 = \int_{-\infty}^{\infty} (x-\mu)^2 f(x)dx \geq \int_{|x-\mu|\geq\varepsilon} (x-\mu)^2 f(x)dx \geq \varepsilon^2 \int_{|x-\mu|\geq\varepsilon} f(x)dx$$

$$= \varepsilon^2 P\left(|X-\mu| \geq \varepsilon\right) \quad \text{(from Equation I.9)}$$

$$\Rightarrow P\left(|X-\mu| \geq \varepsilon\right) \leq \frac{\sigma^2}{\varepsilon^2} \tag{I.10}$$

6. **Law of large numbers**

The law of large numbers is concerned with the convergence of estimates of the mean of a random variable. There exist two versions of the law of large numbers. One is the weak law of large numbers and the other, the strong law of large numbers. While both of these refer to the estimate of sample average from a large number of observations/experiments, they differ only in the mode of convergence of the estimate.

The weak law of large numbers: Suppose that X_j, $j = 1, 2, ..,$ be a sequence of independent and identically distributed random variables, each having finite mean, $E[X_j] = \mu$. Then, for any $\varepsilon > 0$:

$$P\left(\left|\frac{X_1 + X_2 + \ldots + X_n}{n} - \mu\right| \geq \varepsilon\right) \to 0 \text{ as } n \to \infty \tag{I.11}$$

Thus, the above implies convergence in probability. In the case of the strong law of large numbers, however, the convergence is in the almost sure sense as stated below.

The strong law of large numbers: For the sequence of independent and identically distributed random variables X_j, $j = 1, 2, ...,$ each with finite mean μ, the strong law states that:

$$P\left(\lim_{n\to\infty} \frac{X_1 + X_2 + \ldots + X_n}{n} = \mu\right) = 1 \tag{I.12}$$

Proofs of the two laws are straightforward (Ross, 2002)

7. **Central limit theorem (CLT)**
Given n independent random variables X_j, $j = 1, 2, .., n$, with X_j having mean μ_j and variance σ_j^2, define $Y = \sum_{j=1}^{n} X_j$ to be another random variable with mean, $\mu = \sum_{j=1}^{n} \mu_j$ and variance, $\sigma^2 = \sum_{j=1}^{n} \sigma_j^2$. The CLT states that, under certain conditions, the distribution $F_Y(y)$ of Y approaches a normal distribution with mean μ and variance σ^2 as $n \to \infty$. The theorem is an illustration of convergence in distribution. The theorem implies that if X_j are continuous random variables, the density function of Y approaches a normal *pdf*:

$$F_Y(y) = \frac{1}{\sigma\sqrt{2\pi}} \exp\left(-\frac{1}{2}\left(\frac{y-\mu}{\sigma}\right)^2\right) \tag{I.13}$$

In other words, if Z is the normalised random variable $\frac{Y-\mu}{\sigma}$ with zero mean and unit variance, the CLT states that $F_Z(z)$ tends to $N(0, 1)$ as $n \to \infty$.
Proof:
Assume that $\emptyset_j(\omega)$ and $\emptyset_Y(\omega)$ are the characteristic functions of X_j and Y. Since the random variables X_j are independent, we have:

$$\emptyset_Y(\omega) = \prod_{j=1}^{n} \emptyset_j(\omega) = (\emptyset_j(\omega))^n \tag{I.14}$$

Here, ω is the frequency parameter in the definition of the characteristic function (see Equation 9.24) and not a point in the sample set Ω. Thus, the characteristic function of Z, as defined above, is given by:

$$\emptyset_Z(\omega) = \left[\exp\left(-\frac{i\omega m}{\sigma\sqrt{n}}\right)\emptyset_j\left(\frac{\omega}{\sigma\sqrt{n}}\right)\right]^n$$

$$= \left[\exp\left(-\frac{i\omega m}{\sigma\sqrt{n}}\right)\left\{1 + \frac{i\omega m}{\sigma\sqrt{n}} + \frac{(m^2+\sigma^2)}{2}\left(\frac{i\omega m}{\sigma\sqrt{n}}\right)^2 + \ldots..\right\}\right]^n$$

(by MacLaurin's series)

$$= \left[1 - \frac{\omega^2}{2n} + o\left(\frac{\omega^2}{2n}\right)\right]^n$$

$$\to \exp\left(-\frac{\omega^2}{2}\right) \text{ as } n \to \infty \quad \left(\text{using } \left(1+\frac{1}{n}\right)^n = e\right) \tag{I.15}$$

The last result shows that $\emptyset_Z(\omega)$ is in the form of the characteristic function of an $N(0, 1)$ random variable and hence the result.

8. **The vector random variable $X = (X_1, X_2, \ldots..X_n)$ is jointly normal if only if $Y = a_1X_1 + a_2X_2 + \ldots + a_nX_n$ is normal for any $a_j \in \mathbb{R}$, $j = 1, 2..n$.**
Proof:
First, assume that X is (jointly) normal. The characteristic function of Y is:

$$\emptyset_Y\left(y = \sum_{j=1}^{n} a_j x_j\right)$$
$$= E\left[\exp\left(i\omega\left(a_1x_1 + a_2x_2 + \ldots.. + a_nx_n\right)\right)\right]$$

$$= \exp\left(-\frac{1}{2}\sum\nolimits_{j,k}^{n} \omega a_j \sigma_{jk}^2 \omega a_k + i\sum\nolimits_{j}^{n} \omega a_j m_j\right)$$

$$= \exp\left(-\frac{1}{2}\omega^2\sum\nolimits_{j,k}^{n} a_j a_k \sigma_{jk}^2 + i\omega\sum\nolimits_{j}^{n} a_j m_j\right) \tag{I.16}$$

The equation above shows that Y is normal with $E[Y] = \sum_j^n a_j m_j$ and $\sigma_Y^2 = \sum_{j,k}^n a_j a_k \sigma_{jk}^2$, where $m_j = E(X_j)$ and $\sigma_{jk} = E\left(\left(X_j - m_j\right)\left(X_k - m_k\right)\right)$. Conversely, suppose that $Y = \sum_{i=1}^n a_i x_i$ is normal with mean m and variance σ^2. By the definition of a characteristic function of an $N(m, \sigma)$ random variable, we have:

$$\exp(-\frac{1}{2}\omega^2\sigma^2 + i\omega m) = E\left[exp\left(i\omega\left(a_1x_1 + a_2x_2 + \ldots.. + a_nx_n\right)\right)\right] \tag{I.17}$$

Here:

$$m = \sum\nolimits_{j}^{n} a_j E[X_j] \text{ and}$$

$$\sigma^2 = E\left[\left(\sum\nolimits_{j=1}^{n} a_j X_j - \sum\nolimits_{j}^{n} a_j E[X_j]\right)^2\right]$$

$$= E\left[\left(\sum\nolimits_{j=1}^{n} a_j\left(X_j - m_j\right)\right)^2\right], \text{ where } m_j = E[X_j]$$

$$= \sum\nolimits_{j,k}^{n} a_j a_k E\left[\left(X_j - m_j\right)\left(X_k - m_k\right)\right] \tag{I.18}$$

Hence $\boldsymbol{X}$ is normal.

References

Ross. S. (2002) *A first Course in Probability*, 6th Edition, Pearson Education, Inc., New Delhi, India.

Index

Absolute displacement, 186, 188, 191, 203
Absolute response, 188, 194, 196
Absolute (ABS) summation rule, 198
Accelerating rotor, 301
Accelerogram, 179
Adjoint operator, 42, 69, 70, 72–3
Admissible function, 80–1, 375
Aleatory type of uncertainties, 309
Algebraic multiplicity, 142, 389
Algorithmic damping (in integration schemes), 281, 296, 300
Amplification matrix, 279, 281, 285, 287 294, 300
Amplitude error (in time integration schemes), 295, 300
Amplitude plot, 168, 193
Asymptotically stable / asymptotic stability, 279, 282, 285, 287, 290
Autocorrelation function, 324–5, 327
Auxiliary equation, 37, 73
Average acceleration scheme, 295
Axially loaded bar (or rod), 132, 139
Axial (dual) vector, 241, 395
Axially moving string, 243, 246
Axially vibrating bar (or rod), 109, 112, 125

Backward (implicit) Euler method:
- amplification matrix, 285
- asymptotic stability, 285
- definition, 277
- finite difference equations, 285
- local truncation error, 286
- spectral radius, 285

Backward PDE, 357
Banach fixed point theorem, 280
Band limited signal, 278
Bandwidth, 171–3
Base excitation (motion), 179, 181, 186, 191, 196
Basis function, 100–1, 118, 124, 371
Basis set, 100, 101, 136, 209, 231, 387
Basis vectors, 231
Bearing:
- isotropic, 257–8
- orthotropic, 258–9

Bearing damping, 263
Beam theory:
- Euler-Bernoulli, 53
- Timoshenko, 27

Bearing stiffness, 253, 255
Bearing element, 249, 252
Beat phenomenon, 152–3
Bernoulli trial, 311
Best approximation 57–8
Bessel's inequality, 58
Biharmonic equation, 59, 120
Biharmonic wave equation, 23, 26, 48, 50
Bilinear form, 124
- coercivity, 105–6
- continuity, 99, 104–6, 118
- ellipticity, 105
- positive definite, 100, 103, 382
- positive semi-definite, 103, 382
- symmetric, 103, 382

Binomial distribution, 320
Bi-orthogonality property, 70, 209, 254–5, 265–6
Bisection method, 141–2

Elements of Structural Dynamics: A New Perspective, First Edition. Debasish Roy and G Visweswara Rao.

Block Lanczos, 236
shifted, 272
Body force, 6, 240
Borel σ-algebra, 313
Boundary conditions:
Cauchy type, 38, 369
Dirchlet, 47, 102, 369
essential (geometric), 26, 80, 85–6, 102, 375
fixed-free, 47
free free, 135
homogeneous, 41, 182, 200
inhomogeneous (non-homogeneous), 68, 181
natural (force), 26, 80, 86, 99, 102, 111, 117, 119, 375
Neumann, 46, 102, 113, 127, 369
time dependent, 199, 200
Boundary value problem (BVP), 77
self-adjoint, 44, 83
Sturm–Liouville, 48, 56
Brownian filtration, 340
Brownian increments, 339
Brownian motion, 329, 339
Bubnov-Galerkin method, 106
Buckling instability, 73

Cable sag, 90–1
Cable tension, 89, 97
Campbell diagram, 257, 259 (of rotor bearing system)
Canonical basis, 213
Canonical transformation, 37
Cardinality (of sample space), 313
Cauchy's equations of motion, 16
Cauchy-Schwartz inequality, 103, 105, 125, 380, 382–3, 386
Cauchy (or fundamental) sequence, 58, 280, 370–1, 377
Cauchy stress tensor, 8
Cea's lemma, 125
Central difference method:
definition, 277
Jury's stability criterion, 287
relation to Newmark-β method, 295
time stepping map, 286
Central limit theorem, 321, 410
Central moments (of a random variable):
m^{th} order moments, 316
variance, 316
Centrifugal force, 242
Characteristic function (of a random variable), 317, 410–11
Characteristic polynomial, 137, 141, 287
Characteristic roots (eigenvalues), 55, 73, 141, 173
Chebychev inequality, 409
Cholesky decomposition, 144, 211
Closed ball, 368, 373
Closed set 312, 368, 407
Coercivity (ellipticity) of bilinear form (*see bilinear form*)
Combination rules:
ABS, 198
CQC, 198
SRSS, 198
Compactly supported function, 383
Comparison function, 86, 375
Complementary function, 38, 61, 66
Completeness, 375
of a set of functions, 377
Complex frequency response, 163, 165, 192, 335
Complex stiffness, 175
Conditional distribution function, 317–18
Conditional expectation:
definition, 319
existence and uniqueness, 319
Conditional probability:
random events, 317
random variables, 317–18
Conditional stability / conditionally stable, (for integration scheme), 290–2, 294
Conditional variance, 319
Congruent transformation, 216
Conservation of mass, 13
Continuity of a bilinear / linear form (*see bilinear form and linear form*)
Continuity requirements (of functions), 104, 118, 121
Continuous system (*see System*), 21–2
deterministic response of, 36
classical methods, 79
eigensolution, 69,

forced vibration, 59
support motion, 68,
stochastic response of, 331
moments, 332–3
Continuous version:
of a stochastic process, 329, 330
Contraction mapping theorem (for stability of an integration scheme), 279–80, 306
Convergence of eigenvalues, 224, 236
inverse and power iteration, 213–14, 216
Jacobi method, 221
Lanczos method, 236
QR method, 228
subspace iteration, 232
Convergence of random variables, 408
Convex function, 379, 380
Convex functional, 103, 379
strictly convex, 379
Convolution integral, 66, 146, 160, 183, 196, 391
Coordinates:
cartesian, 34
curvilinear, 34
generalized, 60, 131, 135, 145, 156
modal, 145, 159
physical, 150, 153
polar, 75
spatial, 181
Coordinate system (reference), 239
Coordinate transformation, 34
Coriolis effects, 69
Coriolis force, 238, 242
Correlation of random variables, 322
Countable set, 372
Covariance matrix, 316, 321–2
Cramer's spectral representation:
stochastic process, 326
Critical speed, 63, 246, 258–9, 301
Critical velocity, 73
Cumulative (probability) distribution function, 313

D'Alembert's solution, 38–9
Damping:
hysteretic, 174–5, 177
non-proportional, 178
proportional, 159–60
viscous, 157, 167, 174
Damping coefficient, 176
viscous, 157, 174–5
Damping force, 157, 163, 174
fictitious, 242
Damping ratio, 158, 171, 194
modal, 159, 160, 172, 184, 201
viscous, 174, 184, 189, 194, 196
Deflation, 215–6
Deformation gradient, 7, 13, 18, 239, 367, 394
Degrees of freedom (*dof* s):
master, 262–3
rotational, 188, 249
slave, 262–3
support, 186
tanslational, 249
Delta property (of the shape functions), 112, 114, 119
DeMoivre–Laplace limit theorem, 321
Dense space, 56, 372
(See Space also)
Denumerable, 46
Determinant search (for eigensolution), 84
Diagonalisability, 142, 388
Differential operator, 136, 142, 209, 388
linear, 135, 173
non-self adjoint, 253
self-adjoint, 83, 90, 375
time-invariant, 136
Diffusion (noise) coefficient (in SDEs) 344–5, 349
additive, 338–9, 357
multiplicative, 349, 351
Diffusion equation, 36
Diffusion (or Noise) term, 344, 346, 350
Dimensional descent, 115
Dirac delta function, 59
Direct integration method (or scheme), 131, 268, 275
explicit, 276
implicit, 276
Discretization (spatial), 45, 100, 107, 117, 124, 275
Discretization (temporal), 275, 277–8, 281

Discretization error (in direct integration methods), 277, 282
Disk element, 249, 251
(in rotor bearing system)
Disk unbalance force, 252
Displacement:
axial, 109
transverse, 188, 241
vertical, 204, 246
Displacement vector:
absolute, 186, 188, 203
pseudostatic, 186
relative, 186
Divergence theorem, 15–16
Domain discretization, 127
Domain integrals (integration), 107, 111, 113
Duhamel integral, 66
Dynamic component, 68–9, 89, 181, 186, 203
Dynamic condensation, 260, 262, 264
2-dof model, 301
Dynamic displacement (component), 182, 199
Dynkin's formula, 352–3, 355

Effective loading, 69, 182, 189, 201
Effective reaction force, 183
Eigenexpansion, 56, 59, 183, 275
of initial system state, 279
Eigenfunction:
orthogonal, 46
orthonormal, 56
Eigensolution procedures, 209
Eigenspace, 137, 142, 215, 232, 389
Eigenvalue:
complex, 159, 256, 259, 260
convergence of, 224, 236
distinct, 143, 212–13, 216
imaginary, 140
largest, 210–11
minimax characterisation, 376
multiple, 215
positive, 393
real, 145, 209
repeated, 142, 215, 245
smallest (lowest), 156, 211
zero, 142, 210, 216
Eigenvalue problem:
adjoint, 73, 253
generalized, 93, 144, 147, 211, 215, 223
quadratic, 94, 157, 159
reduced (order), 156, 231, 259
shifted, 218
standard (canonical), 90, 137, 144, 220, 235, 376, 388
Sturm-Liouville, 21
symmetric, 88, 145, 245, 389
unsymmetric, 158–9, 253–4
Eigenvalue separation property, 84, 376
Eigenvector, 137, 142
convergence of, 214, 216
left, 254
orthogonal property, 145, 154
right, 254
Einstein summation, 6
El Centro earthquake, 179
Elements:
line (1-dimesional), 101
quadrilateral, 113, 117, 121–2
triangular, 101, 113, 117
Elliptic PDE, 36
Ellipticity (coercivity) of bilinear form (*see bilinear form*)
Energy:
potential (or strain), 6, 25, 27, 31, 84, 143, 242, 249
kinetic, 4, 6, 25, 27, 31, 84, 143, 242, 249, 251
Energy dissipation, 132, 156–7, 174, 175
Energy dissipation function, 157
Energy functional, 80, 85, 102
Energy dissipation per cycle, 175
Energy integral, 25, 28
Epistemic type of uncertainties, 309
Equilibrium:
static, 89
dynamic, 2, 5, 10, 11, 89
Error analysis (Lanczos method), 235
Error norm, 235–6
Error estimates in FEM, 124
A-priori, 124
Essential boundary conditions, 26
(*see* also boundary condition)

Estimation of modal damping, 172
Euler-Bernoulli beam, 24, 48, 53, 56, 59, 62
Euler's formula, 139
Euler–Maruyama (EM) method:
  time integration of linear SDEs, 358, 360
Euler version, 15
  (of the conservation of mass)
Extensibility (of cable), 90
Expectation (or mean) of a random variable, 315
Expected (average) value, 315
Explicit Euler map of test equation, 282
Extremum of a norm, 100
Extremum of a functional, 103, 379

Field variable, 13, 14
Finite difference approximations:
  backward Euler, 277
  central difference, 277
  forward Euler, 277
Finite element method (scheme), 79, 99, 100, 129
First variation of energy integral, 25, 28, 32, 35
Flexural rigidity, 12, 18
Force:
  body, 6, 81
  damping, 157, 163, 174
  inertia, 163
  spring, 163
  surface (traction), 6, 16, 81
Force balance, 11, 16
  principle of, 1
Force diagram, 163
Force equilibrium, 133, 165
Forward Euler method, 277–8, 281
Fourier coefficients, 57–8
Fourier sine series, 46
Fourier transform:
  of sync function, 278
Frame (of reference):
  body-fixed (inertial), 238, 396
  moving, 239–40
  rotating, 246, 250, 396
Frame-indifferent (quantities), 238–9, 242
Free vibration, 157, 167, 169, 171
Frequency (or mode) cross-over, 91–2
Frequency equation, 51
Frequency parameter, 46, 47, 50
Frequency ratio, 163, 167, 169, 173
Frequency (harmonic) response, 152, 154, 162–3, 170, 172, 192
Friedrich's inequality, 105
Frobenius norm, 221
Function:
  compactly supported, 383
  locally integrable, 383
  weak derivate of, 383
Functional:
  convex, 103, 379
  energy, 2, 80, 85
  positive definite, 79
  strictly convex, 379
Functional discretization, 79, 95, 101
Fundamental frequency, 155, 191
Fundamental solution matrix, 357

Galerkin method, 80, 86, 90, 191
Galerkin orthogonality, 124
Gauss quadrature, 107
Gaussian distribution (*see Normal distribution*)
Generalized coordinates, 2, 4, 5, 60, 80, 100, 145, 156
Generealised derivative, 330
Generalized force, 183–4
Generalized mass and stiffness, 183
Generalized velocities, 4
Generalized-α method:
  algorithmic damping, 300
  amplification matrix, 298
  characteristic equation, 298
  comparison with Newmark-β method, 303, 304
  faster convergence, 304
  Jury's stability criteria, 298
  spectral radius, 300
  time stepping map, 297
  unconditionally stable solution, 300
Geometric multiplicity, 142, 389
Global discretization error, 282
Gram-Schimdt orthogonalization, 215–16
Green's function, 64
Green's identity, 43, 51

Green strain tensor, 30
Green's theorem, 33, 54
Ground motion, 173, 182, 196
  spatial variability of, 198
  uniform, 199
Gyroscopic coupling, 253
Gyroscopic damping matrix, 250
Gyroscopic effects, 69
Gyroscopic moment, 249
Gyroscopic system, 73, 243

Hamilton's principle, 3–5, 16, 242, 244, 252
Hamiltonian, 243
Harmonic motion, 84
Harmonic solution, 48, 51, 54, 72, 83
Heaviside function, 66, 313
Helmholtz equation, 55
Hermitian, 45
Hessenberg form:
  lower, 224
  upper, 224
HHT-α method, 296–7, 299
High frequency dissipation:
  of Newmark-β method, 296
  of HHT-α and generalised-α methods, 297
Hilbert space, 42
Hooke's law, 13
Homogeneous material, 29
Homogeneous PDE, 41
Householder transformation, 220, 224–5, 227
Hyperbolic, 36
Hysteresis loop, 174–5
Hysteretic damping (or structural damping) factor, 174
Hysteretic damping model, 174, 177

Identity matrix, 136
Impulse response function, 147
Incremental summation procedure, 185, 391
Independent:
  random events, 317
  random variables, 319
Inequality:
  Chebychev, 409
  Markov, 409
Influence vectors, 186
Inhomogeneous boundary conditions (see Boundary conditions)
Inhomogeneous (nonhomogeneous) PDE, 59, 61, 62, 64, 69
Initial conditions, 23, 38, 41, 46
Initial displacement function, 23
Initial velocity function, 23
Inner product, 42, 63, 72, 86, 103
Inner product space, 56, 370, 372, 380, 399
Integration scheme (or method):
  asymptotic stability, 290
  conditionally stable or unstable, 290
  feature of algorithmic damping, 296
  stable, 279
  unconditionally stable, 289, 299
  unstable, 279
Interpolants, 117
Interpolating bases (functions), 101
Interpolation error, 125
  bound by Taylor's theorem, 384
Invariant subspace, 231, 393
Invertible, 142
Invertibility, 142, 388
Isolation system, 167
Isolator efficiency, 196
Isotropic, 29, 54, 120
Iteration methods (for eigensolution), 210
  inverse, 210, 213, 215, 219, 231
  power, 210, 212, 231
  subspace, 231–2, 246, 273
Ito calculus:
  introduction, 338
Ito's formula, 344
  higher-dimensional, 349
  of a scalar function, 347
  response moments p-dimensional SDE, 352
  solution of SDE (of MDOF system), 352
  solution of SDE (of SDOF system), 349
Ito integral process, 341, 346
Ito isometry, 341

Jacobian, 37–8

Jacobi method, 220
convergence of, 221, 223–5
Joint random variables:
probability distribution and density functions, 314
Jury's stability criterion, 287, 403–4

Kernel, 64
Kinetic energy, 4, 6, 25, 27, 31, 71, 84, 143, 242, 249
Kirchhoff's plate theory, 26, 29
Kirchhoff shear force, 36
Kolmogorov's backward equation, 356, 357
Kolmogorov's continuity condition, 329, 338
Kolmogorov's extension theorem:
consistency requirement, 323
existence of a stochastic process, 323
Kronecker delta, 7, 56, 90, 395

Lagrange polynomials, 107
Lagrangian, 4, 15, 32, 36, 71, 244, 250
Lagrange (Lagrange's) equations of motion, 4, 132, 157
Lame parameters, 369
Lanczos (transformation) method, 233, 237–8
two-sided, 254
Lanczos method and error analysis, 235
Lanczos vectors, 233, 254
Laplace's equation, 36
Laplace transform, 61, 63, 66–7, 400–2
Laplacian operator, 22, 26
Law of large numbers:
strong, 360, 409
weak, 409
Lebesgue-integral, 316
Lebesgue measure, 312
Liebnitz rule, 14
Linear differential operator, 135–6
self-adjoint, 331
Linear form, 102–3, 105, 111, 382
continuity of, 106, 382
Linear homogeneous equation, 36
Linear homogeneous PDE, 40
Linear mapping:
equilibria of the SDOF oscillator, 290
Linearly independent, 80, 136, 142, 377, 387
Linear momentum balance (conservation), 16, 239
Linear span, 56
Linear time-invariant (LTI) system, 131, 136, 147, 163, 209, 275
Linear transformation (or operator), 136, 137, 373, 387–8, 393
Lipschitz continuous function, 367
Lipschitz domain, 367
Lipschitz boundary, 6, 367
Lipschitz condition:
first order equation, 276
Lipschitz constant, 8
Local (physical) coordinates, 114, 119
Local truncation error, 282
Localization theorem, 15, 16, 368
Longitudinal vibration, 22, 47
LR method, 228
Lumped parameter approach, 132
Lumped parameter model, 132, 186

MDOF system:
damped, 156, 158
undamped, 132
Magnification factor, 163–4
Markov chain, 328
Markov inequality, 409
Markov process:
statistical property, 328
strongly Markov, 357
Markov property, 357
Marginal (probability) density functions, 315
Marginal (probability) distribution function, 315
Martingale (stochastic process), 342–3, 356
submartingale, 342
supermartingale, 342
Mass balance, 13, 239
(law of)
Mass conservation, 15
Mass imbalance, 250, 252
Mass-orthonormalization, 219, 231, 233

Material:
homogeneous, 29
isotropic, 29
St. Venant Kirchhoff type, 29
Material coordinate, 15
Material constants, 13, 80, 116
Material point, 238, 244, 246
Material volume, 13
Matrix:
banded, 111
constitutive, 24
damping, 157, 159, 160, 177
diagonal, 142, 216, 218, 228
gyroscopic, 243
Hessenberg, 224
lower triangular, 144, 393
mass, 83, 107, 144, 211, 235
non-singular, 134, 144, 388
participation factor, 189
positive definite, 134, 218, 393
positive semi-definite stiffness, 135
reflection, 227
singular, 135, 137
skew-symmetric, 242–3, 253, 394–5
sparse, 111
stiffness, 83, 107, 134–5, 230, 242, 250
symmetric, 84, 134, 216, 389
trace of, 229–30, 394
triangular, 224, 227, 229, 259, 261
tri-diagonal, 111
unitary, 227
unsymmetric, 253–4
Matrix-vector equations, 132
Maximal orthonormal set, 57
Mean square error, 126
Mean square value, 316–17
root mean square (RMS) value, 327
Measurable set, 312, 407
Measurable space, 313
Measurability, 340
Measurement of damping, 167, 177
half power method, 170
logarithmic decrement method, 167
Membrane vibrations, 113
Method:
classical, 79, 95
direct integration, 131, 275
finite element, 79, 99, 129, 131
Newmark, 289
Rayleigh-Ritz, 80
separation of variables, 21, 41, 47–8
Metric space, 280, 367, 399
Minimization, least square, 94, 379
Minimum energy norm, 128
Minimum residual norm, 86
Modal:
coordinates, 61, 145, 159
damping ratio, 159, 160, 172
mass, 148
matrix, 146, 148, 159, 165
participation factor, 183
response:
deterministic input, 147, 185, 303
stochastic input, 332, 337
stiffness, 148
superposition, 146, 174, 188, 209, 246
Mode:
anti-symmetric, 91
fundamental, 150
symmetric, 91
Mode (or frequency) crossover, 91
Mode shape, 148, 153, 184, 190, 209, 246, 248
Moment generating function:
of vector random variable, 317
Momentum balance, 13, 16, 239, 241
Momentum equation, 16
Moment-curvature relationship, 33
Moment/s of inertia:
diametrical, 246
polar, 246
Monomial, 82
Monte Carlo integration approach, 360
Moving load, 60, 62, 69, 77
Multi-degree-of-freedom system:
eigenvalue problem, 131
response for deterministic input, 179, 180
frequency response, 150–1
general loading, 145, 147, 161
support motion, 185
transient response, 149, 151, 152, 165
response for stochastic input, 337
moments, 337, 352–3, 356

PSDs, 338
Multi-support excitation (motion), 199, 202, 206
Multiple stochastic integral (MSI), 359, 360
Multiplicity:
algebraic, 142, 389
geometric, 142, 389

Nanson's relation, 9
(between deformed and undeformed areas)
Natural boundary condition, 26
(*see* also boundary condition)
Natural (normalized) coordinates, 121, 122, 124
Natural frequency, 47, 52, 53, 61, 73, 91, 111, 146
damped, 158, 184
undamped, 146
Newmark-β method:
amplification matrix, 290
amplitude and periodicity errors, 295
characteristic equation, 290
comparison with generalised-α, 303, 304
conditional stability, 291–2
degenerate cases, 292, 294
implicit method, 289
linear mapping, 290
second order accuracy, 289
unconditional stability, 290
Newton's force balance, 11, 16, 115
Newton's second law, 2, 22, 132
Nodes, 100–1, 106, 109
Noise term (in SDEs):
additive, 338, 347
multiplicative, 338
Non-elliptic (system), 85
Non parabolic equation, 37
Non-proportionl damping, 178, 206
Non-self (or nonself-) adjoint, 69, 71, 132, 209, 237, 243, 253
Norm:
energy, 125
euclidean, 370
Frobenius, 221
functional, 100, 370
L^2, 370
on inner product space, 370
on normed space, 370
residual, 86
Sobolev, 380
vector, 370
Normal (probability) distribution:
scalar valued random variable, 321
vector random variable, 321–2
simulation, 407
Null space or Kernel, 44, 136–7, 373, 388
Nullity, 373
Numerical dissipation, 281
generalized-α method, 307
Newmark-β method, 296
time integration algorithms, 307
Numerical integration, 107
Nyquist frequency (or sampling frequency), 278

Odd periodic expansion, 40
Open ball, 407
Open set, 313, 407
measure of, 312
Order and rate of convergence of eigenvalues, 212
Orders of approximation, 399
(of a function)
Ordinary differential equations (ODEs):
autonomous, 136
coupled, 131–2, 136, 146, 157
coupled first orde, 136, 144, 159
coupled second order, 132–3, 147, 159
non-autonomous, 136
second order, 183
stability, 402–3
uncoupled, 145
uncoupled first order, 142
uncoupled second order, 159
Orthogonal, 44, 46, 49, 54, 57, 69, 70, 86, 145, 148, 389
Orthogonal projection, 86
Orthogonal (orthogonality) property, 145
Orthogonal property of the eigenfunctions, 44, 61, 63
Orthogonal transformation, 221, 233, 394

Orthonormal basis, 56–8, 63, 71
Orthonormalization, 231–2, 390

Parabolic equation, 36
Parabolicity, 38
Partial derivative, 37
Partial differential equation (PDE), 21
  homogeneous, 41, 369
  inhomogeneous (nonhomogeneous), 59, 61, 64, 69, 201, 369
Partial differential operator, 24
Partial fractions expansion, 162
Partial sum, 57
Particular integral (solution), 38–9, 61
Peak response, 150, 163, 196, 198
Peak response estimate, 197–8
Periodicity error, 291, 295
Permutation symbol, 367
Phase plot, 165, 193
Piecewise polynomials, 100
Planar oscillation, 89
Plane (2D) elasticity problem, 115
Plane strain, 115–16
Plane stress, 115–16
Plate:
  rectangular, 34, 54, 121
  with non-smooth boundaries, 34
Poisson distribution, 320
Poisson's equation, 36, 104
Poisson's ratio, 12, 27, 116, 369
Polynomial:
  cubic, 119, 121
  interpolating, 127
  linear, 112, 123, 125
  quadratic, 119, 123
Polynomial basis (set / functions), 101, 118
  higher order, 118
Polynomial interpolant, 125
Positive definite, 100, 103, 210, 218, 228, 238, 253
Positive definite energy functional, 100
Positive definite functional, 79, 103
Positive semi-definite (or semidefinite), 103, 211
Potential energy 25, 28, 32, 71, 80, 84, 143, 242, 244, 249
Power set, 320
Power spectral density:
  mean energy density, 327
  stationary stochastic process, 326
Probability density function, 314
Probability distribution function, 313, 315
Probability mass function, 313
Probability measure, 312–13, 323, 344
Probability space, 312, 317–18, 323
Probability theory:
  basic concepts, 311
Proportional damping, 159, 160, 178, 188
Pseudo-absolute acceleration, 197
Pseudo-static component (displacement), 69, 181–2, 186, 194, 199, 203
Pseudovelocity response, 196

Q–factor, 172–3
QL method, 268
QR decomposition, 228
QR method, 224, 228, 230, 259, 268
Quadrature points, 107–9
Quadratic covariation, 347–8
  of Brownian motion components, 351
Quadratic eigenvalue problem, 94, 157, 159
Quadratic (scalar) function, 134
  homogeneous, 242
Quadratic variation:
  Brownian motion, 346
  stochastic process, 346

Radius of curvature, 36
Radius of gyration, 27
Radon-Nikodym theorem, 319
Random process (*see Stochastic process*)
Random field, 309, 323
Random variables:
  as an invertible mapping, 313
  continuous, 313, 316, 318
  discrete, 313–14, 321
  independent, 319
  independent and identically distributed (i.i.d), 321, 361, 407
  joint, 314
  scalar, 313–15, 320–22
  uncorrelated, 317, 319, 322
  vector, 317, 410

Rate of convergence (of an eigenvalue), 212–3, 216, 232
Rate of convergence (of the eigenvector), 214, 216
Rayleigh (probability) distribution, 322, 407
Rayleigh's (or Rayleigh) quotient, 84, 153
Rayleigh's (minimum) principle, 84, 153
Rayleigh Ritz method, 155, 375
Rectangular plate, 34, 54
(*see plate also*)
Recursive relationship (Lanczos method), 234
Reduced order model, 260, 262, 265, 267
Relative frequency concept, 312
Residual, 85, 100
weighted, 85, 97, 100, 102
Response amplitude, 156, 168
peak, 150
Resonance, 151, 153, 158, 170, 172
Resonance condition, 61
Response spectra, 196–7, 207
Reimann integrable, 315
Riemann-Stieltjes integral, 315
Riesz representation theorem, 104–5
Rigid body motion, 103
Rotor bearing system, 246, 248–9, 253, 255, 257, 260, 263, 267
2 dof model and time integration, 301–3
Rotary inertia, 27, 29, 52–3
Rotary inertia effect, 12, 31
Rotating shaft, 241, 249, 250
Round-off error, 216, 234
Ruth–Hurwitz criterion, 404

Sagging cable, 89
Sample space:
continuous, 311–12
definition, 311
discrete, 311–12
Sampled data and sampling period, 278, 400, 401
Sampling frequency (or Nyquist frequency), 278, 400
Sampling theorem, Shannon, 277
Scalar field, 238
SDOF oscillator (or system), 60, 132
damped, 157–8, 162
impulse response of:
damped, 160
undamped, 147
response for deterministic input:
frequency response,163
general loading, 61
support motion, 191
transient response, 150
response for stochastic input:
moments, 333
PSDs, 335
undamped, 65
Seismic excitation, 179
Seismic qualification, 179, 196
Self-adjoint, 42–44, 83, 90, 132, 209, 237
Self-adjoint property, 42, 44, 132
Semi-discretization, 100, 105, 107, 109, 121–2, 124, 131, 268
equations of motion, general 275
Semi-discretized (or semidiscretized) systems, 157, 200, 209
Separable (*see* space)
Separation of variables, 21, 41, 47
Series solution (or expansion), 46, 58, 90
Shaft eccentricities, 250
Shaft unbalance force, 250
Shannon's sampling theorem, 278
Shape functions:
compactly supported, 111
definition, 101
element, 111–12, 114, 116, 118–19, 121–23
global, 109–11, 113–14
linear, 114
Shear deformation, 26–9, 52
Shear force, 25, 27, 28, 33, 35, 50
Kirchhoff, 36
Shear frame, 165, 172
Shear modulus, 12, 24, 28, 31
Shear strain, 24, 27
Shift, Shifting, 213, 216
Similarity transformation, 142, 146, 389, 394
Simulation (of a random variable), 407
Simpson's rule (of integration), 185, 391, 392

Singularity, 50, 52, 65
Smoothed average earthquake response spectra, 197
Sobolev space, 104
Space:
closed 57, 373
complete, 56, 371
convex, 57, 373
dense, 372
euclidean space, 367, 399
finite dimensional, 56, 111, 387
Hilbert, 42, 86, 372, 379
infinite dimensional (function), 56–7
inner product, 56, 370, 372, 380, 399
L^2, 379
L^p, 379
linear vector, 369
metric 280, 399
normed, 369–70, 380, 399
null, 44, 373
separable, 56, 372, 377
Sobolev, 104, 380–2
solution, 104
test, 104, 119
vector, 44, 86, 103, 136, 142, 144, 173
Span, 56–7, 124, 138–40, 155, 387
Spatial coordinate, 13, 15
Spectral decomposition, 132
Spectral pseudo-velocity, 196
Spectral radius:
explicit Euler, 282
implicit Euller, 285
Newmark-β method, 295
Spectral relative displacement, 196
Spurious high frequency mode, 275, 281, 296
Square integrable, 103
Square integrable function, 57, 372
Square root mapping, 228, 230
Square root of sum of the squares (SRSS), 198
St. Venant–Kirchhoff type material, 369
Stability, of solution to weak form, 104
Standard deviation, 316
Standard uniform variate, 322, 407
State of resonance, 151
State space, 253, 260, 264, 268
State-space vector, 135
Stationary stochastic process:
of order m, 324
properties, 325
weakly or wide-sense, 324
Stationary standing wave, 39
Stationary wave solution, 39
Steady state absolute response, 194
Steady state relative displacement, 194
Steady state solution (or response), 162, 163, 165, 174, 192
Step function input, 149
Stochastic (ordinary) differential equation (SDE), 311, 337–8
solution by Ito's formula, 349–50
Stochastic finite element, 310–11, 364
Stochasticity, 309
Stochastic calculus:
Ito, 338–40
Stratonovich, 340
Stochastic integral, 339
Stochastic process;
adapted, 340
continuous time, 329
definition, 323
existence of, 323
finite-dimensional joint distribution, 323
Martingale, 342
spectral representation of, 325
stationary, 323, 326–7, 331
nonstationary, 334, 336
Stochastic steady state, 334
Stochastic Taylor expansion (STE), 359
Stone-Weierstrass theorem, 107
Strain:
normal, 24, 27
shear, 27
Strain energy, 6, 7, 9, 31
internal, 24–5, 27, 28
Strain tensor, Green, 30
Stress tensor (*see* Tensor also), 9
Stress resultant, 25, 32
Stress-strain relationships, 24, 30, 31
Strong form, 79, 81
Structural damping (hysteretic damping) factor, 174
Sturm-Liouville operator, 42–3, 373
Sturm sequence property, 216–8

Subspace, 56–7
  invariant, 209, 231
Sufficiently differentiable, 86
Support excitation (or motion), 68–9
  differential (or nonsynchronous), 69, 201
  Multi-, 198, 201–2
Sylvester's law of inertia, 216, 218
System:
  conservative, 132, 143, 145–6, 159
  continuous, 6, 21, 131, 135, 181, 199, 209, 237
  damped, 157–60, 169
  discrete, 5, 81, 131–2, 135, 155
  discretized, 101, 125, 157
  homogeneous, 137
  non-classically damped, 273
  non-conservative, 85
  non-self adjoint, 132, 237
  positive definite, 134
  self-adjoint, 132
  semi-definite, 134, 135, 173
  semi-descretized (semi-discrete), 157, 200, 209
  shifted, 214
  time invariant, 136, 163
  torsional, 174
  undamped, 132, 156
  unrestrained, 103

Taut string, 22, 38, 90
Taylor series:
  family of direct integration methods, 276
Temporal discretization (for time integration), 275
Tensor: 7, 367
  second order, 394
  skew-symmetric, 241
  strain (Green-Lagrange), 7
  stress (Cauchy, 1st and 2nd Poila Kirchoff), 8, 9
Time history analysis, 196
Time integration of SDEs, 357, 358
Time-invariant, 61, 132, 136, 144, 147, 163, 209, 241
Time stepping maps, 280
Traction forces, 6, 16
Trace of a square matrix, 229
Transfer function, 161
Transformation:
  congruent, 216
  Householder, 225
  Jacobi, 220
  linear, 231, 253
  nonunitary, 228
  orthogonal, 221, 233
  QR, 230, 255, 259
  similarity, 210, 216, 220, 224–5, 230, 259, 260
  unimodular, 220
  unitary, 227
Transient solution (response), 158, 160, 191
Translation speed, 238
Transmissibility:
  force, 165
  motion, 194
Transformation methods:
  (for eigensolution)
    Householder, 225–6, 230
    Jacobi, 210, 220, 222
    Lanczos, 233, 235, 237, 254
Transport theorem, Reynolds, 13, 14
Travelling seismic waves, 198
Travelling speed, 71
Trial function, 80, 82, 84, 85
Triangle inequality, 369
Tridiagonal form, 224, 227, 230, 233, 254 259
Truncation error (in direct integration methods), 277

Uncertainties:
  aleatory type, 309
  epistemic, 309
Uncorrelated:
  random variables, 319
Uniform probability distribution, 322
Uniform ground motion, 199
Uniqueness, of solution to weak form, 104, 106
Unit circle (concept of stability), 279
Unit-step function, 313
  Heavyside, 408
Unit vector, 242

Upper bound, 83–4, 375
Upper triangular matrix, 227, 229

Variable separable form, 41, 46, 48
Variance, 316, 319, 320–22
Variational calculus, 2, 19, 81
Variational formulation, 36, 80, 102
Variational principle, 25, 28, 80
Vector field, 238
Vector iteration method, 211
Vector space (*see Space also*):
  finite dimensional, 86
  infinite dimensional, 56
  linear, 103
Vector stochastic process, 311
Velocity gradient, 394
Vibration (or oscillation):
  axial (longitudinal), 22, 47, 56, 112
  flexural, 12
  torsional, 12, 22
  transverse, 12, 114, 120
Vibration isolator, 194
Vibrating membrane, 104, 113
Vibration solution (or response):
  forced, 83, 173, 183
  free, 89, 167, 170, 184
  steady state, 86
Virtual displacement, 3, 10, 25, 32, 72
Virtual work, 2
Viscous damping, 157, 167, 174
Voigt convention, 9

Wave equation, 36, 38–9, 41–2, 45–6, 48, 56, 59
  bi-harmonic, 23
  one-dimensional, 45
Weak derivative, 104, 126, 380–1, 384
Weak form, 45, 102–5, 107, 114, 117, 119, 121, 124, 126, 199
  element-wise representation, 111
Weak formulation of PDE, 101–2
Weak solution, 104, 106, 124
Weight (weighting) function, 80, 102, 104, 106, 117, 199
Weighted integral, 80, 86
Weighted residual, 85
Weighted residual methods, 79, 85, 102
Weighted sum, 132
Wiener–Kienchine relations:
  stationary stochastic process, 326
Wiener process:
  limit of a random walk, 329
  mathematical model for Brownian motion, 329
  statistical characteristics, 329–30
Well-posedness of the weak form, 103
Whirl:
  backward, 258, 397
  forward, 258, 397
Whirl speeds, 257–8
White noise process:
  autocorrelation function of, 328
  band limited, 328
  definition, 327, 338
  response of an SDOF oscillator, 334, 352
  response of an MDOF system, 352

Young's modulus of
  elasticity, 12, 369

Z-transform (in stability of time stepping maps), 287, 400–2
σ-algebra, 311–15, 317, 320, 340